SEEING SOCIOLOGY
AN INTRODUCTION
Second Edition

Joan Ferrante
Northern Kentucky University

With contributions from
Chris Caldeira
University of California, Davis

Australia • Brazil • Japan • Korea • Mexico • Singapore • Spain • United Kingdom • United States

WADSWORTH
CENGAGE Learning

***Seeing Sociology: An Introduction*, Second Edition**
Joan Ferrante

Executive Editor: Mark Kerr
Acquiring Sponsoring Editor: Seth Dobrin
Developmental Editors: Kristin Makarewycz and Chris Caldeira
Assistant Editor: Mallory Ortberg
Editorial Assistant: Nicole Bator
Media Editor: John Chell
Brand Manager: Liz Rhoden
Market Development Manager: Michelle Williams
Content Project Manager: Cheri Palmer
Art Director: Caryl Gorska
Manufacturing Planner: Judy Inouye
Rights Acquisitions Specialist: Dean Dauphinais
Production Service and Composition: Lachina Publishing Services
Photo Researcher: Janice Yi, Q2a/Bill Smith
Text Researcher: Jill Krupnik, Q2a/Bill Smith
Copy Editor: Lachina Publishing Services
Illustrator: Lachina Publishing Services
Text Designer: RHDG
Cover Designer: RHDG
Cover Image: Alamy

Library of Congress Control Number: 2012943395
Student Edition:
ISBN-13: 978-1-133-93523-0
ISBN-10: 1-133-93523-0

Loose-leaf Edition:
ISBN-13: 978-1-133-95193-3
ISBN-10: 1-133-95193-7

Wadsworth
20 Davis Drive
Belmont, CA 94002-3098
USA

Cengage Learning is a leading provider of customized learning solutions with office locations around the globe, including Singapore, the United Kingdom, Australia, Mexico, Brazil, and Japan. Locate your local office at **www.cengage.com/global.**

Cengage Learning products are represented in Canada by Nelson Education, Ltd.

To learn more about Wadsworth, visit **www.cengage.com/wadsworth**

Purchase any of our products at your local college store or at our preferred online store **www.CengageBrain.com.**

Printed in Canada
1 2 3 4 5 6 7 16 15 14 13 12

About the Author

Doug Hume

JOAN FERRANTE is a professor of sociology at Northern Kentucky University (NKU). She received her PhD from the University of Cincinnati in 1984. Joan decided early in her career that she wanted to focus her publishing efforts on introducing students to the discipline of sociology. She believes it is important for that introduction to cultivate an appreciation for the methods of social research and for sociological theory beyond the three major perspectives. As a professor, she teaches sociology from an applied perspective so that students come to understand the various career options that the serious student of sociology can pursue. Joan is the author of "Careers in Sociology" (a Wadsworth sociology module), a guide to making the most of an undergraduate degree in sociology. She also teaches sociology in a way that emphasizes the value and power of the sociological framework for making a difference in the world. With the support of the Mayerson Family Foundations, Joan designed the curriculum for a student philanthropy project at NKU in which students, as part of their course work, must decide how to use $4,000 in a way that addresses some community need. That curriculum has now been adopted by dozens of universities across the United States. For the past decade, she has also supported a study abroad scholarship called "Beyond the Classroom" for which any NKU student who has used her sociology texts (new or used) can apply. Joan's university has twice recognized her as an outstanding professor with the Frank Milburn Sinton Outstanding Professor Award and the Outstanding Junior Faculty Award.

To Dr. Horatio C Wood, IV, MD

Brief Contents

Contents

Tony Rotundo

Tony Rotundo

Chris Caldeira

Chris Caldeira

Chris Caldeira

Courtesy of Linda Sarna, UCLA School of Nursing

Missy Gish

Photo by Chris Caldeira; Mural by Paul Ygartua/www.ygartua.com

Courtesy of Asia and Pacific Transgender Network Development

Chris Caldeira

Courtesy of Terra Schulz

Lisa Southwick

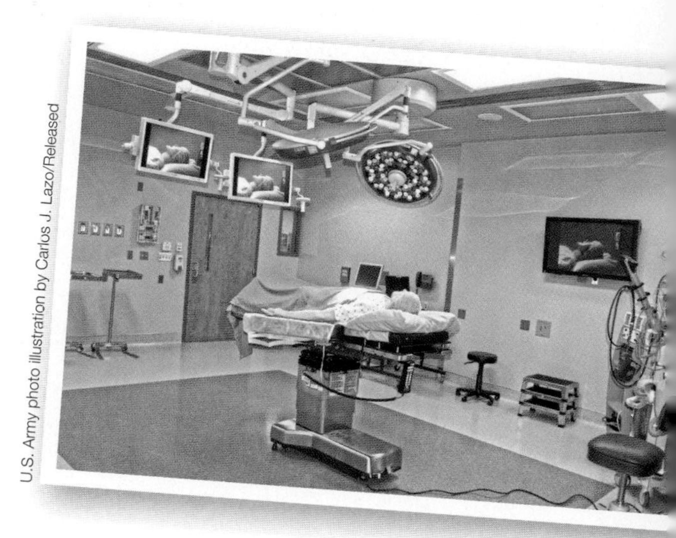

U.S. Army photo illustration by Carlos J. Lazo/Released

Chris Caldeira

Preface

We live in an age when we must force ourselves to concentrate—to sit still, focus, and shut out all that is going on around us or might be going on if we just check our e-mail, Facebook, or text messages. How can we immerse ourselves in ideas when the technology that surrounds us makes it so difficult to carve out a space that allows us to read, reflect on what we read, and do the hard work of making ideas our own? In writing the second edition of *Seeing Sociology*, I thought long and hard about how students experience reading a textbook, given the realities of their lives today.

I realize that even good students find it difficult to engage in a sustained course of reading. This textbook is consciously structured to support an uninterrupted and coherent reading experience, even if the reading is done in short bursts of time. Specifically, there are 14 chapters, each broken into 5 to 8 pedagogically meaningful modules, or chunks. In addition, the book capitalizes on the instructional value of photographs as tools for provoking thought and clarifying abstract concepts.

Modular Format

Each chapter consists of five to eight self-contained modules, three to seven pages in length. The modular format allows instructors to assign all or specific modules within a chapter. Each module

- is assigned a title and opens with a basic objective in bold print alerting readers to the organizing concept.
- poses a question that prompts readers to react, respond, or recall something that has direct personal relevance to the material covered.
- ends with a critical thinking question that prompts students to think about what was just read and apply that material to personal relationships and events that matter to them.
- challenges students with a Write-a-Caption exercise inviting them to study a photograph and then use sociological language and concepts to analyze what they see.

Photographs and Captions Seamlessly Integrated with Text

Photographs, captions, and text are seamlessly integrated and have equal significance for showcasing how sociologists observe, interpret, and analyze the world around them. Specifically,

- the photographs function as objects of sociological analysis,
- the text presents the vocabulary and concepts that guide interpretation and analysis, and
- the captions apply the vocabulary and concepts to clearly convey the photograph's sociological significance.

The adage "a picture is worth a thousand words" helps to explain the pedagogical value of this seamless integration. While pictures can help us to convey a point in short order, we must also realize that the picture per se does not convey meaning; rather, the photo evokes varying meanings in the minds of those viewing it. In other words, a photo can facilitate many interpretations, but the interpretations this book cultivates and facilitates are sociological.

Retained Features

Each chapter (except the last) includes an Applying Theory module. These modules are written for those instructors who wish to expose students to theoretical traditions within sociology (rational choice, post-structural, feminist) beyond the "big" three perspectives.

There are also Research Question modules (available online). These modules were written with the purpose of demonstrating the various kinds of questions sociologists ask and the data-gathering strategies they employ to answer questions such as: "With whom do Americans talk about important matters?" "How does online dating shape intimate relationships?" "Do gangs have a purpose?" "Why do some people give inconsistent answers to questions asking their race?" For the most part, the Research Question modules describe a study published in a major sociological journal within the past decade. Students who read them will come to understand how sociologists observe and explain human behavior and activities.

Each chapter includes Write-a-Caption exercises, which ask students to write a photo caption that describes the sociological relevance of a given image, reinforcing the idea that key course concepts can be observed in everyday interactions and activities. The first edition of *Seeing Sociology* included Write–a-Caption exercises at the end of the first module in every chapter, but in the second edition I have included one at the end of every module.

Changes to This Edition

In revising *Seeing Sociology*, my strategy was simple: I tried to make the second edition better than the first. As I reviewed every photograph, I asked, "Can I find a more effective photograph to represent a given sociological idea?" In the end, I replaced approximately 330 photos with new, hopefully more effective ones. I also studied the questions that open each module and asked, "Can this question engage students so that they will want to read what follows?" In response, I revised or changed the opening question in 56 of 100 modules. Finally, I read each word, sentence, paragraph, and module, asking, "Is this the best word? Can this sentence be made more clear and concise? Does this paragraph stay on task or does it deviate from a point to be made? Is the module a coherent whole?" These questions prompted hundreds of minor changes, as evidenced by the fact that I have added 250 new references and eliminated some 200. In addition, I also updated and revised all tables, charts, and maps, adding new ones and dropping others to reflect the issues of the day.

Each chapter now opens with a video related to the content in the chapter, for instructors to use as lecture-launchers to pique student interest. Video topics include how the recession has altered spending culture in the United States

(Chapter 2, Culture), a study of two-year-old twins and the role that nature and nurture play in their development (Chapter 3, Socialization), and the mechanisms of social control that North Korea uses to get its people to conform and behave in ways that correspond with government ideology (Chapter 6, Deviance).

New to this edition is Aplia, an online interactive homework solution. Aplia helps students understand sociology as a science through fresh and compelling content, brief engagement activities that illustrate key concepts, and thought-provoking questions. Key features include:

- Engagement activities pique student interest and motivate students to learn about a concept. Short experiments, videos, and surveys provide a range of experiential learning opportunities.
- Questions about real-world situations hone students' critical thinking skills.
- Auto-assigned and auto-graded assignments hold students accountable for the material before they come to class, increasing their effort and preparation.
- Immediate, detailed explanations for every answer enhance student comprehension.
- Gradebook Analytics allow instructors to monitor and address performance on a student-by-student and topic-by-topic basis.

Below are module-by-module changes that I consider major:

- In the first edition, there were two modules dedicated to the methods of social research. In the second edition, the two have been incorporated into one (Module 1.6—"Research Methods"). As in the first edition, the module describes a research study to illustrate the steps of research. For this edition, the featured study has changed to one guided by the following question: Do graduates who majored in degrees with clear career paths such as nursing and education do better in the job market initially and later than graduates who majored in, say, sociology or history, degrees with no clear career path?
- Module 2.2, "Material and Nonmaterial Culture," now incorporates a series of new examples that compare and contrast consumption- and conservation-oriented behaviors in Cuba and the United States.
- Module 2.3, "Cultural Diversity," has been completely revised to give special emphasis to exploring what constitutes a culturally diverse setting and to introducing concepts that capture the complexity of cultural diversity, including cultural capital, cultural anchors, subcultures, and countercultures.
- The "Resocialization" module (3.4) has been thoroughly revised to consider the process of resocialization as interactive, during which the affected party reconstructs his or her identity and renegotiates relationships with significant others who must also adjust to the changing person and circumstances.
- The modules titled "Reference Groups" and "Ingroups and Outgroups" in the first edition have been incorporated into one module, "Constructing Identities" (Module 5.5). This newly titled module considers how the groups we belong to (ingroups), the groups we value and to which we aspire to belong (reference groups), and the groups we do not belong to (outgroups) are important to one's identity and sense of self.
- Module 8.3, "Race and Ethnicity in Brazil and the United States," now offers a comparative analysis of racial/ethnic classification in Brazil versus the United States. The different ways that the Brazilian and U.S. governments

define race highlight the social processes by which people are divided into racial-ethnic categories that are implicitly or explicitly ranked on a scale of social worth.

- Module 9.2, "Life Chances and Structural Constraints," considers the concept of gendered institutions and now gives emphasis to the institution of sport, where it is commonplace for men to coach females, but not for females to coach men. Simply consider that 58 percent of women's college teams have a male head coach, while fewer than 2 percent of men's teams have a female head coach.
- Module 9.3, "Gender Stratification," updates the World Economic Forum's annual rankings of the world's countries with regard to gender equality. The latest rankings name Iceland as the country with the greatest gender equality and Yemen with the greatest gender inequality. We consider why this is the case.
- The module on "The U.S. Economy and Jobs" (Module 10.2) now identifies the fastest-growing and fastest-declining occupations and what these trends say about the forces shaping career opportunities. This module also gives special attention to structural factors underlying job loss, including outsourcing and technological innovations like automation.
- Module 11.1, "Defining Family," places new attention on the social forces giving rise to what is popularly known as the "modern family."
- Module 11.7, "Applying Theory: Feminist Theory," now applies feminist theory to the study of family. Feminists argue that the family must be studied in the context of the larger economic and political structures in which they are embedded. In this module, caregiving and work–family balance issues are analyzed in this larger context.
- Module 12.1, "Education and Schooling," significantly strengthens the application of the symbolic and conflict perspectives in an analysis of the processes inside and outside schools that affect academic outcomes.
- Module 12.3, which examines the costs and rewards of education, now includes an analysis of the four models for paying for college and also considers college loan programs in 70 countries, with emphasis on the portion of the money borrowed that students are required to repay and on the percentage of money loaned that is paid back or recovered.
- The "Social Change" module (13.1) still identifies some of the major social forces contributing to societal change. However, robotics is now the object of analysis. Robotics is changing virtually every area of life. It is one innovation in a long line of technologies that has fundamentally changed the way people relate to each other and the environment.
- Module 13.3, "Medical Sociology," is a new module. It describes the specialty of medical sociology, which emphasizes the social context of health care and gives special emphasis to the ways in which health, disease, and illness are defined and experienced. This specialty also focuses on how medical care is organized and delivered, and on the relationships among health care providers and other stakeholders in the delivery of health care.
- Module 13.6, "Applying Theory: Rational Choice Theory and Health Risk," still features rational choice theory but now applies it to health risk—specifically, why people take risks that endanger their health and that of others—and to policies that could change risky behavior.
- Module 14.5, "Ageism and the Anti-Aging Industry," is new. It focuses on the meaning of ageism and the historical factors that have gradually eroded the status of the elderly over time and contributed to widespread ageism. One

manifestation of ageism is the anti-aging industry that promises to fight the physical changes that come with age.

Ancillary Materials

I believe that a textbook is only as good as its supplements. For this reason, I have written the Test Bank, PowerPoint Slides, and Instructor's Manual, with assistance from Kristie Vise, my colleague at NKU, to accompany *Seeing Sociology*. We have tried to create ancillary materials that support the vision of this textbook.

INSTRUCTOR'S MANUAL The Instructor's Manual includes standard offerings such as Learning Objectives and a Key Terms glossary. It also includes several unique features:

- Write-a-Caption answers: A possible caption for each photograph in the Write-a-Caption exercise that ends each module.
- Multiple-choice questions to accompany the short videos that open each chapter. These questions test the students' ability to apply the material in the video to concepts covered in the chapter.
- Sample answers to Critical Thinking questions: Each module ends with a Critical Thinking question, the purpose of which is to get students to reflect on key ideas, concepts, and theories covered. Typically, the questions can be answered in 250 to 400 words. The Instructor's Manual includes a sample answer from an actual sociology student to each Critical Thinking question. The sample answer can serve as an example to share with the students as a way of stimulating thoughts about how to answer these questions. Instructors may also want to read sample answers as a way to prepare for questions students may have about them.

TEST BANK Like most textbooks, the ancillary materials for instructors include a test bank with multiple-choice and true-false questions. In addition to test questions about the textbook material, there are several multiple-choice questions relating to the short film clips. These questions can be found at the end of the multiple-choice questions for each chapter and are labeled by topic ("TOP").

POWERPOINT SLIDES These slides highlight key ideas and points covered in each module. They are useful if instructors want to give students a quick overview of material covered or post online as a review. Included on the PowerPoint® slides are Write-a-Caption exercises as described above.

POWERLECTURE™ WITH EXAMVIEW® This one-stop digital library and presentation tool includes preassembled Microsoft® PowerPoint® lecture slides. In addition to a full Instructor's Manual and Test Bank, PowerLecture™ also includes ExamView® testing software with all the test items from the printed Test Bank in electronic format, enabling you to create customized tests in print or online, and to keep all of your media resources in one place, including an image library with graphics from the book itself and videos.

APLIA™ Aplia helps students understand sociology as a science through fresh and compelling content, brief engagement activities that illustrate key concepts, and thought-provoking questions. Key features include:

- Engagement activities pique student interest and motivate students to learn about a concept. Short experiments, videos, and surveys provide a range of experiential learning opportunities.
- Questions about real-world situations hone students' critical thinking skills.
- Auto-assigned, auto-graded assignments hold students accountable for the material before they come to class, increasing their effort and preparation.
- Immediate, detailed explanations for every answer enhance student comprehension.
- Gradebook Analytics allow instructors to monitor and address performance on a student-by-student and topic-by-topic basis.

SOCIOLOGY COURSEMATE Cengage Learning's Sociology CourseMate brings course concepts to life with interactive learning, study, and exam preparation tools that support the printed textbook. CourseMate includes an integrated eBook, glossaries, flash cards, quizzes, videos, and more—as well as EngagementTracker, a first-of-its-kind tool that monitors student engagement in the course. The accompanying instructor website, available through login .cengage.com, offers access to password-protected resources such as an electronic version of the instructor's manual, test bank files, and PowerPoint® slides. CourseMate can be bundled with the student text. Contact your Cengage sales representative for information on getting access to CourseMate.

Acknowledgements

The acknowledgement section—the place to recognize and give credit to those who have influenced the ideas in this book and its creation—is the most difficult part of the book to write. I have always struggled to find the words to capture the essence and depth of the various relationships that are special to my intellectual and, by extension, personal life. I find myself using clichés like "this book could not have been written without. . .," "I wish to extend my deepest gratitude to. . .," "I acknowledge the profound influence of. . .," and so on. I am never satisfied that I have captured and conveyed the unique relationships and contributions I value so highly. Here, I will simply state the names and acts for which I am most thankful and leave it at that.

Chris Caldeira, my former editor and now a graduate student at the University of California, Davis, conceived the book's structure and approach. She is the lead photographer, contributing 98 photographs to this edition, and she is the person with whom I talk most about this book. Her role is so large that her intellectual and photographic contributions are acknowledged on the title page of this text.

Phillip (deceased) and Annalee Ferrante, my parents, whom I most admire for their work ethic, their optimism and perseverance in the face of difficulties, and their belief that the best effort matters.

Missy Gish, who manages the overwhelming number of details associated with writing a textbook and preparing it for production, including taking photographs.

Kristin Makarewycz, who read the pages of this book in preparation for production and handled the many details of moving the book from my hands to production.

Robert K. Wallace, my husband and colleague, who offers unwavering support.

There are also the colleagues and students (former and current) who contributed one or more photographs to this edition. They include Serina Beauparlant, Kyle Cowgill, Jason Eric Dustin, Katie Englert, Jeremiah Evans, Barbara Houghton, Boni Li, Scott T. McLaren, Tony Rotundo (34 photos), Billy Santos, and Lisa Southwick (50 photos).

Behind the scenes there is a team of people who worked to make this book a reality. You can find their names listed in an unassuming manner on the copyright page of this book. As one measure of the human effort expended, consider that there were dozens of people reading, copyediting, designing, and proofing the pages of the book for at least six months before it reached the market.

I dedicate this book to Dr. Horatio C Wood, IV, M.D. Our relationship goes back to my days in graduate school. Over the decades I have always made a point of formally acknowledging the tremendous influence he has had on my intellectual life, academic career, and philosophy of education. Dr. Wood died on May 28, 2009, but his influence remains as important and strong as ever today.

Reviewers

I am grateful to the many colleagues from universities and colleges across the United States who reviewed the text and provided insightful comments: Annette Allen, Troy University; Andre Arceneaux, Saint Louis University; Arnold Arluke, Northeastern University; Aurora Bautista, Bunker Hill Community College; David P. Caddell, Mount Vernon Nazarene University; Gregg Carter, Bryant University; Andrew Cho, Tacoma Community College; Mirelle Cohen, Olympic College; Pamela Cooper, Santa Monica College; Janet Cosbey, Eastern University; Gayle D'Andrea, J. Sargeant Reynolds Community College; Karen Dawes, Wake Technical Community College; Kay Decker, Northwestern Oklahoma State University; Melanie Deffendall, Delgado Community College; David Dickens, University of Nevada, Las Vegas; Dennis Downey, University of Utah; Angela Durante, Lewis University; Keith Durkin, Ohio Northern University; Murray A. Fortner, Tarrant County College; Matt Gregory, Tufts University and Boston College; Derrick Griffey, Gadsden State Community College; Laura Gruntmeir, Redlands Community College; Kellie J. Hagewen, University of Nebraska; Anna Hall, Delgado Community College; Laura Hansen, University of Massachusetts; James Harris, Dallas Community College; Garrison Henderson, Tarrant County College; Melissa Holtzman, Ball State University; Xuemei Hu, Union County College; Jeanne Humble, Bluegrass Community and Technical College; Allan Hunchuk, Thiel College; Hui M. Huo, Highline Community College; Faye Jones, Mississippi Gulf Coast Community College; Rachel Tolbert Kimbro, Rice University; Philip Lewis, Queens College; Carolyn Liebler, University of Minnesota; Beth Mabry, Indiana University of Pennsylvania; Gerardo Marti, Davidson College; Tina Martinez, Blue Mountain Community College; Donna Maurer, University of Maryland, University College; Marcella Mazzarelli, Massachusetts Bay Community College; Jeff McAlpin, Northwestern Oklahoma State University; Douglas McConatha, West Chester University of Pennsylvania; Janis McCoy, Itawamba Community College; Elizabeth McEneaney, California State University, Long Beach; Melinda Messineo, Ball State University; Arman Mgeryan, Los Angeles Pierce College; Cathy Miller, Minneapolis Community and Technical College; Krista Lynn Minnotte, Utah State University; Lisa Speicher Muñoz, Hawkeye Community College; Elizabeth Pare, Wayne State University; Denise Reiling, Eastern Michigan University; Robert Reynolds, Weber State University;

Judith Richlin-Klonsky, Santa Rosa Junior College; Lisa Riley, Creighton University; Luis Salinas, University of Houston; Alan Spector, Purdue University, Calumet; Rose A. Suggett, Southeast Community College; Don Stewart, University of Nevada, Las Vegas; Toby Ten Eyck, Michigan State University; Katherine Trelstad, Bellevue College; Sharon Wettengel, Tarrant Community College; Robert Wood, Rutgers University; and James L. Wright, Chattanooga State Technical Community College.

I also thank the focus group participants whose valuable feedback helped us to shape this book: Ghyasuddin Ahmed, Virginia State University; Rob Benford, Southern Illinois University; Ralph Brown, Brigham Young University; Tawny Brown, Warren, Columbia College Online; Kay Coder, Richland College; Jodi Cohen, Bridgewater State College; Sharon Cullity, California State University, San Marcos; Anne Eisenberg, State University of New York, Geneseo; Dana Fenton, Lehman College; Rhonda Fisher, Drake University; Lara Foley, University of Tulsa; Glenn Goodwin, University of La Verne; Rebecca Hatch, Mt. San Antonio College; Idolina Hernandez, Lone Star College, CyFair; Joceyln Hollander, University of Oregon; Jennifer Holsinger, Whitworth University; Amy Holzgang, Cerritos College; Michelle Inderbitzin, Oregon State University; Mike Itashiki, Collin County Community College; Greg Jacobs, Dallas Community College District; Kevin Lamarr James, Indiana University, South Bend; Krista Jenkins, Fontbonne University; Elizabeth Jenner, Gustavus Adolphus College; Art Jipson, University of Dayton; Maksim Kokushkin, University of Missouri, Columbia; Amy Lane, University of Missouri, Columbia; Marci Littlefield, Indiana University–Purdue University, Indianapolis; Belinda Lum, University of San Diego; Ali Akbar Mahdi, Ohio Wesleyan University; Tina Martinez, Blue Mountain Community College; Lori Maida, State University of New York, Westchester Community College; Linda McCarthy, Greenfield Community College; Richard McCarthy and Elizabeth McEneaney, California State University, Long Beach; Julianne McNalley, Pacific Lutheran University; Krista McQueeney, Salem College; Angela Mertig, Middle Tennessee State University; Melinda Miceli, University of Hartford; Anna Muraco, Loyola Marymount University; Aurea Osgood, Winona State University; Rebecca Plante, Ithaca College; Dwaine Plaza, Oregon State University; Jennifer Raymond, Bridgewater State College; Cynthia Reed, Tarrant County College, Northeast; Nicholas Rowland, Pennsylvania State University; Michael Ryan, Dodge City Community College; Martin Sheumaker, Southern Illinois University, Carbondale; Carlene Sipma-Dysico, North Central College; Juyeon Son, University of Wisconsin, Oshkosh; Kathy Stolley, Virginia Wesleyan College; Stacie Stoutmeyer, North Central Texas College, Corinth Campus; Ann Strahm, California State University, Stanislaus; Carrie Summers-Nomura, Clackamas Community College; Zaynep Tufekci, University of Maryland; Deidre Tyler, Salt Lake Community College; Georgie Ann Weatherby, Gonzaga College; Sharon Wettangel, Tarrant County College; Rowan Wolf, Portland Community College; James Wood, Dallas County Community College; LaQueta Wright, Dallas Community College District; and Lori Zottarelli, Texas Woman's University.

I am grateful to the student focus group participants who provided helpful input: Greg Arney, Pauline Barr, Wesley Chiu, John Liolos, Jessica St. Louis, Fengyi (Andy) Tang, and Rojay Wagner.

I also appreciate the useful feedback of the survey participants: Mike Abel, Brigham Young University, Idaho; Wed Abercrombie, Midlands Technical College, Airport; Dwight Adams, Salt Lake Community College; Chris

Adamski-Mietus, Western Illinois University; Isaac Addaii, Lansing Community College; Pat Allen, Los Angeles Valley College; Robert Aponte, Indiana University–Purdue University, Indianapolis; David Arizmendi, South Texas College, McAllen; Yvonne Barry, John Tyler Community College; Nancy Bartkowski, Kalamazoo Oakley, University of Cincinnati; Kirsten Olsen, Anoka Ramsey Community College; Roby Page, Radford University; Richard Perry, Wake Technical Community College, Raleigh; Kenya Pierce, College of Southern Nevada, Cheyenne; Sarah Pitcher, San Diego City College; Cynthia Reed, Tarrant County College, Northeast; Paul Renger, Saint Philip's College; Melissa Rifino-Juarez, Rio Hondo College; Barbara Ryan, Widener University; Rita Sakitt, Suffolk Community College, Ammerman; Luis Salinas, University of Houston; Robert Saute, William Paterson University; Terri Slonaker, San Antonio College; Kay Snyder, Indiana University of Pennsylvania; Julie Song, Chaffey College; Andrew Spivak, University of Nevada, Las Vegas; Kathleen Stanley, Oregon State University; Rachel Stehle, Cuyahoga Community College, Western; Susan St. John-Jarvis, Corning Community College; Tanja St. Pierre, Pennsylvania State University, State College; Rose Suggett, Southeast Community College, Lincoln; Becky Trigg, University of Alabama, Birmingham; Tim Tuinstra, Kalamazoo Valley Community College; David Van Aken, Hudson Valley Community College; Vu Duc Vuong, De Anza College; Tricia Lynn Wachtendorf, University of Delaware; Kristen Wallingford, Wake Technical Community College, Raleigh; Margaret Weinberg, Bowling Green State University; Robyn White, Cuyahoga Community College; Beate Wilson, Western Illinois University; Sue Wortmann, University of Nebraska, Lincoln; Bonnie Wright, Ferris State University; Sue Wright, Eastern Washington University; Delores Wunder, College of DuPage; and Meifang Zhang, Midlands Technical College, Beltline.

Theme Index

INDEX OF HEALTH CARE COVERAGE

INDEX OF GLOBALIZATION/GLOBAL INTERDEPENDENCE COVERAGE

INDEX OF DIVERSITY COVERAGE

INDEX OF ENVIRONMENT COVERAGE

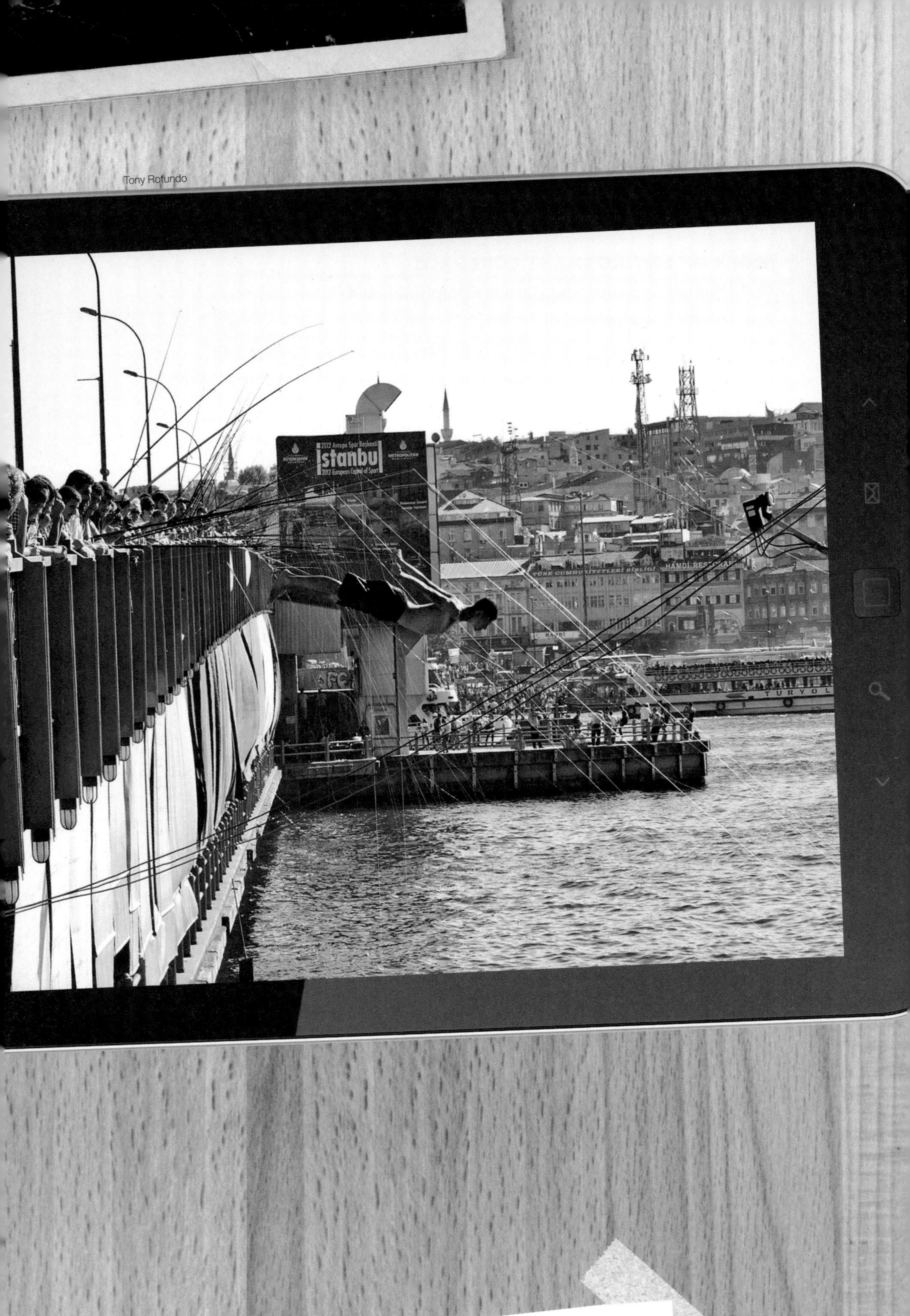

Tony Rotundo

CHAPTER 1

THE SOCIOLOGICAL PERSPECTIVE

Sociology is a field of study that pushes you to see the world around you in new ways. This journey requires you to leave your comfort zone, to be curious about events taking place around you, and to wonder and care about those who live nearby and far away. By taking this journey, you will come to see how social experiences—the time you are born in history, the place you live, and the countless number of people known and unknown with whom you interact and depend—profoundly shape what you think and do. You will come to understand that "things are not what they seem" (Berger 1963, 21). You will also come to see that, for better or worse, what you think and do affects those with whom you live, work, and share the planet. Knowing these things will help you to be a positive force in your own life and others' lives as well.

Go to Sociology CourseMate on cengagebrain.com to watch a video on how smartphones have become a social force, transforming human interaction and activities.

MODULE 1.1 What Is Sociology?

Objective

You will learn that sociology is a field of study that pushes its students to ask questions about the way human activities are organized. The answers almost always reveal that things are not what they seem.

Chris Caldeira

Do you sometimes wonder what is going on inside buildings as you walk past them on the street? Given the opportunity, would you take the time to look inside?

If you answered yes, then you will certainly appreciate the sociological perspective. We can think of sociologists as curious observers walking the streets, fascinated with what they cannot see taking place behind the building walls. The buildings themselves offer few clues about the people who built them and who may no longer live there. The drive to understand how and why these structures were built, to look inside and to learn more, captures the sociological mind-set (Berger 1963).

Sociology: A Definition

Sociology is the scientific study of human activity in society. **Human activity** involves all the things people do with, to, and for one another and what they think and do as a result of others' influence. The activities sociologists study are age-old and too many to name, but they may include people searching for work, securing food, adorning the body, competing for some desired outcome (a scholarship, a job, love), celebrating something, communicating, assigning people to categories, consuming products, burying the dead, and so on.

When sociologists study human activities, they focus on social forces that shape the way human relationships and activities are organized and the effects that organization has on people's opportunities, sense of self, and relationships with others and the larger environment.

Social forces are anything human-created that influence, pressure, or push people to interact, behave, or think in specified ways. People can embrace or challenge social forces, be swept along or be passed by. The mobile phone is one example of a social force. It is a human-created technology that has transformed and is still transforming human activity in countless ways.

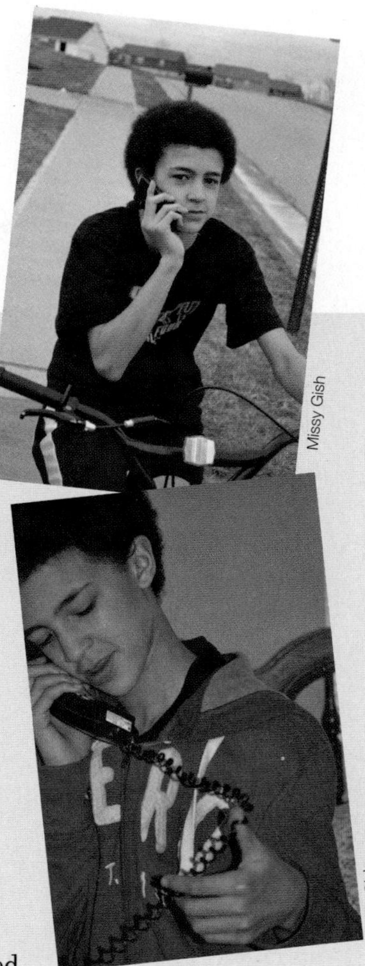
Missy Gish
Missy Gish

➤ Keeping in touch with others is a human activity. The mobile phone is a revolutionary social force that has changed the way we keep in touch. Notably it frees people from being in a specific physical space when they communicate with others. While the landline phone allowed people to communicate with others in faraway places, all parties had to arrange to be in a fixed location—an office, at home, in a telephone booth. In addition, people waited until they got home or to some other fixed location to call someone. With the mobile phone people can contact others at will. While most people have embraced the mobile phone as a necessity, many also have been swept along by this technology lamenting that it seems to control them. A shrinking minority across the globe have been bypassed, having no access to this technology. The point is that sociologists focus on the social forces or things human beings create and then examine the meanings assigned, the consequences (intended and unintended), and issues surrounding who controls and benefits from those creations.

The following are just three examples of the many human activities covered in this textbook. Each is accompanied by some of the questions that sociologists ask in their quest to learn how human activities are organized and analyze their associated effects. At this point we will not look to answer the questions posed; the answers can be found in relevant modules identified in parentheses. This is simply a preview of the kinds of social activities sociologists study and the kinds of questions they ask. In time you will learn that there is a pattern to these questions and that the discipline offers a rich vocabulary to assist in answering them.

Lisa Southwick

❮ Humans everywhere create and assign people to categories they consider important. These categories may relate to gender, race, and age. In the United States, most would see this mother and her baby daughter as belonging to different racial categories. Sociologists ask, What social forces help to explain the decision to classify biologically related people as distinct races? How did this practice come to be? Who has the power to put such a practice in place? How might classifying a mother and daughter as different races affect the relationship between the two? What effect might this practice have on race relations in general? (See Chapter 8, Module 8.1.)

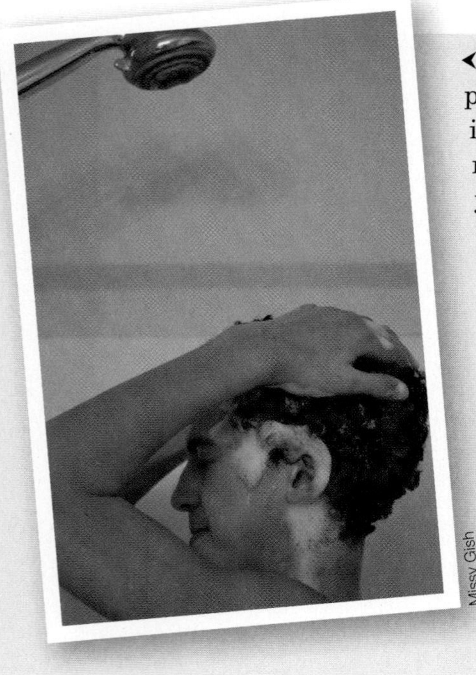

Missy Gish

The human activity depicted in this photograph shows a way of washing hair while in the shower. Notice that the water is not running. When you shampoo your hair, would you think to first wet it, turn the water off as you lathered, then turn the water back on to rinse? If you live in the United States, you would likely keep the water running. If you live in Germany, you would likely turn the water off while you lather and wash your hair, then turn it back on to rinse. Sociologists ask, What does this difference suggest about German and American societies? What social forces help to explain people's "decision" to use water in this way? What meanings do Germans and Americans assign to natural resources such as water? (See Chapter 2, Module 2.4.)

AP Photo/Koji Sasahara

The South Korean government is investing $100 million to create robots that can teach students English. Eventually, robots will perform the labor-intensive activity of teaching a language. They are being built to ease a shortage of English-language teachers in the country and to offset expenses associated with hiring instructors from countries such as the United States (J. Young 2010). The robots are expected to be in classrooms by 2013. Sociologists ask, What are the intended and unintended consequences of using robots to teach students a language? Who benefits from this arrangement and at whose expense? What will interaction between students and robots be like? In what ways, if any, will it be like human interaction? In what ways will it differ?

Why Study Sociology

Sociology offers a framework to help us understand how human activity is organized. The sociological framework is especially relevant today, if only because many are questioning the ways some human activities have been organized. For example, many are questioning our dependence on fossil fuels to power the trains, planes, cars, and buses that move people and goods from one place to another, and to power the batteries that run computers, mobile phones, and other smart technologies. The fact that most human activity depends on fossil fuels is something that concerns those who believe fossil fuels are too costly—environmentally, economically, and politically. Some have responded by turning to wind and solar power.

➤ This laundromat in San Francisco suggests that there are other ways to organize human activity, but of course such efforts are resisted by the millions of people worldwide who make their living extracting, refining, delivering, and selling fossil fuels. Still, sociologists are interested in understanding the processes by which change is initiated and resisted.

Chris Caldeira

As we will see, sociology offers a perspective that prompts us to identify social forces shaping any human activity and to ask questions about that activity's consequences. In the process we are positioned to recognize flaws in the way an activity has been organized. Real change leaders understand there are many ways to organize human activities, and they are committed to taking risks, thinking outside the box, and motivating and inspiring others to organize things differently (Katzenbach et al. 1997).

(Write a Caption)

Write a caption that addresses one way in which the mobile phone is a social force affecting how human relationships and activities are organized.

Lisa Southwick

Hints: In writing this caption

- think about the age of the child and how the mobile phone affects his relationship with his parents;
- consider the likelihood that his parents will contact him directly or go through a third party, such as another parent who could verify his location; and
- reflect on what is lost or gained as a result of this direct connection.

Critical Thinking

Name a social force that has affected the way you relate to others. Explain. To date, how have you embraced, challenged, and/or resisted that force?

Key Terms

human activity
sociology
social forces

MODULE

1.2 The Sociological Imagination

Objective

You will learn that the sociological imagination allows us to see how the larger forces related to time and place shape our personal lives and the society in which we live.

Prints & Photographs Division, Library of Congress, LC-USZ62-92188

Lisa Southwick

The photo to the left was taken in the early 1900s in the United States; the contemporary photo to the right was also taken in the United States. In what ways might each baby's expectations about what to do with his or her time be shaped by each kind of learning experience?

Sociologist C. Wright Mills (1963) wrote about the **sociological imagination**, a perspective that allows us to consider how outside forces, especially the time in history and society (place) in which we live, shape our life story or biography. A **biography** consists of all the events and day-to-day interactions from birth to death that make up a person's life. The photos opening this module illustrate the connection between biography and the forces of time and place. Notice that the child born in 1900 (left) is focused primarily on a specific task—walking. Compared with the baby on the right, there are few things to distract his attention from the task at hand. The child to the right is learning to stand, and for the moment her attention is on the flowers but could shift at any instant. We might argue that these experiences work to create children with different expectations and tolerances about how the time in their day should be filled and the importance of material items to occupy their attention.

The forces of history and society affect our most personal experiences. For example, how might the forces of society shape expressions of affection and closeness? It seems that the rules governing touch in the United States dictate that "one must be in a romantic relationship to get much touch, that touch has sexual connotations, and that daily interpersonal interaction tends not to involve touch" (Traina 2005).

➤ Many societies such as Papua New Guinea (left) allow men to walk arm-in-arm in public without assuming romantic involvement. Not so in the United States, where for the most part physical contact between same-sex friends is discouraged.

The sociological imagination is empowering because it allows those who possess it to distinguish between what sociologist C. Wright Mills (1959) called troubles and issues. Mills (1959) defined **troubles** as individual problems, or difficulties, that are attributed to personal shortcomings related to motivation, attitude, ability, character, or bad judgment. The resolution of a trouble, if it can indeed be resolved, lies in changing the person in some way. For example, Mills states that when only one person is unemployed in a city of 100,000, that situation is a trouble. For its relief, we focus on that person's shortcomings—"She is lazy," "He has a bad attitude," "He didn't try very hard in school," or "She had the opportunity but didn't take it."

An **issue**, on the other hand, is a societal matter that affects many people and that can only be explained by larger social forces that transcend the individuals affected. When 24 million men and women are unemployed or underemployed in a nation with a workforce of 156 million, that situation is an issue. Clearly, we cannot hope to solve widespread unemployment by focusing on the character flaws of 24 million individuals. A constructive assessment of this crisis requires us to think beyond personal shortcomings and to consider the underlying social forces that created it. For example, the economy is structured so that corporate success is measured by ever-increasing profit margins. Under such an arrangement, profits are increased by lowering labor costs, which can be achieved through laying off employees, downsizing, transferring jobs from high-wage to low-wage areas, and otherwise eliminating the amount of human labor needed to produce a product or deliver a service.

Chris Caldeira

When you drive through neighborhoods where houses have been abandoned or boarded up, do you think the former residents irresponsible? Or do you wonder whether these homes once belonged to those who worked at a factory or other place of employment that shut down or moved to a location where labor is less expensive? If you see the residents as the source of the problem, then you are framing neighborhood deterioration as a trouble; if you place greater emphasis on economic forces that leave people without jobs or with lowered wages, then you are framing neighborhood deterioration as an issue.

The ability to distinguish between troubles and issues allows us to think more deeply about the cause of and potential solutions to problems that seem, on the surface, to be entirely personal. Arguably the best-known effort to connect personal troubles to larger social issues was that of sociologist Émile Durkheim, who wrote *Suicide* in 1897 and who is still regarded as an authority on that subject today.

Suicide

When we think about who commits suicide, we often think of people who are deeply troubled. In *Suicide*, Durkheim argued that it is futile to study the immediate circumstances that lead people to kill themselves, because an infinite number of such circumstances exist. For example, one person may kill herself in the midst of newly acquired wealth, whereas another kills herself in the midst of poverty. One person may kill himself because he is unhappy in his marriage and feels trapped, whereas another kills himself because his unhappy marriage has just ended in divorce. We can find cases of people who kill themselves after losing a business; in other cases a lottery winner kills himself because he cannot tolerate family and friends fighting one another to share in the newfound fortune. Because almost any personal circumstance can serve as a pretext for suicide, Durkheim concluded that any situation could serve as an occasion for someone's suicide.

Given these conceptual difficulties, Durkheim offered a definition of suicide that goes beyond its popular meaning (the act of intentionally killing oneself). His definition—the severing of relationships—takes the spotlight off the victim and points it outward toward relationships that have been severed. To make his case that relationships are key to understanding suicide, he argued that every group has a greater or lesser propensity for suicide. The suicide rates for various age, sex, and race groups in the United States, for example, show that suicide is more prevalent for some categories of people—males in general and especially males age 65 and older. From a sociological perspective, these differences in suicide rates cannot be explained by pointing to each victim's immediate circumstances. Durkheim believed that comparing suicide rates by group yields important insights about the larger social forces that push people to take their own lives. To grasp this point, we consider suicide rates for men and women of different ages and ask, What do rates suggest about men's and women's relationships to others? (See Chart 1.2a.)

▼ Chart 1.2a: Male-Female Differences in Suicide Rates (per 100,000), United States
The chart shows the annual number of suicides per 100,000 people for eight age and sex groups. Note that males age 85 and over have the highest suicide rate—each year 45 of every 100,000 men commit suicide. Females in that age category have a suicide rate of 4 of every 100,000 suicides. Is there any age category where females have the higher rate relative to males?

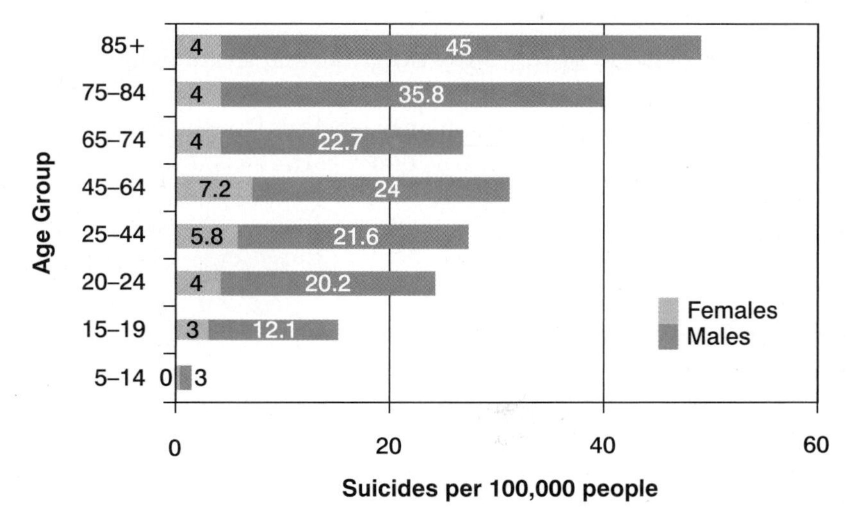

Source: Centers for Disease Control (2009)
Note: This data is based on deaths that have been deemed suicide by a medical examiner. The numbers likely underestimate the actual number of suicides committed each year.

In reviewing these rates, it is important to point out that women attempt suicide about three times more often than men. In addition, the most common method of suicide among women is poisoning with drugs or other chemicals, a method that is more likely to fail; suicide among men most commonly involves firearms (Centers for Disease Control 2009). Durkheim would maintain that the different suicide rates reflect the pressures the group places on men to succeed at suicide and not use it as a cry for help, as well as males' greater access to guns and knowledge of how to use them (guns are viewed as an expression of masculinity and as something males, and not females, are expected to possess and know how to use).

In thinking about the character of relationships, Durkheim identified four types of relationships or social ties that bind the individual too weakly or strongly to others: egoistic, altruistic, anomic, and fatalistic.

Egoistic describes a state in which the ties attaching the individual to the others in society are weak. When individuals are weakly attached to others, they encounter less resistance to suicide. Relative to men, society offers women more opportunities to form meaningful relationships with others; women are disproportionately assigned nurturing roles and men are disproportionately assigned roles that distance them from others. These differences in the ties that bind men and women to the society offer insights about why males have a much higher rate of suicide.

Altruistic describes a state in which the ties attaching the individual to the group are such that a person's sense of self cannot be separated from the group. When such people commit suicide, it is on behalf of a group they love more than themselves. The classic example is soldiers willing to sacrifice their lives to advance the ideals and cause of their unit.

can someone aquire all of these states?

Anomic describes a state in which the ties attaching the individual to the group are disrupted due to dramatic changes in circumstances. Durkheim gave particular emphasis to economic circumstances, such as recession, depression, or economic boom. In all cases, a reclassification occurs that suddenly casts individuals into lower or higher statuses. Those cast into a lower status must reduce their desires, restrain their needs, and practice self-control. Those cast into a higher status must adjust to increased prosperity. Aspirations and desires may be unleashed, and an insatiable thirst to acquire more goods and services, and even to feel that one needs still more, may arise.

Fatalistic describes a state in which the ties attaching the individual to the group are so oppressive that there is no hope of release. Under such conditions, individuals see their futures as permanently blocked. Durkheim asked, "Do not the suicides of slaves, said to be frequent under certain conditions, belong to this type?" (1897, 276).

(Write a Caption)

Write a caption that applies the terms *troubles* and *issues* to this unemployed person's situation.

Hints: In writing this caption

- review the distinction between troubles and issues,
- think about the underlying causes to a trouble versus an issue, and
- consider a solution to unemployment when it is framed as a trouble versus when it is framed as an issue.

© iStockphoto.com/Catherine Lane

Critical Thinking

Describe something in your biography that is shaped by the time in history in which you were born or by the society (place) in which you live.

Key Terms

altruistic	egoistic	sociological imagination
anomic	fatalistic	troubles
biography	issue	

The Emergence of Sociology

MODULE 1.3

Objective

You will learn about a major historical event that triggered the birth of sociology.

Chris Caldeira

Barbara Houghton

What would it be like to live in a society where mechanization did not exist—that is, a society where human or animal muscle, not fossil fuels, was the source of power?

Throughout much of history, human and animal muscle were the key sources of power in all labor-intensive activities. This changed with the Industrial Revolution, the name given to the changes in manufacturing, agriculture, transportation, and mining that transformed virtually every aspect of society from the 1300s on. The defining feature of the Industrial Revolution was mechanization, the process of replacing human and animal muscle with machines powered by burning wood, coal, oil, and natural gas. The new energy sources eventually replaced hand tools with power tools, sailboats with steamships and then freighters, and horse-drawn carriages with trains. Mechanization changed how goods were produced and how people worked. It turned workshops into factories, skilled workers into machine operators, and handmade goods into machine-made products.

Prints & Photographs Division, Library of Congress, LC-DIG-nclc-05247

❮ Consider the effort required to make bread before mechanization. Bakers plunged fists into gluey dough and massaged it with their fingers until their muscles hurt (Zuboff 1988).

Prints & Photographs Division, Library of Congress, LC-DIG-nclc-05178

❯ People also took their dough to small local bakeries, where it was shaped and baked in wood- or coal-heated brick ovens. This baker and his apprentice used long-handled wooden shovels to move bread in and out of the oven (Advameg, Inc. 2007).

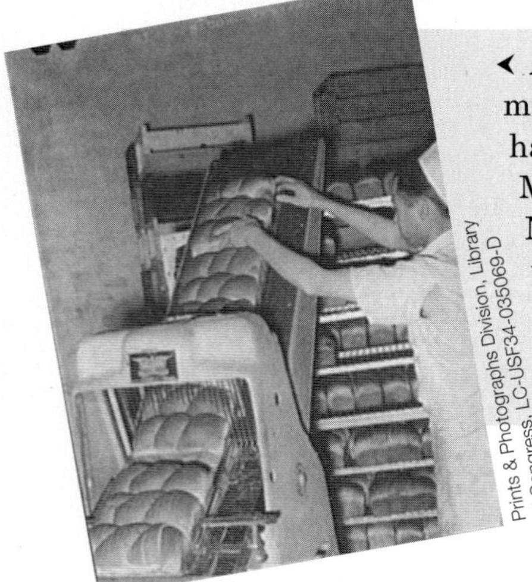

Prints & Photographs Division, Library of Congress, LC-USF34-035069-D

❮ After mechanization, the human muscle needed to make bread was largely eliminated. Bakers no longer had to spend seven or more years as apprentices. Machines rendered the time-consuming skills obsolete. Now people with little or no skill could do the baker's work, and at a faster pace. Before mechanization, customers knew the person who baked their bread. With mechanization, they came to depend on "strangers" to sustain them.

Changes to Society

Bread baking eventually moved out of the home and small bakery shops, and by the 1940s commercial bakeries were stocking grocery shelves. While this may seem unimportant, it is just one example of the way the Industrial Revolution weakened people's ties to others in their community, their workplace, and their home. Other innovations that changed the ways people related to each other included the railroad, the steamship, running water, central heating, electricity, and the telegraph. Monthlong trips by stagecoach, for example, became daylong trips by coal-powered trains. These trains permitted people and goods to travel day and night—in rain, snow, or sleet—and to previously unconnected areas.

The railroad and other innovations in transportation facilitated economic competition, interdependence, and upheaval. Now people in one area could be priced out of a livelihood if people in another area could provide lower-cost labor, goods, and/or materials (J. Gordon 1989). In the case of bread, the innovations in transportation allowed bread to be made outside the community and then shipped long distances for sale to people the makers did not know. The nature of bread changed to accommodate this new reality; over time, dozens of additives gave commercial bread a standard texture, shape, taste, and most importantly, a shelf life that allowed it to be shipped long distances and sit on a store shelf for weeks.

The Industrial Revolution, centered in Europe and the United States, pulled together, often by force, people from even the most remote parts of the planet. The resulting division of labor meant that people who did not know one another were now interconnected.

▲ The Industrial Revolution included the assembly line. To produce a car, Ford workers depended on more than those working in the factory. They also depended on workers in remote corners of the world, such as this rubber tapper (right) in Indonesia (a Dutch colony), to supply the material for tires. A strike in the U.S. factory or a disease to the latex-producing rubber plants would affect workers in both countries. (Both photographs were taken in 1923.)

In sum, the Industrial Revolution changed everything—the ways in which goods were produced, the relationships between what were once geographically separated peoples, the ways people made their livings, the density of human populations, the relative importance and influence of the home in people's lives, and the emergence of a consumption-oriented economy and culture. The accumulation of wealth became a valued pursuit. In *The Wealth of Nations*, Adam Smith (1776) argued that the invisible hand of the free market via private ownership and self-interested competition held the key to progress and prosperity. These unprecedented changes caught the attention of the early sociologists. In fact, sociology emerged out of their effort to understand the effects of the Industrial Revolution on society.

(Write a Caption)

Write a caption that applies the defining feature of the Industrial Revolution to the photograph and describe at least one change that defining feature ushered in.

Hints: In writing this caption

- determine what is the defining feature of the Industrial Revolution, and
- consider at least one consequence of shifting from horse-drawn carriages to coal-powered trains.

Critical Thinking

Identify an activity in your life that depends on burning fossil fuels (electricity, oil consumption). Explain its significance.

MODULE

1.4 The Early Sociologists

Objective

You will learn how the writings of the early sociologists shape our understanding of society today.

Auguste Comte · Karl Marx · Émile Durkheim · Max Weber · W.E.B. DuBois · Jane Addams

Do you feel any connection to the past when you look at these photos of people, considered great thinkers in their time, who still influence thinking today?

In this module we consider the transforming ideas of six early theorists. Three of the six—Karl Marx, Émile Durkheim, and Max Weber—are nicknamed the "big three" because their writings form the heart of the discipline. We also consider three other central figures: Auguste Comte because he gave sociology its name, Jane Addams for her methods of applying knowledge, and W.E.B. DuBois for his work on the color line.

Auguste Comte (1798–1857)

French philosopher and father of positivism Auguste Comte gave sociology its name in 1839. **Positivism** holds that valid knowledge about the world can be derived only from using the scientific method. Comte maintained that sociologists were scientists who studied the results of the human intellect (DeGrange 1939). What did he mean by this?

First, sociology is a science, and only those sociologists who follow the scientific method can presume to have a voice in describing and guiding human affairs. Second, sociologists study the things humans have created and their effects on society. Comte recommended that sociologists study **social statics**, the forces that hold societies together and give them endurance over time, and **social dynamics**, the forces that cause societies to change. Comte's preoccupation with order and change is not surprising, given that he was writing at a time when the Industrial Revolution was transforming society in unprecedented ways.

Chris Caldeira

➤ Comte gave us concepts for thinking about the social forces that hold any town, city, or place together over time even as new residents are born or move in and as other residents move out or die. Those concepts also encourage us to focus on how a place changes over time and on the social forces responsible. As a case in point, Eureka, Nevada, was established as a mining town (gold, silver, and lead) in 1864, reached its peak population in 1878 (10,000 pop.), and today is home to about 1,100 people. From the beginning the town's identity centered on mines, but the worth of the ore was determined by global demand. As the value of what Eureka mined declined in the marketplace and residents and those in surrounding towns moved away, Eureka's identity still centered on its mines as tourist attractions and its location along the "Loneliest Road in America."

Karl Marx (1818–1883)

Karl Marx was born in Germany but spent much of his professional life in London working and writing in collaboration with Friedrich Engels. His most well-known writing is *The Communist Manifesto*, a 23-page pamphlet issued in 1848 and translated into more than 30 languages (Marcus 1998). Upon reading it today, one is "struck by the eerie way in which its 1848 description of capitalism resembles the restless, anxious and competitive world of today's global economy" (Lewis 1998, A17).

According to Marx, the sociologist's task is to analyze and explain **conflict**, the major force that drives social change. Specifically, Marx saw class conflict as the vehicle that propelled society from one historical epoch to another. He described **class conflict** as an antagonism growing out of the opposing interests held by exploiting and exploited classes. The nature of that conflict is shaped by each class's relationship to the **means of production**, the resources such as land, tools, equipment, factories, transportation, and labor that are essential to the production and distribution of goods and services.

The Industrial Revolution was accompanied by the rise of two distinct classes: the **bourgeoisie**, the owners of the means of production; and the **proletariat**, those individuals who must sell their labor to the bourgeoisie. The bourgeoisie's interest lies with making a profit and the proletariat's with increasing wages. To maximize profit, the bourgeoisie search for labor-saving technologies, employ the lowest-cost workers, and find the cheapest materials to make products.

Marx believed that the pursuit of profit drove the explosion of technological innovation and the unprecedented levels of production during the Industrial Revolution. He felt that capitalism was the first economic system capable of maximizing human ingenuity and productive potential. He also maintained that capitalism ignored too many human needs, that too many workers could not afford to buy the products of their labor, and that relentless efforts to reduce labor costs left the worker vulnerable and insecure. Marx called the drive for profit a "boundless thirst—a werewolf-like hunger"—that takes no account of the health and the length of life of the worker unless society forces it to do so" (Marx 1887, 142). Marx believed that if this economic system were in the right hands—the hands of the workers, or the proletariat—public wealth would be abundant and distributed according to need.

➤ Although the workers in these photographs are separated by more than a hundred years, Marx would argue that they have something in common: they are among the proletariat because they own only their labor, which they must sell to the bourgeoisie, who are always looking to cut labor costs.

Lou Linwei/Alamy

© CORBIS

Émile Durkheim (1858–1918)

To describe the Industrial Revolution and its effects, Frenchman Émile Durkheim focused on the division of labor and solidarity. Division of labor is the way a society divides up and assigns day-to-day tasks. Durkheim was interested in how the division of labor affected **solidarity**, the system of social ties that acts as a cement bonding people to one another and to the wider society. Durkheim observed that industrialization changed the division of labor and, by extension, the nature of solidarity from mechanical to organic.

Preindustrial societies are characterized by **mechanical solidarity**, a system of social ties based on uniform thinking and behavior. Durkheim believed that uniformity is common in societies with a simple division of labor where, for the most part, everyone performs the same tasks needed to maintain their livelihood. This sameness gives rise to common experiences, skills, and beliefs. In preindustrial societies, religion and family are extremely important. As a result, the social ties that bind are grounded in tradition, obligation, and duty. These societies do not have the technology or resources to mass-produce a variety of products that people can buy to distinguish themselves from others.

The Industrial Revolution fueled a different kind of solidarity that resulted from a complex division of labor where the workers needed to create a product did not have to know or live near one another; in fact, they often lived in different parts of a country or the world. In addition, the materials needed to make products came from many locations around the globe.

Industrialization gave rise to **organic solidarity**, a system of social ties such that for the most part people relate to others in terms of their specialized roles in the division of labor and as customers. Customers buy tires from a dealer, travel by airplane flown by a pilot they might never see, and drink coffee made from beans harvested and roasted by people they've never met. Customers do not need to know someone personally to interact with him or her. In industrial societies most day-to-day interactions are short-lived, impersonal, and instrumental (i.e., we interact with most people for a specific reason). In addition, few individuals possess the knowledge, skills, and resources to be self-sufficient. Consequently, social ties are still strong, not because people know one another, but because strangers depend upon one another to survive.

➤ This Eskimo family (top) likely has the skills to survive on its own. Thus the family lives in an environment characterized by mechanical solidarity. It is unlikely that the family below possesses the skills to survive on its own. This family's way of life—the clothes its members wear, the food they eat, the house they live in—depends on the labor of strangers, most of whom do not live in the immediate community, but are spread across the world.

Prints & Photographs Division, Library of Congress, LC-USZ62-135995

Lisa Southwick

Max Weber (1864–1920)

The German scholar Max Weber made it his task to analyze and explain how the Industrial Revolution affected **social action**—actions people take in response to others—with emphasis on the forces that motivate people to act. Weber suggested that sociologists should focus on the meanings guiding thought and action. He believed that social action is motivated in one of four ways. In reality, motives are not so clear-cut but involve some mixture of the four.

1. *Traditional*—a goal is pursued because it was pursued in the past (i.e., "that is the way it has always been").
2. *Affectional*—a goal is pursued in response to an emotion such as revenge, love, or loyalty (a soldier throws him- or herself on a grenade out of love and sense of duty for those in the unit).
3. *Value-rational*—a desired goal is pursued with a deep and abiding awareness that there can be no shortcuts or compromises made in reaching it. Instead, the actions taken to reach a desired goal are guided by a set of standards or codes of conduct (Weintraub and Soares 2005).
4. *Instrumental rational*—a valued goal is pursued by the most efficient means irrespective of the consequences. In the context of the Industrial Revolution, the valued goal is profit, and the most efficient means are the most cost-effective ones, no matter the costs to workers' health, society, or the environment. In contrast to value-rational action, no code of conduct governs the pursuit of goals. One might equate this type of action with that of an addict who will seek a drug at any cost to self or to others. There is an inevitable self-destructive quality to this form of action (Henrik 2000).

Weber believed that instrumental rational action could lead to **disenchantment**, a great spiritual void accompanied by a crisis of meaning in which the natural world becomes less mysterious and revered and becomes the object of human control and manipulation. In the case of raising chickens for human consumption, disenchantment results when the goal of profit outweighs any moral responsibility to treat the animals with kindness and when the means used to turn a profit are such that "we no longer recognize the animals in a factory farm as living creatures capable of feeling pain and fear" (Angier 2002, 9).

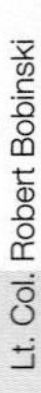

Lt. Col. Robert Bobinski

Navy Petty Officer 2nd Class Manhea Kim

▲ Disenchantment is an outcome of instrumental rational action. This kind of action is embodied in the way factory farms raise baby chicks for eventual slaughter (left). From birth on they are destined to live in overcrowded conditions never seeing the light of day. This treatment represents the most cost-efficient means to reach the valued goal of making a profit. It ignores the code of good animal husbandry (right), which involves an obligation to care for animals' well-being by providing food, protection, and shelter and the chance to be "chickens" that can move freely and enjoy the outdoors (Scully 2002).

W.E.B. DuBois (1868–1963)

A voice that was initially ignored but later "discovered" as important to sociology is that of American-born W.E.B. DuBois. In trying to describe the Industrial Revolution and its effects on society, DuBois offered the concept of the **color line**, a barrier supported by customs and laws separating nonwhites from whites, especially with regard to their roles in the division of labor. DuBois (1919 [1970]) traced the color line's origin to the scramble for Africa's resources, beginning with the slave trade upon which the British Empire and American Republic were built, costing black Africa "no less than 100,000,000 souls" (246). The end of slave trade was followed by the colonial expansion that accompanied the Industrial Revolution. That expansion involved rival European powers competing to secure labor and natural resources. By 1914 all of Africa had been divided into European colonies. DuBois maintained that the world was able "to endure this horrible tragedy by deliberately stopping its ears and changing the subject in conversation" (246). He felt that an honest review of Africa's history could only show that Western governments and corporations coveted Africa for its natural resources and for the cheap labor needed to extract them.

Courtesy of Urbain Ureel

AP Photo/Ben Curtis

➤ The color line reflected the deep social divisions between Europeans and Africans that was solidified by slave trade and by colonization. DuBois's observations are reflected in these scenes in which African labor (top) moved goods through the continent for export to Europe and the United States around the turn of the 20th century. If DuBois were alive today, he would most certainly emphasize that the West continues to exploit African labor and resources. One example of ongoing exploitation, diamond mining (bottom), was the subject of the 2006 film *Blood Diamond*.

Jane Addams (1860–1935)

In 1889 Jane Addams (with Ellen Gates Starr) cofounded one of the first settlement houses in the United States, the Hull House. Settlement houses were community centers that served the poor and other marginalized populations. Wealthy donors supported them, and university faculty and college students lived with the clients, serving and learning from them. The Chicago Hull House was one of the largest and most influential settlements in the United States. At the time of its founding, immigrants constituted almost half of Chicago's population. In addition, the city was industrializing and experiencing unprecedented population growth. These dramatic changes were accompanied by a variety of social problems, including homelessness, substandard housing, and unemployment.

Hull House facilities contained a school, boys' and girls' clubs, recreation facilities, a library, and much more. Hull House had strong ties with the University of Chicago School of Sociology. Jane Addams was the forerunner of what is today called public sociology. Addams worked to give her clients a voice and to address issues related to child labor, worker safety, and other areas (Hamington 2007).

Addams maintained that the settlement houses were equivalent to an applied university where knowledge about how to change people's situations could be applied and tested. Addams advocated for **sympathetic knowledge**, firsthand knowledge gained by living and working among those being studied, because knowing others increases the "potential for caring and empathetic moral actions" (Addams 1912, 7). Addams made a point of never addressing an audience about the Hull House without bringing a member who knew its conditions more intimately than she "to act as an auditor" of her words (Addams 1910, 80).

Chris Caldeira

◀ Jane Addams's ideas are alive today in any center that serves the needs of a community. The Salvation Army represents such a place. It has 119 adult rehabilitation centers addressing "issues of substance misuse, legal problems, relational conflicts, homelessness, unemployment, and most importantly, a need for spiritual awakening and restoration" (Salvation Army 2012).

(Write a Caption)

Write a caption that relates the 1940s photograph of home-grown and -canned vegetables to a type of solidarity.

Hints: In writing this caption

- review the concept of solidarity,
- identify which type of solidarity the photograph best represents, and
- consider who prepared it and whether the preparer is likely to personally know the people who will eat it.

Prints & Photographs Division, Library of Congress, LC-DIG-fsac-1a35476

Critical Thinking

Which one of the six theorists' ideas resonate the most with you? Explain why.

Key Terms

bourgeoisie	means of production	social action
class conflict	mechanical solidarity	social dynamics
color line	organic solidarity	social statics
conflict	positivism	solidarity
disenchantment	proletariat	sympathetic knowledge

Sociological Perspectives

MODULE

(1.5)

Objective

You will learn how sociological perspectives guide research and analysis of any topic.

Lisa Southwick

Think of a time you did research for a teacher-assigned essay, research paper, or speech. Did you struggle with how to incorporate the information available to you on the topic?

Imagine the chosen topic is mobile phones. The search engine Google Scholar yields 1.5 million hits for the search term *mobile phone*. As you browse though the titles, you wonder how to select and organize the information. This writing assignment can be less stressful if your research and writing is informed by at least one of the three major sociological perspectives. This is because each perspective offers a vision of society, a key question that guides readings and analysis, as well as a vocabulary to answer that question.

Sociological Perspectives

A **sociological perspective** is a conceptual framework for thinking about and explaining how human activities are organized and/or how people relate to one another and respond to their surroundings. There are three major sociological perspectives, each of which focuses our attention on different dimensions of human activity. These perspectives are functionalist, conflict, and symbolic interaction.

Functionalist Perspective

Functionalists ask, How is an existing social order in society maintained? Their answer is that social order is possible because all the parts of society contribute to order. Functionalists see society as a system of interdependent parts. To illustrate, they use the human body as an analogy. The human body is composed of parts such as bones, ligaments, muscles, a brain, a spinal cord, a heart, and lungs, all working together in impressive harmony. Society, like the human body, is made up of interdependent parts, such as schools, automobiles, sports teams, mobile phones, funeral rites, laws, and languages. Like the various body parts, each of society's parts performs a function.

A **function** is the contribution a part makes to maintain the stability of an existing social order. In the most controversial form of this perspective, functionalists argue that all parts of society, even something like poverty, contribute in some way to maintaining order and stability in the larger society. They strive to identify how parts—even seemingly problematic ones—contribute to the existing order.

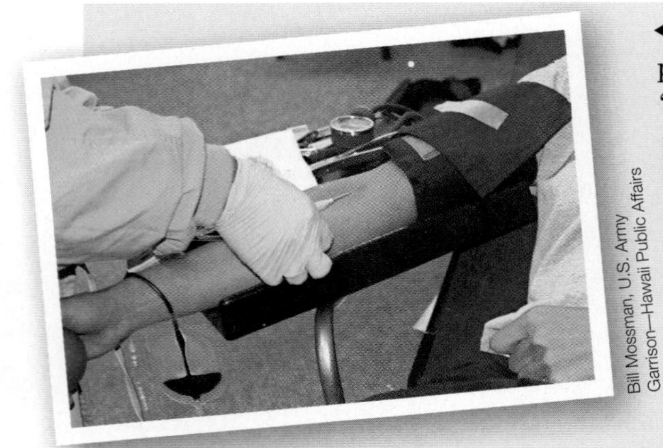

Bill Mossman, U.S. Army Garrison—Hawaii Public Affairs

◀ Consider just one function of poverty: many poor people often "volunteer" for over-the-counter and prescription drug tests. Most new drugs, from AIDS vaccines to allergy medicines, must eventually be tried on healthy human subjects to determine their potential side effects (e.g., rashes, headaches, vomiting, and drowsiness) and appropriate dosages. The chance to earn money motivates people to volunteer for clinical trials. Because payment is relatively low and the risks are often high, however, the tests attract a disproportionate share of low-income, unemployed, or underemployed people as "volunteers." Therefore, from a functionalist perspective, the pharmaceutical and medical systems would be seriously strained if poverty were eliminated.

Sociologist Robert K. Merton realized that it is not realistic to think just about a part's functions because a part can also contribute to disorder and instability. Merton introduced the concept of dysfunction to focus attention on a part's disruptive effects to an existing social order. Merton also recognized that both functions and dysfunctions can be manifest (anticipated/intended) or latent (unanticipated/unintended).

MANIFEST AND LATENT FUNCTIONS. **Manifest functions** are a part's anticipated, recognized, or intended effects on maintaining some social order. **Latent functions** are a part's *un*anticipated, *un*recognized, and *un*intended effects on an existing social order. To illustrate this distinction, consider the manifest and latent functions of the mobile phone. One obvious manifest function of the mobile phone is that it offers people a tool for communicating with others without being confined to some fixed space such as an office, a home, or a phone booth.

A latent function of the mobile phone is that few people anticipated that it would become a "smart" phone, allowing people to take, send, and receive photos; watch television; send text messages; listen to and share music; hear radio shows; and access the Internet. This anytime-anywhere access eliminates unnecessary delays in connecting people and information sources.

MANIFEST AND LATENT DYSFUNCTIONS. A part can also have manifest or latent disruptive consequences. **Manifest dysfunctions** are a part's anticipated disruptions to an existing social order. A manifest dysfunction of mobile phones relates to drivers who become distracted when dialing, talking, and texting, increasing their chances of being involved in or causing an accident. It is not just drivers who become distracted. In a survey of 439 doctors who perform cardiopulmonary bypass (CPB) surgery, 55.6 percent reported using their mobile phones while performing surgery to send or check text messages, access e-mail, check postings on social networking sites, or otherwise use the Internet (Smith, Darling, and Searles 2011; Richtel 2011). **Latent dysfunctions** are the unanticipated disruptions to the existing social order. One latent dysfunction of the cell phone is that it can be used to report on and document events as they happen. This capability allows people to disrupt the existing order by bypassing the news media, government censors, and others who try to control or suppress the flow of information.

AP Photo/Paul Sancya

➤ Another latent dysfunction relates to the toxic waste generated from the improper disposal of mobile phones. The adverse consequences of such disposal to the environment and human health will only increase as disposable mobile phones, designed to allow for 60 minutes of conversation, enter the market in mass.

The strength of the functionalist perspective is that it offers a balanced overview of a part's anticipated and unanticipated effects on an existing social order. One weakness is that it leaves us wondering about a part's overall positive or negative effect on that order. This perspective provides no formula for evaluating whether functions outweigh dysfunctions or about whether a social order needs to be disrupted.

Conflict Perspective

Conflict theorists, who draw inspiration from Karl Marx, see conflict over scarce and valued resources as an inevitable fact of life. In any society, advantaged and disadvantaged groups compete for scarce and valued resources, such as well-paying jobs, admission to a chosen college, water, and even bodily organs for transplant. Not surprisingly, the advantaged seek to protect their positions and increase their advantage, whereas the disadvantaged seek to improve their position. Conflict theorists ask this basic question: Who benefits from a particular social arrangement and at whose expense? In answering this question, these theorists seek to identify the groups that benefit and the strategies employed to maintain their advantage. Conflict theorists also seek to identify which groups are at a disadvantage (relative to the advantaged). With regard to mobile phones, the advantaged groups would include mobile phone manufacturers (Motorola, Nokia, Apple) that control device production, upgrades, and use, and the providers (Verizon, AT&T) that control terms of use. The disadvantaged groups would include factory workers who assemble mobile phones and those addicted to mobile phones, whose sleep, work, and relationships are disrupted or otherwise negatively affected.

Missy Gish

◀ Smart phones, through barcode scanner apps, connect potential consumers to a website that offers information about a product, service, or location. From a conflict point of view, corporations use this technology to influence and manage people. Specifically the technology allows corporations to connect directly with customers, learn about their needs and preferences, customize ads, eliminate the middleman, and sell products directly.

The concept of ideology helps explain one way dominant groups protect their advantages. **Ideologies** are seemingly commonsense views justifying the existing state of affairs. Those who embrace an ideology, including the disadvantaged, view the existing arrangements as normal and legitimate. But upon close analysis, ideologies reflect a viewpoint that benefits the dominant groups and disguises their advantages. The ideology supporting barcode scanner apps on mobile phones is that people benefit when they have access to technologies that allow them to learn about products on demand. But closer analysis reveals the real purpose is to manipulate people's needs and desires for commercial gain.

The strength of the conflict perspective is that it forces us to consider how advantaged groups benefit from the way human activity is organized and how they control access to scarce and valued resources. The weakness is that it simplistically portrays the advantaged group as all-powerful and the disadvantaged as victims incapable of changing their circumstances. In the case of the cell phone, conflict theorists might overlook the millions of previously unconnected people in the world now connected and ignore the empowering elements of the mobile phone to circumvent power structures and deliver news at a grassroots level.

Symbolic Interactionist Perspective

Sociologist Herbert Blumer coined the term *symbolic interaction* and outlined its essential principles, which follow. Symbolic interactionists focus on **social interaction**, everyday encounters in which people communicate, interpret, and respond to each other's words and actions. Symbolic interactionsists ask, How do people involved in interaction and other human activity "take account of what each other is doing or is about to do" and then direct their own conduct accordingly (Blumer 1969)? The process depends on (1) self-awareness, (2) shared symbols, and (3) negotiated order.

SELF-AWARENESS. All interaction and human activity depends on the parties being self-aware. **Self-awareness** occurs when a person is able to observe and evaluate the self from another's viewpoint. People are self-aware when they imagine how others are viewing, evaluating, and interpreting their words and actions. Through this imaginative process people become objects to themselves; they come to recognize that others see them, for instance, "as being a man, young in age, a student, in debt, trying to become a doctor, coming from an undistinguished family and so forth" (Blumer 1969, 172). They also come to recognize that others have expectations about what they should say and do. In imagining others' reactions, people respond and make adjustments (apologize, change facial expressions, lash out, and so on).

Lisa Southwick

➤ This facial expression conveys the message "Oops, I shouldn't have done that." Such a reaction requires self-awareness; that is, this woman had to imagine another's evaluation and judgment of something she did, which prompted her to respond with such a look.

SHARED SYMBOLS. A **symbol** is any kind of object or idea to which people assign a name, meaning, or value (Blumer 1969). The things assigned meaning may be physical (chairs, cell phones, cars, books, a color, the wave of a hand), social (a friend, a parent, the president of the United States, a bus driver), or abstract (freedom, greed, justice, empathy). In the context of driving, for example, the color red has come to symbolize *stop*. Objects can take on different meanings depending on audience and context: a tree can have different meanings to an urban dweller, a farmer, a poet, a home builder, an environmentalist, or a lumberjack (Blumer 1969). People derive meaning from observing others. Through observation, people learn such things as a wave of the hand means good-bye, that true friends are rare, or when it's okay to respond to text messages in front of others.

NEGOTIATED ORDER. When we enter into interaction with others, we take for granted that a system of expected behaviors and shared meanings is already in place to guide the interaction. Although expectations are in place, symbolic interactionists emphasize that those expectations can be ignored, challenged, or changed (Blumer 1969). Room for negotiation exists in most interactions; that is, the parties involved have the option of negotiating other expectations and meanings. The **negotiated order**, then, is the sum of existing expectations and newly negotiated ones (Strauss 1978).

As we have seen, symbolic interactionists focus on self-awareness, shared symbols, and negotiated order. Thus, when studying mobile phones they examine the ways in which this technology shapes sense of self, the meanings assigned to mobile phones (e.g., for what reasons people call each other, when is it appropriate to call), and how people negotiate cellular use and interactions. With regard to the mobile phone's effect on interaction, research suggests that, while people may have many phone numbers stored in their directory, for the most part, they regularly contact only a few people with whom they have strongest ties—family, close friends, and partners.

➤ The mobile phone also makes it easy to record (audio and/or visual) events for replay later. This recording can allow people to review and scrutinize the behavior and words spoken. Symbolic interactionists would be interested in how the ability to "relive" the past allows people " to 'see' things about their own and others' behaviors and words they previously could not perceive" and how that review changes (or verifies) the viewers' understanding of the event (Humphreys 2011, 578).

Lance Cpl. Jody Lee Smith

The strength of the symbolic interactionist perspective is that it views people as active agents rather than as passive participants shaped by outside forces. While outside forces certainly shape interpretations and the course of interaction, people still have power to assess the situation and ultimately decide on the actions to take. One weakness is that symbolic interactionists may overestimate that power.

(Write a Caption)

Write a caption that uses one of the three theoretical perspectives to analyze the mobile phone and its use in public settings (look closely and you can see two people using phones).

Hints: When writing this caption

- review each of the three theoretical perspectives, and
- think about what you want to emphasize—intended and unintended consequences; advantaged and disadvantaged groups or meanings assigned (by others, to the mobile phone, to some relationship).

C. Todd Lopez

Critical Thinking

Which of the three perspectives do you find most compelling? Explain.

Key Terms

function	manifest dysfunctions	social interaction
ideologies	manifest functions	sociological perspective
latent dysfunctions	negotiated order	symbol
latent functions	self-awareness	

MODULE 1.6

Research Methods

Objective

You will learn the process by which sociologists ask questions and research answers.

How do you think your choice of college major will affect your job prospects upon graduation and beyond?

Cristina Willard

Sociologists Josipa Roksa and Tania Levey (2010) recognized that some programs of study such as education and nursing offer training for specific occupations. Other programs of study such as sociology and literature focus on general skills such as critical thinking, complex reasoning, and verbal and written communication but offer no clear occupational path. The researchers wondered to what extent the occupational specificity of a college major affected college graduates' transition into the labor market and their evolving occupational status. The two sociologists designed a study in which they were able to follow the transitions from college to first job and beyond with specific emphasis on the role the college major plays in shaping job opportunities. In this module we draw upon Roksa and Levey's research to illustrate the steps of the social research process.

The Methods of Social Research

Research methods are the various techniques that sociologists and other investigators use to formulate and answer meaningful questions and to collect, analyze, and interpret data. Sociologists adhere to the **scientific method**, a carefully planned research process with the goal of generating observations and data that can be verified by others. The research process involves at least six interdependent steps:

- determining the topic or research question,
- reviewing the literature,
- choosing a research design,
- identifying variables and specifying hypotheses,
- collecting and analyzing the data, and
- drawing conclusions.

It is important to know that researchers do not always follow the six steps in sequence. For example, they may not decide on a specific research question until they have familiarized themselves with the literature. Sometimes an opportunity arises to collect specific data, and a research question is defined to fit that opportunity. Although the six steps need not be followed in any particular order, all must be completed at some point in the research process to ensure the quality of the project.

Choosing a Topic/Reviewing the Literature

It is impossible to list all the topics that sociologists study, because almost any subject involving human activity represents a potential subject for research. Sociology is distinguished from other disciplines not by the topics it investigates but by the perspectives it takes in framing, studying, and drawing conclusions about human activity. As sociologists, Roksa and Levey were interested in the ways in which their college major presents graduates with opportunities and disadvantages in the labor market.

Chuck Cannon, *Fort Polk Guardian* staff writer

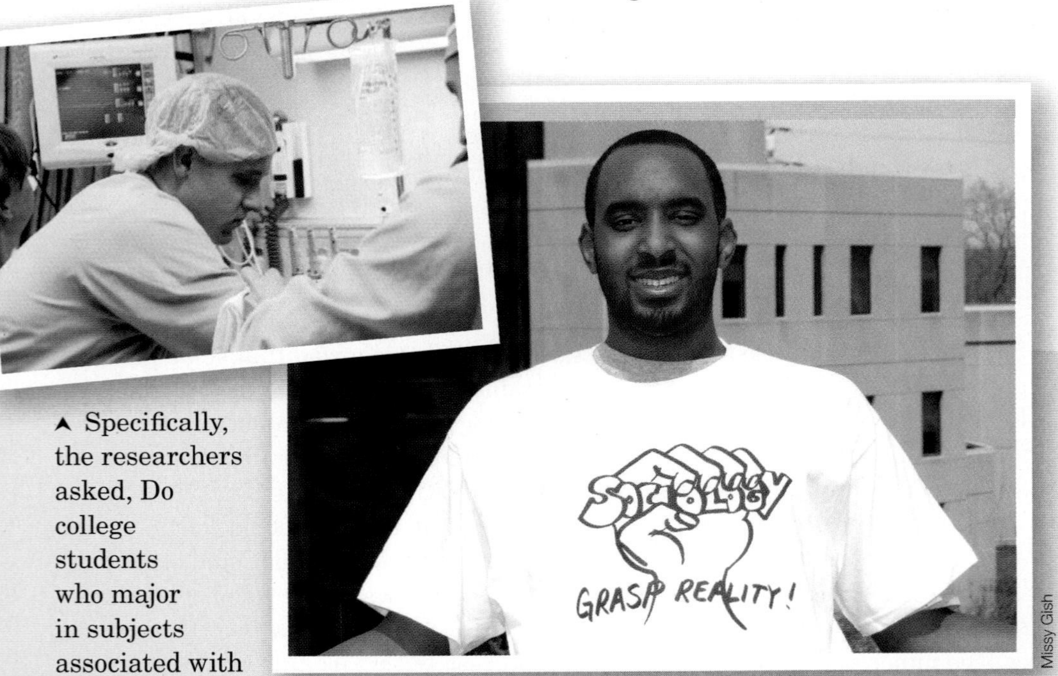

Missy Gish

▲ Specifically, the researchers asked, Do college students who major in subjects associated with defined career paths such as nursing have a career advantage in the labor market over those students who choose majors such as sociology with no certain career path? In the case of the students pictured, the nursing student plans to enter the nursing profession upon graduation. The sociology undergrad has accepted a full-tuition scholarship, a stipend, and free housing to attend graduate school to pursue a master's degree in student affairs.

When doing research, sociologists consider what knowledgeable authorities have written on the chosen topic, if only to avoid reinventing the wheel. Reading the relevant literature can also generate insights that researchers may not have considered.

Lisa Southwick

◂ Even if researchers believe that they have a new idea, they still must consider the works of other thinkers and show how their research verifies, advances, or corrects existing research. Today, most people can access the equivalent of a library and much more via computer search engines.

In writing up an account of their research study, researchers cite those who have influenced their work. Roksa and Levey cited about 50 references to scholarly books and articles. Ideally, these materials are written by credentialed authors who carefully document their sources of information and/or who actually conducted research studies. Scholarly writings are also often reviewed by qualified experts who assess quality, recommend whether they deserve to be published, and offer suggestions for improvement.

In choosing the literature to review, researchers certainly read the most current materials, but the publication date should not be the only criterion. Classic and groundbreaking articles written decades or even centuries ago can be very important resources. In reviewing the literature, researchers learn not only what is known about the topic of interest but also about things not yet known. By identifying what is not known, researchers also establish what they can do in their research to advance knowledge on the chosen topic. In this regard, Roksa and Levey learned that a number of published research studies suggest that college graduates who have majored in subjects with clear career options have an advantage (392). The authors found that these studies "focused largely on entry into the labor market," and while entry is an important phase of a career path, so too are the years that follow. Roksa and Levey decided to focus not just on entry but the decades following that first job as well.

Choosing a Research Design

After determining the research question, researchers typically decide upon a **research design**, which involves deciding who or what to study and the method of gathering data. Roksa and Levey chose to examine the occupational careers of 1,970 college graduates who were part of the National Longitudinal Survey of Youth (NLS). Conducted by the U.S. Department of Labor, Bureau of Labor Statistics, this survey is an important data source for social scientists. This survey interviewed participants at multiple points in time about the jobs they held and other significant life events. For this NLS survey, respondents were first interviewed in 1979 when they were 14–22 years old and then every year until 1994. They were interviewed again in 1996 and for the last time in 1998. Ideally, Roksa and Levey would have liked the survey to have continued until the present, but the Bureau of Labor Statistics ended the interview cycle started over 30 years ago to begin a new longitudinal survey following a new cohort. Given that Roksa and Levey were interested in the long-term effects of choice of major, data generated from the first longitudinal survey offered the best option.

Specifying Variables, Operational Definitions, and Hypotheses

A **variable** is any behavior or characteristic that consists of more than one category. Age is a variable as people can range in age from seconds old (at birth) to 100-plus years old. Grade point average is a variable ranging from 0 to 4.0. All variables used in a research study must be **operationalized**; that is, the researcher must give clear, precise instructions about how they observed or measured them. Two key variables in the Roksa and Levey study were occupation specificity of major and economic/occupational status of jobs held.

"Occupation specificity of college major" was established by calculating the percentage of graduates in a particular major who obtain a job related to their field of study. Roksa and Levey classified each major as one of three levels:

1. *High occupational specificity*—examples include education majors (74 percent of whom enter the job market as educators) and health care–related majors (82 percent of whom enter the job market as medical professionals);
2. *Moderate occupational specificity*—examples include business majors (56 percent of whom enter the job market taking business or management-related careers) and social work/protective services (60 percent of whom enter the job market as human/protective service professionals); and
3. *Low occupational specificity*—examples include English, philosophy, and sociology majors for whom no dominant occupational category stands out.

Since Roksa and Levey were interested in the economic and occupational status of a graduate's first and subsequent jobs, they calculated an occupational and economic status value for each job respondents held over their observed work life.

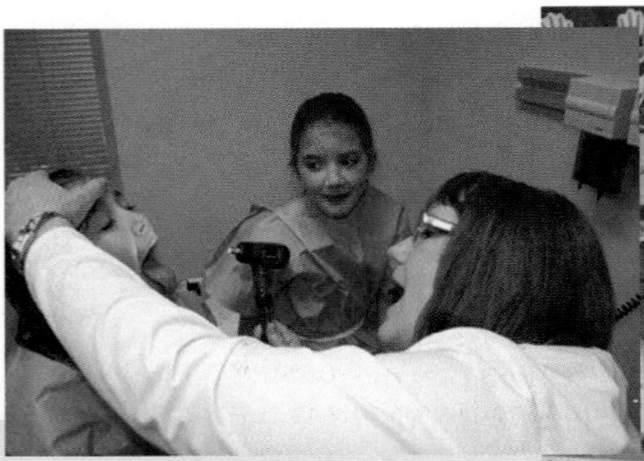

Department of the Army

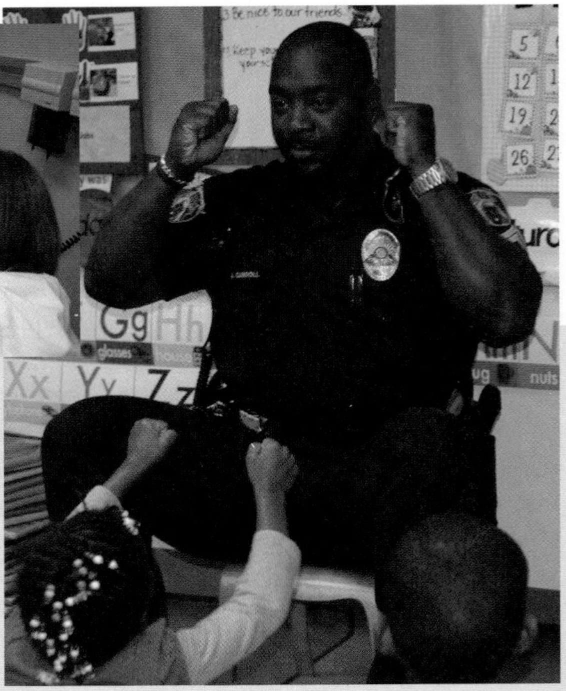

Robin Brown

▲ How would you measure occupational status of, say, a nurse versus that of a police officer? Roksa and Levey measured a job's *occupational status* as the percentage of incumbents in a particular occupation who have completed at least some college. Presumably, the greater the proportion of incumbents with some college, the higher that job's occupational status. The percentage of nurses with some college is 55 percent; the percentage of police officers with some college is 15 percent.

Roksa and Levey determined a job's economic status by calculating the proportion of incumbents holding that job who earned $14.30 or more per hour. In 1998, $14.30 was the median hourly income of all workers in the United States. Presumably, the greater the proportion of incumbents who earned more than $14.30 per hour, the higher a job's economic status.

With a study's variables established, operational definitions must be reliable and valid. An operational definition is considered **reliable** if someone using the operational definition as described repeats the measure and gets essentially the same results. Consider the reliability of Roksa and Levey's measure of the occupational specificity of a college major. They first calculated the percentage of graduates in a particular major who obtained a job related to their field of study. But, the researchers did not establish what percentage distinguishes majors high in occupational specificity from those majors of moderate specificity, nor did they establish what percentage distinguished majors moderate in occupational specificity from majors that were low. Without clear instructions specifying the cutoffs, the measure of occupational specificity cannot be reliable. In this case, the researchers could have defined high occupational specificity as at least 70 percent of college graduates in a major take a job associated with that major; moderate occupational specificity as 40–60 percent of college graduates in a major take a job associated with that major; and low occupational specificity as less than 40 percent of college graduates in a major take a job associated with that major.

An operational definition must also be assessed for its **validity**, that is, the extent to which it claims to measure. In the case of Roksa and Levey's study, is the percentage of graduates in a particular major who obtain a job related to their field of study a valid measure of occupational specificity? This measure does seem valid. However, we might question the validity of the measure Roksa and Levey used to calculate a specific job's occupational status—the percentage of incumbents in the occupation who have completed at least some college. Specifically, is "some college," which could be as little as one credit hour, a valid measure of occupational status? We might argue that the percentage of job holders with at least two years of college, or even a four-year college degree, constitutes a better measure of an occupation's status.

INDEPENDENT AND DEPENDENT VARIABLES. Researchers strive to find associations between variables to explain or predict behavior or certain outcomes. The behavior to be explained or predicted is the **dependent variable**. The variable that explains or predicts the dependent variable is the **independent variable**. In the Roksa and Levey study, the dependent variable—the variable to be explained—is economic and occupational status of a graduate's first and subsequent jobs. The independent variable—the variable that Roksa and Levey believes explains this dependent variable—is occupational specificity of the major.

The relationship between independent and dependent variables is described in a **hypothesis**, or a prediction about the relationship between the independent and dependent variables. Specifically, if the hypothesis is supported by the data, then researchers can claim that if they know the value of an independent variable (whether a major is low, moderate, or high in occupational specificity), then they can predict the independent variable—the occupational and economic status of the first job after graduation (e.g., the percentage of people holding a job who have some college or who earn more than the median hourly wage). Roksa and Levey hypothesized that:

- the greater the occupational specificity of a major, the higher the economic and occupational status of the first job obtained after graduation relative to graduates with moderate and low occupational specificity; and
- as a career path evolves over time, the advantages graduates with high occupational specificity initially hold over graduates with moderate and low occupational specificity will narrow.

In addition to identifying independent and dependent variables, researchers also identify **control variables**, other variables that may affect the dependent variable and that researchers hold constant so they can focus on just the relationship between the independent variable and the dependent variable. Think about it this way—there are a number of variables that could affect the major one chooses that could also affect the occupational and economic status of the job one holds. One such variable is a graduate's sex/gender (females tend to dominate health care and education majors, and female-dominated occupations often have lower status and pay relative to occupations dominated by males). How do researchers hold variables such as sex/gender constant? They compare how occupational specificity of major (independent variable) affects graduates' economic and occupational status (independent variable) for males and females as separate groups. If the effect of a major's occupational specificity on economic and occupational status is essentially the same for males as a group as for females, then sex/gender does not play a role in determining economic and occupational status. However, if males' economic and occupational status (as a group) is higher than that of their female counterparts with the same majors, then sex/gender plays an important role in determining occupational and economic status.

Collecting and Analyzing the Data

Researchers collect data that they then analyze to see if there is support for their hypotheses. Recall that Roksa and Levey used data collected by the U.S. Department of Labor, Bureau of Labor Statistics (2011b) on the occupational careers of 1,970 respondents who graduated from college. Among other things, the data set included information about each respondent's major, the job secured upon graduation, salary associated with that job, any jobs secured at later points in career, and associated salaries.

When researchers analyze the collected data, they search for common themes and meaningful patterns. In presenting their findings, researchers may use graphs, tables, photos, statistical data, quotes from interviews, and so on. Roksa and Levey performed a statistical analysis on the data that is beyond the scope of this module. It should be noted, however, that their statistical analysis supported their hypotheses. Specifically, Roksa and Levey found that their research supported the following.

- Upon entering the labor market after graduation, college graduates with majors considered high in occupational specificity had an advantage over those graduates with majors considered of moderate and low occupational specificity.
- Those with majors considered high in occupational specificity tend to experience the lowest *growth* in occupational status over their work careers.
- Over the course of their career, college graduates with a major considered low in occupational specificity experience the greatest increase in occupational status.
- There are significant gender differences that favor males regardless of major at point of entry in the labor market and beyond. Perhaps the most interesting finding is that, as a group, males with majors considered low in occupational

specificity equaled or surpassed the income earned by women as a group regardless of major.

Drawing Conclusions

Generally, researchers conclude by discussing the broader implications of their research. In their conclusion, Roksa and Levey maintained that their findings have important implications for the recent push to vocationalize higher education. They argue that while it is important for universities to have programs of study that offer clear career paths, that should not be the exclusive focus of postsecondary education. Roksa and Levey's findings suggest that majors that cultivate general skills offer considerable opportunity for occupational mobility.

The conclusion section also addresses **generalizability**, the extent to which findings can be applied to a larger population beyond those studied. In the case of surveys, if a sample is randomly chosen, if almost all subjects agree to participate, and if the response rate for every question is high, we can say that the sample is representative of the larger population from which the sample was drawn and thus the findings are generalizable to that larger population. In Roksa and Levey's study, the college graduates studied were randomly chosen and entered the labor market in the 1980s. But can the research findings be generalized to those graduates who entered the labor market in the 1990s and 2000s? One also wonders if male advantage in the labor market would exist today at the level they found. Still, Roksa and Levey's findings support arguments against reducing, even abandoning, general education requirements or withdrawing support for majors with low occupational specificity.

Other Data-Gathering Strategies

As we will see, Roksa and Levey's research design involved two types of data-gathering strategies—secondary sources and structured interviews. Researchers can choose from a variety of other data-gathering strategies, including interviews, observation, secondary sources, case studies, and experimental design.

SURVEYS AND INTERVIEWS. Self-administered surveys and interviews are two ways in which sociologists gather data. A **self-administered survey** is a set of questions that respondents read and answer on their own. Respondents may be asked to write out answers to open-ended questions or to choose the best response from a list of potential responses (forced choice). The self-administered survey is one of the most common methods of data collection. Self-administered surveys have a number of advantages. No interviewers are needed to ask respondents questions, the surveys can be given to large numbers of people at one time, and respondents are less likely to feel pressure to give the socially acceptable response (especially when surveys are anonymous).

U.S. Census Bureau

➤ Some drawbacks of self-administered surveys are that respondents can misunderstand the meaning of a question, skip over some questions, or just stop answering them.

The results of a questionnaire depend not only on respondents' decisions to fill it out, answer questions conscientiously and honestly, and return it, but also on the quality of the survey questions asked and a host of other considerations. In comparison to self-administered surveys, interviews are more personal. The interviewer asks questions and records the respondent's answers.

➤ When respondents give answers during interviews, the interviewers must avoid pauses, expressions of surprise, inflections in their voice, or body language that reflect value judgments. Refraining from such conduct helps respondents feel comfortable and encourages them to give honest answers.

U.S. Census Bureau

Interviews can be structured or unstructured, or a combination of the two. In a structured interview, the wording and sequence of questions, and sometimes response choices, are set in advance and cannot be altered during the course of the interview. Recall that Roksa and Levey used responses from a structured survey conducted by the U.S. Bureau of Labor Statistics that asked respondents a series of in-depth questions about their job history. The survey was considered longitudinal because the survey was repeated each year for decades.

In contrast to the structured interview, an unstructured interview is flexible and open-ended. The question-answer sequence is spontaneous and resembles a conversation in that the interviewer allows respondents to take the conversation in an unplanned direction. The interviewer's role is to give focus to the interview, ask for further explanation or clarification, and probe and follow up on interesting ideas expressed by respondents.

No matter what type of interview, questions have to be clear and meaningful. Remember that writing good questions is much more difficult than it appears. Consider issues related to income. Researchers cannot simply ask, "What is your income?" but must specify if they mean hourly, weekly, monthly, or annual income; pretax or after-tax income; household or individual income; this year's or last year's income; or income from employment or other sources.

OBSERVATION. **Observation** involves watching, listening to, and recording human activity as it happens. This technique may sound easy, but the challenge lies in knowing how to observe and what is significant while still remaining open to other considerations. Good observation techniques are developed through practice and involve being alert, taking detailed notes, and making associations between observed behaviors. Observation is useful for (1) learning things that cannot be surveyed easily, and (2) experiencing the situation as those being observed experience it. One disadvantage is that observation research is time-consuming and specific to a particular setting. Observation can take two forms: nonparticipant and participant. **Nonparticipant observation** consists of detached watching and listening: the researcher only observes and does not become part of group life.

Courtesy of T. Scott McLaren

◂ Researchers engage in **participant observation** when they join a group, interact directly with those they are studying, assume a role critical to the group's purpose, and/or live in a community under study as did this man, who lived with people in a village in Mauritania (Africa).

In participant and nonparticipant observation, researchers must decide whether to hide their identity and purpose. One reason for choosing concealment is to avoid the **Hawthorne effect**, a phenomenon in which research subjects alter their behavior simply because they are being observed. This term originated from a series of worker productivity studies conducted in the 1920s and 1930s involving female employees of the Western Electric plant in Hawthorne, Illinois. Researchers found that no matter how they varied working conditions—bright versus dim lighting, long versus short breaks, piecework pay versus fixed salary—productivity increased. One explanation is that workers were responding positively to having been singled out for study (Roethlisberger and Dickson 1939).

SECONDARY SOURCES. **Secondary sources or archival data** has been collected for a purpose not related to the research study. This kind of data includes that gathered by census bureaus, research centers, and survey companies such as Gallup. Sociologists may use this already collected data to do their research. The advantages of secondary data are that it is often free or at least costs less to obtain than it would to collect. Governments and other large agencies such as the U.S. Bureau of Labor Statistics have the resources to execute large-scale surveys of randomly chosen populations, something few researchers can do on their own. Because Roksa and Levey used survey data collected by the Bureau of Labor Statistics, they used secondary sources.

Secondary data goes beyond surveys and includes biographies, photographs, letters, e-mails, websites, films, advertisements, graffiti, and so on. With this kind of data, sociologists often do what is called **content analysis**. That is, they identify themes, sometimes counting the number of times something occurs or specifying categories in which to place observations. A researcher who studies family photographs over time may look to see the extent to which pets are included.

CASE STUDIES. **Case studies** are objective accounts intended to educate readers about a person, group, or situation. Well-written case studies shed light through in-depth descriptions of an individual, an event, a group, or an institution. Case studies should tell a story with a beginning, a middle, and an ending. Researchers interested in the occupational careers of college graduates may choose to do three case studies—one of a college graduate with a high occupational specificity major, a second of a graduate with a moderate occupational specificity major, and a third of a graduate with a low occupational specificity major.

(Write a Caption)

Imagine you had a chance to study a casino and its patrons. Write a caption that describes a data-collection technique you might employ.

Chris Caldeira

Hints: In writing this caption

- review the various methods of gathering data,
- identify the data-gathering method that fits best with what you would like to study, and
- highlight the essential features of the selected strategy.

Critical Thinking

After reviewing Roksa and Levey's study, what part of the research process would you label the *most* critical to the quality of the findings? Explain.

Key Terms

case studies
content analysis
control variables
dependent variable
generalizability
Hawthorne effect
hypothesis
independent variable
nonparticipant observation
observation
operationalized
participant observation
reliable
research design
research methods
scientific method
secondary sources or archival data
self-administered survey
validity
variable

Applying Sociology: Stepping Outside Comfort Zones

MODULE **1.7**

Objective

You will learn how stepping outside our comfort zones helps us to realize the promise of sociology.

Chris Caldeira

Do you like to travel alone to places where the food, language, people, and transportation are different from what you know?

That is what Chris Caldeira (2009) did the summer before she started graduate school in sociology at University of California–Davis. She traveled across the United States by car, from Vermont to San Francisco, avoiding the interstate highway system whenever possible, and after that she traveled to Thailand, Vietnam, and Laos.

We began this textbook with a promise: that sociology offers a perspective that allows us to see our world in new ways. But sociology cannot deliver on that promise if students of the discipline feel no urge to know what is going on behind closed doors, lack curiosity about people, or never wonder about those who live in distant places. Sociologists do not just open books; they also open doors, because behind each door are facets of human life to be perceived and understood.

In addition to learning the vocabulary and theories of the discipline, the best sociologists feel the urge to leave their *comfort zone*, a state of being grounded in habit such that a person does not have to think too much about how to respond or what to do next. When people step outside their comfort zones, they first must figure out the situation in which they find themselves and then decide how to act without yet knowing how others will respond. That uncertainty creates a heightened sense of awareness, not just about the new situation, but about old habits that are not applicable to the new settings. Traveling can help us realize the promise of sociology because it allows us to see in new ways the world in which we live our day-to-day lives and to open our eyes to the possibility that things can be different. Three examples from Chris's trip illustrate how insight can come from allowing the worlds we choose to explore shape us rather than forcing those worlds to conform to our expectations (Steves 2009).

Example 1: Language

In e-mails Chris wrote about crossing paths with two girls in Thailand, both of whom were studying English on their own:

> I told the girls I knew English and that we could talk. So I sat with them and asked questions to give them some experience talking and then I talked to give them experience listening. I explained that they were not finishing their sentences. I learned that both were studying English, but in a self-paced course in which they study on their own and go into school only to take the exams. Both said how hard it is and I said, "that's because you need a teacher to help with things you don't understand! I mean how can you learn to speak a language without a teacher?" Also I noticed that the book that one was studying had sentences like "Some plants are biennials" and "Buffaloes are almost extinct." I told them that most Americans would be confused about words like biennial—it all seemed way too advanced, which is why they were having trouble.

But the transaction did not end here. The experience of this transaction pushed Chris to learn something about the Thai language. Chris learned that the Thai language orders the world differently than English. She learned that Thai is what is known as a topic-prominent language—it demands that the topic or object of a sentence always appear at the very beginning of a sentence (e.g., *That book*, I read). The topic (that book) is followed by the subject or actor (I) and then the action taken (read). The English language demands the reverse—that the subject or actor begin the sentence and the topic end it (e.g., I read that book). It is quite possible that in trying to learn English on her own, this girl knew not to start her sentences with the object but forgot to put it at the end, giving her sentences an unfinished quality (e.g., I read). This encounter also caused Chris to reflect about the way the girls were learning language on their own and without a teacher to explain what they could not understand. She wondered whether American and other students taking online courses also felt so alone and disconnected.

Example 2: Constructing Reality

Chris encountered many situations in which she had to figure out what was going on. That process caused her to think about the social construction of reality, the process by which people make sense of the world. At those points she realized that people approach any situation with preconceived notions or assumptions about what things are and how things should be organized. Chris wrote, "I had a lot of questions about what was real or not real on that trip." When things did not correspond to "her reality," Chris found herself doubting her interpretations and seeking an interpretation that made sense. Until she learned the culturally appropriate interpretation, she was not sure what was real.

➤ For example, while in Thailand, Chris thought this rack of clothing was part of a yard sale. She walked up to the rack and began looking through the clothes and then it hit her. This was not a yard sale; these were newly washed clothes someone had hung out to dry. Chris realized that she was looking through someone's wardrobe.

Chris Caldeira

Example 3: The Vastness of the United States

Perhaps the biggest lesson Chris learned from her travels across the United States is that it is not easy to generalize about life in the United States or its people. In fact, "America is so vast that almost everything said about it is likely to be true, and the opposite is probably equally true" (Farrell 2010). As she traveled from Vermont to California, Chris did not stay at Marriotts or Holiday Inns, hotels typically located off the interstate, but chose instead to stay in campsites or local motels. She walked through largely abandoned neighborhoods near downtown Detroit, visited cowboy bars in Wyoming, drove through trailer parks, and walked past gun shops. At night she read online to learn what sociologists had written about things she had experienced that day. She found that some sociologist at one time or another had written something on every experience she researched.

➤ Sociologist Katarzyna Celinska's (2007) research revealed that gun owners tend to place high value on self-reliance—a value that has deep roots in American culture but that nevertheless undermines feelings of connection to a larger community.

Chris Caldeira

Chris Caldeira

◀ For most Americans the cowboy on the bucking bronco is an iconic image that evokes strong symbolic associations with the so-called taming of the Wild West. Sociologist Gene L. Theodori (2007) found that the cowboy on the bucking bronco symbolizes rodeo, the preferred sport of college students from rural America. In fact, some 3,000 student-athletes attending about 200 colleges participate in this intercollegiate sport. Rodeo draws 82 percent of its athletes from small towns across the United States.

Chris Caldeira

▶ The research of sociologist William Julius Wilson offered Chris the historical context for understanding what she saw in many largely abandoned neighborhoods of Detroit. Wilson (1983) describes the structural or economic transformations that created such neighborhoods as the one pictured, including the shift from a manufacturing-based economy to a service- and information-based economy and the transfer of millions of manufacturing jobs out of the United States. Detroit has been experiencing a structural transformation for decades, and the personal toll of the transformation was painfully evident in this and other neighborhoods.

One of the most important things Chris learned from her travels is that sociology helped her to frame her experiences. At the same time, her experiences gave sociology a deeper meaning and importance. Chris realized that reading and experiencing together gave her a more complete and lasting learning experience.

Chris returned home with a better sense of what it means to be an American and a global citizen. The people Chris met are now part of her life, and she finds that she pays attention to news coming from the places she visited. As strange as this may seem, after Chris shared her adventures with me through photographs, e-mails, and discussion, I find myself more connected to the places she visited as well. In fact, when a typhoon hit Vietnam a few months after Chris returned home, I remember having a strong urge to call Chris, because I knew that she would be worried.

Critical Thinking

Describe a time when you left your comfort zone. What did you learn about yourself and others?

Summary: Putting It All Together

CHAPTER 1

Sociology is the scientific study of human activity in society. More specifically, it is the study of the social forces that affect that activity. Social forces are any human-created ways of doing things that influence, pressure, or make people behave and think in specified ways. People can embrace or challenge social forces, be swept along or be bypassed. Sociology offers a framework to help us understand the social forces that affect our sense of self and our relationships with and connections to others. That framework includes the sociological imagination, a perspective that allows us to consider how outside forces shape our life stories or biographies and helps us to distinguish between troubles and issues.

Sociology emerged out of an effort to document and to explain a transformative social force, the Industrial Revolution, on society. In an effort to understand this event, the early sociologists—Auguste Comte, Karl Marx, Émile Durkheim, Max Weber, W.E.B. DuBois, and Jane Addams—gave us conceptual tools (1) to examine how important social forces such as the division of labor, the means of production, solidarity, and the color line connect or divide us from others in our community and beyond; (2) to think about the reasons we pursue goals, the means we use to achieve them, and their consequences; and (3) to achieve sympathetic understanding. In addition, sociology offers three broad perspectives that guide an analysis of any human activity or social issue. Those perspectives are functionalist, conflict, and symbolic interaction.

It is impossible to compile a list of the topics that sociologists study, because almost any topic involving humans is a potential area of scrutiny. Sociology is distinguished not by the topics studied but by the perspectives and questions used to frame research and analysis—questions like, What are the intended and unintended consequences of X? Who benefits from a social arrangement and at whose expense? How are meanings assigned, accepted, sustained, and challenged?

Sociologists adhere to the scientific method, a carefully planned research process with the goal of generating observations and data that can be verified by others. When sociologists do research, they decide on a research question, establish a research design, make observations, carefully collect and analyze data to test hypotheses, and draw conclusions.

The best sociologists leave their *comfort zone* to study other ways of organizing social life. The new environment creates a heightened sense of awareness and a realization that old ways of doing things are not applicable to the new setting. When sociologists leave their comfort zones it opens their eyes to the possibility that things can be different.

Go to cengagebrain.com to link to Aplia and CourseMate for the chapter quiz and other activities.

Tony Rotundo

Missy Gish

CHAPTER 2
CULTURE

Sociologists view culture as a key concept for capturing the human capacity for devising ways to interact and live together and for negotiating the surrounding environment. Sociologists are interested in how culture shapes human experiences and how human experiences shape culture. The photographs that open this chapter—one showing a family in Thailand traveling by motorbike, the other showing a family in the United States traveling by car—offer some insights. The comparison prompts us to ask how an automobile-centered and a motorbike-centered culture affect the ways in which family members relate to and interact with each other. We can also see that the experience of traveling with family members by car or by motorbike "creates" a different kind of person—in the case of the car a more autonomous individual and in the case of a motorbike a more group-oriented individual.

Go to Sociology CourseMate on cengagebrain.com to view a video about how the recession has altered the spending culture in the U.S.

MODULE 2.1

Culture

Objective

You will learn the meaning of culture and the challenges of defining a culture's boundaries.

Spc. Tobey White

Have you ever interacted with someone from another culture? What did you learn about that person's culture? Do you think that person learned something about your culture?

Defining Culture

Sociologists define **culture** as the way of life of a people. To be more specific, culture includes the shared and human-created strategies for adapting and responding to one's surroundings, including the people and other creatures that are part of those surroundings. The list of human-created strategies is endless: it includes the invention of physical objects such as cars and motorbikes to transport people from one place to another, values defining what is right and good, beliefs about the world and how things in it operate, a language with which to communicate, and rules guiding behavior in almost any situation.

A culture can be something as vast as U.S. culture or much smaller in scale, such as the culture of a family, a school, a community center, or a coffee shop. In our everyday use of the word *culture*, we often think of culture as something with clear boundaries and as something that explains differences among groups. You may be surprised to learn that sociologists face at least three challenges in defining a culture's boundaries:

- *Describing a culture*—Is it possible to describe something that encompasses the way of life of a people? How would you describe U.S. culture or Thai culture? Or on a smaller scale, how might you describe the culture of a school or a coffee shop?

- *Determining who belongs to a culture*—Is everyone who lives in the United States part of American culture? Does a person have to look Thai or live in Thailand to be considered part of that culture?
- *Identifying the distinguishing markers that set one culture apart from others*—Is riding a motorbike a marker of Thai culture? Does owning an automobile make someone American?

Given these challenges, is *culture* a useful term? First, there is no question that cultural differences exist. For example, anyone who travels to Thailand will note that people largely depend on motorbikes to get around, while people in the United States largely depend on cars. Likewise, anyone traveling to Saudi Arabia and then perhaps to Brazil will notice that women in the two countries dress quite differently when out in public. But once you think you have identified a clear marker, you can always find exceptions to the rule and see that the marker is not unique to one culture (Wallerstein 1990).

Culture as a Rough Blueprint

On some level culture is a blueprint that guides and, in some cases, even determines behavior. Think about it: for the most part, people do not usually question the origin of the objects around them, the beliefs they hold, or the words they use to communicate and think about the world "any more than a baby analyzes the atmosphere before it begins to breathe it" (Sumner 1907, 76). Much of the time people think and behave as they do simply because it seems natural and they know of no other way.

Although culture is a blueprint of sorts, you will notice that people who are purported to share a culture are not replicas of one another. For example, Americans typically eat three meals per day—breakfast, lunch, and dinner associated with morning, noon, and evening. Still, every American does not eat at the same time or eat the same food. Of course, we can find people in the United States who skip breakfast or eat rice for breakfast, but they readily admit that in doing so they are not following what is considered typical in American culture. The fact that people are not cultural replicas makes it especially difficult to describe a culture and determine who belongs.

Cultural Universals and Particulars

Anthropologist George Murdock (1945) distinguished between cultural universals and particulars. **Cultural universals** are those things that all cultures have in common. Every culture has natural resources such as trees, plants, and rocks that people put to some use. In addition, every culture has developed responses to the challenges of being human and living with others. Those challenges include the need to interact with others, to be mentally stimulated, to satisfy hunger, and to face mortality. In every culture, people have established specific ways of meeting these universal challenges.

Cultural particulars include the *specific* practices that distinguish cultures from one another. For example, all people become hungry and all cultures have defined certain items and objects as edible. But the potential food sources defined as edible vary across cultures. That is, what is appealing to eat in one society may be considered repulsive or is simply unavailable in another.

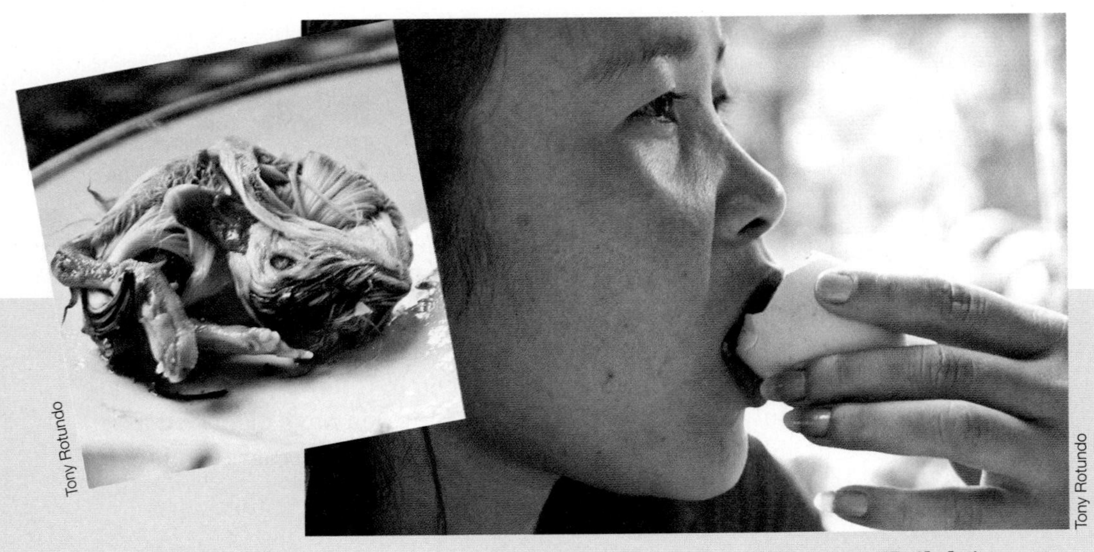

▲ This woman is eating a balut, or *kailuk* as it is called in Laos. Kailuk is a duck embryo that is boiled alive and then eaten from the shell. Popular in Southeast Asian countries, kailuk is considered a snack food and a good source of protein.

All cultures provide formulas for expressing **social emotions**, feelings that we experience as we relate to other people, such as empathy, grief, love, guilt, jealousy, and embarrassment. In that sense, such formulas for expressing social emotions are universal. Grief, for instance, is felt at the loss of a relationship; love reflects the strong attachment that one person feels for another person; jealousy can arise from fear of losing the affection of someone to another (S. Gordon 1981). People do not simply express a social emotion directly; they evaluate and can modify their feelings to fit with culturally established rules about how to express them. With regard to grief, some cultures embrace its open expression such that the bereaved are encouraged to wail loudly. Other cultures embrace self-control of grief to the point of encouraging their members to suppress any visible expression of grief (Galginaitis 2007).

Passing on Culture

The process by which people create and pass on culture suggests that it is more than a blueprint. Babies enter the world, and virtually everything they experience—being born, being bathed, being toilet trained, learning to talk, playing, and so on—involves others facilitating the experience. The people present in a child's life at any one time—father, mother, grandparents, brothers, sisters, playmates, caregivers, and others—expose the child to their "versions" of culture. In this sense people are carriers and transmitters of culture with a capacity to accept, modify, and reject cultural experiences to which they have been exposed and to choose cultural experiences to pass on to the next generations. As a case in point, consider that Christmas is celebrated in the United States as if everyone in the country participates in the festivities. Businesses close on Christmas Eve and Day; public schools give children the week of Christmas off; stores, houses, and streets are decorated as early as October; and Christmas-themed television commercials run for a month or more. Yet despite this exposure, many people in the United States reject this cultural option, as this reflection from one of my students illustrates:

> I grew up in a religion that did not celebrate Christmas or Easter even though we called ourselves Christian . . . While growing up I never challenged this; I just did what I was told.

Yet this same student as an adult decided to celebrate Christmas and other holidays that her family rejected in her youth.

> Now that I am older and have a child . . . [t]his year was the first year I put up a Christmas tree. I didn't know how to decorate it and I didn't know any Christmas carols but I learned. I decided that I don't want my children growing up the way I did.

When sociologists study cultures, they do not get caught up in identifying distinct markers that set people of one culture apart from those of another (e.g., all Christians celebrate Christmas). Nor do they assume that physical cues (eye shape, hair texture, skin shade) qualify someone for membership in a culture. Instead, they are most interested in how culture shapes human behavior and in how people create, share, pass on, resist, change, and even abandon culture.

Chris Caldeira

▲ The mobile phone is an element of culture because it is something people have created to help them communicate with others. People around the world have embraced the mobile phone and incorporated it into an already existing culture. Sociologists are interested in how the mobile phone is shaping human behavior and changing existing cultures now that people can be contacted anytime and anywhere and distracted from what they are doing.

(Write a Caption)

Write a caption that makes use of the concepts of cultural universals and cultural particulars. This profile of George Washington is one of five images of U.S. presidents carved into the stone of Mount Rushmore.

Chris Caldeira

Hints: In writing this caption

- review the concepts of cultural universals and cultural particulars,
- consider that every country is likely to have some iconic leaders who guided people through events considered critical to national identity, and
- think about what George Washington represents that is central or particular to American identity.

Critical Thinking

Do you think of yourself as belonging to a culture? In what ways does that culture function as a rough blueprint guiding your behavior and thinking?

Key Terms

culture	cultural universals
cultural particulars	social emotions

Material and Nonmaterial Culture

MODULE 2.2

Objective

You will learn that culture consists of two components: material and nonmaterial.

Chris Caldeira

Missy Gish

What messages does each lawn mower convey about the larger culture of which it is a part?

The lawn mower on the left would elicit little, if any, attention in Cuba (the place where it was photographed) but would certainly elicit a reaction from the typical American. What does that lawn mower say about Cuban culture? What does the lawn mower on the right say about U.S. culture? The lawn mowers are part of what sociologists call material culture. But lawn mowers were not created in a vacuum. As we will learn, a number of larger social forces account for their creation.

Material Culture

Material culture consists of all the natural and human-created objects to which people have assigned a name and attached meaning. Examples of material culture are endless and include iPods, cars, clothing, tattoos, trees, diamonds, and much more. In order to grasp the social significance of material culture, sociologists strive to understand the larger context in which an object exists. They also work to identify the meanings people assign to that object and the ways it is used. From a sociological point of view, material objects are windows into a culture because they offer clues about how people relate to one another and about what is valued.

Chris Caldeira

◄ The photo of the Cuban lawn mower that opens this module and this photo of a man repairing a more than 50-year-old automobile suggest that Cubans throw away nothing. These photos prompt us to ask how Cuba became a nation of recyclers. In 1960, after Fidel Castro seized U.S. assets and subsequently declared Cuba a socialist country, the United States imposed an economic embargo on Cuba and broke diplomatic relations. Cuba formed close ties with the then Soviet Union, receiving $4 billion to $6 billion in foreign aid each year from its ally. When the Soviet Union collapsed in 1989, Cuba lost an important revenue source. This loss, in conjunction with more than 50 years of embargo, has made economic hardship a way of life for the Cuban people—a hardship that supported the creation of a culture where almost nothing is thrown away.

Nonmaterial Culture

In contrast to material culture, **nonmaterial culture** is intangible human creations. Intangible means that these creations are things that cannot be touched with the hands. Nonmaterial culture includes values, beliefs, norms, and symbols.

VALUES. **Values** are general, shared conceptions of what is good, right, desirable, or important with regard to personal characteristics/ways of conducting the self and other desired states of being. While it is impossible to make a complete list of values, examples include individual freedom, happiness, consumption, conservation, sharing, cleanliness, obedience, independence, salvation, responsibility to others, and national security. Cultures are distinguished from one another, not according to which values are present in one society and absent in another, but rather according to which values are evoked as reasons for taking some action (like going to war or deciding to give a 6-year-old his own smartphone) and which values are most cherished and dominant (Rokeach 1973). One might argue, for example, that American culture values consumption over conservation and Cuban culture values conservation over consumption.

BELIEFS. **Beliefs** are conceptions that people accept as true concerning how the world operates and the place of the individual in relationship to others. In contrast to values, beliefs are about what is or is not true or real. People hold a variety of beliefs that can be accurate or inaccurate. Beliefs may be descriptive (I believe the earth is round), causal (I believe that fluoride prevents cavities), or prescriptive (The government should not intervene in people's lives) (Rokeach 1973). Beliefs can be rooted in blind faith, experience, tradition, or science. With regard to happiness, many Americans hold the belief that happiness can be achieved by accumulating wealth and consuming products. In contrast, many Cubans hold the belief that happiness can be achieved by getting the most use out of what one has. It is not that Cubans or Americans cannot see how happiness could be achieved in other ways; rather the beliefs each tends to hold about what actions result in happiness are derived, in part, from the opportunities open to them and the kinds of behavior each society encourages.

◀ For the most part, in the United States people are encouraged to throw away things without thinking of other uses to which they might be put. Americans are not encouraged, for example, to think about how they might use an empty toilet paper roll. People living in Cuba, on the other hand, are encouraged to think about such things—empty toilet paper rolls can function as hair curlers.

NORMS. A third type of nonmaterial culture is **norms**, written and unwritten rules that specify behaviors, thoughts, and appearances that are appropriate and inappropriate to a particular social situation. Examples of written norms are rules that appear in college student handbooks (e.g., to be in good academic standing, maintain a 2.0 GPA), on signs (smoke-free area), and on garage doors of automobile repair centers (Honk Horn to Open). Unwritten norms exist for virtually every kind of situation: wash your hands before preparing food, raise your hand to indicate that you have something to say, and do not throw anything away. Sometimes norms are formalized into laws (see Figure 2.2a).

▼ **Figure 2.2a: Minimum Drinking Age by Country**
One must be at least 21 years old to legally drink alcoholic beverages in the United States, which has one of the highest minimum-age drinking laws in the world. In Mexico the minimum age is 18, and in Canada the age is 19. In some countries, such as Italy, the drinking age is 16. In still other countries, such as Albania, Cambodia, and Ghana, there is no minimum age.

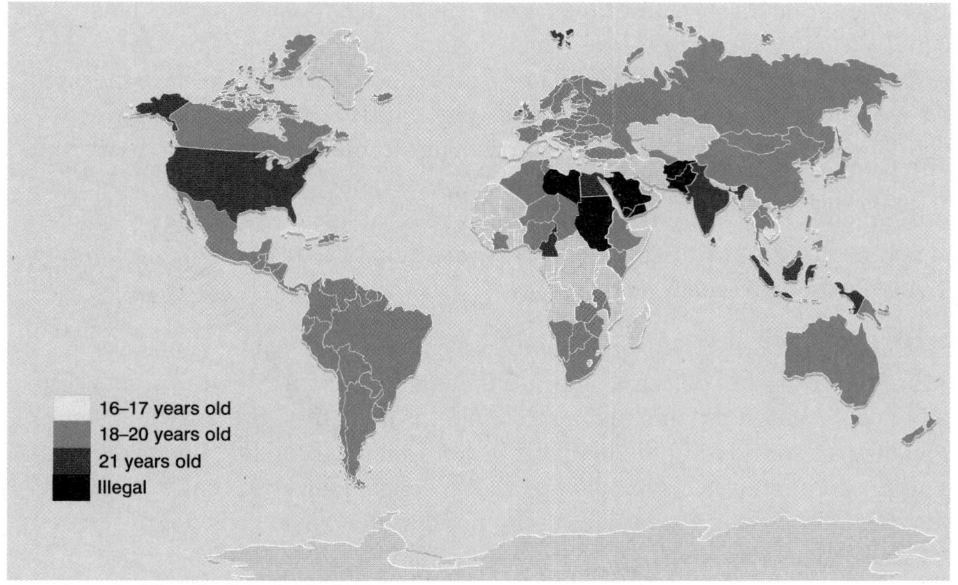

Source: International Center for Alcohol Policies (2010)

Depending on the importance of the norm, punishment for violations can range from a frown to the death penalty. In this regard, we can distinguish between folkways and mores. **Folkways** are norms that apply to the details of daily life: when and what to eat, how to greet someone, and how to dress for a school event such as a prom. As sociologist William Graham Sumner noted, "Folkways give us discipline and support of routine and habit"; if we were forced constantly to make decisions about these details, "the burden would be unbearable" (1907, 92). Generally, we go about everyday life without asking why until something reminds us or forces us to see that other ways are possible.

Whereas folkways apply to day-to-day details, **mores** are norms that people define as essential to the well-being of a group. People who violate mores are usually punished severely: they may be ostracized, institutionalized, or condemned to die. Mores are regarded as the only way.

➤ Folkways and mores are not always clear-cut. On first thought, we may think it is a folkway to "knock before opening a door." However, failure to knock on the door could be considered a violation of a more in the United States as residents may feel they have the right to shoot to kill if they think a person is invading their property.

Missy Gish

Mores in the United States reflect the high value placed on individual freedom. Mores in Cuba reflect high value placed on social responsibility. For example, in the early years of the HIV/AIDS epidemic (1986–1989) when little was known about the condition, Cuban health officials quarantined those who tested HIV-positive. They also strategically tested some groups considered especially vulnerable to the disease and thus key to containing an epidemic, such as pregnant women, health workers, blood donors, and those who had traveled to high-incidence areas of the world. The sexual contacts of HIV-positive individuals were traced and tracked down for testing and treatment. In time, these strict policies were relaxed to reflect knowledge gained about how to best treat the condition and protect public health. The emergency response gave Cuban health officials time to think through the best strategies. The Cuban response was resisted in the United States because such tactics violate mores related to individual rights, freedom, and privacy.

SYMBOLS. Another type of nonmaterial culture is **symbols**, which are anything—a word, an object, a sound, a feeling, an odor, a gesture, an idea—to which people assign a name and a meaning. In the United States, when someone makes a fist and then holds up his or her index and middle finger, depending on the context, it is a symbol of peace or victory. However, that meaning is not self-evident, because positioning the hand as described does not universally elicit these meanings.

➤ In some cultures a hand held in the pictured way can convey eventual victory in the midst of a long struggle, defiance of authority, an insult equivalent to the middle finger, the number 2, a symbol for the letter V, or "everything is okay or cool with me."

Chris Caldeira

In the broadest sense of the word, **language** is a symbol system that assigns meaning to particular sounds, gestures, characters, and specific combinations of letters. The complexity of human language is believed to set people apart from animals. Arguably language is the most important symbol system people have created. When we learn the words of a language, we acquire a tool that enables us to establish and maintain relationships, convey information, and interpret experiences. Learning a language includes an expectation that we will communicate and organize our thoughts in a particular way (Whorf 1956, 212–214). For example, some languages are structured so that speakers have no choice but to address people using special age-based hierarchical titles. For example, in Korea age is an exceedingly important measure of status: the older the people, the more status they can assume. As another example, in the United States the word *my* is used to express ownership of persons or things over which the speaker does not have exclusive rights: *my* mother, *my* school, *my* school bus, *my* country. The use of *my* reflects an emphasis on the individual and not the group. In contrast, the Korean language expresses possession as shared: *our* mother, *our* school, *our* school bus, *our* country.

Linguists Edward Sapir and Benjamin Whorf (1956) advanced the **linguistic relativity hypothesis**, also known as the Sapir-Whorf hypothesis, which states that "No two languages are ever sufficiently similar to be considered as representing the same social reality." The worlds of those who speak different languages "are distinct worlds, not merely the same world with different labels attached" (Sapir 1949, 162). Although languages channel thinking in distinct ways, do not assume that those speaking different languages cannot communicate. It may take some work, but it is possible to translate essential meanings from one language into another. For example, it is certainly possible to translate Korean words into English—a translator can emphasize *our* teacher, not *my* teacher—but lost in translation will be a Korean worldview that actually thinks in terms of *our* teacher. It is difficult for an English-speaking American who has no firsthand experience with Korean culture, which makes the group the central point of reference (rather than the individual), to completely grasp the meaning of "our."

Chris Caldeira

◀ How might the meaning of *Venceremos!* be translated and conveyed to English speakers in the United States so that they understand its meaning in the same way people living in Cuba do? For Cubans, *Venceremos!* means "We shall overcome," and it is used in reference to the revolutionary socialist goal of achieving equality and dignity for all. Just as Americans view the word *freedom* as an ideal worth fighting for, Cubans see the word *Venceremos* as something worth fighting for even as the country adopts some capitalist principles. *Venceremos* has driven the Cuban revolution since 1960 to resist U.S. (and capitalist) influence and interference, which they believe undermines equality and dignity.

(Write a Caption)

Write a caption that uses concepts from this module to describe a cultural practice shown in this photo.

Chris Caldeira

Hints: In writing this caption

- note that these Vietnamese school girls are using umbrellas to shield themselves from the sun's rays,
- think about what a tan symbolizes to Americans considered white and how a tan likely symbolizes something different in Vietnam, and
- consider what this practice suggests about how a tan is viewed in Vietnam.

Critical Thinking

Have you ever interacted with (or observed) someone you believed was from another culture? Using the concepts from this module, describe some features of that person's culture.

Key Terms

beliefs
folkways
language
linguistic relativity hypothesis
material culture
mores
nonmaterial culture
norms
symbols
values

Cultural Diversity

MODULE

2.3

Objective

You will learn the meaning of cultural diversity and how to think about cultural variety.

Pvt. Choi Keun-woo (USAG Yongsan)

Would you consider this graduating class to be culturally diverse? What might a truly diverse graduating class look like?

Sociologists use the term **cultural diversity** to capture the cultural variety that exists among people who share some physical or virtual space. That space may be as large as the planet or as small as a household. When the focus is on *cultural* diversity, sociologists examine the extent to which people in a particular setting vary with regard to the

1. material components of culture—the objects they possess and have access to and the meanings (positive or negative) assigned to those objects; and
2. nonmaterial components of culture—the values, beliefs, norms, symbolic meanings, and language guiding behavior and thinking.

Sociologists often look at the people who occupy a particular setting, such as a university campus, and then try to establish the extent to which cultural diversity is present. In glancing at the photo of the graduating class, we might consider whether the people pictured reflect the racial, age, and gender diversity of the United States. It is not that everyone classified as the same age, race, and gender shares a specific culture; rather they are likely to share similar cultural experiences related to being classified and perceived as such. That is, they experience and struggle with norms, values, beliefs, and symbolic meanings that have come to be associated with specific age, race, and/or gender categories. For example, people believed to be a particular age often face intense pressures to dress, live, and act in certain ways.

As you might think, we have to go beyond physical appearances to make a judgment about cultural diversity. Cultural capital is a useful concept for thinking about the diversity that exists in a particular setting. In the most general sense, **cultural capital** includes all the material and nonmaterial resources a person possesses or has access to that are considered useful and desirable (or not) in a particular social setting. We can think of cultural capital as objectified, embodied, and institutionalized (Bourdieu 1986).

Objectified cultural capital consists of physical and material objects that a person owns outright or has direct access to. These objects have a monetary value that is tied to others' willingness to buy, sell, own, and hold on to them. These objects also have symbolic value in that they convey meaning about the owner/seller/buyers' status. Finally, objectified cultural capital includes the ability to understand, appreciate, and explain an object's meaning and value.

➤ If we use the concept of objectified cultural capital to think about this homemade board game and its value in a specific setting, say American society, we might conclude that it would bring the owner little, if any, acclaim, nor would it be an object that most American children would covet (relative to electronic games or videos or the latest toy). In another setting, Cuba, this board game may be considered an innovative and clever way to recycle materials.

Embodied cultural capital consists of everything that has been consciously and unconsciously internalized through the socialization process. The socialization process shapes a person's character and outlook. Embodied cultural capital includes the words and languages one hears, has acquired, and has used to communicate with others, think about the world, and present oneself to others.

◀ Imagine that parents manage to instill in these toddlers an enduring interest in dinosaurs and in boxing. How might these respective interests shape each child's character and outlook? What words will each child learn as he pursues the cultivated interest? How might the respective experiences shape the way each child presents himself to others?

Institutionalized cultural capital consists of anything (material or nonmaterial) recognized as important to success in a particular social setting. Examples include academic credentials for a job search, the ability to dress the part (to look like a doctor, professor, or carpenter), physical qualities such as straight white teeth noticed by clients/customers, a youthful appearance when entering a bar, and so on. Whatever the attribute or item—a professional degree, nice clothes, a nice smile—the institutionalized backing gives those who possess it advantages beyond the attribute or item itself.

➤ In the United States (and elsewhere) teeth have important symbolic meanings that go beyond the teeth per se. This is a photo of the same person with his front teeth intact and with two front teeth missing. Do assumptions about his intelligence, wealth, occupation, and marital status change depending on the presence or absence of front teeth? The perfect smile, institutionalized through commercials and dental hygiene products, has become a desired attribute.

Chris Caldeira

Chris Caldeira

The three concepts—objectified, embodied, and institutionalized cultural capital—help us to think about what people bring to a setting. Given the endless variety of cultural experiences people can have, you see that establishing the degree to which a social setting is culturally diverse is not an easy task. Universities often highlight campus diversity by profiling their student body according to geographic location (in-state versus out-of-state residents, international students), sex, race, ethnicity, and age. In addition, they highlight the number of student clubs and organizations to showcase the variety of cultural experiences available. Because universities present diversity as something that "enhances the quality of the living and learning environment," many offer diversity scholarships and leave it to the applicant to make a case for how "aspects of your identity, your life experiences, special skills or values equip you to make a positive contribution to help ensure that the [campus] is rich with diversity, yet inclusive of all its members" (Northern Kentucky University 2012).

If you were to try to catalog cultural diversity on a college campus (or in any setting, for that matter), it is likely that as you visited with various campus groups and populations you would encounter what Cynthia K. Mahmood and Sharon Armstrong (1992) did when the two traveled to Friesland, a province in the Netherlands, to study the Frisian people's reactions to a book published on their culture. The researchers found that the Frisian people could not agree on a single "truth" about them as described in the book. At the same time, the villagers could not come up with a list of cultural characteristics that would distinguish them from other people living in the area. Still, the Frisians remained "convinced of their singularity" and reacted emotionally to the suggestion that they might not constitute a culture. Mahmood and Armstrong's findings suggest that "many things one would want to call cultural are not completely or even generally shared" (D'Andrade 1984, 90). What, then, holds people who believe they constitute a culture together? One answer is **cultural anchors**, some cultural component—material (a color, a mascot, a type of clothing, a book) or nonmaterial (a belief, value, norms, language)—that elicits broad consensus among members of its importance but also allows debate and dissent about its meaning (Ghaziani and Baldassarri 2011).

Chris Caldeira

▲ Given the diversity that exists in the United States, is there such a thing as American culture? We would be hard-pressed to describe a single culture that everyone shares in all its aspects. It is much easier to identify the cultural anchors that unite Americans even in the face of bitter debate and dissent. Notice that whenever there is protest in the United States, the protesting groups (in this case, LGBT [lesbian, gay, bisexual, and transgender] activists) wave the American flag to rally support. The flag is a cultural anchor because it symbolizes a value considered important to all Americans—"freedom"—whether it be freedom to pursue happiness or live a certain lifestyle free from interference. Likewise, LGBT activists come from all walks of life. Their cultural anchors include their flag and the belief that people should be free to choose who they love.

Subcultures and Countercultures

When thinking about cultural variety, the concepts of *subcultures* and *countercultures* are especially useful. Every society has **subcultures** that share in certain parts of the mainstream culture but that possess cultural anchors—values, norms, beliefs, symbols, language, and/or material culture—not shared by those in the so-called mainstream. The anchors may be associated with a physical setting (church, community center, neighborhood) or some selected aspects of life, such as work, school, recreation and leisure, marriage, friendships, fashion, or housing.

Chris Caldeira

◂ Cowboy churches are one example of a subculture. There are 775 such listings in the Cowboy Church Directory (2011). The Cowboy Church movement began in the 1990s in the southwest part of the United States, especially those areas that identify with the cultural anchor known as the cowboy. These churches are generally nondenominational and welcome all faiths and beliefs (Hyslop 2011). Members come to church dressed in attire we associate with cowboys (hats, boots, jeans). Pastors may preach sitting on a horse. No collection plates are passed, although there may be a boot, hat, or wooden birdhouse to drop money in when leaving (Associated Press 2009).

Sociologists use the term **countercultures** in reference to subcultures that challenge, contradict, or outright reject the mainstream culture that surrounds them. Sociologist Milton Yinger maintains that members of countercultures feel strongly that the society as structured cannot bring them satisfaction; some believe that "they have been caught in very bad bargains, others that they are being exploited," and still others think the system is broken (1977, 834). Because countercultures emerge in response to an existing order, Yinger argues that "every society gets the countercultures it deserves." Countercultures express themselves by deploring society's contradictions, "caricaturing its weaknesses, and drawing on its neglected and underground traditions" (850). In response, some countercultures "attack, strongly or weakly, violently or symbolically, the frustrating social order" (834). Yinger presents three broad, and at times overlapping, categories of countercultures:

- **Communitarian utopians** withdraw into a separate community where they can live with minimum interference from the larger society, which they view as evil, materialistic, wasteful, or self-centered. In the United States, the Old Order Amish constitute a communitarian counterculture in that they remain largely separate from the rest of the world, organizing their life so that they do not even draw power from electrical grids.
- **Mystics** search for "truth and for themselves" and in the process turn inward. "They do not so much attack society as disregard it, insofar as they can, and float above it in search of enlightenment" (838). Buddhist monks constitute such a counterculture because they make a point of rejecting the material trappings of capitalistic society. As monks, they are committed to simple living, modest dress, and vegetarian diet—ways of living that run counter to the values of capitalist-driven societies.
- **Radical activists** preach, create, or demand a new order with new obligations to others. They stay engaged hoping to change society and its values. Strategies to bring about change can include violent and nonviolent protest; Ghandi and Martin Luther King Jr. both famously used nonviolent protest to effect societal change. The Occupy Wall Street movement represents one example as, among other things, protestors seek to limit corporate influence and address the ever-widening inequality between the wealthiest 1 percent and the remaining 99 percent.

(Write a Caption)

Write a caption that relates women's roller derby to the concept of subculture and describes potential cultural anchors of that subculture.

Tony Rotundo

Hints: In writing this caption

- visit the Women's Flat Track Derby Association (WFTDA) website and other sites to learn about who is attracted to this sport, and
- consider what cultural anchors unite such a diverse population of athletes.

Critical Thinking

Are you a member of a subculture or counterculture? Describe any cultural anchors that unite that subculture or counterculture.

Key Terms

communitarian utopians
countercultures
cultural anchors
cultural capital
cultural diversity
embodied cultural capital
institutionalized cultural capital
mystics
objectified cultural capital
radical activists
subcultures

MODULE

Encountering Cultures (2.4)

Objective

You will learn concepts for describing people's reactions to foreign cultures.

When you shampoo your hair in the shower, do you step back from running water to lather your hair? Would you think to first wet your hair, turn the water off as you lathered, then turn the water back on to rinse it?

Missy Gish

If you lived in Germany all your life, you would very likely turn the water off while lathering. I learned about this practice from a German exchange student studying in the United States. Apparently, the student's host parents, after noticing that the water did not run continuously while she showered, told her that this was not necessary and that she should let the water run the entire time.

The Home Culture as the Standard

If you uncritically accept the idea that water should run the entire time one showers, and that it is absurd to conserve water in the way mentioned above, then your point of view can be considered ethnocentric. **Ethnocentrism** is a point of view in which people use their home culture as the standard for judging the worth of another culture's ways. Sociologist Everett Hughes describes ethnocentric thought in this way: "One can think so exclusively in terms of his own social world that he simply has no set of concepts for comparing one social world with another. He can believe so deeply in the ways and the ideas of his own world that he has no point of reference for discussing those of other peoples, times, and places. Or he can be so engrossed in his own world that he lacks curiosity about any other; others simply do not concern him" (1984, 474).

Ethnocentrism puts one culture at the center of everything, and all other ways are "scaled and rated with reference to it" (Sumner 1907, 13). Thus, other cultures are seen as strange or inferior. Often the words people choose to describe cultural differences offer clues as to whether they are ethnocentric (Culbertson 2006).

Sgt. Steven King

➤ Notice that this woman is finishing her notes on the left-hand side of the white board. If, upon learning that Arabic is written from right to left, you conclude that this is backward, then you have described the writing process in an ethnocentric way. From the point of view of the Arabic reader, writing English left to right could be considered backward.

Using one's home culture as the frame of reference to describe another culture's ways distorts perceptions. Many Americans, for example, are appalled when they learn that some cultures eat dog and evaluate this practice from the point of view that dogs possess special qualities that other animals commonly eaten by Americans, such as pigs, cows, and chickens, do not.

Mr. Devin L. Fisher (IMCOM)

➤ Keep in mind that people from cultures where dog is defined as a source of food don't eat *pet* dogs; rather, they eat a "special breed of large tan-colored dogs raised especially for canine cuisine" (Kang 1995, 267). In fact, those who eat dog would argue that Americans such as this man slicing meat from a roasted pig are in no position to judge them (Kang 1995). For some reason, many Americans protest that they could never eat Fido but seem to have no reservations about eating Porky Pig or Babe the Pig.

The most extreme and destructive form of ethnocentrism is one in which people feel such revulsion toward another culture that they act to destroy it. Unfortunately, in human history there are many examples where one group acts to destroy another by banning the targeted culture's language; changing the first and/or last names of those in the targeted group; miseducating children by using a special curriculum to indoctrinate them; destroying important symbols such as flags, places of worship, and museums; and even killing those who resist. In the United States between 1870 and 1920, Native American children were sent to boarding schools to ensure total immersion in white culture. This process separated them from their families and home cultures. Upon arrival children were issued uniforms, assigned names considered white, and given haircuts. They were punished for speaking their native languages.

➤ Native Americans enrolled in these boarding schools were also introduced to sports such as football to socialize them to American ideas about competition, winning, status, and teamwork. Taken in 1899, this photograph is of the Carlisle (PA) Indian Industrial School football team.

Prints & Photographs Division, Library of Congress, LC-USZ62-125089

Another type of ethnocentrism is **reverse ethnocentrism**, in which a home culture is regarded as inferior to a foreign culture. People who engage in this kind of thinking often idealize other cultures as utopias. For example, they might label Asian cultures as a model of harmony, American culture as a model of economic opportunity, and Native American cultures as a model of environmental sustainability (Hannerz 1992).

Cultural Relativism

Cultural relativism is an antidote to ethnocentrism. **Cultural relativism** means two things: (1) that a foreign culture should *not* be judged by the standards of a home culture, and (2) that a behavior or way of thinking must be examined in its cultural context—that is, in terms of that culture's values, norms, beliefs, environmental challenges, and history. For example, to understand the German propensity to conserve water when showering, we have to place that practice in the context of Germany's decades-long efforts to encourage environmentally conscious behavior. For one, the German government taxes household water consumption at a high rate. Likewise, to understand why people in some countries eat dog, we must consider that there is likely a shortage of grazing land to support large-scale cattle production.

Critics of cultural relativism maintain that this perspective encourages an anything-goes point of view, discourages critical assessment, and portrays all cultures as equal in value regardless of obviously cruel practices (Geertz 1984, 265). In response to this criticism, there is no question that notions of rightness and wrongness vary across cultures, and if we look hard enough we can probably find a "culture in which just about any idea or behavior exists and can be made to seem right" (Redfield 1962, 451). But that is not the purpose of cultural relativism. Ideally, cultural relativism is a perspective that aims to understand a culture on its own terms; the primary aim is not to condone or discredit it. More than anything, cultural relativism is a point of view that acts as a check against an uncritical and overvalued acceptance of the home culture, thereby constricting thinking and narrowing sympathies (Geertz 1984).

Tony Rotundo

➤ When using their home culture as the standard, Americans often make fun of sumo wrestling and have a hard time appreciating it as a serious sport. An ethnocentric American might describe a sumo wrestling match as two fat guys trying to push each other out of the ring. Cultural relativism reminds us to place the sport in the context of Japanese culture and history. A serious study of the sport would reveal that it is "rich with tradition, pageantry, and elegance and filled with action, excitement, and heroes dedicated to an almost impossible standard of excellence down to the last detail" (Thayer 1983, 271).

Culture Shock

Most people come to learn and accept the ways of their home culture as natural. Thus, when they encounter foreign cultures, they may experience **culture shock**, a mental and physical strain that people can experience as they adjust to the ways of a new culture. In particular, newcomers find that many of the behaviors and responses they learned in their home culture, and have come to take for granted, do not apply in the foreign setting. Newcomers may have to learn to squat when using the toilet and learn to eat new foods. Or, newcomers may experience "invisible" pressures they do not feel in their home culture, as described by this man who moved to the United States from a rural community in Mexico: "I lived in a laid-back, close knit community. The survival of the fittest attitude that exists in the United States made me feel that everyone was running from some sort of monster" (Anonymous [NKU student] 2012).

It is not that any one incident generates culture shock; rather, it is the cumulative effect of a series of such adjustments that can trigger an all-encompassing disorientation. The intensity of culture shock depends on several factors: (1) the extent to which the home and foreign cultures differ; (2) the level of preparation for living in a new culture; and (3) the circumstances—vacation, job transfer, or war—surrounding the encounter. Some cases of culture shock can be so intense and unsettling that people experience "obsessive concern with cleanliness, depression, compulsive eating and drinking, excessive sleeping, irritability, lack of self-confidence, fits of weeping, nausea" (Lamb 1987, 270).

A person does not have to live in a foreign culture to experience culture shock. People who move from West or East Coast cities to towns in the Midwest, and vice versa, can experience it. Children who attend private faith-based schools for eight years and then transfer to a public high school (and the reverse) can experience culture shock. When Hurricane Katrina devastated the Gulf Coast in 2005, 1 million people were forced from this region to live in scattered locations across the United States, resulting in the largest mass dislocation of people in

U.S. history. Many from urban New Orleans found themselves living in very small towns and reported experiences of culture shock. As one example, 100 black residents of New Orleans evacuated to Seguin, Texas, a town of 23,000 with 60 percent of residents considered Hispanic and 31 percent classified as white. They had nothing but the clothes on their backs and had to rely on the goodwill of strangers. Understandably they experienced cultural shock:

> Where were the city buses that they had relied on to get around in New Orleans? Where were all the black people? Where were folks sitting on porches, in a festive tangle of music and gossip? . . . a sign in front of the courthouse bragged that Seguin was "Home of the World's Largest Pecan." . . . Wrangler jeans and cowboy hats were among the donated items. . . . Heifers grazed nearby. . . . In the afternoon the breeze smelled like burned chicken from the nearby Tyson's poultry plant. (Hull 2005)

Reentry Shock

Do not assume that culture shock is limited to experiences with cultures outside of a home country. People can also experience **reentry shock**, or culture shock in reverse, upon returning home after living in another culture (Koehler 1986).

Chris Caldeira

➤ Many people find it surprisingly difficult to readjust when they return home after spending a significant amount of time elsewhere. As with culture shock, returnees face a situation in which differences jump to the forefront. Imagine, for example, that you have lived for years in a culture such as a rural village in Vietnam where the pace of life was much slower and where parents are less worried about danger lurking around the corner that might harm their children.

As with culture shock, the intensity of reentry shock depends on an array of factors, including the length of time someone has lived in the host culture and the extent to which the returnee has internalized the ways of the host culture. Symptoms of reentry shock mirror those of culture shock. They include panic attacks ("I thought I was going crazy"), glorification of and nostalgia for the foreign ways, a sense of isolation, and a feeling of being misunderstood. This comment by one American returning from abroad illustrates: "Many things about the United States made me angry. Why did Americans have such big gas-guzzling cars? Why were all the commercials telling me I had to buy this product in order to be liked?" (Sobie 1986, 96; Thebodo 2011.) In fact, many people become anxious and feel guilty about having problems with readjustment. Upon returning home after a long stay in a foreign culture, many worry about how family, friends, and other acquaintances will react to their sometimes critical views of the home culture; they may be afraid that others will view them as unpatriotic.

The experience of reentry shock points to the transforming effect of an encounter with another culture (Sobie 1986). That the returnees go through reentry shock means that they have experienced up close another way of life and that they have come to accept the host culture's norms, values, and beliefs. Consequently, when they come home, they see things in a new light.

Prints & Photographs Division, Library of Congress, LC-USZ62-100543

(Write a Caption)

Write a caption that applies the concept of ethnocentrism to the 1884 *Illustrated Newspaper* cover showing a female Native American student who attended the government school at Carlisle visiting her home at Pine Ridge Indian Reservation.

Hints: In writing this caption

- review the purpose of boarding schools for Native Americans,
- consider how this student was taught to view her home culture and how she might now hold an ethnocentric view toward it, and
- think about the clues the image holds for things she may now reject about her birth culture and see as inferior.

Critical Thinking

Describe a time when you reacted to a different culture's way of thinking or behaving with an ethnocentric (or reverse ethnocentric) response.

Key Terms

cultural relativism

culture shock

ethnocentrism

reentry shock

reverse ethnocentrism

MODULE 2.5

Cultural Diffusion

Objective

You will learn that many of the items we take for granted in our daily lives originated in foreign settings.

Chris Caldeira

Have you ever taken an interest in something that originated in a foreign culture, such as this man who had a Chinese proverb tattooed on his forearm?

Cultural Borrowing

The process by which an idea, invention, or way of behaving is borrowed from a foreign source and then adopted by the borrowing people is called **cultural diffusion**. The term *borrowed* is used in the broadest sense; it can mean steal, imitate, purchase, or copy. The opportunity to borrow occurs whenever people from different cultures make contact, whether face-to-face, by phone, via the Internet, or through other media channels. Instances of cultural diffusion are endless.

View Stock/Alamy

Courtesy Photo, U.S. Department of Defense

➤ The compass is a Chinese invention of the third century BC. It is an important navigational instrument that helps people know where they are on the earth. Countless cultures have borrowed the compass and improved on its features over the past 23 hundred or so years. The Global Positioning System, known as GPS, is perhaps the most recent variation.

Selective Borrowing

People in one culture do not borrow ideas or inventions indiscriminately from another culture. Instead, borrowing is often selective. That is, even if people in one culture accept a foreign idea or invention, they are, nevertheless, choosy about which features of the item they adopt. Even the simplest invention is really a combination of complex elements, including various associations and ideas of how it should be used. Not surprisingly, people borrow the most concrete and most tangible elements and then shape the item to fit in with their larger culture (Linton 1936).

Christopher Morris/Associated Press

◀ As a result of diffusion, the first McDonald's restaurant in Saudi Arabia opened in 1996. There are now several hundred restaurants in that country, and the Saudi diet includes Big Macs and Chicken McNuggets. The Saudis have not adopted the McDonald's model in total, however. As Saudi Arabia is a sex-segregated society, it did not adopt a queuing system in which any customer can step into one line to place orders. In Saudi Arabia, women queue to the right and men to the left to place orders. There are also many special items on the menu that cater to Saudi tastes, including date pies and spring rolls (McDonald's Saudi Arabia 2012).

The Diffusion Process

Keep in mind that cultural diffusion is a process that generates change in the borrowing society. The introduction of fast foods changed people's eating habits in many ways, including decreasing the amount of time devoted to eating a meal, increasing the likelihood of eating away from home, and altering the types and amounts of food consumed.

Sociologists who study cultural diffusion are interested in the rate at which the borrowing people come to use or apply a new idea, behavior, or invention. Upon making its debut, sociologists ask how quickly others in the culture acquire, learn about, and/or come to use or consume it. The answer depends on a variety of factors, including (1) the extent to which that borrowing causes people to change ways of thinking and behaving, (2) the existence of a media structure that lets people learn about it, and (3) the social status of the first adopters. For example, first use and acceptance by a lower-status group may cause higher-status people to reject it as disrespectful, dangerous, or subversive. Conversely, first use by higher-status groups may cause those in lower status groups to reject it as elitist and condescending.

Cpl. Kim Kyu-ho, Eighth Army Public Affairs

➤ Hip-hop is believed to have emerged in the 1970s in black–Puerto Rican communities of South Bronx, New York. Since that time, hip-hop has diffused across the globe. While Korean artists have borrowed the hip-hop style, the political issues confronted through the music concern Korean youth, such as the intense pressures put on them to succeed in school and the reunification of North and South Korea (K. Lee 2006).

Sociologist William F. Ogburn (1968) believes that one of the most urgent challenges facing people today is the need to adapt to new products and inventions such as smartphones and medical technologies in thoughtful and constructive ways. Needed adjustments are often resisted—people text while driving and overuse the latest medical technologies. Clearly, failure to adjust can have disruptive consequences to society. He uses the term **adaptive culture** in reference to the role that norms, values, and beliefs of the borrowing culture play in adjusting to a new product or innovation, specifically adjusting to the associated changes in society. In this regard, one can argue that Americans adapted easily to things like mobile phones, the automobile, and computers because they save people time and support deeply rooted norms, values, and beliefs regarding individualism and personal freedom.

On the other hand, many point to problems these inventions have created. Some are advocating for laws banning use of mobile phones while driving, increasing automobile gas mileage, and regulating the disposal of obsolete computers. These examples suggest that needed adjustments are not always or easily made.

Chief Warrant Officer 4 Seth Rossman

◀ Think about the emergence of robotic technologies that are capable of doing low-skill and repetitive tasks such as filling prescriptions. This photo is of a prescription-filling robot known as Autoscript III. Some of these robots now work in pharmacies as technicians. According to the Bureau of Labor Statistics (2012), there are an estimated 330,000 pharmacy technicians earning a median pay of $13.60 per hour. As this technology is diffused into many pharmacies, what effect will it have on the people who occupy the jobs robots can now do? Should we think about the impact on populations of workers before robots such as Autoscript become commonplace?

(Write a Caption)

Write a caption that relates the effects of cultural diffusion on these two youth baseball teams, one made up of players from Japan in white uniforms (left) and the other team made up of players from the United States in red and gray uniforms.

Pfc. Terence L. Yancey

Hints: In writing this caption

- think about in which country baseball is believed to have originated,
- think about in which country the cultural practice of greeting the umpire with a bow and the practice of tipping one's cap originated, and
- consider what the Japanese players may have borrowed from the United States beside the game itself (at least on this occasion).

Critical Thinking

Identify something you do, use, say, eat, or think that has been borrowed from a foreign source. Explain its significance to your life.

Key Term

adaptive culture

Applying Theory: Global Society Theories

MODULE 2.6

Objective

You will learn how global society theorists frame the diffusion process.

How did Coca-Cola become a global brand?

pcruciatti/Shutterstock.com

The first glass of Coca-Cola was sold in Atlanta, Georgia, in 1886. Today, Coke is sold in more than 200 countries. Sociologist Leslie Sklair (2002) asks how a drink that is "nutritionally worthless" came to be a truly global product. Sklair argues that Coke (and Pepsi, for that matter) are marketed on a global scale, not as soft drinks per se, but as a lifestyle promising that when anyone, even the most poor, buy the product, they "can join in the great project of global consumerism, if only for a few moments" (611). Sklair's theory of global consumption offers insights about the broader phenomenon of cultural diffusion. First, we place Sklair's ideas in the context of global society theory.

The Connections between Local and Global Contexts

Global society theorists see human activity as embedded in a larger global context. They emphasize that seemingly local events are interconnected with events taking place within other countries and regions of the world. These interconnections are part of a phenomenon known as **global interdependence**, a situation in which human relationships and activities transcend national borders and in which social problems—such as unemployment, drug addiction, water shortages, natural disasters, or the search for national security—are experienced locally but are shaped by events taking place outside the country. Global interdependence is part of a dynamic process known as **globalization**—the ever-increasing flow of goods, services, money, people, technology, information, and other cultural items across political borders. This flow has become more dense and quick-moving as constraints of space and time break down.

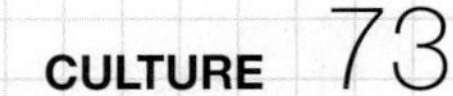

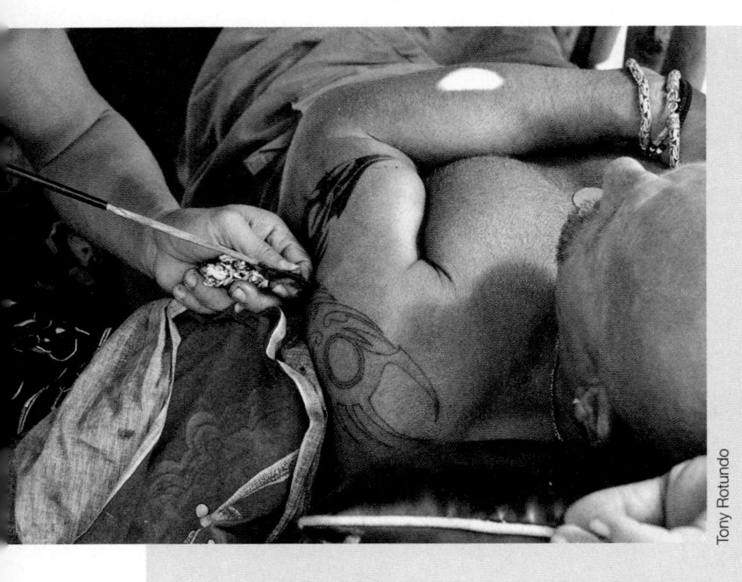

Tony Rotundo

‹ Ultimately global interdependence is enacted through relationships in which at least one of the parties has crossed national borders to interact directly or indirectly with others, as did this tourist, who got a tattoo while visiting Thailand. The tourist represents but one of the 14.1 million tourists from all over the world who visit Thailand each year (World Bank 2011b). This number is one rough measure of global interdependence as it relates to Thailand and its connections to others around the world. Thailand is also part of the globalization process in that tourism has increased steadily over the past two decades from 5.3 million tourists to 14.1 million (Noypayak 2001).

Four Positions on Globalization

Globalization is an ongoing process that involves economic, political, and cultural transformations between and within countries. Global society theorists differ with regard to the character of those transformations and which of the four dominate the globalization process (Appelrouth and Edles 2007). Regardless, when taken together, the four positions capture the nature of globalization.

Position 1: Globalization is producing a homogeneous world that fuses distinct cultural practices into a new world culture as embodied in trends such as worldbeat, world cuisine, and world cinema.

Yonhap News/YNA/Newscom

Sgt. Charles Brice

› Hip-hop music is a global phenomenon. Artists in many countries throughout the world have imported American hip-hop from rappers such as Kaine and D-Roc (of the Ying Yang Twins) but have also refashioned it to suit local sensibilities. From a global perspective, hip-hop gives the disenfranchised a voice, an identity, and a platform to articulate social issues important to them (Edlund 2005; Nawotka 2004).

Position 2: Globalization is producing a homogeneous world by destroying variety or, more specifically, the local cultures that get in the way of progress or cannot compete against large corporations. The engines of this homogeneity—sometimes referred to as McWorld and Coca-colonization—are consumerism and corporate capitalism.

Chris Caldeira

Chris Caldeira

⮝ How can this food vendor compete against a fast-food establishment that is open 24 hours a day and that is committed to delivering food to customers within minutes of ordering? When people eat a Big Mac or drink a Coke, they consume more than a burger or a beverage; they also consume associated values. Those values relate to the time one should spend preparing and eating food, the nature of the relationship between the cook and the person eating, and the place of the individual in relationship to the group (i.e., I can eat whatever I want, whenever I want; I do not have to eat what others are eating at set times of the day).

Position 3: Globalization actually brings value to and appreciation for local products and ways of doing things. Consumption is not a one-way exchange in which the buying culture simply accepts a foreign product as is. Although the products of corporate capitalism penetrate local markets, local ingredients, tastes, and preferences do not disappear.

Ted Aljibe/AFP/Getty Images

➤ In Japan, McDonald's has adapted to Japanese sensibilities in a number of ways. For one, the Japanese believe that bread is not filling and thus see hamburgers as a snack or something to be eaten between meals. For meals, the Japanese prefer rice burgers—a slice of meat served between rice patties. Second, in Japan, food has a social function, allowing those eating together to share food and in the process cement social relationships and create a feeling of community. To support this social function, McDonald's serves rice in common containers to be shared by everyone at the table (Megan 2009).

Position 4: Globalization and its interconnections intensify cultural differences by sparking conflicts as people fight (1) to preserve their identity and way of life, (2) to resist outside influences that clash with cultural ideals, or (3) to protect and enforce boundaries even as they are opened (Appelrouth and Edles 2007).

James R. Tourtellotte

⋀ As travel across borders increases, concerns about national security also have increased. Specifically, airport security and border patrol agents must work to process visitors as quickly as possible, but at the same time they also work to identify and stop real (and sometimes imagined) threats from crossing borders. The officer and drug-sniffing dog pictured here are searching for drugs in a vehicle about to enter the United States.

Capitalism and Globalization

In the spirit of Position 2, Sklair (2002) argues that capitalism is *the* force shaping the character of globalization. He describes the force of capitalism in this way:

- the vehicle of global capitalism is the transnational corporation,
- the driver of global capitalism is the transnational capitalist class, and
- the cultural ideology of consumerism is the fuel that powers the motor of global capitalism.

TRANSNATIONAL CORPORATIONS. The global economy is dominated by a few gigantic transnational corporations (TNCs) marketing their brands all over the world. Sklair does not say what "a few" means, but it would probably include the Global 500 companies. One reason the largest multinational organizations have great influence on the societies in which they operate is their size. Taken together, the annual revenues of the top 10 global corporations are about $2.3 trillion. Only five countries in the world—the United States, China, Japan, India, and Germany—possess a gross national product that exceeds that amount. As one measure of how a relatively small number of TNCs dominate the world economy, consider that in the United States alone there are 275,843 exporting companies, 50 of which account for one-third of all exports (U.S. Census Bureau 2010b).

TRANSNATIONAL CAPITALIST CLASS. Globalization "does not just happen: It is engineered and promoted by identifiable groups of people within identifiable organizations" (Sklair 2002, 601). Obviously, the transnational capitalist class (TCC) includes the top executives of the largest TNCs, as well as executives associated with the major media outlets who advertise their products and services. Sklair includes the federal and state politicians who make and pass laws that support corporations and their profit-making activities as part of this capitalist class.

TRANSNATIONAL CULTURAL IDEOLOGY. The transnational cultural ideology (TCI) is **consumerism**, an ideology that "proclaims that the meaning of life is to be found in the things that we possess." According to this ideology, when one consumes one feels alive, and "to remain fully alive, we must continuously consume, discard, and consume" (Sklair 2002, 601). According to Sklair, the expansion of capitalism depends on creating legions of consumers who buy, not to satisfy real needs, but because they believe that buying things will allow them to become something they could not be otherwise. The ideology of consumerism persuades people to consume not simply to satisfy real need, but to satisfy "artificially created desires" (601).

➤ Consider this cosmetic counter located in the People's Republic of China. Here, Lancôme features photographs of models who appear white to sell its products to Asian women. Lancôme and other cosmetic companies send the message that the typical Chinese woman, no matter how beautiful, cannot achieve this appearance without these products.

Boni Li

Beginning in the 1980s, the media dramatically increased its ability through TV, cable, satellite, and Internet technologies to deliver messages and images on an unprecedented scale. Sklair argues that few people in the world can avoid the reach of advertisers and marketers, whose messages "are beginning to penetrate deeply into the countryside even in the poorest places" (607). Sklair believes that there is "nothing in human experience that has prepared men, women, and children for the modern television techniques of fixing human attention and creating the uncritical mood required to sell goods, many of which are marginal at best to human needs" (605).

(Write a Caption)

Write a caption that ties the concept of consumerism to the child's sense of self.

Chris Caldeira

Hints: In writing this caption

- review the concept of consumption,
- speculate how long the typical child will remain interested in any one toy,
- think of the "need" driving the desire to purchase toys, and
- ask what the child might anticipate feeling from owning a coveted toy.

Critical Thinking

Review the four positions or themes describing globalization effects. Which of the four best captures globalization as you have experienced it? Give an example to support your point.

Key Terms

consumerism global interdependence globalization

Summary: Putting It All Together

CHAPTER 2

Sociologists define culture as the way of life of a people. Culture includes the shared and human-created strategies for adapting and responding to the surrounding environment. On some level, culture is a blueprint that guides and, in some cases, even determines behavior. People of the same culture are not replicas of each other, because they possess the ability to accept, create, reject, revive, and change culture.

Culture consists of material and nonmaterial components. Material culture is the physical objects that people have invented or borrowed from other cultures. Sociologists seek to understand the ways people use these physical objects and the meanings they assign to them. Nonmaterial culture is the intangible aspects of culture, including beliefs, values, norms, and symbols.

People borrow ideas and objects from other cultures through a process known as cultural diffusion. Borrowing is usually selective in that people are choosy about which features of the item they adopt, and they reshape the item to fit the core values of their own cultures. Cultural diffusion is a process that generates change in the borrowing society. One key challenge facing people today is the need to adapt to new products and inventions in thoughtful and constructive ways.

Sociologists use the term *cultural diversity* to capture the cultural variety that exists among people who find themselves sharing some designated physical or virtual space. Cultural capital in all its forms (objectified, embodied, and institutionalized) is a useful concept for thinking about the diversity that exists in a particular setting (Bourdieu 1986). Sociologists use the concepts of subcultures and countercultures to further describe diversity. The difficult part of describing cultures is capturing what holds together people who believe they constitute a culture. One answer is cultural anchors, some cultural component—material (a color, a mascot, a sacred book) or nonmaterial (a belief, value, norms, language)—that elicits broad consensus among members of its importance but also allows debate and dissent about its meaning.

When we encounter different cultures, the natural tendency is to judge them using our home culture as the standard. When people do this, they are engaging in ethnocentrism. Sometimes people find themselves in situations where they must leave their home culture and experience foreign ways of thinking and behaving. Culture shock is the mental and physical strain that people from one culture experience when they must reorient themselves to the ways of a new culture. People can also experience reentry shock upon returning home after living in and adapting to the ways of another culture. Cultural relativism is an antidote to ethnocentrism, in that it acts as a check against an uncritical and overvalued acceptance of the home culture that constricts thinking and narrows sympathies.

Go to cengagebrain.com to link to Aplia and CourseMate for the chapter quiz and other activities.

Chris Caldeira

CHAPTER 3
SOCIALIZATION

Socialization is the link between society and the individual—a link so crucial that neither individual nor society could exist without it (I. Robertson 1988). To understand this link, think of society as analogous to a game with rules and expectations about how it is played. For individuals to be part of the game, there must be mechanisms in place (socialization) by which they learn the rules and expectations. If there are no such mechanisms, the "game" ceases to exist. But the remarkable aspect is that people created the game and pass rules and expectations on to others. Those rules and expectations adapt to challenges and change introduced by new "players" and subsequent generations.

Go to Sociology CourseMate on cengagebrain.com to watch a video featuring a study of two-year-old twins and the role that nature and nurture play in their development.

MODULE

3.1 Socialization

Objective

You will learn that socialization is a process that prepares people to live in society and to interact with others.

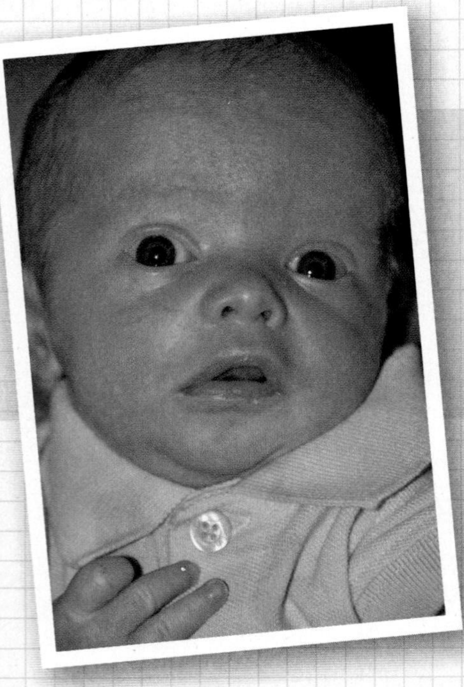
Chris Caldeira

Do you believe that newborn babies should sleep alone in a crib or with a parent? Why do you believe what you do?

Some parents believe that sleeping alone creates an independent child. Others argue that infants sleeping with the parent, especially the mother, is universal to all species and creates an emotionally secure child. Whatever your opinion, it is likely informed by the kind of child you believe each socialization experience will create.

Socialization is the lifelong process by which people learn the ways of the society in which they live. More specifically, it is the process by which humans

- acquire a sense of self or a social identity,
- develop their human capacities,
- learn the culture(s) of the society in which they live, and
- learn expectations for behavior.

Socialization is not a one-way process such that people simply absorb these lessons. It is a process by which people negotiate, resist, ignore, and even challenge those lessons.

Acquiring a Sense of Self

What does it mean to acquire a **sense of self**? From a sociological point of view, children acquire a sense of self when they can step outside the self and see it from another's point of view and also imagine the effects their words and actions have on others. Having a sense of self also means that children have acquired a set of standards about how others expect them to behave and look in a given situation.

Researchers have devised an ingenious method for determining when a child has acquired a sense of self. A researcher puts a spot of blush on the child's nose and then places the child in front of a mirror. Presumably, a child who ignores the blush does so because he or she is not yet able to imaginatively step outside the self and evaluate it from another's point of view. Moreover, that child has not yet acquired a standard about how he or she is expected to look (Kagan 1989).

Lisa Southwick

Lisa Southwick

◀ The child in the top photo does not seem bothered by the spot of blush on her nose. In fact, she does not even appear to know she is the baby reflected in the mirror. This is because she has not yet developed a sense of self. The child in the bottom photo appears to be concerned about the blush. Her facial expression tells us that she has acquired a sense of self, which means that she evaluates herself using a set of standards about how she ought to look.

Acquiring a sense of self also involves learning about the groups to which we belong and, by extension, do not belong. The importance of groups to identity is best illustrated when we think about how we get to know someone. We ask questions like "What is your name?" "Where do you live?" "How many brothers and sisters do you have?" "Are you in school?" "Do you play sports?" The answers to these questions reference group memberships in a family (last name), in a town, in a school, and so on. Consider how much time we spend teaching young children about the groups to which they belong, including working with them to know their family name, their age, their sex, their race, and the country in which they live. Such ideas do not come easily. We have to go over them many times before children get it.

Bill Mossman, U.S. Army Garrison—Hawaii Public Affairs

S.Sgt Vanessa Valentine

▶ Through the socialization process, children come to learn about and to identify with various groups, including the country in which they live. Through rituals involving flags, these children are learning to see themselves as Americans and Iraqis, respectively.

Developing Human Capacities

Socialization is also a process by which human capacities are developed. Our parents transmit, by way of their genes, a biological heritage common to all humans but unique to each person. Our genetic heritage gives us, among other things, a capacity to speak or sign a language, an upright stance that frees arms and hands to carry and manipulate objects, and hands that can grasp objects. If these traits seem too obvious to mention, consider that, among other things, they allow humans to speak innumerable languages and to use their hands to create and use inventions. Babies cannot realize these biological potentials unless caregivers support and encourage their development. Likewise, babies born with physical impairments find ways to compensate if given the opportunity and support.

Thomas Witte/Sports Illustrated/Getty Images

➤ The story of Bobby Martin, a high school football player born without legs, captivated the sports world. One reporter described Martin's situation this way: "Never knowing a life with legs, Bobby from an early age just adapted to using his arms and the pendulum motion of his body for movement. And after perfecting this method of locomotion for 17 years—and seeing it first hand—I can tell you this kid can move with the best of them" (T. Witte 2005).

Learning Culture and Expectations for Behavior

By the time children are two years old, most are biologically ready to pay close attention to social expectations, or the "rules of life." They are bothered when things do not match their learned expectations: paint peeling from a table, broken toys, small holes in clothing, and persons in distress all raise troubling questions for children (Kagan 1989). To show this kind of concern, two-year-olds must first be exposed to information that leads them to expect people and objects in their world to be a certain way (Kagan 1989). They develop these expectations as they interact with others.

To understand how children learn the ways of their culture, keep in mind that they learn through the things we have them do and say. As children learn a language, they learn their culture's names for things in the world and ways of verbally expressing their thoughts and feelings. The clothes they wear and toys they play with convey messages about their culture and its values. In addition, routines—the repeated and predictable activities, such as eating at certain times of the day or taking baths, that make up day-to-day existence—teach children about what is important to the culture (Corsaro and Fingerson 2003).

Internalization

Socialization takes hold through **internalization**, a process by which people accept as binding learned ways of thinking, appearing, and behaving. We know socialization has taken hold if people suffer guilt when they violate expectations. We also know expectations have been internalized when a person conforms even when no one is watching. In these instances an inner voice urges conformity (Campbell 1964).

Lisa Southwick

➤ This boy may have been taught not to drink juice out of the carton, but he has not internalized that rule. When no one is looking, he violates that expectation. It is clear from his glance that he knows this expectation.

Anyone who observes young children and their caretakers understands the effort the two parties expend to make living together "workable." As one example, parents and children negotiate the shopping experience. Parents let their children know through glances, words, and physical contact (e.g., firmly grabbing their hands) that they can't have everything they want. The child often responds by whining or throwing a tantrum. Now it is the parent's move, and the child responds. The two negotiate an understanding about how they will shop together (Corsaro and Fingerson 2003). But their attempts to make shopping together workable cannot be understood apart from the larger social context of consumerism. The U.S. government has relatively few laws limiting children's exposure to advertisements. The typical American child views between 25,000 and 40,000 television commercials per year. By contrast, Sweden, Norway, and Finland ban commercial sponsorship of television programming aimed at children 12 and under (Rabin 2008).

(Write a Caption)

Write a caption that relates this scene to the socialization process.

Missy Gish

Hints: In writing this caption

- ask yourself which of the four dimensions of socialization are being realized when a parent or other caretaker gives a child medicine whenever he or she experiences a physical ailment such as the sniffles, and
- remember that children learn about their culture by the things we ask them to do.

Critical Thinking

Give an example of expectations you learned but did not internalize.

Key Terms

internalization | sense of self | socialization

MODULE

3.2 Nature and Nurture

Objective

You will learn about the importance of social interaction to physical and social development.

If you had to choose, which of the two factors—genetic potential or dedicated training—do you think is more important for achieving success at an activity like gymnastics?

Andy Nelson/*The Christian Science Monitor* via Getty Images

The People's Republic of China is often criticized for how it selects and develops athletic talent. Between the ages of six and nine, gifted children are chosen to join government-funded sports programs. In determining giftedness, officials look for some exceptional physical attribute, such as a wingspan that exceeds height as one sign of a potentially gifted boxer, above-average height as one sign of a potentially gifted basketball player, or exceptional flexibility as one sign of a potentially gifted gymnast (Krovatin 2008). Chinese sports programs develop athletic talent by subjecting those chosen to rigorous training. It seems that the Chinese system places equal weight on innate potential (nature) and dedicated training (nurture) as important factors in creating the best athletes.

Nature and Nurture

Nature is human genetic makeup, or biological inheritance. **Nurture** refers to the social environment, or to the interaction experiences that make up every person's life. Both nature and nurture are essential to socialization. Some scientists debate the relative importance of genetic makeup and the social experiences, arguing that one is ultimately more important than the other. Such a debate is futile, because it is impossible to separate the influence of the two factors or to say that one is more critical. In fact, it is difficult to identify any human trait that can be explained by nature or nurture alone. One might argue, for example, that height is genetically determined. But height depends, in part, on nutrition; children who are undernourished are considerably shorter than they might otherwise be. Likewise, one might argue that certain personality traits such as shyness are inherited. But no scientist has found a specific gene responsible for shyness. While it might seem that babies are born with personalities—that is, some babies seem to be born fussy, active, or serene—we must remember that the babies' experiences began in the womb nine months earlier and were shaped by the mother's life, nutritional health, and emotional experiences.

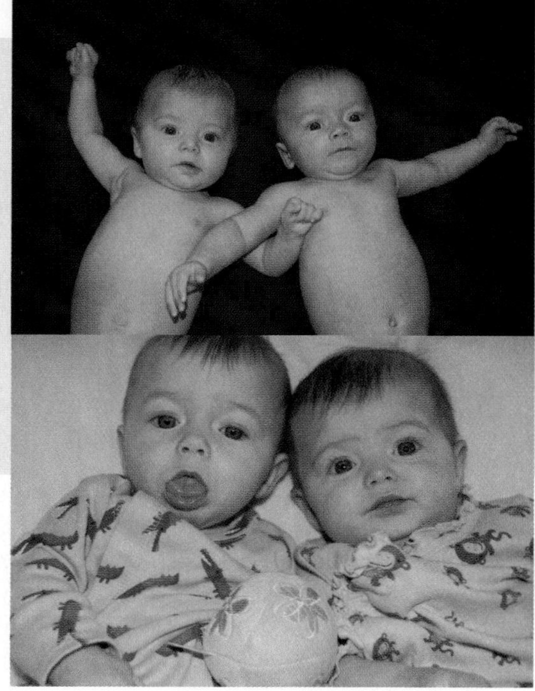

➤ These twins are male and female (nature). Their lives will take different paths as a result of nurture. Nurture is a factor when caretakers dress babies in pink and blue clothes. Those colors signal others to treat each child according to society's ideas about what boys and girls are and can be. But human development is a complex process, and some children, when they get old enough, resist efforts to push them to behave in what are considered sex-appropriate ways.

Trying to separate the effects of nature and nurture is like trying to determine whether length or width accounts for the shape of a picture frame (Kolb 2007). The latest research suggests that nurture and nature collaborate to shape people's lives. To grasp this relationship, consider how language is learned. As part of our human genetic makeup (nature), we possess a cerebral cortex, which allows us to organize, remember, communicate, understand, and create. In the first months of life, all babies are biologically capable of babbling the essential sounds needed to speak any language. As children grow, however, this enormous potential is reduced by the language or languages that the baby hears spoken and eventually learns (nurture).

Research tells us that in order for children to realize their biological potential, they must establish an emotional attachment with a caring adult. In other words, there must be at least one person who knows a baby well enough to understand his or her needs and feelings and who will act to satisfy them. Under such conditions, a bond of mutual expectation between caregiver and baby emerges, and the child learns how to elicit predictable responses in his or her caretakers: smiling causes the caretaker to smile; crying prompts the caretaker to soothe the child.

➤ When there is a bond of mutual expectation, a child can expect his or her caretakers to respond in predictable ways. This baby comes to learn and trust that when she opens her mouth wide, for example, her mother does the same.

Lisa Southwick

When researchers set up experimental situations in which parents failed to respond to their infants in expected ways, even for a few moments, they found that the babies suffered considerable tension and distress. Cases of children raised in situations of extreme isolation offer even more dramatic illustrations of the importance of caring adults in children's lives.

The Effect of Social Isolation

Sociologist Kingsley Davis (1940, 1947) did some of the earliest and most systematic work on the consequences of extreme isolation. His research shows how neglect and lack of social contact (that is, nurture) can delay the development of human potential (that is, nature).

Davis documented and compared the separate yet similar lives of two girls: Anna and Isabelle. During the first six years of their lives, the girls received only minimal care. Both children lived in the United States in the 1940s, and because their mothers did not marry the fathers, the babies were viewed and treated as illegitimate. Anna was forced into seclusion and shut off from her family and their daily activities. Isabelle was shut off in a dark room with her mother, who was deaf and could not articulate speech. Both girls were six years old when authorities intervened. At that time, they exhibited behavior comparable to that of six-month-olds. Anna "had no glimmering of speech, absolutely no ability to walk, no sense of gesture, not the least capacity to feed herself even when food was put in front of her, and no comprehension of cleanliness. She was so apathetic that it was hard to tell whether or not she could hear" (K. Davis 1947, 434). Anna was placed in a private home for mentally disabled children until she died four years later. At the time of her death, she behaved and thought at the level of a two-year-old.

While Isabelle had not developed speech, she did use gestures and croaks to communicate. Because of a lack of sunshine and a poor diet, she had developed rickets: "Her legs in particular were affected; they 'were so bowed that as she stood erect the soles of her shoes came nearly flat together, and she got about with a skittering gait'" (K. Davis 1947, 436). Isabelle also exhibited extreme fear of and hostility toward strangers. Her case shows how the "gene" for rickets is turned on by difficult social experiences.

Isabelle entered into a special needs program designed to help her master speech, reading, and other important skills. After two years in that program, she achieved a level of thought and behavior normal for someone her age. Isabelle's success may be partly attributed to her establishing an important bond with her deaf-mute mother, who taught her how to communicate through gestures and croaks. Although the bond was formed under less than ideal circumstances, it gave her an advantage over Anna.

Other evidence showing the importance of social contact comes from less extreme cases of neglect. Psychiatrist Rene Spitz (1951) studied 91 infants who were raised by their parents during their first three to four months of life but who were later placed in orphanages. When they were admitted to the orphanages, the infants were physically and emotionally normal. Orphanage staff provided adequate care for their bodily needs—good food, clothing, diaper changes, clean nurseries—but gave the children little personal attention. Because only one nurse was available for every 8 to 12 children, the children were starved emotionally. The emotional starvation caused by the lack of social contact resulted in such rapid physical and developmental deterioration that a significant number of the children died. Others became completely passive, lying on their backs in their cots. Many were unable to stand, walk, or talk (Spitz 1951). These cases and the cases of Anna and Isabelle teach us that children need close contact with and stimulation from others if they are to develop normally.

(Write a Caption)

Write a caption that describes how nature and nurture work together to channel a child's behavior in a certain direction.

Hints: In writing this caption

- review the concepts of nature and nurture,
- think about the human potential these children possess, and
- consider the role that nurture plays in determining activities a child might become good at.

MC Spc. 1st Class Chad J. McNeeley

Tina Osterman, EFMP

Critical Thinking

Give a specific example from your life that illustrates how you are a product of both *nature* and *nurture*.

Key Terms

nature

nurture

MODULE

3.3 The Social Self

Objective

You will learn the process by which children acquire a sense of self.

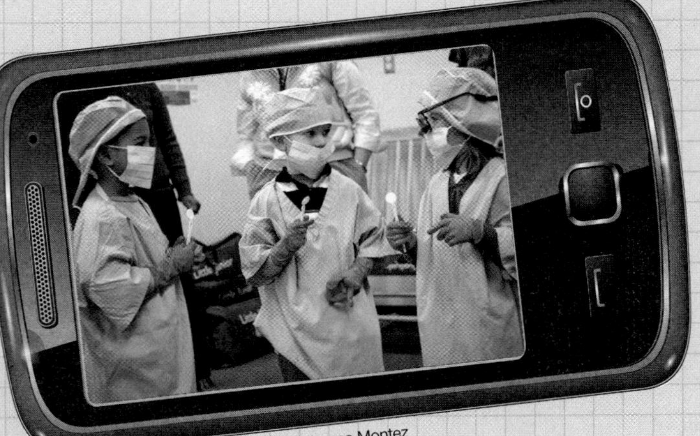

U.S. Air Force photo/Airman Rebecca Montez

Have you ever watched small children playing at being someone significant in their lives, like a doctor?

Sociologists maintain that this kind of play fills a larger social purpose. It gives children practice at seeing how the status they are pretending to be fits in with other statuses in the society. Before children pretend to be another, they must have acquired a sense of self. A sense of self exists when people can

1. imaginatively step outside the self to take the role of the other; and
2. name, classify, and categorize the self.

Role-Taking

How do people learn to take the role of the other? George Herbert Mead (1934) maintained that we learn to do this through a three-stage interactive process that involves **role-taking**, or stepping into another person's shoes and imaging how they view us and others around them. Those stages are (1) preparatory, (2) play, and (3) games. Each stage involves a progressively more sophisticated level of role-taking.

THE PREPARATORY STAGE (UNDER AGE 2). In the preparatory stage, children have not yet developed the cognitive ability to role-take. They mimic or imitate people in their environment but often do not know the meaning of what they are imitating. In this stage children mimic what others around them are doing or repeat things they hear. In the process, young children learn to function symbolically; that is, they are in the beginning stage of learning to name things and learn that particular actions and words have meanings that arouse predictable responses from others.

➤ These small children are in the preparatory stage, as they are imitating the behavior of those around them without understanding the meaning of what they are imitating. The two have noticed that their grandfather is wearing something over his eyes. At this age, neither has a name for what they will one day call sunglasses, nor are they aware of the way sunglasses should be worn or their purpose. We know this because one grandchild is putting them on upside down.

Lisa Southwick

Lisa Southwick

THE PLAY STAGE (ABOUT AGES 2 TO 6). Mead saw children's play as the mechanism by which they practice role-taking. **Play** is a voluntary, spontaneous activity with few or no formal rules. Play is *not* subject to constraints of time (e.g., 20-minute halves) or place (e.g., a regulation-size field). Children, in particular, play whenever and wherever the urge strikes. In play children make rules as they go; they are not imposed by rulebooks or referees. Children undertake play for entertainment or pleasure. These characteristics make play less socially complicated than organized games, such as baseball.

In the play stage, children pretend to be **significant others**—people or characters such as cartoon characters, a parent, or the family pet—who are important in a child's life, in that they greatly influence the child's self-evaluation and way of behaving. When a little girl plays with a doll and pretends to be the doll's mother, she talks and acts toward the doll the same way her mother talks and acts toward her. By pretending to be a mother, the child gains a sense of the mother's expectations and perspective and learns to see herself as an object (through the doll) from her mother's point of view.

Pfc. Cory D. Polom

➤ A child playing doctor with his stuffed animal is seeing the world from viewpoints other than his own. Such play allows him to play two roles simultaneously and, in the process, to think about how a significant other in his life (a doctor) sees the patient and how the patient is supposed to behave relative to the doctor.

THE GAME STAGE (AGE 7 AND OLDER). In Mead's theory, the play stage is followed by the game stage. **Games** are structured, organized activities that involve more than one person. They are characterized by a number of constraints, such as already established roles and rules and a purpose toward which all activity is directed. When people gather to play a game of baseball, for example, they do not have to decide the positions or the rules of the game. The rules are already in place. To be a successful pitcher, for example, one must understand not only how to play that position but how the position of pitcher relates to the other positions.

Rachel Parks, III Corps & Fort Hood Public Affairs

➤ When young children first take part in organized games such as soccer, baseball, or basketball, their efforts seem chaotic. Everyone runs toward the ball and forgets to play his or her position. This chaos exists because children have not acquired the mental capacity to see how each role fits with others.

As they learn to play games, children also learn to (1) follow established rules, (2) imaginatively take the roles of all participants, and (3) see how their role fits in relation to an established system of expectations. In particular, children learn that what is happening with other positions affects the position they occupy. They learn that under some circumstances their position takes on added significance or assumes lesser significance.

Through games, children learn to organize their behavior around the **generalized other**—a system of expected behaviors and meanings that transcend the people participating. An understanding of the generalized other is achieved by imaginatively relating the self to many others playing the game. Through this imaginative process, the generalized other becomes incorporated into a person's sense of self (Cuzzort and King 2002). When children play games such as baseball, they practice fitting their behavior into an already established system of expectations. This ability is the key to living in society because most of the time, whether it is at school, work, or home, we are expected to learn and then fit into an already established system of roles and expectations.

Significant Symbols

We have learned that a sense of self involves the ability to role-take. In order to role-take, children must learn the meaning of **significant symbols**, gestures that ideally convey the same meaning to the persons transmitting them and receiving them. Mead (1934) defined **gesture** as any action that requires people to interpret its meaning before responding. Language is a particularly important gesture because people interpret the meaning of words before they react. In addition to words, gestures also include nonverbal cues, such as tone of voice, inflection, facial expression, posture, and other body movements or positions that convey meaning.

Lisa Southwick

Lisa Southwick

⮝ Through gestures, others convey how they are evaluating our appearance and behavior. Look at the two facial expressions in the photographs. Imagine that you just asked this woman's advice about declaring a particular college major. What meaning do you attach to each facial expression?

A sense of self emerges the moment children internalize **self-referent terms**, words and other symbols used to distinguish the self (including I, me, mine, first name, and last name) and to specify the statuses one holds in society (I am an athlete, doctor, child, and so on). Mead maintains that the self is always recognized in relationship to others. That is, one can be a student only in relationship to teachers and fellow students. Similarly, one can be an athlete only in relationship to other athletes, fans, a referee, and so on.

Mead described the self as having two parts—the me and the I. The **me** is the social self—the part of the self that is the product of interaction with others and that knows the rules and expectations. More specifically, the me is the sense of self that emerges out of role-taking experiences. The **I** is the active and creative aspect of the self. It is the part of the self that questions the expectations and rules. The I is capable of acting in unconventional, inappropriate, or unexpected ways. The I takes chances that can pay off if others label the risk-takers as unique or one-of-a-kind. At other times such risk-taking backfires. The existence of the me and the I suggests that the self is dynamic and complex.

Looking-Glass Self

Like Mead, sociologist Charles Horton Cooley (1961) assumed that the self is a product of interaction experiences. Cooley coined the term **looking-glass self** to describe the way in which a sense of self develops: specifically, people act as mirrors for one another. We see ourselves reflected in others' real or imagined reactions to our appearance and behaviors. We acquire a sense of self by being sensitive to the appraisals that we perceive others to have of us. As we interact, we imagine how we appear to others, we imagine a judgment of that appearance, and we develop a feeling about ourselves somewhere between pride and shame: "The thing that moves us to pride or shame is not the mere mechanical reflection of ourselves but . . . the imagined effect of this reflection upon another's mind" (Cooley 1961, 824).

Cooley went so far as to argue that "the solid facts of social life are the facts of the imagination." Because Cooley defined the looking-glass self as critical to self-assessment and awareness, he believed that people are affected deeply by what they imagine others' reactions to be, even if they perceive reactions incorrectly. The student who dominates class discussion may think that classmates are fascinated by his or her comments, when in fact they are rolling their eyes whenever that student speaks. This student continues dominating discussion because he or she misinterprets others' assessment.

We have learned that self-awareness and self-identity emerge when people can role-take, understand meanings of significant symbols, and use self-referential terms. The question that still remains is how people develop the levels of cognitive sophistication to do these things. For the answer, we turn to the work of Swiss psychologist Jean Piaget.

Cognitive Development

Assessing the impact of Piaget on the study of cognitive development is "like assessing the impact of Shakespeare on English literature, or Aristotle in Philosophy—impossible" (Beilin, 1992, 191). Piaget's ideas stemmed, in part, from his study of water snails, which spend their early life in calm waters. When transferred to tidal water, the size and shape of the snails' shells develop to help them cling to rocks and avoid being swept away (Satterly 1987). Building on this observation, Piaget arrived at the concept of active adaptation, a biologically based tendency to adjust to and resolve environmental challenges. Piaget believed that learning and reasoning are important cognitive tools that help children to adapt to environmental challenges. These tools emerge according to a gradually unfolding genetic timetable in conjunction with direct experiences with people and objects.

Piaget's model of cognitive development includes four broad stages: (1) sensorimotor, (2) preoperational, (3) concrete operational, and (4) formal operational. According to Piaget's model, children cannot proceed from one stage to the next until they master the reasoning challenges of earlier stages. Piaget maintained that reasoning abilities cannot be hurried; a more sophisticated level of reasoning will not show itself until the brain is ready.

SENSORIMOTOR STAGE (BIRTH TO ABOUT AGE 2). In the sensorimotor stage, children are driven to explore the world through their senses (tasting, touching, seeing, hearing, and smelling).

Missy Gish

◀ You might notice that very young children put into their mouths just about anything they pick up. This baby uses his mouth to explore things in the world around him.

The cognitive accomplishments of this first stage include an understanding of the self as separate from other persons and the realization that objects and persons exist even when they are moved out of sight. Before this notion takes hold, children act as if an object does not exist when they can no longer see it. At about eight months, children begin to actively look for an object that was once there.

PREOPERATIONAL STAGE (AGES 2 TO ABOUT 7). Children in the preoperational stage think anthropomorphically; that is, they assign human feelings to inanimate objects. They believe that objects such as the sun, the moon, nails, marbles, trees, and clouds have feelings and intentions (for example, dark clouds are angry; a nail that sinks to the bottom of a glass filled with water is tired). Children in the preoperational stage cannot grasp the fact that matter can change form but still remain the same in quantity. For example, they believe a 12-ounce cup that is tall and narrow holds more than a 12-ounce cup that is short and wide; height is the criterion by which they judge the amount, failing to consider diameter. In addition, children in this second stage cannot conceive how the world looks from another person's point of view. Thus, if a child facing a roomful of people, all of whom are looking in his or her direction, is asked to draw a picture of how someone in the back of the room sees the group, the child will draw the picture as he or she sees the people, showing faces rather than backs of heads. Finally, children in the preoperational stage tend to center their attention on one detail and fail to process information that challenges that detail. That is, they may believe that women have long hair, and when they meet a male with long hair they classify him as a woman and ignore the detail that he also has a beard, something society attributes to a male.

CONCRETE OPERATIONAL STAGE (ABOUT AGES 7–11). By the concrete operational stage, children can take the role of the other. But they have difficulty thinking abstractly without reference to a concrete event or image. For example, children in this stage have trouble envisioning death or that life could go on without them. One 11-year-old struggling to grasp this idea said to me, "I am the beginning and the end; the world begins with and ends with me." Children in the concrete operational stage can think deductively (using general principles to see the errors of flawed conclusions such as in this statement: "something able to swim is a fish, so a frog in water is a fish"). During this stage they also learn to think inductively, applying a specific situation to a general principle. For example, the child comes to truly understand the concept of brother so that he knows he is his brother's brother and that his dad and other men can be someone's brother, too.

FORMAL OPERATIONAL STAGE. In this stage (from the onset of adolescence forward), children learn to think abstractly. They plan for the future, are able to think through hypothetical situations, and can entertain moral dilemmas. In this stage children can conceptualize their life as being part of a much larger system. The world is not so black and white now, but shades of gray.

NASA

➤ Because children in the formal operational stage can think abstractly, they are able to imagine Earth in relationship to the universe and, by extension, grasp the notion that they are one among billions who live on the planet.

Piaget is criticized for, among other things, making conservative assessments of children's cognitive abilities at certain stages (Lourenço and Machado 1996). Some researchers, for example, have found that children at the age associated with the preoperational stage can do tasks associated with the concrete operational stage. Piaget would respond by saying that he was not so much interested in the ages at which children move from one developmental task to another but the *sequence* through which they move (Lourenço and Machado 1996).

(Write a Caption)

Write a caption that uses the concept of role-taking to frame this interaction between mother and daughter.

Lisa Southwick

Hints: In writing this caption

- identify which of the three stages of role-taking this scene best represents,
- think about whether the term *significant* or *generalized other* applies to the mother, and
- consider what the little girl is learning by playing at putting on makeup.

Critical Thinking

Name five self-referent terms that you use to describe yourself. Which one is most important to your sense of self?

Key Terms

games	looking-glass self	self-referent terms
generalized other	me	significant other
gesture	play	significant symbols
I	role-taking	

Resocialization

MODULE 3.4

Objective

You will learn that resocialization is an interactive process during which the affected party reconstructs his or her identity and associated relationships.

Courtesy of J.R. Martinez

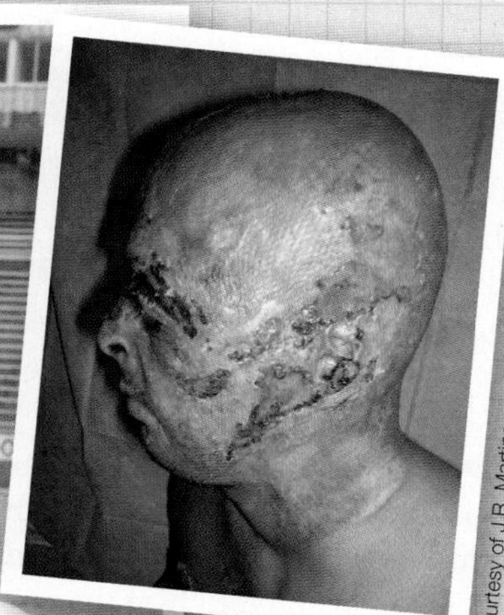

Courtesy of J.R. Martinez

Can you name a key event in your life that changed the way you thought about yourself and the way others viewed and related to you?

For J.R. Martinez that event occurred in 2003 while he was serving in Iraq. The left tire of the Humvee he was driving hit a landmine. The resulting explosion left J.R. with severe burns over at least 40 percent of his body; he also suffered damage to internal organs from inhaling the hot air and smoke. Martinez spent 3 weeks in a coma and 34 months in recovery. He endured dozens of surgeries. When Martinez first looked at himself in the mirror, he remembers that he "just froze for a couple seconds" and thought that he would have been better off had he not survived (Martinez 2011b). But Martinez did more than just survive, he went on to thrive as a soap opera star, a motivational speaker, and a contestant who went on to win *Dancing with the Stars*. In addition to the medical processes, Martinez went through another process that sociologists call resocialization.

Resocialization is an interactive process during which the affected party reconstructs his or her identity. The resocialization process may be voluntary or involuntary, dramatic or subtle, and involve a loss or gain in status and associated relationships. In addition, resocialization is a process by which the affected party renegotiates relationships with significant others who must also adjust to the changing person and circumstances.

➤ This is a photograph of Ellen DeGeneres's mother marching in a gay pride celebration. The organization Parents, Families, and Friends of Lesbians and Gays (PFLAG) is a support group for those who declare nonheterosexual identities and for those who care about them. Keep in mind that when people announce an orientation other than heterosexual, they must renegotiate relationships with others in their life to accommodate that social identity.

Chris Caldeira

Significant others play an important role in the resocialization process because how they react is important to how the affected party reconstructs his or her identity and relationships (K. Daly 1992). For example, Martinez's mother played an important role when she told him that looks weren't everything and that "people are going to be in your life for who you are as a person and not what you look like." Martinez's mother told her son that when she was younger, "everyone told me I was pretty and gave me compliments. No one tells me that now." Somehow these words resonated with Martinez, and he responded, "You know what, Mom? You're right. And now, I'm actually glad this happened to me . . . Now I get to see who liked me as a person, versus who liked me for being the popular guy in school, being the athlete, being the handsome young man" (E. Collins 2011).

The resocialization process can be triggered by a crisis, such as the one Martinez experienced, or it can be triggered by a less dramatic event. Regardless, the event alters an existing and internalized identity. Examples of identity- and relationship-altering events include these situations:

- a woman begins to seriously date someone,
- a significant person in someone's life dies,
- a man who prides himself on being physically fit is diagnosed with a degenerative illness,
- someone retires from a 40-year career,
- a student graduates from high school and enrolls in college,
- a person inherits a significant amount of money, and
- a financially independent person becomes unemployed.

In all these cases the affected party must relinquish an existing identity and come to terms with a new one. Such identify transformations often involve a review of the life that once was and a protest against or celebration over what has changed, followed by a period of mourning (of varying intensity) over what has been lost. Martinez grieved for the handsome man he once was and his mother grieved over his near death and disfigurement. Then the affected parties must renegotiate their identities and relationships. "To do less is to remain mired in loss" (Fein 2011). For Martinez, that negotiation process was pushed further along when a nurse taking care of him asked if he would speak to a fellow burn patient who had become withdrawn upon seeing his body for the first time. After visiting with that patient for 45 minutes and gaining a positive response, Martinez realized that he could impact the lives of others simply by sharing experiences (Martinez 2011a).

The resocialization process and experience varies depending on the extent to which the change is welcomed, the extent to which relationships are adversely affected, and a number of other factors outlined in the life event model.

Life Event Model

The life event model considers how factors such as age and time in history affect life transitions. In addition to these factors, the life event model considers things like prior life experiences, temporal orientation, and whether the transformed identity is considered normative or nonnormative (Lutfey and Mortimer 2003).

Prior life experiences can shape how social transitions are experienced. For example, the resocialization process triggered by a new romantic relationship is affected by whether one has had previous romantic relationships and how significant others such as a best friend responded to those relationships in the past. The experience of childbirth is affected by the number of children a woman has already had, the age at which she gives birth, and the level of support she can anticipate from significant others in her life.

Temporal orientation encompasses ideas about the possible self, including projections of what someone hopes to become or fears becoming. The possible selves for which one hopes "might include the successful self, the creative self, the rich self, the thin self, or the loved and admired self, whereas the dreaded possible selves could be the alone self, the depressed self, the incompetent self, the unemployed self, or the bag lady self" (Markus and Nurius 1986, 954). Martinez's view of his possible self included the belief that once people talked with him they would "not notice the scars anymore." They would see "a human being" with a "sense of humor" who likes to have a good time (Martinez 2004).

Normative / nonnormative transitions relate to whether a social transition is expected/unexpected of people in light of their age, race, sex, social class, and so on. For the most part, in the United States we do not expect 85-year-olds to divorce, but are not too surprised if a 45-year-old divorces.

Much resocialization happens naturally and involves no formal training; people simply learn as they go. For example, people who marry; those who decide to live openly as gay, lesbian, or transgendered; and those who lose their job or make some other transition must incorporate the new status into their social identity and learn new ways to relate to others as they transition from single to married, from assumed heterosexual to another orientation, and from gainfully employed to unemployed. Sometimes circumstances are such that people voluntarily or involuntarily participate in programs created with the purpose of guiding or forcing them through a resocialization process.

Voluntary versus Imposed Resocialization

Resocialization can be voluntary or imposed. Voluntary resocialization occurs when people choose to participate in a process or program designed to remake them. Examples of voluntary resocialization are wide-ranging: the unemployed youth who enlists in the army to bring discipline to her life, the alcoholic who joins Alcoholics Anonymous to transform himself into a "recovering alcoholic," and the unemployed person who enrolls in college to begin training for an anticipated career. Imposed resocialization occurs when people are forced into a program designed to train them, rehabilitate them, or correct some supposed deficiency. People drafted into the military, sentenced to prisons, court-ordered to attend parenting classes, or committed to mental institutions represent examples of those who undergo resocialization that is forced upon them.

In *Asylums*, sociologist Erving Goffman (1961) wrote about a particular type of setting called **total institutions** in which people are isolated from the rest of society to undergo systematic resocialization. Total institutions include homes for the blind, the elderly, the orphaned, and the indigent; mental hospitals; prisons; concentration camps; army barracks; boarding schools; and monasteries and convents. In total institutions, people surrender control of their lives, voluntarily or involuntarily. As "inmates," they carry out daily activities such as eating, sleeping, and recreation in the "immediate company of a large batch of others, all of whom are treated alike and required to do the same thing together" (6).

U.S. Marine Corps photo by Sgt. Mallory S. VanderSchans/Released

➤ This guard checks on an inmate locked in a cell. Total institutions such as prisons control inmates' social interactions with the outside world. This control is symbolized by the use of locked doors, high walls, fences, and security systems.

People confined to total institutions participate in activities (lining up, attending classes, making crafts, showering, running, attending group therapy sessions, praying) that fit in with institutionally planned goals. Those goals may be to care for the incapable, to keep inmates out of the community, or to teach people new roles (for example, to be a soldier, priest, or nun).

Goffman (1961) identified standard procedures that total institutions employ to resocialize inmates. The inmates-to-be arrive with a sense of self. Upon entering a total institution, that self undergoes **mortification**, a process by which the self is stripped of all its supports and "shaped and coded." Specifically, staff members may take inmates' life histories and photograph, weigh, and fingerprint them. The staff strips inmates of any personal possessions through which they present themselves on the outside. Clothes and jewelry are taken and uniforms issued. Access to services such as a hairdresser or workout facilities are denied. Communication with those in the outside world is curtailed, and if restored, closely monitored. "Family, occupational, and educational career lines are chopped off" (17). Inmates have no say in how to spend the day; they follow a schedule of activities to which all adhere. Taken together, these policies enforce a clean break with the past and mark the beginning of a process that, if successful, will result in a new identity and way of thinking and behaving.

(Write a Caption)

Write a caption that relates this father's resocialization or adjustment to being a man without legs.

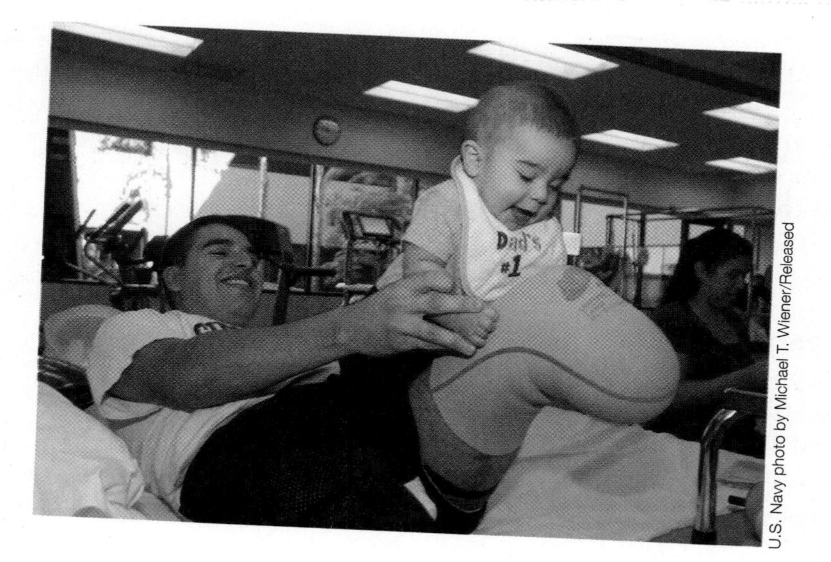

U.S. Navy photo by Michael T. Wiener/Released

Hint: In writing this caption

- think about the broad elements of the resocialization process that revolve around identity and relationships with others.

Critical Thinking

Describe a time when your social identity changed voluntarily or involuntarily. Explain how your relationships changed (and adjusted or not) as a result.

Key Terms

mortification resocialization total institutions

MODULE

3.5 Agents of Socialization

Objective

You will learn how agents of socialization shape behavior, thinking, and social identities.

Think back to when you were a child. Did someone or something in your life shape your identity, thinking, or behavior in any way?

Chris Caldeira

If you answered yes, then you have acknowledged a socializing agent in your life. **Agents of socialization** are significant people, groups, and institutions that shape our sense of self and social identity, help us realize our human capacities, and teach us to negotiate the world in which we live. It is impossible to list all the agents of socialization; they include family, friends, peers, coaches, day cares, schools, governments, religions, and all forms of media. In this module we consider three agents of socialization: primary groups, peer groups, and mass media.

Primary Groups

A **primary group** is a social group whose members share an identity, have face-to-face or other intimate contact, and feel strong emotions for each other. These emotions are not always based on love; they can be based on insecurity and need. Whatever the case, the ties are emotional. Primary groups are "fundamental in forming the social nature and ideals of the individual" (Cooley 1909, 23). Families and military units are two examples of primary groups.

FAMILY. The family is an important agent of socialization because it gives individuals their deepest and earliest experiences with relationships and their first exposure to the rules of life. In addition, the family teaches its members about the world in which they live and ways to respond to it. During difficult times the family can buffer its members against the ill effects; alternatively, it can increase stress. Sociologists Amith Ben-David and Yoav Lavee (1992) offer a specific example. The two sociologists interviewed Israelis to learn how their families responded to missile attacks Iraq launched on Israel during the Persian Gulf War in 1991. During these attacks, families gathered in sealed rooms and put on gas masks. The researchers found that families varied in their responses to this life-threatening situation. Some respondents reported the interaction during that time as positive and supportive: "We laughed and we took pictures of each other with the gas masks on" or "We talked about different things, about the war, we told jokes, we heard the announcements on the radio" (39).

Other respondents reported little interaction among family members, but a feeling of togetherness prevailed: "I was quiet, immersed in my thoughts. We were all around the radio. . . . Nobody talked much" (40). Some respondents reported that interaction was tense: "We fought with the kids about putting on their masks, and also between us about whether the kids should put on their masks. There was much shouting and noise" (39). As this research illustrates, even under extremely stressful circumstances, such as war, the family can teach responses that increase or decrease that stress. Clearly, children in families that emphasize constructive responses to stressful events have an advantage over children whose parents respond in destructive ways.

➤ The family can buffer its members against the effects of stressful events, or it can magnify the stress. This family is riding a boat to safety after its home and community were flooded. How do you think your family would react in such a situation?

U.S. Navy Photo

While a family's income and wealth will surely shape how it responds to stressful events, there are many examples of families that, lacking even the most basic resources, still manage to respond constructively to stressful situations. Save the Children (2007) staff member Jerry Sternin offers one example. He was charged with a seemingly impossible assignment: to help save starving children in Vietnam. He drew inspiration from mothers he termed positive deviants—individuals "whose exceptional behaviors and practices enable them to get better results than their neighbors with the exact same resources."

Sternin identified those few Vietnamese children whose weight suggested they were well nourished and compared their situations with peers who were underweight and malnourished. He learned that the mothers of well-nourished children were behaving in ways that defied conventional wisdom. Among other things, these mothers were (1) using alternative food sources available to everyone. They were going to the rice paddies to harvest the tiniest shrimp and crabs, and they were picking sweet potato greens—considered low-class food—and mixing both food sources with rice; (2) feeding their children when the children had diarrhea, contrary to traditional practice; and (3) making sure their children ate, rather than "hoping children would take it upon themselves to eat." Save the Children successfully introduced these strategies to 2.2 million Vietnamese in 276 villages and to people in at least 20 other countries where malnutrition is widespread (Dorsey and Leon 2000).

MILITARY UNIT. A military unit is a primary group. A unit's success in battle depends in part on the existence of strong ties among its members. Military units train their recruits always to think of the group before themselves. In fact, the overriding goal of military training is to create a primary group and make individuals feel inseparable from their unit.

Scott A. Thornbloom

➤ Military units such as this one are considered primary groups because recruits are trained to think of themselves as a unit and not as individuals. Strategies used to make recruits feel inseparable from their unit include dressing and looking as much alike as possible, assigning a number to the unit, and doing specific tasks together.

Sgt. Virgil P. Richardson

◀ Other strategies include ordering recruits to march in unison; to live, sleep, and eat together; and to work together to accomplish some task. Still other key strategies are to punish the entire unit when one member fails and to focus the unit's attention on a common enemy. Soldiers become so close that they fight for one another, rather than for victory per se, in the heat of battle (Dyer 1985).

Peer Groups

A **peer group** consists of people who are approximately the same age, participate in the same day-to-day activities, and share a similar overall social status in society. Examples of that status include middle school student, adolescent, teenager, or retired. Sociologists are especially interested in the process known as **peer pressure**, those instances in which people feel directly or indirectly pressured to engage in behavior that meets the approval and expectations of peers and/or to fit in with what peers are doing. That pressure may be to smoke (or not smoke) cigarettes, to drink (or not drink) alcohol, and to engage (or not engage) in sexual activities.

Mr. Robert Dozier (FMWRC)

Among other things, peer groups are especially important to children and adolescents as they work to establish an identity and to test their effectiveness at interacting with others. Peer groups can be a source of both emotional support and personal anguish (Atwater 1988).

It is through peer groups that children and others learn what it means to be male or female or to be classified into a racial/ethnic category. In the process, children are influenced by overall societal conceptions of race and gender but also create and integrate their own meanings. Sociologist Amira Proweller (1998) found that white middle-class students tended to perceive black counterparts as being more exotic and sensual in their styles, whereas blacks tended to perceive white middle-class styles as more accepted in school settings. Both blacks and whites adopted each other's styles but criticized each other when acting "out of race," labeling those blacks who did so as "acting white" and those whites who did so as "acting black."

Peer groups are important agents of socialization regarding gender as well. Among other things, peer groups negotiate what constitutes appropriate and inappropriate physical expression toward same- and other-sex persons. Close same-sex friendships are often labeled as gay, and the fear of being labeled as such can restrict how same-sex friends relate to one another. Peers discuss and comment on physical changes accompanying puberty that revolve around girls wearing bras, menstruation, boys growing facial hair, girls removing (or failing to remove) body and facial hair, sexual development, sexual activity, rumors regarding romantic connections, and breakups. Praise, insults, teasing, rumors, and storytelling inform them about the meaning of being male, female, or something in between (Proweller 1998).

Mass Media

Another source of socialization is **mass media**, forms of communication designed to reach large audiences without direct face-to-face contact between those creating/conveying and receiving messages. The tools of mass media—such as the printing press, television, radio, the Internet, and iPods—deliver content in the form of magazines, movies, commercials, songs, and video games to audiences and expose them to a variety of real and imaginary people, including sports figures, animated characters, politicians, actors, disc jockeys, and musicians.

➤ We can argue that any exposure to a cartoon, recorded music, movie, video game, or other content presents an opportunity for socialization if only because it introduces viewers to possible ways to act, appear, and think. Here Bert and Ernie teach viewers about the dangers of playing with matches.

Because media is so pervasive, it is difficult to make any definitive statements about its effects as a whole, except perhaps to say that a very significant but unquantified portion of what we know about the world and people in it is acquired through popular media sources. Since the 1950s (when televisions became widespread), there has been concern about the effects watching violence has on viewers, especially young children. A number of research studies have shown the following:

- The average child age 6 and under spends almost 2 hours per day in front of some type of screen media. Those between the ages of 8 and 18 spend 7 hours, 38 minutes in front of some form of media. Because youth are often engaged with two or more media at once (e.g., surfing the Internet and listening to music), exposure time increases to the equivalent of 11 hours per day (Kaiser Family Foundation 2010).

- If children were not watching, listening to, or playing with some media, it is likely they would be engaged in play or other physical activity, socializing with friends, reading, or doing homework (University of Michigan Health System 2010).
- A substantial proportion of children's programming includes violence, with the highest incidence found in animated programming, music videos, video games, and PG-13 rated movies (University of Michigan Health System 2010).
- Content analyses of media violence reveals that violence is often glamorized and is disproportionately aimed at women and racial minorities. In addition, it is something that goes unpunished, is often accompanied by humor, and fails to depict human suffering and loss as a consequence. Finally, violence is presented as an appropriate method for addressing problems and achieving goals (University of Michigan Health System 2010).
- A review of more than 2,000 research studies reveals a clear association between viewing violence and (1) aggressive behavior and thoughts, (2) a desensitization to violence, (3) nightmares, and (4) fear of being a victim (University of Michigan Health System 2010; Harvard Pediatric 2010).

Missy Gish

◀ Some video games allow players to simulate aggressive and violent responses. These games embed players in virtual environments where they assume the role of the aggressor and are rewarded for injuring or killing others. Players are not simply *watching* violent scenes; they are strategizing and actively engaging in violent acts, albeit aimed at characters who are part of a virtual reality.

Clearly, viewing violence and other aggressive actions does not in itself lead viewers to engage in violent behavior such as bullying, abuse, homicide, or assault. Otherwise, everyone who viewed media violence would engage in such behavior. To understand the reasons people engage in violent and aggressive behavior, one must look beyond the media and consider other factors such as the viewer's gender, employment opportunities, age, and so on. Moreover, to understand the media's role in perpetuating violence and other aggressions, one must consider a multitude of factors that mute or accentuate its effects. Those factors include the type and amount of violence being viewed; age of viewer; and quality of a viewer's relationships with family, friends, and others. The fact that media violence is only one of many factors that contribute to violent and aggressive behavior does not mean it does not play an important role. Moreover, it is a factor worth studying because it is something to which exposure can be monitored (University of Michigan Health System 2010; Harvard Pediatric 2010).

(Write a Caption)

Write a caption that addresses the importance of the peer group in establishing a sense of self but that also recognizes the influence of dominant images in the larger society.

Hints: In writing this caption

- think about the characteristics that make these teens, who have formed a group, a peer group; and
- consider the relative contributions of a peer group and the media (e.g., celebrities) in setting a dance style.

Cpl. Kristin E. Moreno

Critical Thinking

Name an agent to socialization that is currently key in shaping your sense of self and social identity. Note that the agent can be a video game, a celebrity, or someone personally involved in your life such as a parent or coach.

Key Terms

agents of socialization
peer group
primary group
mass media
peer pressure

Applying Theory: The Three Perspectives

MODULE 3.6

Objective

You will learn how the three theoretical perspectives help us think about how people use social media to present themselves and to connect with others.

Missy Gish

Missy Gish

If you saw these photographs on Facebook or some other social networking site, what impression would each make on you?

If you are one of the hundreds of millions of people who use a social networking service such as Facebook, you have had to make decisions about how to present yourself online. Did you post a photograph? Did you share your age, sexual orientation, sex, race, income, or occupation? Who did you invite into your circle of friends or connections? What kinds of things do you reveal about your life? Do you keep your space private or did you open it up to the Internet public?

Sociologists are interested in these questions because they offer a window into how people present and distinguish themselves. The sociological perspectives—functionalist, conflict, and symbolic interactionist—help us to think about Facebook and other online social networking platforms as a facilitator in this process.

Functionalist Perspective

Functionalists focus on how parts of society function in expected and unexpected ways to maintain existing social order. They also pay attention to how parts disrupt the existing order in expected and unexpected ways. Of course, the use of social networking sites as a platform for presenting the self to others is the part of society we are analyzing.

Some expected, or *manifest*, functions of social networking sites are that they facilitate connections with family, friends, and other parties; allow members to share photos and videos; support discussions with like-minded people about hobbies and other interests; and help users plan face-to-face meetings with friends. They also allow users to establish and maintain contacts with a far greater number of people than is typically possible using nondigital means.

Lisa Southwick

➤ It is impossible to manage connections with everyone we have come to know or hope to know via phone calls and face-to-face meetings. There is simply not enough time. Social networking sites allow individuals to manage a potentially infinite number of contacts.

An unexpected, or *latent*, function of social networking sites is that they give people a tool for rekindling relationships with those they have lost contact with, including lost relatives. In addition, police can study a suspect's Facebook or other networking pages to find incriminating evidence about a person's activities and interactions (Hodge 2006).

One *manifest dysfunction* of such websites is that there is no way to tell whether people are presenting real or fabricated selves or if they are hiding some information that they are sharing with others in their network. A unexpected, or *latent*, dysfunction of social networking sites is that once something is posted for others to access, there is no way to control how it will be shared or used. Many people create a Facebook page for the purpose of meeting like-minded friends or staying in touch, not thinking that potential employers may view postings for clues about someone's character apart from the resume and interview.

Conflict Perspective

Conflict theorists seek to identify advantaged and disadvantaged groups, document unequal access to scarce and valued resources, and describe the ways in which advantaged groups promote and protect their interests. With regard to online social networking websites, conflict theorists ask, Who ultimately controls the information conveyed about the self, and who ultimately benefits from this arrangement and at whose expense? Conflict theorists maintain that the advantaged groups include those who own the social networking websites, advertisers, potential employers, and other parties interested in selling products. No matter how much users think they benefit from social networking, in the final analysis they are the disadvantaged groups, especially if they mistakenly believe that they control the information they have posted about themselves and their lives. Although most social networking platforms offer users ways to control those who have access to the information shared, many fail to activate them because they are confusing or because there are too many privacy options to manage. On close analysis, we see that Facebook and other social networking companies use "like" buttons to place cookies on a browser's computer even when a person ignores invitations to "like" or connect with the inviting entity. Even non-Facebook members' activities can be tracked. Depending on the type of cookie issued, it allows Facebook and other companies to track people's Internet browsing activities, access the contents of shopping carts, and collect other kinds of personal data (Roosendaal 2010).

Facebook, for example, makes it clear that it has access to

> data about you whenever you interact with Facebook, such as when you look at another person's profile, send someone a message, search for a friend or a Page, click on an ad, or purchase Facebook Credits . . . . This may include your IP address, location, the type of browser you use, or the pages you visit. . . . Sometimes we get data from our advertising partners. . . . For example, an advertiser may tell us how you responded to an ad on Facebook or on another site in order to measure the effectiveness of—and improve the quality of—those ads. (Facebook 2011)

Symbolic Interactionist Perspective

Symbolic interactionists study social interaction and focus on self-awareness, symbols, and negotiated order. Symbolic interactionists ask, How do people experience, interpret, influence, and respond to what they and others are saying and doing? Symbolic interactionists are interested in how people present the self to others in face-to-face and other settings such as online environments. These theorists would consider what effect social networking has on a person's sense of self. In this vein, they would ask a number of questions. Why do some people become addicted to keeping up with and responding to postings? What does it mean that many teens (and adults, for that matter) access a social networking site several times a day? Why do some keep the site open while working, check in between classes or on lunch breaks, and even disrupt their sleep so as not to miss a posting (Boyd and Jenkins 2006)? How do users tailor their presentations of self in general and for particular audiences? These theorists would also consider what effect social networking has on a person's sense of self.

Symbolic interactionists studying Facebook and other social media sites familiarize themselves with the associated vocabulary and symbols such as "admin," "fan," "like," "status," and "event." Symbolic interactionists would also apply the concept of negotiated order to think about the dynamics surrounding questions related to how to handle friend requests from people you really don't want to friend, whether to respond to an invitation to an event, how long to wait to respond to a posting, and so on.

➤ Symbolic interactionists are interested in the number of people a person friends on Facebook and the proportion of those friends who regularly interact face-to-face versus the proportion who are online acquaintances.

A summary of each of the three perspectives is presented in Table 3.6a. For each theory, the table shows the question that guides analysis and the vocabulary used to answer that question.

Chris Caldeira

▼ **Table 3.6a: Summary of Three Theoretical Perspectives on Social Networking Theory**

	Functionalist	Conflict	Symbolic Interaction
Question	What are the anticipated and unanticipated functions and dysfunctions of digital social networking?	Who controls social networking platforms? Who benefits from that arrangement?	How do people experience, interpret, influence, and respond to social networking?
Answer	**Manifest function**: a way of connecting with a potentially infinite number of people **Manifest dysfunction**: lacks checks on accuracy of postings **Latent function**: a way of finding lost relatives; a crime-investigation tool **Latent dysfunction:** information posted may be used for purposes the user did not anticipate	**Advantaged groups**: owners of social networking websites, employers, and others interested in selling products **Disadvantaged groups**: social networkers who do not really control postings **Scarce and valued resource**: information about and access to social networkers	**Self-awareness**: social networking platforms offer feedback **Shared symbols**: vocabulary of social networking terms; symbols used to convey mood **Negotiated order**: the process by which people work out dynamics of interaction with "friends"

Critical Thinking

Which of the three theoretical perspectives best captures how you think about Facebook and other social networking tools? Explain.

Summary: Putting It All Together

CHAPTER 3

Socialization is the process by which people learn to live in society. We are socialized (and resocialized) by agents of socialization. Three important agents are primary groups, peer groups, and mass media. Socialization takes hold through internalization. Any discussion of socialization must take into account nature and nurture, both of which are essential to human and social development. Social interaction is critical to the socialization experience. Cases of children raised in extreme isolation show that people need meaningful social contact with others to realize their human potential.

Sociologists maintain that a sense of self emerges out of interaction experiences. Children acquire a sense of self when they (1) take the role of the other, and (2) name, classify, and categorize the self relative to other social categories. Children learn to take the role of others through three stages: preparatory, play, and game.

The concept of the looking-glass self sheds light on the role of the imagination in role-taking. The social experiences of the looking-glass self and the three stages of role-taking prepare the child to establish his or her place in a wider system of rules and expectations. Psychologist Jean Piaget's theory of cognitive development explains how children gain the conceptual abilities to role-take.

Socialization is a lifelong process because people learn new expectations that come with making the many transitions in life. All social transitions involve resocialization, an interactive process during which the affected party reconstructs his or her identity. That process may be voluntary or involuntary, dramatic or subtle, and involve a loss or gain in status and relationships with significant others that must be renegotiated. Significant others play an important role in the resocialization process because how they react is important to the outcome of the reconstruction process for the affected party (K. Daly 1992). Total institutions specialize in resocialization. They are a setting in which people are isolated from the rest of society to undergo systematic resocialization.

Go to cengagebrain.com to link to Aplia and CourseMate for the chapter quiz and other activities.

Chris Caldeira
Silver State
High School
CA$H LOANS
456-3724
CA$H LOANS
456-3724
TAX
SERVICE
CHECK
CASHING
SUBWAY
SUBWAY
BUS

4 CHAPTER SOCIAL STRUCTURES

Most people could walk into any of these buildings and interact with those inside even though they have never met. This is because a largely invisible social structure coordinates interaction in predictable ways between those who enter as "customers" and those who provide services. To gain some idea of the power of social structure to coordinate human interactions, consider what happens when someone does not behave as expected. That person may be stared at, asked to leave, or reported to the police or other authority. As we will learn, social structures prompt people to assume a social identity, shape the content and course of interaction, and present people with opportunities and obstacles.

Go to Sociology CourseMate on cengagebrain.com to watch a video about how factory closures in small and midsized towns in the United States disrupt the division of labor.

MODULE

4.1 Defining Social Structure

Objective

You will learn about how the largely invisible system sociologists call social structure coordinates human activity.

Think of the last time you walked into a doctor's office. What did you expect would happen? Did you expect to wait? To pay the co-pay before seeing the doctor? To get a prescription?

Lisa Southwick

When you entered the doctor's office, you stepped into what sociologists call a **social structure**, a largely invisible system that coordinates human activities in broadly predictable ways. Social structures

- shape relationships with and opportunities to connect to others,
- prompt the people involved to assume a social identity, and
- can erect barriers to achieving goals.

Sociologists study social structures that involve as few as two people (doctor–patient, customer–store clerk, parent–child), that are global in scale (the food industry, the pharmaceutical industry), and that are every size in-between (a local bar, a college dorm, an extended family, a hospital). Sociologists may study the social structure of a particular store, such as a Walmart Supercenter in China, or they may simply study the social structure guiding interactions between a cashier and customer. The supercenter and the cashier–customer relationship are embedded in other social structures.

Boni Li

Journalist 2nd Class Jim Williams

⮝ For example, the cashier-customer relationship is embedded within a particular Walmart store. And each Walmart store is part of the Walmart Corporation (a global enterprise with 8,000 stores in 15 countries). In addition, the Walmart store is part of a community that may embrace or resent its presence.

Social structures encompass at least four interrelated components: statuses, roles, groups, and institutions.

Status and Roles

Sociologists use the term *social status* in a very broad way. **Social status** is a human-created position in society. Examples are endless but include female, teenager, doctor, patient, retiree, sister, homosexual, heterosexual, employer, employee, and unemployed. A social status has meaning only in relation to other social statuses. For instance, the status of a physician takes on quite different meanings depending on whether the physician is interacting with another physician, a patient, or a nurse. Thus, a physician's behavior varies depending on the social status of the person with whom he or she is interacting.

Note that some statuses, such as sister or teenager, are ascribed. **Ascribed statuses** are the result of chance in that people exert no effort to obtain them. By chance a person is born in a particular year and inherits certain physical characteristics. So birth order, race, sex, and age qualify as ascribed statuses. Other statuses, such as nurse's aide and college student, are **achieved statuses**; that is, they are acquired through some combination of personal choice, effort, and ability. The distinction between ascribed and achieved statuses is not clear-cut. One can always think of cases in which people take extreme measures to achieve a status typically thought of as ascribed; a person may undergo sex transformation surgery, lighten his or her skin to appear to be another race, or hire a plastic surgeon to create a younger appearance. Likewise, ascribed statuses can play a role in determining achieved statuses, as when women seemingly choose to enter a female-dominated career such as elementary school teacher, or when men "choose" to enter a male-dominated career such as car repair.

People usually occupy more than one social status. Sociologists use the term **status set** to capture all the statuses any one person assumes (see Figure 4.1a). Sometimes one status takes on such great importance that it overshadows all other statuses a person occupies. That is, it shapes every aspect of life and dominates social interactions. Such a status is known as a **master status**.

Unemployed, retired, ex-convict, and HIV-infected can qualify as master statuses. The status of physician can be a master status as well, if everyone, no matter the setting (party, church, fitness center), asks health-related questions or seeks health-related advice from the person occupying that status.

Figure 4.1a: Diagram of a Hypothetical Status Set
John was born with Down syndrome, and that status can be considered a master status because it shapes every aspect of his life, dominates his social interactions, and overshadows other statuses that he occupies—such as brother, video gamer, uncle, and hospital employee.

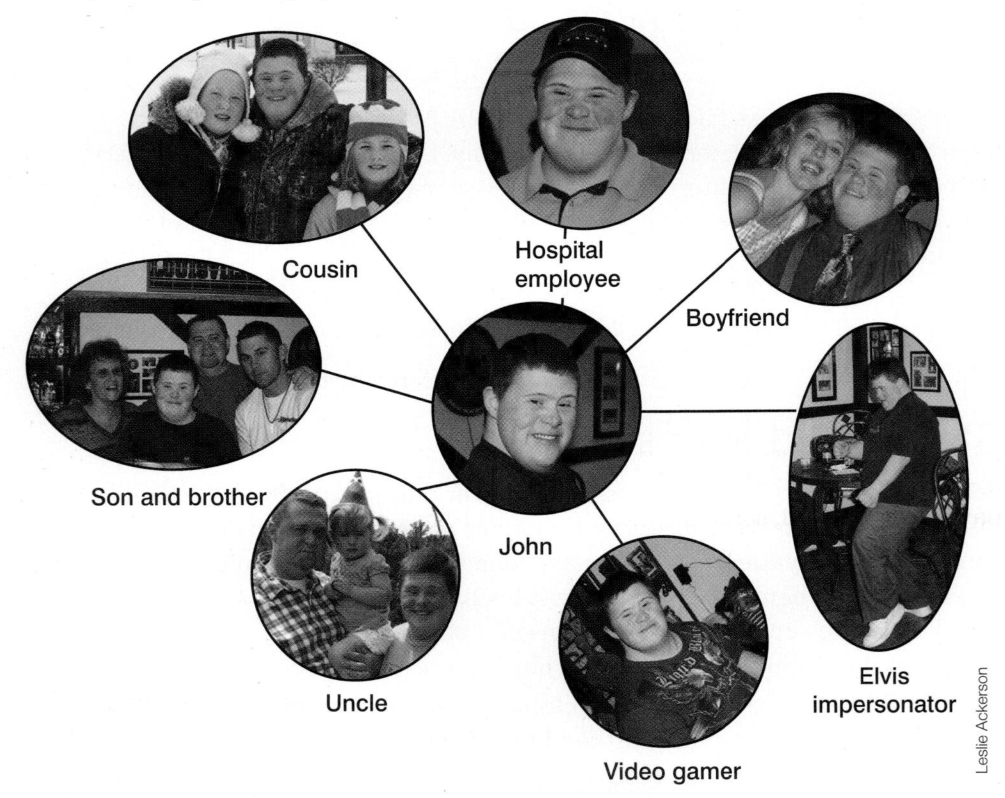

Sociologists use the term **role** to describe the behavior expected of a status in relation to another status—for example, the role of brother in relationship to sister; the role of physician in relationship to patient. Any given social status is associated with an array of roles, called a **role-set**, which are the various role relationships with which someone occupying a status is involved (Merton 1957a). The social status of son is associated with a role-set that can include a relationship with a mother and/or father. The social status of school principal is associated with a role-set that includes relationships with students, parents, teachers, and other school staff members.

The distinction between role and status is this: *people assume or are assigned to statuses* and *they enact or perform roles*. Associated with each social status are **role expectations**, or norms about how a role should be enacted relative to other statuses. The role of physician in relation to a patient specifies that a physician has an obligation to establish a diagnosis, to not overtreat, to respect patient privacy, to work to prevent disease, and to avoid sexual relations with patients (Hippocratic Oath 1943). Relative to a physician, patients have an obligation to answer questions honestly and to cooperate with a treatment plan. Quite often

role performances, the actual behavior of the person occupying a role, do not meet expectations—as when some physicians knowingly perform unnecessary surgery or when some patients fail to comply with treatment plans. The concept of role performance reminds us that people carry out their roles in unique ways. Still, there is predictability in our interactions with others because if people deviate too far from the expected range of behaviors, negative sanctions (penalties)—ranging in severity from a frown to imprisonment and even death—may be applied (Merton 1957a).

If we consider that people hold multiple statuses and that each status is enacted through its corresponding role-set, we can identify at least two potential sources of stress and strain: role conflict and role strain.

ROLE CONFLICT. **Role conflict** is a predicament in which the roles associated with two or more distinct statuses that a person holds conflict in some way. For example, people who occupy the statuses of college student and full-time employee often experience role conflict when professors expect students to attend class and keep up with coursework and employers expect employees to be available to work hours that leave little time for schoolwork. The student–employee must find ways to address this conflict, such as working fewer hours, quitting the job, skipping class, or studying less.

ROLE STRAIN. **Role strain** is a predicament in which there are contradictory or conflicting role expectations associated with a single status. For example, doctors have an obligation to do no harm to their patients. At the same time, doctors have an obligation to pay their office staff. To meet their obligation to pay the office staff, some physicians may feel pressured to recommend unnecessary medical treatments to generate sufficient revenue to stay in business.

Role expectations are embedded within a larger cultural context. For example, in the United States patients are expected to schedule an appointment, whereas in Japan patients are expected to just walk in (Gharib 2009). Doctors in the United States expect to be paid for number of services provided rather than for health outcomes and, as a result, tend to order more tests such as X-rays, CAT scans, and MRIs; prescribe more drugs; and perform more surgeries than their European and Japanese counterparts (see Table 4.1a and Table 4.1b).

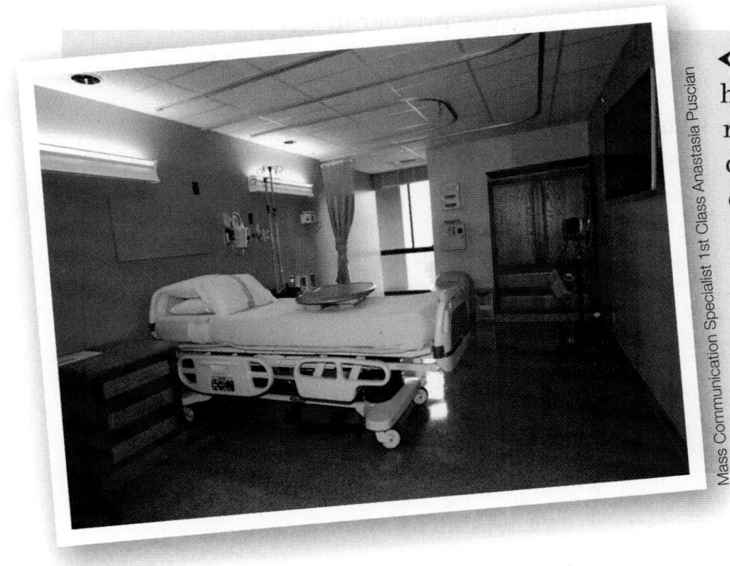

Mass Communication Specialist 1st Class Anastasia Puscian

◀ How do you keep medical hospitals and other facilities running if there is a decrease in the number of patients? Can you see how a hospital might have to keep the patient load constant in order to pay doctors, nurses, and other staff who work there and to pay for the upkeep of equipment and buildings?

▾ Table 4.1a: The Cultural Context of the Physician-Patient Relationship: Health Care Spending and Resources Devoted to Health Care in the World's Richest Countries, 2009-2010

The countries in this table are rank-ordered according to the amount of money each spends on health care, as a percentage of GDP. You can see that the United States spends the most on health care—17.4 percent of GDP. In which country is the per capita spending on health care the highest? Which country most highly subsidizes the cost of health care with taxpayer dollars?

Country	Health Care Spending as a Percentage of GDP*	Per Capita Spending on Health Care ($)	Per Capita Spending on Pharmaceuticals ($)	Percentage of All Health Care Costs Subsidized by Tax Dollars
United States	17.4	7,960	919	47.6
France	11.8	3,978	640	77.9
Germany	11.6	4,281	628	76.0
Denmark	11.5	4,348	318	85.0
Canada	11.4	4,363	743	72.4
Austria	11.0	4,289	554	77.6
United Kingdom	9.8	3,487	381	84.0
Iceland	9.7	3,538	554	82.0
Norway	9.6	5,352	391	84.0
Italy	9.5	3,137	561	77.8
Spain	9.5	3,067	578	73.6
Finland	9.2	3,226	462	75.0
Japan	8.5	2,878	558	81.0

Source of data: OECD (2011d)
*GDP (gross domestic product) is the value of all goods produced and services provided each year in a country.

▾ Table 4.1b: Health Care Outcomes—Life Expectancy, Infant Mortality, C-Sections—in 10 of the World's Wealthiest Countries

The data in this table suggest that some health care systems deliver better outcomes than others. Given the high amount of public and private spending, one might expect the United States to have the best health outcomes—specifically the highest life expectancy and the lowest infant mortality. Statistics show, however, that the United States is not a leader in either category. Which country on this list has the best and worst outcomes related to life expectancy and infant mortality? Which country performs the greatest number of C-sections per 1,000 live births? Why do you think some countries' rates are so high relative to Iceland, Norway, and Sweden?

Country	Life Expectancy at Birth	Infant Mortality (per 1,000 live births)	C-Section (per 100,000 live births)
Austria	80.4	3.8	286
France	81.0	3.9	200.3
Iceland	81.5	1.8	157
Italy	81.8	3.7	383.8
Japan	83.0	2.4	—
Norway	81.0	3.1	171
Spain	81.8	3.3	248
Sweden	81.4	2.5	171
Switzerland	82.3	4.3	324
United States	78.2	6.5	323

Source of data: OECD (2011d)

Groups

Like statuses and roles, groups are an important component of social structure. A **group** consists of two or more people interacting in largely predictable ways who share expectations about their purpose for being. Group members hold statuses and enact roles that relate to the group's purpose. Groups can be classified as primary or secondary. **Primary groups** are characterized by face-to-face contact and/or by strong emotional ties among members who feel an allegiance to one another. Examples of primary groups include the family, military units, cliques, and peer groups.

Mass Comm. Spc. 2nd Class Ron Kuzlik

➤ This child, like all children who see physicians, is part of a primary group (e.g., a family or other caretaking unit). When physicians examine children and recommend treatment and other forms of action that affect health outcomes, they must consider how the family or other caretakers will support or detract from care.

Secondary groups consist of two or more people who interact for a specific purpose. Secondary group relationships are confined to a particular setting, and the involved parties relate to each other in terms of specific roles. People join secondary groups as a means to achieve some agreed-upon end, whether it be to cheer for a sports team, to achieve a status (graduate from college), or to accomplish a specific outcome, such as fund-raising for a good cause. Secondary groups can be small to extremely large in size. They include a work unit, a college classroom, a parent–teacher association, and a church. Larger secondary groups include the staff members who work at a university medical center or fans gathering in a stadium. Certainly, some people who are part of secondary groups form close relationships with each other and can constitute a primary group.

Institutions

A fourth component of social structure is **institutions**, relatively stable and predictable social arrangements created with the purpose of coordinating behavior and interactions to meet some need (e.g., provide medical care or care for young children). Institutions consist of statuses, roles, and groups. Examples of institutions include education, medicine, and sports (P. Martin 2004). We will use examples from medicine as practiced in the United States to illustrate characteristics of institutions.

1. *Institutions have a history*. Institutions have standardized ways of doing things, and over time people come to accept these ways without question. In the United States, Americans have come to expect that the physician will write a prescription whether they actually need one or not. Many patients suggest, even demand, a specific prescription.
2. *Institutions continually change*. Over time ways of doing things become outdated and are replaced by new ways. Change can be planned and orderly, forced, and/or chaotic. Change can come from within or from outside the institution. Sometimes a change is hardly noticed; at other times it revolutionizes ways of doing things.

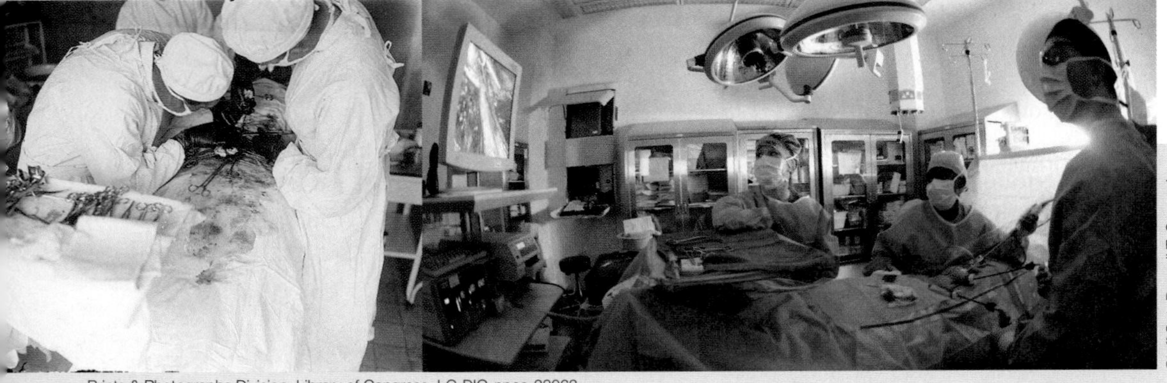

Prints & Photographs Division, Library of Congress, LC-DIG-npcc-23063

Staff Sgt. Darnell T. Cannady

⋏ Before the advent of tiny digital cameras and microchips, surgeons used their fingers to operate and control instruments (left). Today, surgeons may use new technologies (right) to see deep inside the patient's body and organs and direct instruments to do what fingers once did. Such technology means that surgeons now have the ability to operate on a patient from practically any location in the world. This change can potentially revolutionize the delivery of health care and the patient–physician relationship.

3. *Institutions allocate scarce and valued resources in unequal ways.* The U.S. medical system rewards physicians according to specialty, with some specialties earning considerably more than others. As a result, the United States does not produce enough physicians in lower-paying specialties such as general practice to meet the health care needs of its population, especially the least wealthy segments. In the United States, just 2 percent of medical students say they want to be primary care physicians. The most common reasons given for rejecting this vocation are inadequate pay relative to debt incurred to pay for medical school and the costs of liability insurance (*Online Newshour* 2009).
4. *Institutions allocate privileged and disadvantaged status.* These privileges and disadvantages are reflected in individuals' salaries, benefits, degree of autonomy, and amount of prestige.

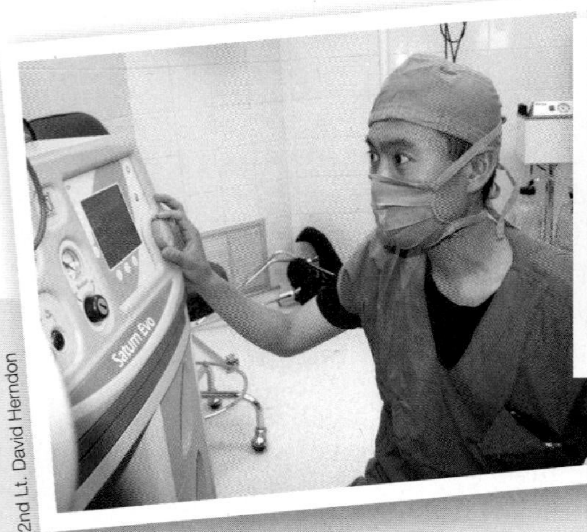

2nd Lt. David Herndon

Ronnie Kaufman/Larry Hirshowitz/ Blend Images/Corbis

⋏ The U.S. Department of Labor (U.S. Bureau of Labor Statistics 2011a, 2011b) lists anesthesiologists as the health care status with the highest annual average salary ($220,000). Home health aides (right) receive the lowest average salary ($21,760). Of course, there are many good reasons for paying some categories of workers more than others. On the other hand, should the lowest-paid employees (many of whom have families) work for poverty-level wages?

5. *Institutions promote ideologies that legitimate their existence.* These legitimating ideologies are largely created and advanced by those occupying the most advantaged statuses or by those who benefit from institutionalized ways of doing things. The masses often accept these ideologies and resist change. For example, many critics of universal health care coverage dismiss other health care systems, even those that provide such coverage with better health outcomes than the United States. They call up images of "big government," "rationed care," and "socialized medicine" ruining the U.S. system that they claim is the best in the world. Table 4.1a shows that taxpayer dollars account for 47 percent of all money spent on health care—an amount equal to what most countries spend on health care.

(Write a Caption)

Write a caption that uses some of the key concepts in this module to describe the social structure underlying this interaction between a drill instructor and recruit.

Hints: In writing this caption

- review the key terms covered in this module, and
- select two or three terms (e.g., statuses, role, role expectations) that capture the dynamics illustrated in this photograph.

Sgt. Jose Nava

Critical Thinking

Use the concepts in this module to describe a social structure that coordinates your activity in predictable ways.

Key Terms

achieved statuses
ascribed statuses
group
institutions
master status
primary groups
role
role conflict
role expectations
role performances
role-set
role strain
secondary groups
social status
social structure
status set

MODULE

4.2 Division of Labor and Social Networks

Objective

You will learn how division of labor and social networks connect people to one another and to society.

How many pairs of hands does it take to make a pair of athletic shoes?

Mass Comm. Spc. 3rd Class Travis K. Mendoza

The Global Assembly Line

In her book *Factory Girls: From Village to City in a Changing China*, Leslie T. Chang describes the incredibly involved process of making a pair of athletic shoes:

> It takes two hundred pairs of hands. . . . Everything begins with a person called a cutter, who stamps a sheet of mesh fabric into curvy irregular pieces, like a child's picture puzzle. Stitches are next. They sew the pieces together into the shoe upper, attaching other things—a plastic logo, shoelace eyelets—as they go. After that sole-workers use infrared ovens to heat pieces of the sole and glue them together. Assemblers—typically men, as the work requires great strength—stretch the upper over a plastic mold, or last, shaped like a human foot. They lace the upper tightly, brush glue on the sole, and press the upper and sole together. A machine applies 90 pounds of pressure to each shoe. Finishers remove the lasts, check each shoe for flaws and place matched pairs into cardboard boxes. The boxes are put in crates, ten shoes to a crate and shipped to the world. (2009, 98)

Although the headquarters of the major athletic footwear corporations are located in places like Oregon, Massachusetts, and Germany, the shoes are most likely made in Asian countries, most notably China, Indonesia, and Vietnam. Keep in mind that the 200 pairs of hands represent only some of the hands involved in getting the shoes to consumers. Other hands navigate the freighters carrying the shoes from China to international ports, where dock workers unload them. Then truck drivers transport them to stores located around the world. And we cannot forget the labor (pairs of hands) that creates the commercials that promise superior athletic prowess when worn.

In *The Division of Labor in Society* (1933), sociologist Émile Durkheim provides a general framework for understanding the forces driving global-scale interactions among workers whose combined efforts bring us products like athletic shoes. The **division of labor** refers to work that is broken down into specialized tasks, each performed by a different set of workers trained to do that task. The labor and resources needed to manufacture products often come from many locations around the world.

A complex division of labor affords different experiences to those doing the work depending on the task with which they are charged. Thus the division of labor creates people with different outlooks and experiences. Furthermore, this labor structure forces people to relate to one another in terms of their specialized roles. When we interact in this manner, we ignore personal differences and treat individuals who perform the same tasks as interchangeable. As a result most day-to-day interactions inside and outside the workplace are largely impersonal and instrumental (that is, we interact with people for a specific reason, not to get to know them). We do not need to know personally the people with whom we work or do business to interact with them. We do not need to know those who work in faraway places who contribute to the overall production process. In spite of this anonymity, the largely invisible ties that bind people to one another and to their society are very strong because in industrial societies few individuals possess the knowledge, skills, and materials to be self-sufficient. Consequently, people need one another to survive (see Figure 4.2a).

REUTERS/Crack Palinggi

Mass Comm. Spc. 2nd Class Joy Marie Kirch

▲ Many Americans do not make a conscious connection between the athletic shoes they wear and those who labor to make them. The chances are quite high that those who sew them are laboring in Asian factories.

▾ Figure 4.2a Countries with Factories That Manufacture Nike Footwear

In societies characterized by organic solidarity, people do not personally know the workers who produce the products they consume, such as footwear. The map illustrates countries where Nike footwear is manufactured (Nike 2010). Nike is the leader in global market share. Other major competitors include Reebok, Adidas, and New Balance.

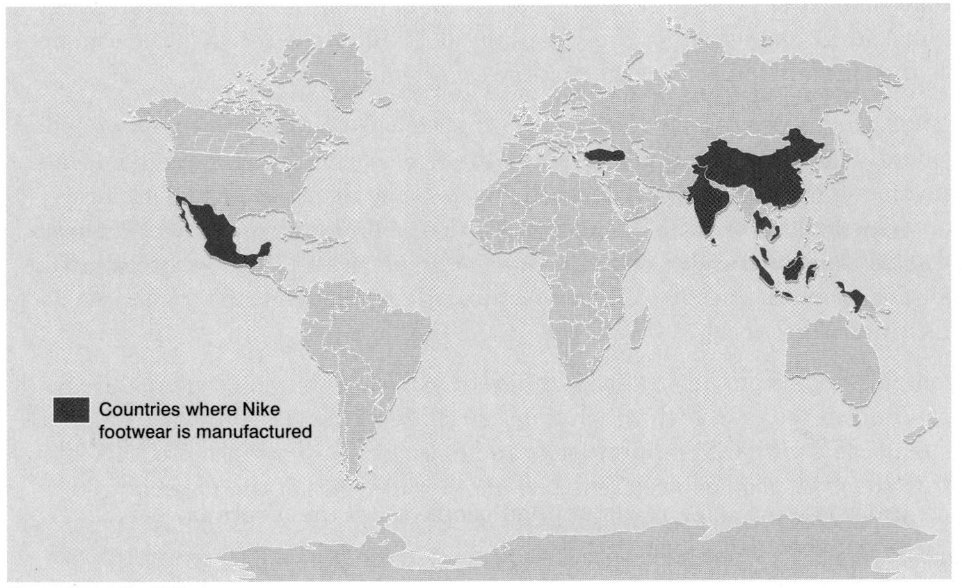

Source of data: Nike 2012

Disruptions to the Division of Labor

Durkheim hypothesized that societies become more vulnerable as the division of labor becomes more complex. The reason is that a disruption to the division of labor eliminates jobs. Durkheim was particularly concerned with the kinds of events that disrupt people's connection to the division of labor. Such disruptions also break down people's ability to connect to each other and their community through their labor. As one example, consider the life of someone who has been laid off from work. After losing a job, daily interactions shift from colleagues at work to contacts with people at the unemployment office. Moreover, the newly unemployed person loses the structure that comes with a job. Now instead of working, the person may watch TV or search the Internet for job openings. Relationships with a partner or children may become strained if the partner or children must work more hours or give up activities to compensate for the lost job. This example illustrates how labor or work is central in people's lives and that "absence of decent and productive work is the primary cause of poverty and social instability" (International Labour Union 2009).

Durkheim identified five key events that disrupt the division of labor and its ability to bind people to each other and their communities:

1. *industrial and commercial crises*, triggered by technological revolutions and outsourcing of labor to locations where wages are lower, that result in plant closings and massive layoffs.
2. *worker strikes* that occur when a group of workers agree to stop working as a way to enforce a demand or express a grievance related to pay, benefits, or

working conditions. In this situation, the employer must find replacement workers or shut down operations. Workers sacrifice pay and benefits with the hope of forcing change.

3. *job specialization* creating a situation in which workers labor in isolation, unaware of how their work relates to the work of others who are part of the production process. Thus, workers lack a complete understanding of the overall enterprise and, by extension, cannot see ways the process might be more efficient or productive.
4. *forced division of labor* where occupations are filled according to nationality, age, race, or sex—rather than ability. There are many examples of what Durkheim calls the forced division of labor, as when some occupations are formally or informally reserved for a particular group—childcare workers are almost always female; airplane pilots are almost always male. The problem is that the division of labor channels some groups (in this case women) toward low-paying jobs and other groups (men) toward higher-paying jobs.
5. *inefficient development or use of workers' talents and abilities* so that work for some is nonexistent, irregular, intermittent, or subject to high turnover. A strong and efficient division of labor is one that cultivates and uses all its available talent. The division of labor is not operating efficiently when, in times of high unemployment, there are job openings in some sectors (like IT or nursing) but not enough qualified people to fill them because needed skills have not been developed.

▲ As early as 1907, Detroit, Michigan—home to GM, Ford, and Chrysler—was considered the automotive capital of the world. Beginning with an oil embargo in 1979 and continuing through today, Detroit has been affected by a series of ongoing industrial and commercial crises. These crises have disrupted the division of labor and have affected the city's ability to connect people to others and to their communities. As a result, on some streets of Detroit and surrounding suburbs, more houses and businesses are abandoned than are occupied because people no longer have the money to maintain the lifestyle they attained when employed.

Social Networks

Imagine you are looking for a job. Do you know someone who might help you find one? If you do, then you are using your social network to gain employment. If no one in your social network can connect you to employment opportunities, you are at a disadvantage in the job market. A **social network** is a web of social relationships linking people to one another.

Figure 4.2b: Example of Social Networks

Think about Maddy's social network as depicted in this image. You can see some of the people who are part of Maddy's social network. Through her dad (Greg), however, Maddy is also connected to Bill's network. Maddy is connected to the networks of all those who are part of her network. Here only Bill's network is shown.

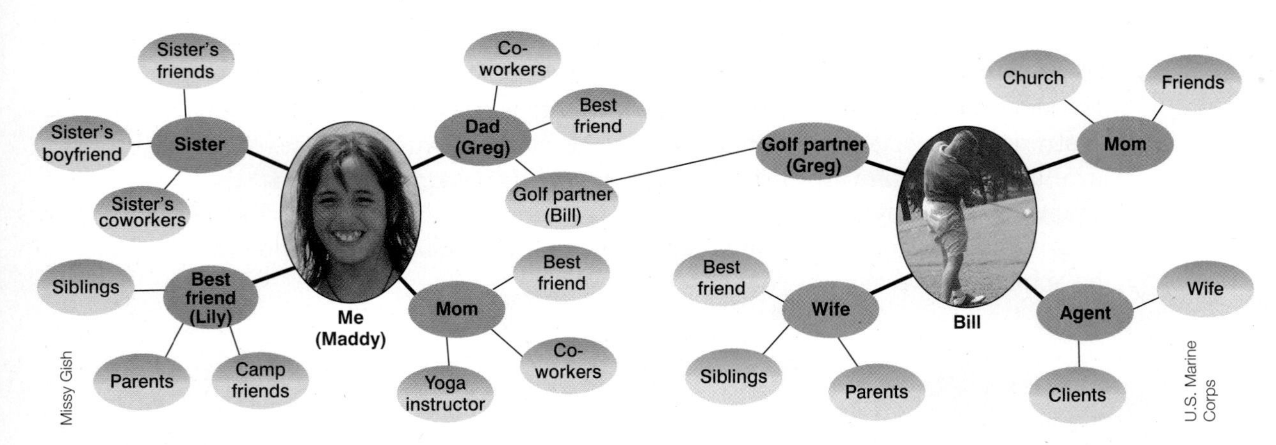

Electronic-Supported Social Networks

Sociologists, of course, are interested in how digital technologies affect the size and character of social networks and people's ability to connect to others through them. Even before digital technologies such as the Internet, e-mail, and cell phones, sociologists realized that it is not easy to map a person's social networks. Digital technologies are not the first innovations to expand the size and reach of social networks. The telephone, airplane, train, and car expanded social networks to the point that it is impossible to manage and sustain connections with everyone we have come to know or hope to know.

In what ways do digital technologies further enhance the ability of people to connect with others? They

- allow people in different time zones and on different schedules to communicate at their convenience,
- increase the speed of communication,
- expand the number of people with whom one communicates to theoretically include anyone with access to the Internet, and
- offer people who live geographically far apart a convenient and inexpensive tool by which to remain in touch.

Strong ties that may have faded without these technologies are more likely to remain strong, and a person's weak ties are maintained and perhaps strengthened.

emily moving away (+) us going to different colleges

Digital technologies do not just expand the reach of social networks but also enhance local connections. Preliminary research suggests that most of the e-mails and text messages we send each day are to those who live nearby—to significant others, work colleagues, and friends—all of whom use the technology to check in, send reminders, or arrange a face-to-face meeting (Wellman and Hampton 1999).

The Importance of Weak Ties

Sociologist Mark Granovetter (1973), who wrote one of the most cited articles in sociology, "The Strength of Weak Ties," studied 100 professional, technical, and managerial workers living in a Boston suburb who had recently changed jobs. He learned that 54 of these workers found their new jobs through a personal contact. To measure the strength of workers' ties to their named contacts, Granovetter asked how often they saw that contact. He found that 27.8 percent saw their contact rarely (once a year or less), 56 percent saw their contact occasionally (more than once a year but less than twice a week), and 16.7 percent saw their contact often (at least twice a week). This led him to conclude that "the skew is clearly in the weak end of the continuum" (1371).

Lisa Southwick

Chris Caldeira

▲ According to Granovetter's measure of weak and strong ties, if a contact is someone the job seeker sees at the gym at least once a week, the relationship is not a strong tie. If the contact is someone the person sees once a year but is a best friend, that relationship is defined as a weak tie. How do these caveats affect your understanding of weak and strong ties?

In interviews with the workers who changed jobs, Granovetter learned that the contacts who helped them land the job were most often an old friend, a former coworker, or former employer. In addition, that contact was likely to be someone with whom that worker had maintained sporadic contact over the years. These findings led Granovetter to conclude that it is "remarkable that people receive such crucial information from individuals whose very existence they have [mostly] forgotten" (1372). In his interviews, Granovetter asked workers how their contacts knew about the job opening. He found that

- 45.0 percent said that their contact knew the prospective employer,
- 39.1 percent said their contact was a prospective employer, and
- 12.2 percent said that their contact knew someone who knew the potential employer.

Granovetter also learned, to his surprise, that the information paths to new employment opportunities involved only one to three linked contacts. He hypothesized that if the paths had involved more linked contacts, larger numbers of people "might have found out about any given job, and no particular tie would have been crucial" (1372). Granovetter's findings do not discount the importance of strong ties and individual initiative in finding a job. But his research does speak to the importance of cultivating weak ties within social networks when searching for employment.

(Write a Caption)

Write a caption that relates Durkheim's disruptions to division of labor to a change in the way X-rays are read: putting film on a light board to viewing X-rays on a computer screen.

Hints: In writing this caption

- review the five disruptions to the division of labor,
- decide which one disruption best applies to these images, and
- explain how this change will disrupt the ties that connect radiologists to others.

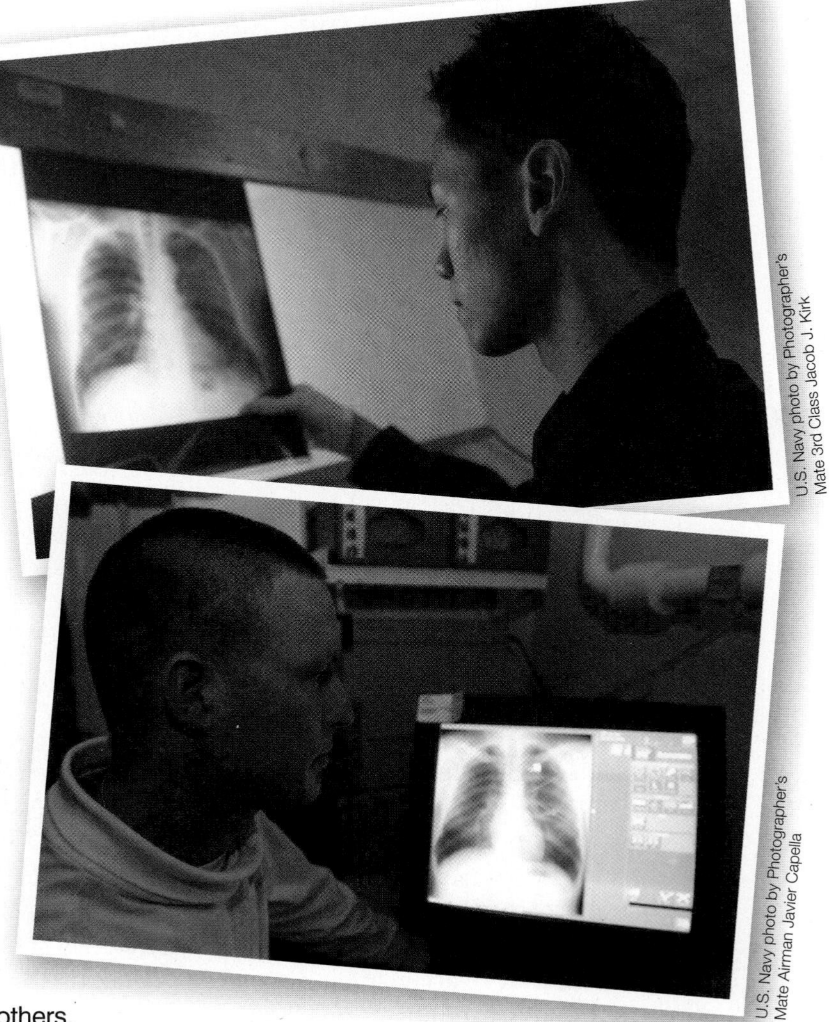

U.S. Navy photo by Photographer's Mate 3rd Class Jacob J. Kirk

U.S. Navy photo by Photographer's Mate Airman Javier Capella

Critical Thinking

Think about jobs you have held over your lifetime. How did you learn about the job opening? If you have never worked, then think about how a friend or family member learned about the job he or she holds.

Key Terms

division of labor

social network

Formal Organizations

MODULE 4.3

Objective

You will learn that formal organizations coordinate human activity with the aim of achieving some valued goal.

Have you ever participated in an organization-sponsored event where you did something for a good cause?

This nine-year-old donated a foot of her hair to Locks of Love, a not-for-profit organization that provides hair pieces to children who have lost their hair as a side effect of medical treatments. The organization claims to receive 4,000 donations of hair from individuals and hair salons each week. If you have been part of such an effort, you have experienced the power of **formal organizations**, coordinating mechanisms that bring together people, resources, and technology and then direct human activity toward achieving a specific outcome. That outcome may be to provide hairpieces to children who have lost their hair to illness, to maintain order in a community (as does a police department), to challenge an established order (as does People for the Ethical Treatment of Animals), or to provide a credentialing service (a university). Formal organizations are a taken-for-granted part of our lives. If you were born in a hospital, attended a school, acquired a driver's license from a state agency, secured a loan from a bank, worked for a corporation, received care at a hospital, or purchased a product at a store, you have been involved with a formal organization (Aldrich and Marsden 1988).

U.S. Army

example

Formal organizations can be voluntary, coercive, or utilitarian, depending on the reason that people participate (Etzioni 1975). **Voluntary organizations** draw in people who give time, talent, or money to address a community need. Voluntary organizations include food pantries, political parties, religious organizations, and fraternities and sororities.

➤ There are millions of voluntary organizations worldwide to which people give time, talent, and money. One example is a homeless coalition that provides winter shelter for those without homes.

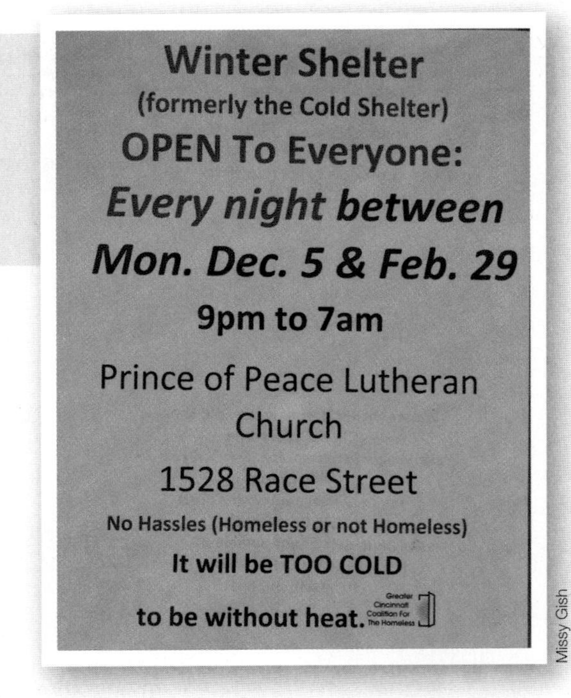

Missy Gish

Coercive organizations draw in people who have no choice but to participate. Examples include public schools that students are required by law to attend, the military when there is a draft, and treatment facilities to which people have been ordered to report because a judge or other authority deems them medically or socially unfit (Spreitzer 1971). Prisons also qualify as coercive organizations.

Utilitarian organizations draw in those seeking to achieve some desired goal in exchange for money. That goal may be to earn an income as an employee, to acquire a skill by enrolling in a special program, or to purchase a desired product at a department store.

From a sociological perspective, formal organizations, whatever their type, have a life that extends beyond the personnel and clients/customers. Indeed, formal organizations prevail even as the people die, quit, retire, or get fired. One concept that sociologists use to think about the ways in which organizations coordinate resources and human activity is bureaucracy.

Bureaucracy

Sociologist Max Weber (1925) defined a **bureaucracy**, in theory, as a completely rational organization—one that uses the most efficient means to achieve a valued goal, whether that goal is feeding people (McDonald's), recruiting soldiers (military), counting people (Bureau of the Census), collecting taxes (IRS), drilling for oil (ExxonMobil), or providing a service (hospitals).

There are at least six means by which a bureaucracy efficiently coordinates human activity to meet organizational goals.

1. *A clear-cut division of labor exists.* Each office or position is assigned a specific task geared toward accomplishing the organizational goals. Google, for example, employs 20,000 people worldwide in occupations including office managers, technicians, laborers, chemists, engineers, and so on. All employees—software engineers, product managers, web developers, ad quality raters, and facilities managers—work toward achieving Google's mission, which is "to organize the world's information and make it universally accessible and useful." (Google 2012).
2. *Authority is hierarchical.* Most bureaucracies publish organizational charts depicting how authority and responsibility are distributed among the positions that make up the organization. Within bureaucracies there may be vice presidents, regional managers, managers, assistant managers, and entry-level employees, all of whom report to someone in the chain of command.

➤ This office building evokes images of a bureaucracy. We can imagine employees, each with a job description, assigned to an office, cubicle, or work space, working to achieve organizational goals. The building suggests a hierarchy, with people holding the most power and authority in the organization occupying offices on the top floors.

Chris Caldeira

3. *Written rules specify the way positions relate to each other and describe the way an organization should operate.* Administrative decisions, rules, procedures, and job descriptions are recorded in operations and training manuals, handbooks, or bylaws.
4. *Positions are filled according to objective criteria.* Criteria include academic degrees, seniority, merit points, or test results, but not emotional considerations, such as family ties or friendship.
5. *Authority belongs to the position.* It does not belong to the particular person who fills a position. This characteristic implies that while on the job, managers, for example, have authority over subordinates because they hold a higher position. Managers cannot demand that subordinates do tasks unrelated to work, such as washing a manager's car or babysitting.
6. *Organizational personnel treat clients or customers as cases.* That is, "without hatred or passion, and hence without affection or enthusiasm" (Weber 1947, 340). To put it another way, no client receives special treatment. This objective approach is believed necessary because emotion and special circumstances can interfere with the efficient delivery of goods and services. Many organizations require employees to greet every customer with standard lines such as "Thank you for shopping at Kmart. May I help you?"

Lisa Southwick

◄ The "take a number" system is designed to ensure that employees treat clients/customers as "cases." This system eliminates emotion and bias in determining who is next. Taking a number sends the message that no person is special or can demand to be served before another.

Taken together, these six characteristics describe a bureaucracy as an **ideal type**—a deliberate simplification or caricature in that it exaggerates essential traits. Ideal does not mean desirable; an ideal is simply a standard against which real cases can be compared. Anyone involved with a bureaucracy realizes that actual behavior departs from this ideal. Still, the six ideal traits of bureaucracy can be used to evaluate the extent to which any bureaucracy follows or departs from these traits. Such an evaluation may reveal problems or strengths in an organization's structure. In this regard, sociologists distinguish between formal and informal dimensions of organizations.

Formal and Informal Dimensions

The **formal dimension** is the official, by-the-book way an organization should operate. The formal dimension is known through an organization's job descriptions and through its written rules, guidelines, and policies. The **informal dimension** encompasses any aspect of the organization's operations that departs from the way it is officially supposed to operate. The informal dimension includes employee-generated responses that violate official policies and regulations. A manager who demands that employees work off the clock, employees who give their friends free food and soft drinks, and servers who spit in a rude customer's drinks are all displaying behavior that departs from official policies.

▲ These photos speak to informal and formal dimensions of organizations. When employees sleep on the job or engage in activities unrelated to work, such as browsing the Internet while on the clock, they are behaving in ways that circumvent official organizational policies. When checkout clerks at Home Depot wear a smock proclaiming "I Put Customers First," they are complying with official policy.

Performance Measures

Many bureaucracies have performance measures in place, quantitative indicators of how well their employees or clients are performing with reference to some valued goal. Managers often evaluate employees using statistics related to sales, customer satisfaction, and production quotas. Such measures can be useful management tools for two reasons: they are considered to be objective, and they permit comparison across individuals, over time, or across departments. Management can tie pay increases and promotions to objective measures of good performance and can use objective measures of poor performance to justify actions such as firings, pay reductions, or demotions.

But performance measures have shortcomings, one of which is that they can encourage employees to concentrate on meeting performance goals and to ignore problems generated by their drive to score well.

Lisa Southwick

< The ongoing mortgage crisis in the United States can be traced in part to performance measures that rewarded loan officers for the number of loans they made and little else. As a result, many loan officers concentrated on making loans regardless of risk to the bank. To meet quotas, loan officers falsely inflated applicants' incomes and created fake W-2s as documentation. If an applicant had no credit record, loan officers created phony loan histories showing timely payments (French and Wilkinson 2007). In the end, loan officers gave many people keys to houses they could not afford. Perhaps one of the greatest incentives for making so many bad loans was that these loans were sold to third-party investors who assumed the risk instead of the bank. Then the third-party investors bought insurance to protect against defaults, so insurance giants such as AIG were left holding the bag.

(Write a Caption)

Write a caption that relates each photograph to one of three broad organizational types—voluntary, coercive, or utilitarian.

Hints: In writing this caption

- review the three broad types of organizations,
- consider into which category a detainee holding cell best fits and the kind of "client" it serves, and
- consider into which category this payday loan enterprise fits for those who use it for the first time. Under what circumstances might it fit a different organizational type?

Master Sgt. Adam M. Stump

Chris Caldeira

Critical Thinking

Name three organizations to which you are connected. Classify each as voluntary, coercive, or utilitarian. Explain why you classified each the way you did.

Key Terms

bureaucracy	formal organizations	utilitarian organizations
coercive organizations	ideal type	voluntary organizations
formal dimension	informal dimension	

MODULE 4.4

Rationalization and McDonaldization

Objective

You will learn about the problems that come with using any means necessary to achieve a desired goal, most notably to achieve a profit.

Why are Korean government and businesses investing in robots to teach children English?

One answer is that the government and businesses see robotic teachers as an efficient way to do the long hard work of teaching someone a language. No longer will teachers get tired and frustrated with students making slow progress. Also, robots have no favorite or least favorite students. Furthermore, robots will not engage in inappropriate behavior, sexual or otherwise, with students.

What does this possibility of robotic teachers in school and elsewhere say about the way society is organized, the way in which people think, and the choices open to them? The answer to these questions lies with a concept sociologist Max Weber called rationalization. Before defining that concept, we will consider the meaning of another related concept: instrumental rational action.

Instrumental Rational Action

Instrumental rational action is result-oriented behavior that emphasizes the most efficient methods or means for achieving some valued goal, without regard for the adverse consequences those methods may have. In the context of an industrial and capitalist society, *efficient* means the most cost-effective and time-saving way to achieve a goal—whether that goal is producing eggs, satisfying hunger, or recovering from an illness. Instrumental rational action, for example, drives the treatment of animals raised on factory farms. Obviously, the more chickens a factory farm can house and the faster it can raise them to egg-laying maturity, the more eggs it can produce and sell.

Jeff Vanuga

➤ Corporations that process pigs are driven by instrumental rational action. Farmers contract with corporations to raise pigs that the corporation has provided. The pigs are confined in crowded and controlled environments, never having the opportunity to move around freely outdoors.

Weber maintained that the rise of instrumental rational action as a dominant way of organizing human activity is a product of industrialization. Through a process he labeled rationalization, instrumental rational action came to replace other ways of thinking and behaving. Before industrialization, thought and action was organized around emotion, mysterious forces, and tradition. When thought and action are driven by emotion, they are accompanied by a physical sensation, such as feelings of love, hate, or fear. When thought and action are driven by respect for mysterious forces, they are grounded in a fear of offending or in a desire to please spirits or the gods. Thought and action driven by tradition are guided by a respect for the ways things were done in the past.

Tony Rotundo

➤ This photo of three little pigs in a barn represents the way many of us imagine or would like to believe pigs are raised. This image suggests that the farmer who raises these pigs is driven by emotion or an obligation to care for them. In reality, most pigs are raised on corporate farms and live under crowded conditions with little room to move or even turn around.

Rationalization, then, is a process in which thought and action organized around emotion, superstition, respect for mysterious forces, or tradition are replaced by instrumental rational action. Keep in mind that Weber used the term to refer to the way daily life has come to be organized so that people are forced to use the "efficient" structures that are already in place to meet their needs (Freund 1968). For example, most food choices on the market are processed and take little time to prepare. People who purchase such food are likely to know home-cooked meals are healthier and to prefer them to processed food, but it is difficult to resist the efficiency of processed foods.

McDonaldization of Society

Sociologist George Ritzer (1993) describes an organizational trend guided by instrumental rational action in which people use the most cost-efficient and quickest means to achieve some valued end. He called that trend the McDonaldization of society. Ritzer defines **McDonaldization of society** as the process by which the principles of the fast-food industry have been applied to other sectors of American society and the world. Those principles are (1) efficiency, (2) calculability, (3) predictability, and (4) control.

EFFICIENCY. **Efficiency** involves using methods that will achieve a desired end in the shortest amount of time. Many organizations advertise products and services touting claims that buying them will help consumers move most quickly from one state of being to another—from hungry to full, from fat to thin, from uneducated to educated, or from sleep-deprived to rested. In some cases, an organization puts a system in place where customers serve themselves. In the name of saving time, customers scan, bag, and pay for purchases; clear their table after eating in a quick-service restaurant; and check themselves into the airport without expecting to be compensated for their labor.

© Kim Kulish/Corbis

➤ Drive-thru flu vaccination centers have applied the efficiency principle of McDonaldization to health care services. According to one nurse who administered the shots, "The drive-through concept is so popular because of its ease and convenience. For many people it is very difficult to get themselves and/or children out of the car and into a doctor's office, or other inside location. It is very convenient for those who just don't want to get out of their car" (*University of Kentucky News* 2003).

CALCULABILITY. The second McDonaldization principle is **calculability**, which emphasizes numerical indicators as a way customers can judge the results of a product or service or the speed with which it is delivered (e.g., delivery within 30 minutes, lose 10 pounds in 10 days, earn a college degree in 24 months, limit menstrual periods to four times a year, or obtain eyeglasses in an hour). These are clear measures that customers can draw upon to assess results. Ritzer argues that the emphasis on quantity promotes the idea that size matters, more is better, and something obtained quickly is superior to a product that takes time to deliver.

PREDICTABILITY. The principle of **predictability** is the expectation that a service or product will be the same no matter where in the world or when (time of year, time of day) it is purchased. With regard to food products, this kind of predictability requires that the product be genetically modified. If the consumer expects strawberries year-round, the produce must be able to withstand shipment from one continent to another. The fruit is genetically modified so that the berry stays firm and does not decay before it reaches its destination (Barrionuevo 2007). Another way to increase the predictability of a food's taste involves adding artificial and natural flavorings. Keep in mind a food's flavor is altered and reduced when it is processed, frozen, and heated. One way to counteract the effects is to add engineered flavorings. Since the smell of food affects how the food tastes, ice creams and candies, for example, made from the same ingredients, can be made to taste dramatically different depending on what scent or fragrance flavorists add.

➤ How do apples get their uniform and unblemished appearance? Notice the "cosmetically challenged" apples are not uniform in size and color like the apples shown in the bottom photo. Among other things, the apples in the bottom photo have been subjected to an aggressive regimen of chemical sprays to maintain the smooth skin, coated with wax for shine, and perhaps stored in a controlled environment for 10 months or more. Apples like those on the bottom are considered to be among the most pesticide-contaminated fruit (Lloyd 2011).

CONTROL. The fourth McDonaldization principle, **control**, involves replacing employee labor with "smart" technologies and/or requiring, even demanding, that employees and customers behave in a certain way. From an employer's point of view, humans are a source of uncertainty and unpredictability. The quality and consistency of people's work, for example, is affected by any number of factors, including how they feel, if they are paying attention, the personal problems they face, and their relationship with the boss. Consider how customers dialing 411 to obtain a phone number are greeted, not by a human voice, but by a computerized voice asking for the city and state and then asking for the name of the person or business they wish to contact. Customers cannot proceed unless they give clear and precise answers to the questions. When a computer-generated voice asks for the city and state, the customer cannot use free-flowing speech expressing uncertainty about a city's name. It is easy to see how smart technology controls the interaction and saves labor costs.

Assessing McDonaldization

Clearly, there are advantages to McDonaldization—people can check bank balances, take college classes, and pay bills at the hours of their choosing. Though McDonaldization has facilitated amazing feats, the model has drawbacks. In the quest to deliver products and services quickly and efficiently, an organization can create dehumanizing structures that can lower the quality of life. Ultimately, we must ask whether so-called rational processes create irrationalities (Weber 1994; Ritzer 2008). Sociologists use the term **iron cage of irrationality** to label the process by which supposedly rational systems produce irrationalities. As one example, consider that the pharmaceutical industry creates medicines for just about every social, physical, and psychological problem people face. It is irrational, however, that this industry seeks to fix the conditions that make us human by offering medications that eliminate monthly menstruation, minimize wrinkles, and alter moods (even normal sadness and grief).

(Write a Caption)

Write a caption that relates the computer-guided drone missiles to the process of rationalization.

Hints: When writing this caption

- review the concept of rationalization,
- consider what motivates those who risk personal harm and death by fighting an enemy in more direct ways (e.g., emotion or tradition), and
- think about how someone who kills from a distance is different from someone who faces the enemy.

Mate Airman Nicholas C. Messina

Critical Thinking

Give an example of McDonaldization that involves something other than fast food.

Key Terms

calculability
control
efficiency
instrumental rational action
iron cage of irrationality
McDonaldization of society
predictability
rationalization

Alienating and Empowering Social Structures

MODULE 4.5

Objective

You will learn about social structures that promote alienating and empowering workplaces.

If you are employed, how satisfied are you in your current job?

In his book *Three Signs of a Miserable Job*, Patrick Lencioni (2007) names three signs of dissatisfaction:

1. *anonymity*—my boss has little interest in me as a person with a unique life, interests, and aspirations;
2. *irrelevance*—my work makes no real difference in people's lives; and
3. *cluelessness about impact*—there is no way of knowing whether I make an impact or contribution through my work.

Comm. Spc. 1st Class Todd Hack

Karl Marx wrote about such issues more than 150 years ago. Marx (1844) wrote: "labor is *alien* to workers when they do not affirm themselves through labor." The alienated worker "does not feel happy, but rather unhappy; he does not grow physically or mentally, but rather tortures his body and ruins his mind."

Alienation

Human control over nature increased with mechanization and the growth of bureaucracies. Both innovations allowed people with the help of machines to extract raw materials from the earth quickly and more efficiently. Mechanization also reduced the amount of human labor needed to complete a task and increased the speed by which necessities such as food, clothing, and shelter could be produced and distributed. Karl Marx believed that increased control over nature is accompanied by **alienation**, a state of being in which humans lose control over the social world they have created and are dominated by the forces of their inventions. The phrase "social world they have created" refers to the way people relate to and interact with each other.

◂ Surveillance via cameras, drones, mobile phones, and computers has given governments, corporations, and individuals the power and ability to watch and monitor others. As a result, people have lost control over private lives and may never know when someone is watching them and for what purpose. In addition, it is hard to know how the behavior caught on surveillance will be interpreted by those watching or reviewing tapes.

Airman 1st Class Betty M. Leonard

Although Marx discussed alienation in general, he wrote more specifically about alienation in the workplace. Marx maintained that workers are alienated on four levels: (1) from the process of production, (2) from the product, (3) from the family and the community of fellow workers, and (4) from the self.

ALIENATION FROM THE PROCESS AND PRODUCT. Workers are alienated from the *process* when they produce not for themselves or for people they know but rather for an abstract, impersonal market. In addition, workers are alienated when they do not own the tools they use to produce things and when what they produce has no individual character and sentimental value to either the worker or the consumer (Marx 1888).

Workers are alienated from the *product* when their roles are rote and limited and when their employers treat them like replaceable machine parts. Marx believed that most jobs do not allow people the chance to be active, creative, and social (T. R. Young 1975).

ALIENATION FROM FAMILY AND COMMUNITY OF FELLOW WORKERS. Workers are alienated from the *family* because households and workplaces are separate spheres. Specifically, the workplace makes few, if any, accommodations to family life unless forced to do so. Workers can lose touch with their families when they work shifts late at night, early in the morning, or on weekends, keeping them from participating in family life. They also lose connection to their families when they must relocate to find employment. An estimated 25 percent of the Philippines' labor force works outside the country in 190 countries, leaving children to be cared for by relatives (Migration Policy Institute 2007). The United States and Western European countries, for example, recruit 85 percent of all nurses trained in the Philippines. The exodus of nurses is so great that the World Health Organization has expressed concerns about its effect on the Philippines' system of medical care (Coonan 2008).

Workers are also alienated from the *community of fellow workers* because they compete for jobs, business, advancement, and salary. Migrant workers, especially the undocumented, are often viewed as taking the jobs of people in the host country. As workers compete for jobs, they fail to consider how they might unite as a force to better control their working conditions.

© David De La Paz/epa/Corbis

➤ These Honduran workers earn approximately 81 cents per hour, about 24 cents less per hour than their Mexican counterparts and about 40 cents less per hour than their counterparts in China.

ALIENATION FROM THE SELF. Finally, workers are alienated from the *self*, or from the human need to realize one's unique talents and creative impulses. When Karl Marx developed his ideas about alienation in the late 1800s, he was describing alienation from self as it relates to industrial society. More recently, sociologist Robin Leidner has described the alienation from self that can occur in service industries when management standardizes virtually every aspect of the service provider–customer relationship, so that neither party feels authentic, autonomous, or sincere: "Employers may try to specify exactly how workers look, exactly what they say, their demeanors, their gestures, even their thoughts" (1993, 8). The means available for standardizing interactions include giving employees scripts to follow, requiring workers to follow detailed dress codes, and issuing specific rules and guidelines for dealing with customers and with coworkers. Employers may use surveillance cameras or computer software to monitor worker performance and enforce compliance.

Mr. Sheldon S. Smith (USARC)

➤ Computer technology is an alienating tool if management uses it strictly as a means for monitoring things like how long customer service employees take to respond to people on hold, the number of calls handled per hour, or the time that lapses between calls.

The Best Work Environments

There are many work environments designed to reduce alienation among employees. Each year *Fortune* magazine lists the 100 best companies for which to work. There are a variety of characteristics that earn a company a place on this list, including

- ethical and collaborative values and principles held by the top leadership, including president;
- open communication with immediate supervisor;

- opportunities for personal growth through challenging work, training, career development, and prospects of advancement;
- programs that help employees manage workplace stresses;
- policies that bring balance to work and personal life;
- quality relationships with colleagues;
- chances to give something back to community and society; and
- fairness as it relates to pay and benefits (*Times Online* 2008).

(Write a Caption)

Write a caption that relates the concept of alienation to this photograph of undocumented workers in a holding pen waiting to be returned to the Mexican side of the border.

Gerald L. Nino

Hints: In writing this caption

- review the four levels of alienation,
- identify which of the four levels of alienation is most relevant to undocumented workers (think about why they leave families and communities to work in the United States), and
- explain your choice(s) in a sentence or two.

Critical Thinking

Think about the work you or someone close to you does for a living. Is the workplace empowering, alienating, or some combination of the two?

Key Term

alienation

The Effects of Size

MODULE 4.6

Objective

You will learn some ways in which group size affects social dynamics.

Think about a time when you and a friend were talking and a third person walked over to join the conversation. How was your conversation affected?

Lisa Southwick

A third party joining a two-person conversation already in progress will change the dynamics between those two people. The third person might direct his or her attention toward one of the two, and the other may feel left out or decide to leave. Sociologist Georg Simmel (1950) argued that one of the most important criteria for understanding group dynamics is size. Size affects the relationships among members and shapes the character of interactions.

Dyads, Triads, and Beyond

The smallest group is a **dyad**, which consists of two people. Sociologists are interested in the reasons dyads form, whether by consequence of birth (mother–son), choice (two friends), or because of social necessity (doctor–patient, teacher–student, author–editor). In assessing a dyad, sociologists ask, How much does each party know about the other? Keep in mind that no person can know everything about another. But we can say that in **comprehensive dyads** the involved parties have more than a superficial knowledge of each other's personality and life; they know each other in a variety of ways. In **segmentalized dyads** the parties know much less of the other's personality and personal life; and what they do know is confined to a specific situation, such as the classroom, a hair salon, or other specialized setting (Becker and Useem 1942).

Lisa Southwick

Chris Caldeira

⮝ Sociologists consider this husband–wife dyad (left) comprehensive, as it is very likely that after decades of marriage each knows a lot about the other as a person. The customer–food vendor dyad is segmentalized because the involved parties interact out of social necessity—one out of a need to purchase food; the other out of a need to make a living. It is likely that the customer and vendor interact only when the customer wants a bite to eat.

If a third member is added to a dyad (a stepfather to a mother–son dyad, a nurse to a doctor–patient dyad, a second student to a teacher–pupil dyad), the result is a **triad**, or a three-person group. The addition of a third person significantly alters the pattern of interaction between the two people who made up the original dyad. Now, two members can form an alliance against the third. Obviously, as more members are added—a fourth, a fifth, and so on—the patterns of interaction and alliance possibilities increase.

Lisa Southwick

➤ At some point—perhaps around seven members—the group breaks down into subgroups because it is impossible to focus everyone's attention on the group per se. That is, it is difficult to have a single, focused conversation in which everyone is listening, taking turns talking, or focusing on the task at hand. Unless someone steps forward and directs communication among the members, the group breaks down into dyads and triads, with each smaller group carrying on its own conversations (Becker and Useem 1942).

Oligarchy

Political analyst Robert Michels (1962) was interested in very large groups involving thousands of people. He believed that large organizations inevitably tend to become oligarchical. **Oligarchy** is rule by the few, or the concentration of decision-making power in the hands of a few persons who hold the top positions in a hierarchy. Michels maintained that one of the most bizarre features of any advanced industrial society is that life-and-death choices are made by a handful of people, usually men, who cannot possibly consider all who will be affected.

Organizations become oligarchical because democratic participation is virtually inconceivable in large organizations. Size alone makes it impossible to discuss matters and settle controversies in a timely and orderly fashion. For example, Walmart is the largest employer on the Global Fortune 500 list with 1.9 million employees and millions of stockholders. Obviously, such a large number of employees cannot engage in direct discussion.

The greater an organization's size, the less likely it is that members can comprehend its workings. As a result, leaders may push employees to advance organizational goals, the full consequences of which no one may be able to know or understand. There is also the danger that key decision makers may become so preoccupied with preserving their own leadership or with the bottom line that they do not consider the greater good or the full implications of their choices.

We can apply the concept of oligarchy to the Great Recession of 2008. Some blame this economic crisis on a small group of powerful executives, running financial institutions that were "too big to fail," for creating and abusing unregulated financial products for personal and institutional gain. Others point to complex financial instruments that few, if any, people understood until it was too late. Regardless of the real source of the crisis, a few people at the top of some of the largest organizations in the world ultimately decided to use these products to build profits, with disastrous consequences for national and global economies (Gross 2009).

(Write a Caption)

Lithographer's Mate Seaman Recruit Shanika Futrell

Write a caption that uses the concepts of dyad and triad to address how this mother–daughter relationship must adjust now that the father has returned from military deployment overseas.

Hints: In writing this caption

- review the concepts of dyad and triad, and
- think about how a dyad must adjust when a third person (even one who is welcomed) becomes part of the relationship.

Critical Thinking

Think of a time a third person was added to a dyad to which you belonged. Describe how that third person changed the interactional dynamics between you and the other person.

Key Terms

comprehensive dyads

dyad

oligarchy

segmentalized dyads

triad

MODULE

4.7 Applying Theory: Emotional Labor and Emotion Work

Objective

You will learn the circumstances under which the organizational social structure requires employees to do emotion work.

Do you hold a job in which you are expected to hide either negative or excessively positive emotions when relating to customers, clients, or the public?

Tech. Sgt. John M. Foster

If your job requires you to serve customers, clients, or the public, your employer is asking that you engage in what sociologists call **emotional labor**, work that requires employees to display and suppress specific emotions and/or manage customer/client emotions.

Dramaturgical Theory

Dramaturgical theory, which is most associated with the work of sociologist Erving Goffman (1959), views social interaction as if it were theater and people as if they were actors giving performances before an audience in a particular setting. In social situations, as on a stage, people manage the setting, their dress, their words, and their gestures so that they correspond to an impression they are trying to make. This process is called **impression management**.

Sociologist Arlie Hochschild (2003a) extended the work of Erving Goffman by presenting actors as not only managing their outer impressions but also working at managing their inner feelings. Hochschild recognized that emotions—whether sadness, anxiety, anger, boredom, or nervousness—are more than simply reflex-like responses; they are shaped by expectations about how one should feel in a particular situation. Thus, people are sad at funerals, not simply because someone has died, but because that is how people are supposed to act (Appelrouth and Edles 2007). Likewise, people suppress feelings of envy at award ceremonies when not chosen for an award as they congratulate a competitor. They suppress this emotion because losers are supposed to be gracious.

Emotion Work

Hochschild argues that people do **emotion work**—that is, they consciously work at managing their feelings by evoking an expected emotional state or suppressing an inappropriate one. That work may involve presenting an outward display of an emotion that does not match inner feelings or convincing oneself to actually feel the expected emotion. For Hochschild, emotion work involves *effort*—with emphasis placed on trying to feel a certain way and not the actual outcome of that effort, "which may or may not be successful" (241). Such phrases as "I psyched myself up," "I squashed my anger down," "I tried hard not to feel disappointed," "I made myself have a good time," "I tried to feel grateful," and "I let myself finally feel sad" represent examples of emotion work (Hochschild 2003a, 241). Hochschild acknowledges that there are often discrepancies between "what one does feel and what one wants to feel," which are further complicated by what one thinks one should feel (242). Even when people fail at managing an emotion, their efforts are still influenced by social expectations.

Emotional labor is a requirement for those jobs in which employees

1. must engage with the public,
2. are directed to produce a specific emotional state in clients/customers, whether it be feelings of satisfaction with a service delivered or a reasonable outlook (such as police officers who calm excited citizens), and
3. are required to follow scripts and are penalized for deviating from them (such as clerks who must greet customers with "Welcome to . . .").

➤ Service jobs—customer service representatives, food servers, sales clerks, funeral directors, front desk clerks, cashiers, doctors, nurses, social workers, teachers—involve emotional labor. People in such jobs must present their emotions in ways that suggest they are out to please and that mask aggressive emotions felt toward customers, especially difficult ones.

In her research on flight attendants, Hochschild found their job required them to manufacture feelings of warmth, caring, and cheerfulness and suppress anger or boredom. Hochschild believes that the emotional labor associated with service work can be an alienating experience in much the same way as the repetitive physical labor we associate with the assembly line. In the case of factory work, employers control workers' bodies and movements; in the case of service work, employers direct workers' emotional states and reactions to customers. Both the service and factory workers are alienated from their authentic (true) self.

Pfc. Vanessa Jimenez

➤ When displaying or eliciting certain emotions is an expected part of a paid job, those emotions are commodified. People who work with children are "paid" to build self-esteem in the children they care for. This expectation becomes a problem when building feelings of self-esteem interferes with a teacher's ability to offer constructive criticism, which is also an important ingredient for learning.

Hochschild writes that "those who perform emotional labor in the course of giving service are like those who perform physical labor in the course of making things—both are subject to the demands of mass production. But when the product is a smile, a mood, a feeling, or a relationship, it comes to belong more to the organization and less to the self" (2003b, 198). Hochschild argues that emotional labor is not performed equally across race, class, gender, and age. She makes the case that women especially face greater pressure to manage their emotions than their male counterparts and to present their emotions in motherly or sexy ways. Why is this the case?

First, there is the deeply rooted cultural idea that women are practiced at managing emotions—that they "have the capacity to premeditate a sigh, an outburst of tears, or a flight of joy. In general, women are thought to manage expression and feelings, not only better than, but more often than men" (248). Second, when women fail to manage emotions in expected ways, they are considered less "feminine." Finally, women feel pressure to manage their emotions, because when they lose control of their emotions they are labeled as "unstable" or "too emotional" to do the job. While it is true that men face cultural pressures to manage their emotions in so-called masculine ways, the specific kinds of cultural expectations women face explain why they are overrepresented in occupations that require emotional labor.

Hochschild found that these cultural pressures specific to women help explain why female flight attendants are more likely to handle the babies, deal with the children, and comfort older passengers. Male flight attendants (when on board) are less likely to engage with passengers in these ways because they are not expected to. As a result, they are more likely to enforce rules about where to stow oversized luggage and to monitor seat belt usage. These cultural pressures may also explain why women dominate teaching and medical occupations (e.g., nurses, nurses aides, home health care workers).

Critical Thinking

Give an example of emotional labor from your own work life or from your observations of someone who works in a customer service job.

Key Terms

emotional labor | emotion work | impression management

Summary: Putting It All Together

CHAPTER 4

A social structure is a largely invisible system that coordinates human activities in broadly predictable ways. Social structures shape people's sense of themselves, their relationships and opportunities to connect with others, and the barriers they will encounter. Social structures encompass at least four interrelated components: statuses, roles, groups, and institutions.

The division of labor is a social structure that connects people to one another and to their society. Durkheim hypothesized that societies become more vulnerable as the division of labor becomes more complex. The reason is that a disruption to the division of labor has the potential to affect so many people whose livelihood (and the livelihood of those they support) depends on the job they hold within that structure. Durkheim was particularly concerned with the kinds of events that break down people's ability to connect to each other and their society through their labor. These events include economic crisis and job specialization.

The concept of formal organization emphasizes how these social structures can operate as coordinating mechanisms bringing together people, resources, and technology to achieve specific outcomes. The concept of bureaucracy describes specific mechanisms employed, including a clear-cut division of labor, an authority structure that is hierarchical, and written rules governing operations. Of course, people embedded in bureaucracies depart from official ways of doing things. To capture these dynamics, sociologists use the concepts of formal and informal dimensions of organizations.

Sociologists employ a variety of other concepts—including dyads, triads, oligarchy, strong and weak ties, and emotional labor—to capture how groups and organizations connect or fail to connect people to one another. The concepts of rationalization and McDonaldization alert us to a process by which organizations channel human behavior in the most efficient ways. Alienation in the workplace, which can result from rationalization and McDonaldization, describes the downside of such efficiency.

Go to cengagebrain.com to link to Aplia and CourseMate for the chapter quiz and other activities.

Chris Caldeira

Chris Caldeira

5 CHAPTER
THE SOCIAL CONSTRUCTION OF REALITY

The process by which people make sense of the world is known as the social construction of reality. Contrast going to a market where the whole chicken is displayed, in open air and feet intact, with shopping in a market where chicken parts are displayed, wrapped in plastic (a package of eight breasts, for example). Would your assumptions about the chicken you eat change? When sociologists study the social construction of reality, they focus on how people go about assigning meaning to what is going on around them. Sociologists also focus on the knowledge people draw upon to create a reality upon which they act. For the most part, people do not question the accuracy of that knowledge, nor do they consider alternative realities until they encounter something that challenges existing assumptions. So, for example, there is little about the way chicken is processed in the United States that reminds us that the meat we buy was once a living creature.

Go to Sociology CourseMate on cengagebrain.com to watch a video that features different reality constructions around the world and the United States regarding the age of consent.

MODULE 5.1

Definition of the Situation

Objective

You will learn that people do not approach a situation with fresh eyes.

Obama For America/Handout/Reuters/Corbis

Reuters/Obama For America/Landov Photographers/Source: Reuters/Landov

Which photograph of President Barack Obama as a child are you most familiar with? Do the photographs affect your perception of who Barack Obama is? Why or why not?

The photo on the left is Barack Obama as a child with his Kansas-born mother. The photo on the right is Obama with his Kenyan-born father. It is also important to note that for three years of his life, from 1969 to 1971, Obama lived in Indonesia. Some who opposed Obama's presidential campaign hoped that foreign connections might raise questions in voters' minds about Obama's loyalty to the United States. Obama supporters, on the other hand, emphasized his Kansas connections. These photographs, and the meanings each evokes for viewers, relate to the concept of definition of the situation.

Definition of the Situation

W.I. Thomas (1923) points out that the human nervous system carries "memories or records of past experiences." When we observe something, like the Obama photographs, we do not see them as something new and fresh. Rather, what we see and how we define it is shaped by past experiences. Those experiences begin accumulating as soon as babies are born (perhaps before), at which point they encounter people who define situations for them. Obviously, infants have no real chance to make their own definitions without interference from caretakers.

Kris Gonzalez

➤ Adults define the situation for young children through the things they have them do. This four-year-old is graduating from a "strong beginnings" preschool program. What do you think the graduation ceremony means to this child? Her teachers likely talk to her about what it means to graduate. What might they say?

Thomas (1923) maintains that before people respond or take action there is a fleeting moment during which they deliberate about the meaning of the situation. That is, they attach a subjective meaning, informed by past experiences, to the situation. Thomas described the link between the meaning assigned and the response as follows: if people "define situations as real they are real in their consequences" (Thomas and Thomas 1928, 572). This is known as the **Thomas theorem**. What happens when people assign false definitions to a situation? It can result in a self-fulfilling prophecy.

Self-Fulfilling Prophecies

Sociologist Robert K. Merton (1957b) defined a **self-fulfilling prophecy** as a false definition of a situation that is assumed to be accurate. People behave, however, as if that false definition is true. In the end, the misguided behavior produces responses that confirm the false definition. Only when that definition is questioned and a new definition is introduced will the situation correct itself. Researchers Robert Rosenthal and Lenore Jacobson (1968) designed an experiment to test Merton's hypothesis to see how teachers' expectations about their students' intellectual growth can become a self-fulfilling prophecy.

Pfc. Ethan Hoaldridge

⮜ Jacobson realized that most schools have some system to sort students into "fast," "average," or "slow" learner categories and consequently expose them to "fast," "average," and "slow" learning environments.

Rosenthal and Jacobson believed that once students were categorized or formally labeled, teachers' perception of academic ability or potential affected the way they taught and interacted with students. The experiment took place in an elementary school the researchers called Oak School. The student body came mostly from lower-income families and was predominantly white (84 percent). Sixteen percent of the students were Mexican Americans. At the end of a school year, Rosenthal and Jacobson gave a test, purported to be a predictor of academic "blooming," to the students who were expected to return the next year. Just before classes began in the fall, all full-time teachers were given the names of the students, assigned to "fast," "average," and "slow" groups the previous school year, who had supposedly scored in the top 20 percent. The teachers were told that these students "will show a more significant inflection or spurt in their learning within the next year than will the remaining 80 percent of the children" (Rosenthal and Jacobson 1968, 66). Teachers were also told not to discuss the scores with the students or their parents. In actuality, the names researchers gave to teachers were chosen randomly; that is, the differences between the children earmarked for intellectual growth and the other children existed only in the teachers' minds. The students were retested after one semester, at the end of the academic year, and after a second academic year.

Overall, intellectual gains, as measured by the difference between successive test scores, were greater for the students who had been identified as "bloomers" than they were for the students who were not identified as such. Although all bloomers benefited in general, some benefited more than others: first- and second-graders, Mexican American children, and children in the middle track showed the largest increases in test scores.

Rosenthal and Jacobson designed the study in such a way that those identified as bloomers received no special instruction or extra attention from teachers. Thus, the only difference between them and the other students was the teacher's belief that the bloomers bore watching. Rosenthal and Jacobson speculated that the teacher, "by what she said, by how and when she said it, by her facial expressions, postures, and perhaps by her touch" may have communicated to the bloomers that she expected their academic performance to improve. The two researchers called for further research to narrow down the specific mechanisms whereby teacher expectations shape students' intellectual growth (180).

(Write a Caption)

Write a caption that relates this child's experience to the concept of definition of situation. The mother is making a "welcome home" sign as her daughter looks on.

Cpl. J.R. Stence

Hints: In writing this caption

- think about W.I. Thomas's thoughts on where memories are stored,
- consider the kind of event this little girl is about to experience, and
- think about the kind of situation to which she will apply this experience in the future.

Critical Thinking

Can you think of a personal experience in which you were affected by a self-fulfilling prophecy? Describe that situation.

Key Terms

self-fulfilling prophecy   Thomas theorem

Reality Construction: Assigning Cause

MODULE 5.2

Objective

You will learn the ways people assign cause and in the process construct a reality and response to that reality.

How did the 30-year-plus global HIV/AIDS epidemic begin?

Chris Caldeira

If your answer assumes people with HIV/AIDS are personally responsible for their disease, then you are emphasizing dispositional factors. If your answer emphasizes larger social forces that created populations vulnerable to that illness, then you are emphasizing situational factors.

Dispositional and Situational Factors

Social life is complex. People need a great deal of historical, cultural, and biographical information if they are to accurately pinpoint the cause of even the most routine happenings. Still, most people form judgments about cause, and take action, without complete information. **Attribution theory** examines the process by which people explain their behavior and that of others. Specifically, people attribute cause to one of two types of factors: (1) dispositional or (2) situational.

Dispositional factors are things that people are believed to control, including personal qualities related to motivation, interest, mood, and effort. **Situational factors** are things believed to be outside a person's control—such as the weather, bad luck, and another's incompetence. Usually, people stress situational factors in explaining their own failures ("I failed the exam because the teacher is terrible") and dispositional factors in explaining their own successes ("I passed the exam because I studied hard"). With regard to others' failures, people tend to emphasize dispositional factors ("She failed the exam because she parties too much"). With regard to others' successes, however, people tend to stress situational factors ("She passed the test because it was easy").

Applying Attribution Theory to AIDS

When I ask my students—most of whom were born after 1981 (the year that what is now known as AIDS was recognized as a disease)—about the origin of AIDS, most point to Africa. Some students mention that they have read that the virus jumped from chimpanzees to humans—specifically that the jump occurred when an African person was exposed to an infected chimp's blood while hunting or preparing it for consumption. My students' views are supported in part by AIDS researchers who believe that the global HIV/AIDS epidemic actually started, not in the early 1980s, but around 1930 in southeastern Cameroon, when the virus "jumped" from a chimp to a human host. Specifically, these researchers speculate that the human host was a male from the Bantu ethnic group who hunted and butchered chimpanzees for meat. AIDS researchers hypothesize that for the virus to survive and reach epidemic proportions, it had to jump from human to human and make its way to a big city, where human hosts were numerous and concentrated. The closest such city was Léopoldville (later renamed Kinshasa), located in the country then known as Belgian Congo (Abraham 2006).

An unspoken assumption of this theory is that dispositional factors are at work: if this Bantu male didn't kill or eat monkeys (something we would never do in the West), this epidemic would never have occurred. But one leading HIV/AIDS researcher, Jim Moore (2004) argues there must be more to the story, because (1) one infected hunter could not trigger a global epidemic, and (2) this hunting practice had been going on for thousands of years. So why would it trigger an epidemic in the 1980s?

Moore maintains that more than a single hunter had to be involved. Suppose a single hunter in 1930 was infected by a chimp and then transmitted the virus to, say, 2 people over 5 years, making a total of 3 infected people by 1935. In turn, each of those 2 and the hunter transmitted the virus to another 2 people each over the next 5 years. We now have a total of 9 people infected between 1930 and 1940. Now imagine how the virus spread over a series of 5-year progressions. The number of cases climbs from 9 to 27 (1940–45) to 81 (1945–50) to 243 (1950–55) to 729 (1955–60) to 2,187 (1960–65) to 6,561 (1965–70) to 19,683 (1970–75) to 59,049 (1975–80). While 59,049 is a large number, this progression as noted cannot account for the tens of millions of people eventually diagnosed with HIV/AIDS. For one, not every sexual or other intimate encounter with an infected partner transmits HIV—depending on the kind of encounter, the probability of transmission ranges from 0.5 to 60 infections per 10,000 exposures to an infected source (Varghese et al. 2002). Moore argues that the virus must have been "helped along" by some large-scale event or situational factors that made the population in what was then Belgian Congo particularly vulnerable.

Situational Factors and AIDS

In 1883 King Leopold II of Belgium claimed as his private property 1 million square miles of land in Central Africa where 15 million people lived. Leopold's personal hold over this land, which would become known as the Belgian Congo, was formalized in 1885 by the leaders of 13 European countries and the United States, who were attending the Berlin West Africa Conference. The purpose of that conference was to carve Africa into colonies and divide the continent's natural resources among competing colonial powers (J. Witte 1992). For 23 years Leopold capitalized on the world's demand for rubber, ivory (used in making piano keys, billiard balls, and snuff boxes), palm oil (a machine-oil lubricant and an ingredient in soaps such as Palmolive), coffee, cocoa, lumber, and diamonds.

➤ One of the most grisly policies of Leopold's rule was to sever a hand of Congolese who refused to gather rubber. Leopold's soldiers were also under orders to turn in a hand for every bullet fired, as proof that they were using bullets to kill people, and not killing animals or using them for target practice. Often soldiers would sever a hand to cover for other uses of bullets (Adam Hochschild 1998; Twain 1905).

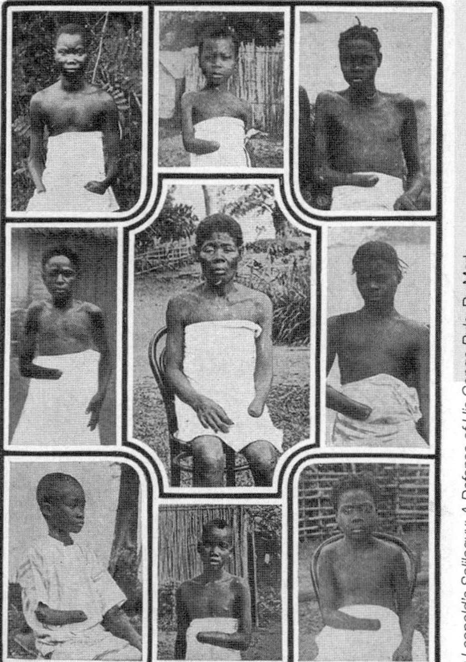

King Leopold's Soliloquy: A Defense of His Congo Rule, By Mark Twain, Boston: The P. R. Warren Co., 1905, Second Edition.

The methods Leopold used to extract rubber for his own personal gain involved atrocities so ghastly that in 1908 international outrage forced the Belgian government to assume control. But like Leopold, the Belgian government also forced the indigenous Africans to leave their villages to work the mines, cultivate and harvest crops, and build and maintain roads and train tracks. By 1930, Belgian colonization had forced enough people together to sustain the HIV/AIDS virus. As Moore describes it:

> Forced labor camps of thousands had poor sanitation, poor diet and exhausting labor demands. It is hard to imagine better conditions for the establishment of an immune-deficiency disease. To care for the health of the laborers, well-meaning, but undersupplied doctors routinely inoculated workers against small pox and dysentery, and they treated sleeping sickness (an infectious disease transmitted by the tsetse fly with symptoms that include severe headache, fever, and lymph node swelling, extreme weakness, sleepiness, and deep coma) with serial injections. The problem was that multiple injections given to arriving gangs of tens or hundreds were administered with only a handful of syringes. The importance of sterile technique was known but not regularly practiced. Transfer of pathogens would have been inevitable. And to appease the laborers, in some of the camps sex workers were officially encouraged. (2004, 545)

The strategies employed to force people to extract resources made them weak and vulnerable to sleeping sickness. The Europeans' approach to diagnosing sleeping sickness involved taking a blood sample through a finger prick or taking spinal taps using unsterilized needles (now a known risk factor in the transmission of HIV). Between 1923 and 1938, an estimated 26 million spinal taps were performed in the Belgian Congo. The risk of transmission by needle sharing is 67 per 10,000 infected sources, and it is 6,000 per 10,000 by exposure to infected blood products (Donegan, Stuart, and Niland et al. 1990; Kaplan and Heimer 1995).

Provided Courtesy of Urbain Ureel

➤ The human muscle required to move this loaded barge into the Congo River speaks to labor demands placed on African workers. Working conditions were so difficult that they left the African workers weak and vulnerable to illnesses. In an effort to escape these conditions, tens of thousands fled into the unfamiliar jungle where, as they searched for and killed animals for food, they encountered chimpanzees, some of which were infected.

The point is that even with "help" from mass testing and inoculation campaigns, it still took five decades for the virus to "burst" onto the global scene (Moore 2004). As long as the disease was confined to central Africa, few, if any, paid attention because AIDS-like symptoms were very common in this environment. In fact, the symptoms of sleeping sickness and AIDS are very similar. As you can see, the story of HIV/AIDS is a complex one that cannot be explained by a hunter's encounter with an infected chimpanzee (dispositional factors). Rather the origin of HIV/AIDS is embedded in a situation (colonization, extreme exploitation of labor) that created a population vulnerable to disease.

Cpl. Noah S. Leffler

(Write a Caption)

Write a caption that relates dispositional and situational factors to this image about weight and what causes people to be overweight.

Hints: In writing this caption

- review the definitions of dispositional and situational factors,
- consider which one of the two factors this image seems to emphasize, and
- counter that factor with an explanation emphasizing the other factor.

Critical Thinking

How does your understanding of HIV/AIDS change, or not change, as a result of reading this module? Explain.

Key Terms

attribution theory   dispositional factors   situational factors

Dramaturgical Model

MODULE **5.3**

Objective

You will learn an approach for thinking about how people manage interactions and the presentation of self.

What about this encounter is like theater? Can you draw an analogy with actors, costumes, and props?

Lisa Southwick

Life as Theater

Sociologist Erving Goffman's writings revolve around the assumption that "life is a dramatically acted thing" (1959, 72). He offered the dramaturgical model for analyzing social encounters. **Dramaturgical sociology** studies social interactions emphasizing the ways in which those involved work (much like actors on stage) to present and manage a shared understanding of reality (Kivisto and Pittman 2007).

Sgt. Christopher M. Gaylord, 5th Mobile Public Affairs Detachment

◀ Goffman uses the analogy of the theater to describe the work of impression management. In the theater, actors use scripted dialogue, costumes, gestures, and settings to convey a particular reality to the audience. Social encounters are like theater and other staged performances, and the involved parties are like actors in costume and on stage performing roles. Like actors, they must work, in conjunction with other cast members, to convince their "audience" that they are who they appear to be or who they say they are.

Lisa Southwick

In social encounters, as on a stage, people manage the setting (stage), their dress (costumes), and their words and their gestures (script) to correspond to the impression they are trying to make. This process is called impression management. This photograph shows two people performing the roles of mechanic and customer. From a dramaturgical point of view, they are in "costume," on "stage" with appropriate props, and following a "script" that we all recognize.

Managing Impressions

When managing impressions some people behave in completely calculating ways with the goal of evoking a particular response (Goffman 1959). An example of such calculation involves someone posting a 10-year-old photograph as an up-to-date likeness of himself on a social networking website. Likewise, thieves posing as utility workers, who knock on doors and request permission to enter under false pretenses, are thoroughly calculating in their attempt to gain entrance (the desired response) so they can rob the occupants.

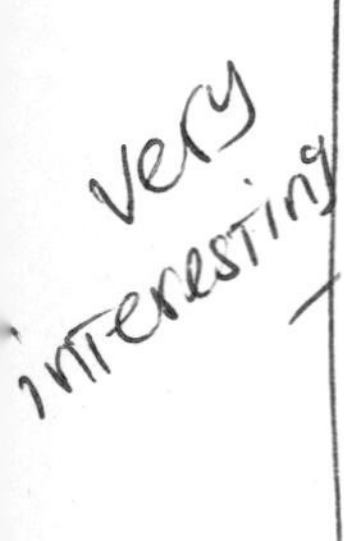

Impression management is not always self-serving; sometimes people have to talk or behave in a particular way because the status they occupy requires them to do so. Coaches work to hide any doubts about whether they think their team can win and instead express great optimism in team meetings. They engage in impression management because that is what coaches are expected to do. At other times people are unaware that they are engaged in impression management because they are simply behaving in ways they regard as natural. Women engage in impression management when they put on makeup, dye their hair, or shave their legs even if they see these activities as things women do. Likewise men engage in impression management when they hide their emotions in stressful situations so that no one questions their masculinity.

Goffman (1959) judges the success of impression management by whether an audience "plays along with the performance." If the audience plays along, the actor has successfully projected a desired definition of the situation or has at least cultivated an understanding among those in the audience that they will uphold that definition even if they doubt an actor's abilities and intentions. According to Goffman, there are times that an audience goes along simply because that actor is "a representative of something" considered important to the society; that is, the audience gives deference, not "because of what they personally think of that actor but in spite of it" (210).

Most students manage to stay awake in class or at least give the appearance that they are paying attention, even when they think the teacher is not very good. They do so not out of respect for the teacher they view as incompetent, but out of respect for the position that the teacher occupies.

Image copyright Marin, 2011. Used under license from Shutterstock.com

◀ Imagine you are a professor speaking in front of this class. What impressions are students conveying through their body language and facial expressions? Why are they not pretending they are interested?

Goffman (1959) recognized that there is a dark side of impression management that occurs when people manipulate their audience in deliberately deceitful and hurtful ways. But impression management can also be constructive. If people said whatever they wanted and behaved entirely as they pleased, social order would break down. According to Goffman, in most social interactions people weigh the costs of losing their audience against the costs of losing their integrity. If keeping the audience seems more important, impression management is viewed as necessary; if being completely honest and upfront seems more important, we may take the risk of losing our audience.

Front and Back Stage

Goffman used a variety of concepts to elaborate on the process by which impressions are managed—including the idea of front and back stage. Just as the theater has a front stage and a back stage, so too does everyday life. The **front stage** is the area visible to the audience, where people feel compelled to present themselves in expected ways. Thus, when people step into an established social role such as a teacher in relation to students or as a doctor in relation to patients, they step onto a front stage such as a classroom or an examining room (Goffman 1959).

The **back stage** is the area out of the audience's sight, where individuals let their guard down and do things that would be inappropriate or unexpected in a front-stage setting. Because backstage behavior frequently contradicts front-stage behavior, we take great care to conceal it from the audience. In the back stage, a "person can relax, drop his front, forgo his speaking lines, and step out of character" (Goffman 1959, 112).

▶ The division between front stage and back stage can be found in nearly every social setting. Often that division is separated by a door or sign signaling that only certain people can enter the back stage without permission or knocking.

Lisa Southwick

Goffman uses a restaurant as an example of a social setting that has clear boundaries between the back stage and front stage. Restaurant employees may do things in the kitchen, pantry, and break room (back stage) that they would never do in the dining areas (front stage), such as eating from customers' plates, dropping food on the floor and putting it back on a plate, and yelling at one another. Once they enter the dining area, however, such behavior stops. Of course, a restaurant is only one example of the many settings in which the concepts of front stage and back stage apply.

Goffman maintains that the self "is a product of a scene that comes off." To put it another way, the self is a performed character, rising out of performances (1959, 252–253). The self cannot be thought of as a lone actor; rather, it depends on a supporting cast and a "stage." For example, a firefighter exists only in relation to a social setting that includes a fire station, a burning building, other firefighters, and victims. Goffman (1959) argues that any analysis of interaction cannot focus on the individual per se (Kivisto and Pittman 2007).

(Write a Caption)

Write a caption that uses the language of front stage and back stage to describe how people respond when they lose their homes. The makeshift tent city was constructed by the people of Haiti after the 2010 earthquake left millions homeless.

Hints: In writing this caption

- review the distinction between front stage and back stage,
- think about how the home functions as a back stage for people, and
- consider what the makeshift tent restores for people in terms of front-stage and backstage dynamics.

Critical Thinking

Describe an anxiety dream you have had. Describe it in terms of the scenes that failed and the "teammates" who did not cooperate.

Key Terms

back stage
front stage
dramaturgical sociology

MODULE 5.4

Ethnomethodology

Objective

You will learn an observational, investigative method of studying how people construct social order.

How would you react if you saw a mother lighting a cigarette while holding a toddler?

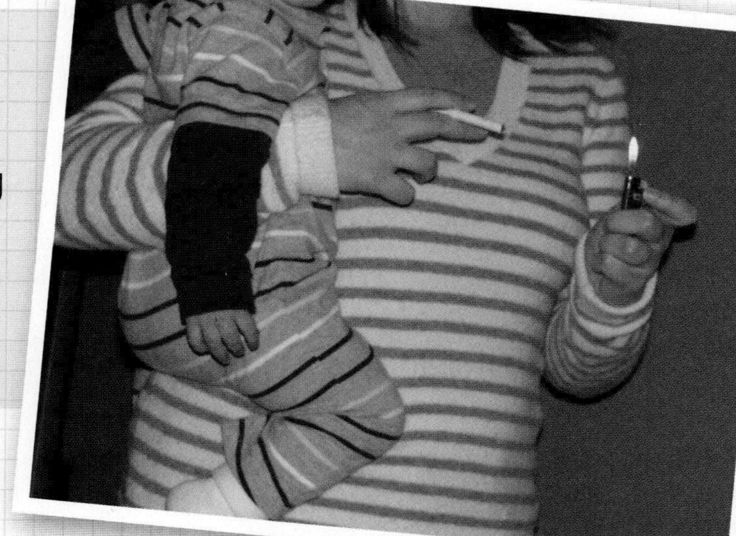

Missy Gish

Most of the time our encounters with others can be considered simply ordinary and routine. That is, we have little reason to question what is going on. Think about the number of times we see a parent carrying a small child and take for granted that parents want to protect their children from harm. We think to question that assumption only when something is out of place, as when we see parents smoking in front of their children. It is the taken-for-granted quality of most social encounters that makes the commonplace "impervious to deeper analysis" (Cuzzort and King 2002, 307). The commonplace lulls us into complacency.

Ethnomethodology

Sociologist Harold Garfinkel coined the term *ethnomethodology* more than 50 years ago. **Ethnomethodology** is an investigative and observational approach that focuses on how people make sense of everyday social activities and experiences. Ethnomethodologists assume that people work at making social encounters meaningful. Social encounters are meaningful when both parties share essentially the same understanding of the situation. Ethnomethodologists seek to penetrate a reality that those caught up in it cannot even begin to question (Cuzzort and King 2002). Garfinkel maintains that the only way we can possibly get at the structure and character of this social order is by disrupting expectations. Once expectations are disrupted, ethnomethodologists observe how people react and/or take action to restore normalcy.

Disrupting Social Order

Disrupting the social order is perhaps the best known investigative technique ethnomethodologists employ. By disrupting expectations, and then observing how the participating parties handle that disruption, ethnomethodologists gain insights into the work people do to maintain the order they know. To reveal such efforts, ethnomethodologists ask, "What can be done to make trouble . . . to produce and sustain bewilderment, consternation, and confusion; to produce the social structured effects of anxiety, shame, guilt, indignation; to produce disorganized interaction?" (Garfinkel 1967, 37–38). As one example, Garfinkel asked his students to engage an acquaintance in conversation and insist that that person clarify everything he or she says by asking "what do you mean?" or "would you please explain?"

> **Acquaintance:** (waving hand cheerily) How are you?
>
> **Student of Garfinkel:** How am I in regard to what? My health, my finances, my school work, my peace of mind, my . . . ?
>
> **Acquaintance:** (red in face and suddenly out of control) Look I was just trying to be polite. Frankly I don't give a damn how you are! (1967, 44)

Garfinkel also instructed his students to engage someone in conversation and to pretend that the chosen person was deliberately misleading them in some way.

Lisa Southwick

➤ Would you get on this bus and then act as if you did not trust the bus driver to take you to the right location? That is what one of Garfinkel's students did. She chose to act as if the bus driver was not forthcoming about whether the bus was on a route that passed by a certain street. After asking the bus driver and then still seeking more reassurance, the bus driver responded, "Look lady I told you once, didn't I? How many times do I have to tell you?" (1967, 52).

In another case, the student chose to question her husband about his reasons for working the evening before and about whether he was really at a poker game several days earlier as he claimed. Garfinkel noted that only 2 of his 35 students chose to engage strangers. "Most admitted being afraid to carry this assignment out with a stranger thinking things could get out of hand." So his students selected "friends, roommates, and family members" as subjects (1967, 51). Even then, students reported rehearsing repercussions in their imagination.

Other assignments Garfinkel devised included asking his students to

- initiate a game of tic-tac-toe with someone. After the chosen opponent made the first move, the student should erase the mark and move it to another square; or
- engage a person in an ordinary conversation. Sometime later in the conversation, the student should show a device recording the conversation and announce "See what I have."

Garfinkel noted that his students found it very difficult to disrupt routine expectations. Depending on the specific experiment, students expressed a range of emotions, including anxiety, distrust, hostility, anger, frustration, and isolation. Garfinkel related his students' reluctance to disrupt expectations to **trust**, the taken-for-granted assumption that in a given social encounter others share the same expectations and definitions of the situation and that they will act to meet those expectations. When one or more of the involved parties, including the student, are put in a position in which they are forced to violate those expectations and/or to mistrust another, the relationship becomes problematic, deteriorates, and eventually collapses.

(Write a Caption)

Write a caption that describes a Garfinkel-inspired field experiment that could be assigned to students enrolled in one of his classes. Give directions for how to go about disrupting the taken-for-granted order between a waitress and her customers.

Lisa Southwick

Hints: In writing this caption

- review the kinds of disruptive interventions Garfinkel's students devised; and
- think about "what can be done to make trouble . . . to produce and sustain bewilderment, consternation, and confusion; . . . to produce disorganized interaction."

Critical Thinking

Imagine you were a student in Garfinkel's class. How would you feel about the kinds of assignments he required? Which of the Garfinkel assignments described in this module would you feel most comfortable carrying out? The least comfortable? Why?

Key Terms

ethnomethodology

trust

MODULE

5.5 Constructing Identities

Objective

You will learn that the groups to which we belong, aspire to belong, and do not belong play important roles in constructing our sense of self and the way we see what is going on in the world.

Stephen Oertwig, U.S. Army, Pacific Public Affairs

Lisa Southwick

Tony Rotundo

In high school which group did you most associate with? Was there a group to which you aspired to belong but, for whatever reason, did not or could not join?

Think about how the groups to which you belong or aspire to belong affect your sense of self and others' views of you. Think also about the groups you do not belong to, would never aspire to join, or are barred from joining. Groups influence a person's sense of self on three levels—a cognitive level (awareness of being or not being a member), an affective level (feelings of belonging or being

rejected), and an evaluative level (perceived value/importance of the group to self-identity and to the larger society) (Christian et al. 2012). In these ways, groups play important roles in constructing identities and social reality. Simply think how one's identity, view of the world, and conception of place in that world varies depending on whether someone is a member of a championship football team, a girls' science club, or the school band. In this module we consider the concepts of reference groups and ingroups and outgroups to understand how groups shape identity and assessments of self and others.

Reference Groups

Sociologists define a **reference group** as any group whose standards people take into account when evaluating something about themselves or others, whether it be personal achievements, aspirations, or their current place in life. In this regard the family is an important reference group, as are classmates, teammates, and coworkers. A person does not have to belong to a reference group to be influenced by its standards. A reference group can also be a group to which a person belonged to in the past or hopes to belong to. There are three broad types of reference groups: normative, comparison, and audience (Kemper 1968).

Normative reference groups provide people with norms that they draw upon or consider when evaluating a behavior, appearance, or a course of action. The mark of a normative reference group's influence is that the person simply takes its norms into consideration; in the end the person may follow those norms or pursue a contradictory course of action (Kemper 1968). Regardless, the person takes the group's norms into account.

Comparison reference groups provide people with a frame of reference for

- judging the fairness of a situation in which they find themselves (i.e., "Everyone at work earns $15 per hour while I only earn $10 per hour. What's up with that?");
- rationalizing or justifying their actions or a way of thinking (i.e., "All taxpayers cheat the government, why should I be the exception?" "Everyone I work with takes home office supplies; it's a way to make up for the low salaries we earn."); and
- assessing the adequacy of their performance relative to others, as when a student earns a 68 on a chemistry test and feels terrible until she assesses that grade in light of her peers' test scores. When she learns that the class average was a 48 and the highest grade was a 70, she feels pretty good. With the curve, a 68 is an A.

Audience reference groups consist of those who are watching, listening, or otherwise paying attention to someone. In addressing an audience, people consider what they believe that audience wants or needs and then adjust the message accordingly (Kemper 1968). Candidates running for elected office often alter their message depending on the audience. Imagine a politician scheduled to speak before an audience of college students and then to an audience of retired adults. The politician considers issues important to each audience group, then tailors the message accordingly. When speaking to college students, the politician is likely to mention efforts to increase access to college loans and grants. When speaking to retired adults, the candidate is sure to mention efforts to safeguard Social Security benefits.

Nonmembers as a Reference Group

When thinking about groups and their importance in shaping social identity, we often overlook the significant role played by those who don't belong—the nonmembers. Nonmembers include those who are eligible to join but, for whatever reasons, do not, as well as those who are ineligible. From the group's point of view, it is those eligible to join and those who have left the group (ex-members), who reduce its completeness. A group is complete when everyone who is eligible joins and no one leaves the group (Simmel 1950).

© keith morris/Alamy

➤ The fact that only 12.3 percent of American workers belong to unions calls attention to those who could join but have not. Nonmembers include those workers who might be afraid to join, those who outright reject union membership, and those who work to dismantle or resist unions.

In the case of unions, the more complete a group, the greater its power and influence (Merton 1957b). Those eligible who outright refuse to join cast doubt on a group's purpose and value and reduce the group's power and influence (Merton 1957b).

Ingroup-Outgroup Dynamics

A group distinguishes itself by the symbolic and physical boundaries its members establish to set it apart from nonmembers. Examples of physical boundaries may be gated communities, special buildings, or other distinct locations. Symbolic boundaries include membership cards, colors, or dress codes such as a uniform. A group also distinguishes itself by establishing criteria for membership—e.g., members must be of a certain race, possess a particular characteristic (interest in the environment), or have achieved some accomplishment (a certain level of education, passing grade on an entrance exam).

Tony Rotundo

Chris Caldeira

▲ Can you identify the symbolic boundaries that distinguish each of these two groups from nonmembers? What do you imagine the membership criteria to be for each? Such criteria are ways groups establish identities and distinguish "us" from "them."

Sociologists use the terms *ingroup* and *outgroup* in reference to intergroup dynamics. An **ingroup** is the group that a person belongs to, identifies with, admires, and/or feels loyalty to. An **outgroup** is any group to which a person does *not* belong. Obviously, one person's ingroup is another person's outgroup. Ingroup formation is built on established boundaries and membership criteria. Ingroup members think of themselves as "us" in relation to some "them" (Brewer 1999).

Michael Heckman, Fort Hood Sentinel

➤ "Us" versus "them" dynamics create a sense of oneness that is especially evident when an ingroup and outgroup compete for some valued outcome—in this case winning a tug-of-war game.

Chris Caldeira

▲ Ingroups and outgroups are everywhere. This gated community is advertising for an ingroup who can afford to buy an executive home. The emphasis on "gated" sends the message that walls, fences, and/or guards control outgroup access to the community.

Depending on the circumstances, ingroup and outgroup dynamics may be characterized as indifferent, cooperative, competitive, or even violent. When sociologists study ingroup and outgroup dynamics they ask, Under what circumstances does the presence of an outgroup unify an ingroup and create an us-versus-them dynamic? Three such circumstances are described.

1. *An ingroup assumes a position of superiority.* **Moral superiority** is the belief that an ingroup's standards represent the only way. In fact, there is no room for negotiation and there is no tolerance for other ways (Brewer 1999). Ingroup members can convey feelings of superiority when they refuse to interact with those who are part of an outgroup, or they establish laws segregating those in an outgroup from those in the ingroup. Such feelings of superiority figure prominently in carrying out genocide against an outgroup.
2. *An ingroup perceives an outgroup as a threat*. In this situation an ingroup believes (rightly or wrongly) that an outgroup threatens its way of life. The ingroup holds real or imagined fear that the outgroup is taking steps to increase its political power or its access to and control over some scarce and valued resource such as jobs.

Chris Caldeira

◂ Real or imagined belief fuels fear and hostility. The motel owner is using an implicit ingroup–outgroup dynamic of American- versus foreign-owned to advertise his motel to potential customers. The suggestion is that "American-owned" is in danger of disappearing. In the United States, an estimated 50 percent of all motels are owned by people of Indian (as in India) origin. Of course, that owner may still be a U.S. citizen or even American-born (Varadarajan 1999; Asian American Hotel Owners Association 2009).

3. *Ingroup–outgroup tensions may be evoked for political gain.* Those with political ambitions may deliberately evoke ingroup–outgroup tensions as a strategy for mobilizing support for some political purpose (Brewer 1999). Thus, a candidate running for elected office may, as a strategy for rallying support for his or her campaign, declare an outgroup such as undocumented workers or gays a threat to the American way of life.

(Write a Caption)

Write a caption that relates this sign to ingroup–outgroup dynamics.

Hints: In writing this caption

- identify the specific ingroup and outgroup involved,
- think about who are the suspicious groups being watched, and
- consider ways in which being on the lookout for an outgroup affects an ingroup's identity.

Chris Caldeira

Critical Thinking

Give a specific example of an ingroup–outgroup conflict of which you are a part.

Key Terms

audience reference groups
comparison reference groups
ingroup
moral superiority
normative reference groups
outgroup
reference group

Applying Theory: Phenomenology

MODULE 5.6

Objective

You will learn to apply the theory of phenomenology to understanding ways people construct reality.

Imagine you are the parents of this transgender child, who has the anatomy of a boy but insists she is a girl. How would you react to and make sense of the situation? More importantly, how does this child make sense of it?

© Jeff Greenberg/Alamy

The social construction of reality (the process by which people make sense of the world) fits within the theoretical tradition known as **phenomenology**, an analytical approach that focuses on the everyday world and how people actively produce and sustain meaning (Appelrouth and Edles 2007).

The Social Construction of Reality

Sociologists Peter L. Berger and Thomas Luckmann wrote what is now considered a classic book, *The Social Construction of Reality.* The two sociologists offer a "sociological analysis of the reality of everyday life" that emphasizes the knowledge people draw on to create the reality upon which they act (1966, 19). Berger and Luckmann are interested in understanding the forces that determine how the world appears to people; it does not matter if that world is "real." In this regard Berger and Luckmann made a number of points about how people construct reality.

1. ASSIGNED MEANINGS. For the most part, everything going on around us was named and assigned meaning before we arrived on the scene. We live in a geographic location that has a name and a reputation, and we use tools with names like "can opener" and "computer." We are part of a web of human relationships, including teacher–student, employee–coworkers, mother–daughter, and so on. Our world is already ordered; the things in it have names and assigned meanings.

Tony Rotundo

➤ We take for granted that most of the things in our day-to-day environment have a name and a meaning, until we encounter something that has a name and/or meaning with which we are not familiar. Most Americans have assigned a meaning to a grasshopper that rarely evokes the meaning "food."

2. ZONE PROXIMITY. Berger and Luckmann divide the reality of everyday life into a continuum of zones. At one end of the continuum are the zones closest to us: the zones in which we actually live our lives—home, work, school, neighborhood—and experience directly. At the other end of the continuum are the zones that are most remote and farthest from our direct experience. People give greatest attention to the zones closest to them because their interest is heightened by "what they are doing, have done or plan to do in it" (1966, 22). Typically, people are less interested in the remote zones because what is going on there does not directly affect them and is perceived as less urgent than what is going on in the zones closest to them.

Tech Sgt. Cecilio M. Ricardo Jr.

⮜ Some critics maintain that because only a handful of political leaders have sons and daughters serving in the military, the consequences of decisions to deploy troops are relegated to remote zones of their lives. Arguably politicians would make different decisions, say about military involvement, if their children were to be deployed. Imagine whether Presidents George Bush or Bill Clinton, both of whom had military-age children while in office, would make the same decisions about combat if their daughters were deployed to carry out those orders.

3. ROUTINE. Everyday life can also be divided into a continuum of sectors, from the most routine to unusual. The **routine** includes the usual ways of thinking and doing things. The unusual includes any unanticipated disruption to routine that challenges "reality." As long as the routine is not disrupted, there is no need to question reality. Berger and Luckmann caution, however, that when the routine is disrupted, people work to keep their sense of reality intact. One illustration comes from the situation of transgender children. Many parents of these children report receiving advice that "if you would stop letting your kid dress like a girl, it wouldn't act like a girl" (Kirchner and Wyker 2009). Such advice fails to recognize that "it's not some obsessive craziness, it's not some mania. It's not something he saw on television and decided to imitate. This is something that every morning he wakes up and he's living it, he's experiencing it. This is his experience; this is what's going on inside his brain and soul" (Kirchner and Wyker 2009).

4. BARRIERS TO CHALLENGING "REALITY." It is not easy to challenge reality if only because the language we use to describe and think about it reinforces that reality. While the existence of transgender people challenges established reality, the language we have to think about the transgender experience supports a binary reality of male and female. So in talking about their situation, transgender children can only draw upon existing words that define the world as being made up of boys or girls. As one transgender child tried to explain the situation, "I'm mad at God because I'm a girl, but I'm not" (Kirchner and Wyker 2009).

5. MENTAL FRAMEWORKS. We draw upon typificatory schemes to organize the world and our relationships. **Typificatory schemes** are mental frameworks that prompt people to place what they observe into preexisting social categories with essential characteristics. For example, we possess a typificatory scheme about what it means to be brothers. Whether we know it or not, when we learn that we are about to meet two people who are brothers, we have some essential characteristics in mind—the two should physically resemble one another, be the offspring of the same parents, share memories of growing up, and be of the same race. These thoughts seem obvious only until we meet brothers who possess characteristics that violate one or more of these assumptions.

Missy Gish

◄ Generally, most people assume that brothers will appear to be the same race. Upon seeing these brothers, most people would immediately question whether they are half brothers or whether one is adopted. They would label the pair as an exception to the rule, as a way of keeping guiding assumptions intact. Thus, it is unlikely that they would question their assumption that brothers should share a race.

6. OBSERVATIONS FROM CLOSE AND FAR. Social reality encompasses a continuum of experiences with people. At one end of the continuum are those with whom we frequently and intensively interact—our inner circle. At the other end of the continuum are those anonymous others with whom we will likely never interact; they have almost no power to alter typifications we hold about them because we observe them from afar primarily through the lens of the media. As one example, young-to-middle-age people often typify those considered elderly as having any number of shortcomings, but their relationship with a grandparent they care deeply about can challenge those typifications.

7. IMAGINING PAST AND FUTURE. Typifications are not just applied to those who exist in the present. We also apply typifications to those who have preceded and will follow us in history. We relate to our predecessors through typifications that are often projections of an imagined reality; that is, we may describe our great-grandparents as immigrants who worked hard to realize the American dream.

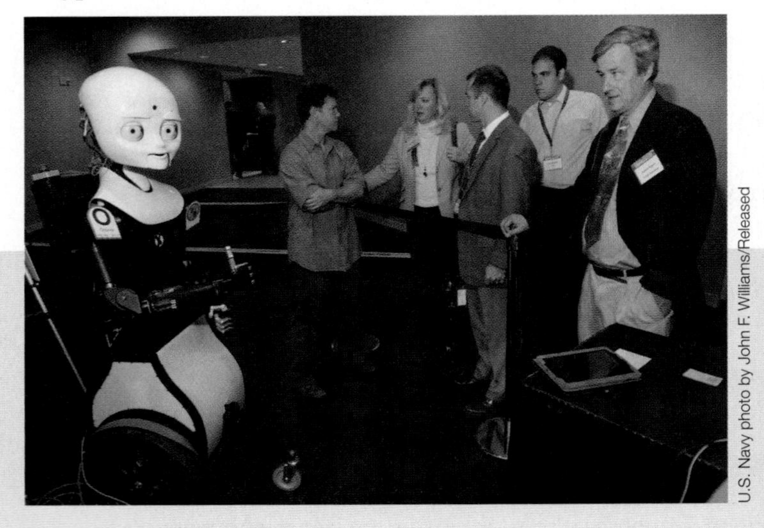
U.S. Navy photo by John F. Williams/Released

◂ Typifications made about future generations are the projections of a current generation's hopes and worries. Some project a future where androids (robots with human qualities) interact seamlessly with human counterparts.

8. TIME. The calendar places us in the context of history (born on a particular day, graduated from high school in a particular year, joined the army at a particular age). Whether we know it or not, our place in history informs our daily reality. Consider how those who wake from a coma are compelled to locate themselves in time and "re-enter the reality of everyday life" by asking what day it is, and what year. Others help by filling them in on key events that occurred.

Critical Thinking

Imagine what you would say to someone who fell into a coma and came to ten years later. What events would you have to fill him or her in on?

Key Terms

phenomenology routine typificatory schemes

Summary: Putting It All Together

CHAPTER 5

The process by which people make sense of the world is known as the social construction of reality. When people observe something, they assign a meaning, informed by past experiences. That meaning becomes the basis for a response. The Thomas theorem specifies the link between assigned meaning and response: if people "define situations as real, they are real in their consequences." However, when people assign false definitions to a situation it can result in a self-fulfilling prophecy.

Most people form judgments about cause and take action without complete information. When people try to explain something, they tend to focus on one of two types of causal factors: dispositional and situational. Usually, people stress situational factors to explain their own failures and dispositional factors to explain others' failures. But with regard to success, people point to dispositional factors to explain their successes and situational factors to explain the success of others.

The dramaturgical model offers insights about the role people play in constructing reality and in managing impressions. That model views social encounters as if they were theater and the involved parties as if they were actors performing roles on a stage. Like actors, they must manage impressions and convince their "audience" that they are who they appear to be or who they say they are. Other related concepts include front stage, back stage, and team.

Ethnomethodology is an investigative and observational approach that gives attention to how people make sense of everyday social activities and experiences and collaborate with others to create, give, and sustain order to social life. To understand what holds social order together, ethnomethodologists disrupt that order and then observe how the participating parties respond to that disruption.

Social groups play important roles in helping people to construct their identities and to assign meaning to what is going on around them. In particular, people use reference groups to evaluate personal achievements, aspirations, or circumstances. Reference groups encompass three types: normative, comparison, and audience. Ingroup and outgroup dynamics also play key roles in helping people to construct reality. When sociologists study ingroup–outgroup dynamics they ask, Under what conditions does the existence of an outgroup create an us-versus-them dynamic?

Within sociology, phenomenology is a theoretical perspective that emphasizes the knowledge people draw on to create the reality upon which they act. Specifically, the perspective emphasizes the factors that determine how the world appears to people; it does not matter if that world is "real." For one, everything going on around us has been assigned a name and meaning before we arrive on the scene. In addition, people do not question the parts of their lives that have become routine until what is routine is disrupted in some way, and even then people find ways to restore the routine.

Go to cengagebrain.com to link to Aplia and CourseMate for the chapter quiz and other activities.

Courtesy of Linda Sarna, UCLA School of Nursing

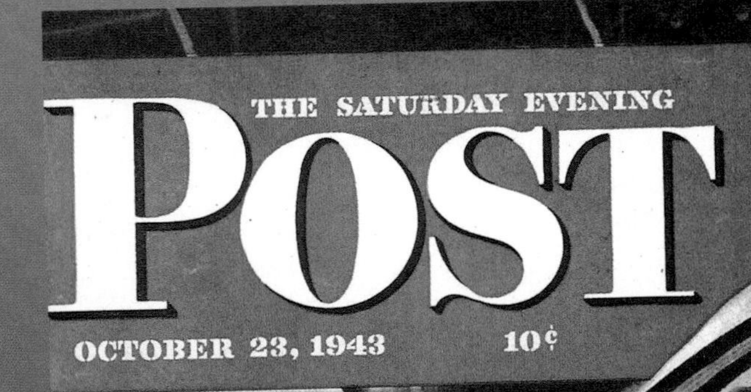

CHAPTER 6

DEVIANCE

This 1943 magazine cover would be unthinkable today because clearly any nurse who encouraged a hospital patient to smoke would be considered deviant. While one might be tempted to argue that smoking is now considered deviant for good reasons, this argument fails to consider that negative health effects of smoking, even secondhand smoke, were known as early as 1900. Yet those warnings were not seriously heeded for almost 80 years. The then-prevailing voices of tobacco advertisers convinced men and eventually women that smoking was healthy, even liberating. Sociologists are interested in the process by which behavior gets defined as deviant; that process is usually not a simple one.

Go to Sociology CourseMate on cengagebrain.com to watch a video on the mechanisms of social control that North Korea uses to get its people to conform and behave in ways that correspond with government ideology.

MODULE

6.1 Defining Deviance

Objective

You will learn how sociologists define deviance and how almost any behavior or appearance can be defined as deviant depending on context.

U.S. Navy photo by Mass Communication Specialist 2nd Class Joshua Mann/Released

When you think of military homecomings, do you think of two women engaged to be married greeting each other with a kiss and embrace?

The U.S. Navy released this photograph on December 21, 2011. This kiss attracted national, even global, attention because it signaled a historical shift in military policy ushered in by the repeal of "Don't Ask, Don't Tell" (DADT) signed into law on December 22, 2010. The law allowed the military some time to adjust. On September 20, 2011, the military indicated that gays could now openly serve in the military without fear of discharge. The United States joined 25 countries that explicitly allow gays to serve in the military (Frank 2010; see Figure 6.1a). The repeal of DADT reveals an important fact about the nature of deviance: what is considered deviant at one time and place may not be considered deviant at another time or place.

▼ **Figure 6.1a: Countries In Which Gays Can Join the Military and Be Open about Their Sexual Orientation**

The countries highlighted on this map explicitly allow gays to serve openly in the military. The United States officially joined this group September 20, 2011. Something that the U.S. military once formally punished is now no longer subject to disciplinary action. One should not conclude that the countries highlighted are uniform in their treatment of gays. As examples, Estonia has never had a ban on gays serving in the military; Germany removed its ban in 2001; and while the Slovenian military does not ban homosexuality, its medical community labels homosexuality a psychiatric disease.

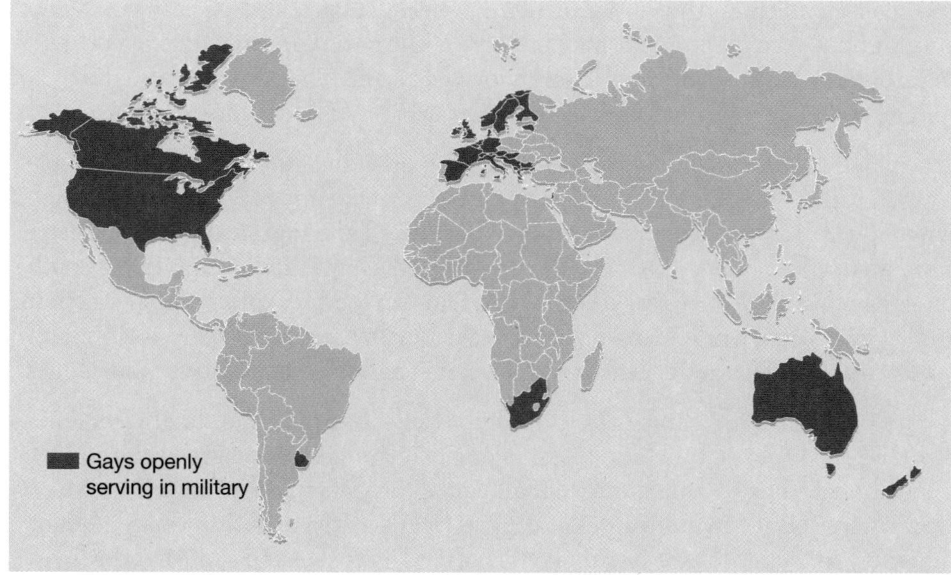

Source of data: Frank (2010)

Defining Deviance

Deviance is any behavior or physical appearance that is socially challenged and/or condemned because it departs from the norms and expectations of some group. **Norms** are rules and expectations for the way people are supposed to behave, feel, and appear in a particular social situation. Norms exist for virtually every kind of situation, including how many times a day to eat, how to greet a friend, what to wear to school, how to handle the American flag, and when to use a gun.

Norms can vary according to whom they apply to and whether people (1) know they exist, (2) accept them, (3) enforce them uniformly, (4) think them important, (5) back them up with the force of law, and (6) adhere to them in their public *and* private lives (Gibbs 1965). Consider speed limits, which are backed by the force of law. Most people know the speed limits from observing posted signs but, depending on the setting, do not find it important to follow them to the letter of the law. In fact, most drivers exceed posted limits by 10 or 15 miles per hour without fear of getting caught and, even when caught, are not always cited for the exact number of miles they were driving over that limit. Norms vary by group. Some people, depending on the groups with which they identify, celebrate tattoos as a normal or expected rite of passage; others treat tattoos as a broad indicator of some character flaw.

Some norms exist for seemingly valid reasons—for example, to prevent harm to self and/or others. But there are just as many examples of norms in place for which there seems to be no valid reason. People often overlook behavior that is objectively deviant, such as corporate crime, but take sharp notice of behavior that is for all practical purposes harmless, such as a certain appearance, way of dressing, or way of expressing sexuality (K. Erikson 1966).

The Sociological Perspective

The definition of deviance given earlier suggests that what makes something deviant is the presence of a social audience that regards a behavior or appearance as deviant and takes some kind of action to discourage it. Deviance is not inherent to a specific behavior. Marrying a first cousin, for example, is not in itself deviant; if so, that behavior would be deviant everywhere in the world. Deviance is something that is conferred. For sociologists the critical factor is not the individual who violates norms per se; it is the social group's response that ultimately determines whether a behavior is deviant. Thus, we can say that deviance exists in relation to norms in effect at a particular time and place.

The sociological contribution to understanding deviance lies not with studying the deviant individuals, but with studying the context under which something is deemed deviant. Sociologists note that almost any behavior or appearance can qualify as deviant under the right circumstances. How is it that anything can be defined as deviant? Émile Durkheim offered an intriguing explanation. Durkheim (1901) argued that while ideas about what is deviant vary, deviance is present in all societies. He defined deviance as those acts that offend collective sentiments.

The fact that there are some acts that offend collective sentiments always and everywhere led him to conclude there is no such thing as a society without deviance. According to Durkheim, deviance will be present even in a "community of saints" (100). Even in a seemingly perfect society, acts that most persons would view as insignificant or minor may offend, create a sense of scandal, or be treated as crimes. To explain this, Durkheim (1901) drew an analogy to the "perfect and upright" person.

Just as "perfect and upright" people judge their own smallest failings with a severity that others reserve for the most serious offenses, so too do those who belong to groups considered exemplary. In such societies, some act will offend, simply because "it is impossible for everyone to be alike if only because each of us cannot stand in the same spot" (100). Thus, what makes an act or appearance deviant, even criminal, is not so much the act itself or its consequences, but rather the fact that the group has defined it as something dangerous or threatening to its well-being.

➤ The military represents a setting in which acts that most people would view as minor offenses, if that, are treated as crimes. Facial hair or a chin strap slightly off angle takes on critical significance for new military recruits in basic training.

U.S. Army

Durkheim argued that the ritual of identifying and exposing a wrongdoing, determining a punishment, and/or carrying it out is an emotional experience that binds together the members of a group and establishes a sense of order and community. Durkheim maintained that a group that went too long without noticing deviance or doing something about it would lose its identity as a group.

Who Defines What Is Deviant?

In answering the question "Who defines what is deviant?" sociologists focus on the ways in which specific groups (such as undocumented workers), behaviors (such as child abuse), conditions (such as teenage pregnancy, infertility, or pollution), or artifacts (such as song lyrics, guns, art, or tattoos) become defined as problems. In particular, sociologists examine claims makers and claims-making activities.

Claims makers are those who articulate and promote claims and who tend to gain in some way if the targeted audience accepts their claims as true. Claims makers include government officials, advertisers, scientists, professors, and other stakeholders. Claims-making activities are actions taken to draw attention to a claim—actions such as "demanding services, filling out forms, lodging complaints, filing lawsuits, calling press conferences, writing letters of protest" (Spector and Kitsuse 1977, 79). Studying claims-making activities can help us understand why smoking in public places has been largely banned in the United States. Before 1980, people could freely smoke in public. In fact, professors and students smoked during class without eliciting a raised eyebrow. Today, such behavior would be unthinkable.

Prints & Photographs Division, Library of Congress, LC-USZ62-89928

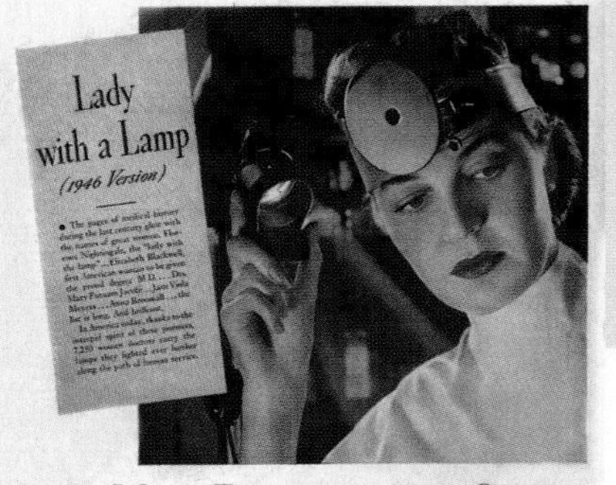

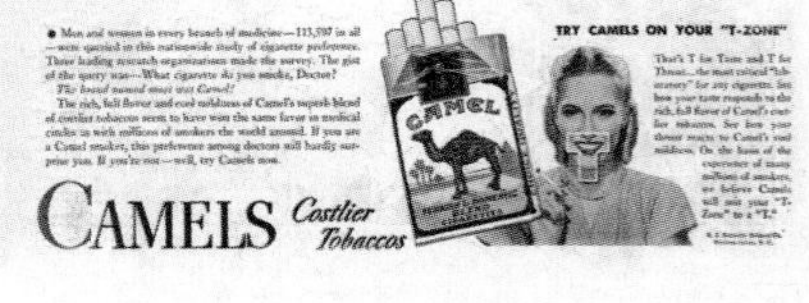

Apic/Getty Images

➤ Negative health effects of smoking were claimed as early as 1900, as evidenced by this poster (top) showing a skeleton rising from the smoke of a burning cigarette. However, other claims makers (bottom) argued successfully for decades that because even doctors smoke, it must be safe.

The success of a claims-making campaign depends on a number of factors: claims makers' access to the media; available resources; position in society; and skill at fund-raising, promotion, and organization. When sociologists study the process by which a group or behavior is defined as deviant, they focus on who makes claims, whose claims are heard, and how audiences respond to them (Best 1989).

Sociologists also pay attention to any labels that claims makers apply because they tend to evoke a specific cause, consequence, and/or solution to a problem (Best 1989). For example, labeling an addiction—whether it be to gambling, credit card use, prescription drugs, or alcohol—as a medical problem is to locate the cause in the biological workings of the body or mind and to suggest that the solution rests with a drug, a vaccine, or surgery. Labeling an addiction as a personal failing, on the other hand, is to locate the cause in the character of the person, such as an inability to delay gratification or a lack of discipline.

(Write a Caption)

Write a caption that relates this "reserved for drunk driving victim" display to the concept of claims making.

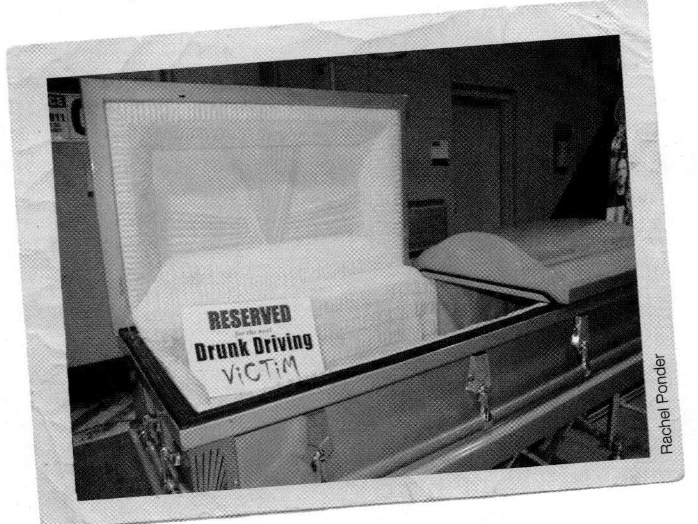

Rachel Ponder

Hints: In writing this caption

- review the concept of claims making, and
- consider the labels employed to bring attention to the problem of drunk driving.

Critical Thinking

Describe something that your parents claim was deviant in their lifetime but is no longer the case today, or something that is deviant today but was not considered so in the past.

Key Terms

claims makers   deviance   norms

Mechanisms of Social Control

MODULE 6.2

Objective

You will learn the mechanisms groups use to elicit conformity and punish deviance.

What is a referee's purpose? Can you think of a nonsport activity that has the equivalent of a referee?

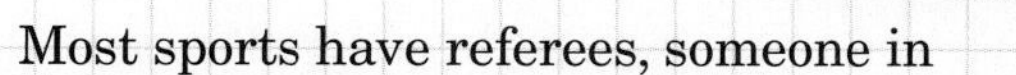

Tony Rotundo

Most sports have referees, someone in a position of authority to watch athletes (and coaches and fans) to make sure they abide by the rules. If you stop and think about it, most human activities have the equivalent of referees. "Referees" include bystanders who stare in a disapproving manner when they observe someone doing something they consider odd. "Referees" also include teachers who watch students as they take exams and police officers who arrest people they believe have committed crimes.

Referees employ what sociologists call **mechanisms of social control**, strategies people use to encourage, often force, others to comply with social norms. For example, a referee may call a foul when an athlete violates rules of the game, or, if the violation is deemed serious enough, even eject that athlete from the game. Ideally, from society's point of view, people should want to conform. When conformity cannot be achieved voluntarily, mechanisms of social control are employed. Methods of social control include positive and negative sanctions, engaging in censorship or surveillance, exercising authority, and applying group pressure.

Sanctions

Sanctions are reactions of approval or disapproval to behavior that conforms to or departs from group norms. Sanctions can be formal or informal and positive or negative. **Formal sanctions** are reactions backed by laws, rules, or policies specifying the conditions under which people should be rewarded or punished for specific behaviors. By contrast, **informal sanctions** are spontaneous, unofficial expressions of approval not backed by the force of law or official policy. **Positive sanctions** are expressions of approval for compliance. In contrast, **negative**

sanctions are expressions of disapproval. Police officers employ negative formal sanctions when they arrest people suspected of breaking the law. Teens employ informal sanctions when they share images via cell phones with the intent of inviting people to praise or ridicule a particular person's behavior.

Jeremiah Evans

➤ From a sociological point of view, this restaurant is announcing that women who breast-feed their babies will not experience negative informal sanctions (staring, whispers, verbal demands to stop) from patrons and the staff. In addition, the sign is essentially inviting in a social audience that approves of and supports public breast-feeding.

Censorship and Surveillance

Censorship is an action taken to prevent information believed to be sensitive, unsuitable, or threatening from reaching some audience (e.g., children, voters, employees, prisoners, or others). Censorship relies on censors—people whose job it is to remove or block access to information deemed problematic in movies, books, letters, e-mail, TV, the Internet, and other media. Reporters Without Borders (2010), which ranks countries according to the level of Internet freedom, named the following governments as engaging in the most extensive Internet censorship: Saudi Arabia, Burma, China, North Korea, Cuba, Egypt, Iran, Uzbekistan, Syria, Tunisia, Turkmenistan, and Vietnam.

Surveillance, another mechanism of social control, involves monitoring movements, conversations, and associations with the intent of catching people in the act of doing something wrong. Surveillance activities include tapping phones; intercepting letters, e-mail, and documents; videotaping/recording; and electronic monitoring.

Cpl. Giovanni Lobello

◄ With the widespread availability of portable technologies that can record voice and capture images, there is no situation in which a person is potentially free from the possibility of surveillance.

Sociologist Kingsley Dennis (2008) describes the rise in new forms of personal surveillance in which people are able to monitor others through portable devices such as mobile phones. Dennis cites 1992 as the year this technology came of age. In that year a bystander videotaped Los Angeles police officers brutally beating Rodney King, a black man who led police on a 100-mile-per-hour car chase and, after being stopped, appeared to disobey police commands to lie down. The tape also showed that officers appeared to use excessive force. The acquittal of three participating officers triggered five days of protest, turned violent, in South Central Los Angeles, resulting in dozens of deaths, thousands of injuries, and $900 million in property damages.

Dennis (2008) points out that in the hands of responsible parties, personal recording devices can be liberating tools allowing the public to record wrongdoing. In the wrong hands, such devices can also be used to harass, stalk, intrude, and otherwise ruin lives. Dennis offers one example of what he calls virtual vigilantism that occurred in South Korea after a girl who became known as the "Dog Shit Girl" failed to pick up after her dog when it defecated on the floor of a subway train. A bystander used his mobile phone to photograph the girl, dog, and droppings and posted them online to mobilize the Internet-based community to humiliate her as punishment. The photo became a globally shared story. The girl eventually issued an apology over the Internet saying, "I know I was wrong, but you guys are so harsh. I regret it, . . . if you keep putting me down on the Internet I will sue all the people and at the worst I will commit suicide" (350).

Obedience to Authority

In his now-classic study, social psychologist Stanley Milgram (1974) wanted to learn why some people obey an authority figure's command to behave in ways that conflict with their conscience. Milgram designed an experiment to see how far people would go before they would refuse to conform. He placed an ad in a local paper asking for volunteers. When participants arrived at the study site—a university—they were greeted by a man in a laboratory jacket who explained to them and another apparent volunteer that the study's purpose was to find out whether the use of punishment improves a person's ability to learn. Unknown to each participant, the apparent volunteer was actually a confederate—someone working in cooperation with the investigator. The participant and the confederate drew lots to determine who would be the teacher and who would be the learner. The draw was fixed, however, so that the confederate was always the learner and the real volunteer was always the teacher. In Milgram's experiments, the confederate-learner was strapped to a chair and electrodes were placed on his wrists. The volunteer-teacher, who could not see the learner, but could hear him moan and scream, was placed in front of an instrument panel containing a line of shock-generating switches. The switches ranged from 15 to 450 volts and were labeled accordingly, from "slight shock" to "danger, severe shock."

The researcher explained that when the learner made a first mistake, a 15-volt shock would be administered. With each subsequent mistake, the teacher should increase the voltage. The "teacher" had no idea that the learner was a confederate and was not actually being shocked.

Missy Gish

➤ At one point during the experiment, the learner would even say that his heart was bothering him and would go silent. If a volunteer-teacher expressed concern, the researcher firmly said to continue administering shocks. In each case, as the strength of the shock increased, the learner expressed greater discomfort. Although many teachers protested, a substantial number (65 percent) obeyed and continued "no matter how painful the shocks seemed to be, and no matter how much the victim pleaded to be let out" (Milgram 1987, 567). These results are especially significant because the volunteers received no penalty if they refused to administer the shocks.

➤ In Milgram's experiment, obedience was founded simply on the firm command of a person with an assumed status (university-based researcher) and supporting symbols (white lab coat) that gave minimal authority over the participant. The authority was rated as minimal because the research participants were under no obligation to be involved in the experiment.

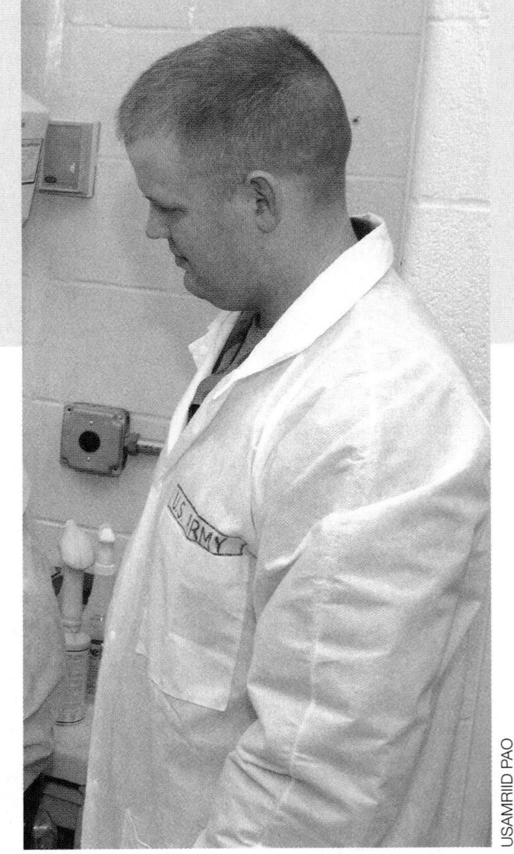

USAMRIID PAO

Milgram's study offers insights as to how events such as the death of 6 million Jews during the Holocaust, the death of 1.7 million Cambodians under Pol Pot, and more recently, prisoner abuse by U.S. military at Abu Ghraib prison in Iraq could have taken place. These kinds of events require the cooperation of many people. This fact raises important questions about people's capacity to obey authority. Milgram concluded that "behavior that is unthinkable in an individual who is acting on his own may be executed without hesitation when carried out under orders" (1974, xi).

As a case in point, Staff Sergeant Ivan L. Frederick II was caught up in the Abu Ghraib prison scandal. Abu Ghraib, one of the world's most notorious prisons under Saddam Hussein, was converted to a U.S. military prison after he was removed from power. Frederick faced charges of cruelty toward prisoners, maltreatment, assault, and indecent acts. He was sentenced to eight years in prison. Letters and e-mail messages that Frederick wrote to family members asserted that he was simply carrying out orders (Hersh 2004).

➤ In January 2004, Frederick wrote, "I questioned some of the things that I saw, . . . such things as leaving inmates in their cell with no clothes or in female underpants, handcuffing them to the door of their cell—and the answer I got was, 'This is how military intelligence [MI] wants it done.'"

AP Photo

Group Pressure

Social psychologist Irving Janis (1972) coined the term **group think**, a phenomenon that occurs when a group under great pressure to take action achieves the illusion of consensus by putting pressure on its members to suppress expression of doubt and ignore the moral consequences of their actions. Group think is most likely to occur when members are from similar backgrounds, do not seek outside opinions, and unquestionably believe in the rightness of their cause. The concept of group think grew out of Janis's research on the group dynamics underlying the making of foreign policies with disastrous consequences and comparing them with the group dynamics underlying the making of foreign policies with successful outcomes.

Chris Caldeira

➤ One disastrous decision Janis investigated was a 1950 decision ordering U.S. soldiers to cross the 38th parallel (marked on highway sign) into North Korea, a line established after WWII separating North from South Korea. (Note that the North Korean army had crossed into South Korea, pushed its way to Pusan, almost taking it, but Americans and South Korean soldiers pushed the North Korean army back to the 38th parallel. But then instead of stopping at the line, Americans decided to invade North Korea.) That decision had the effect of prolonging the Korean War another three years. In the end, 2.5 million civilians and 2 million soldiers on both sides died; North and South Korea were reduced to rubble; and neither side gained territory.

Janis and other researchers have applied group think theory to group dynamics underlying other historical events with disastrous results, including the Watergate break-in, Iran hostage crisis, Kent State massacre, and space shuttle *Challenger* explosion (Esser 1998). There have also been attempts to simulate group think in laboratory settings. Group think is analogous to a syndrome, in that not all (just some) of the antecedents, elements, and symptoms must be present to achieve an illusion of consensus (Esser 1998; see Figure 6.2a).

Master Sgt. Gerold Gamble

➤ Group think–like processes can also be applied to any group under pressure to perform at a high level or maintain a tradition of excellence, as was the case with the elite Florida A&M Marching Band. In 2011, one of the band's drum majors died after being the focus of a hazing ritual where he was repeatedly punched by a small group of band members. The antecedents of group think (see Figure 6.2a) applied to this elite marching band, whose members held celebrity status on campus. That elite reputation and celebrity status gave some members the illusion of invulnerability and suppressed any objective assessment of risk associated with such actions (Alvarez and Brown 2011).

▼ **Figure 6.2a: Group Think: Its Antecedents, Elements, and Symptoms**

Antecedents to Group Think

- Leaders are directive in style
- Members share similar social backgrounds and beliefs
- Group is isolated from outside sources of information and analysis
- Group under pressure to take action

↓

Group Think

- An illusion of invulnerability accompanied by excessive optimism and risk taking
- Guiding assumptions are not questioned
- Certainty about the group's purpose and morality
- Labeling the opposition as weak, evil, biased, vindictive, disloyal, or uninformed
- Internal pressure on members to not raise questions
- An illusion that members are unanimous; interpreting silence to mean agreement
- Dissenting information cannot surface

↓

Symptoms of Group Think

- Alternative strategies are not considered
- Objectives are not clear
- No assessment of potential risks associated with decisions
- Weak effort at gathering information
- No thought given to a contingency plan if things go wrong

(Write a Caption)

Write a caption that describes the type of sanction employed to control behavior. In the photo to the left a skull and bones is used to warn people not to walk past the sign, where there is a minefield. In the other photo a boy holds a certificate celebrating his volunteer work.

Spc. Kelly Lowery

Jim Gordan, CIV, USACE

Hints: In writing this caption

- review positive and negative sanctions,
- think about the kind of sanction captured by each photo and the behavior each seeks to promote or prevent, and
- decide whether each sanction is backed by an official or formal body such as a school or government.

Critical Thinking

Can you think of a setting in which your behavior is routinely monitored? Describe that setting.

Key Terms

censorship

formal sanctions

group think

informal sanctions

mechanisms of social control

negative sanctions

positive sanctions

sanctions

surveillance

MODULE

6.3 Labeling Theory

Objective

You will learn that labeling theorists define a deviant as someone who is noticed and punished.

Have you ever cheated on an exam? Did you get caught?

Lisa Southwick

When I asked my students to answer these questions anonymously, almost everyone indicated that they had cheated on an exam at some point in their educational career, but less than 10 percent said they had been caught. The students caught received a zero on the relevant exam and some failed the course. This example raises questions about the nature of deviance but especially about who gets caught. Such questions are at the heart of labeling theory.

Labeling Theory

In *Outsiders: Studies in the Sociology of Deviance*, sociologist Howard Becker states the central thesis of labeling theory: "All social groups make rules and attempt, at some times and under some circumstances, to enforce them. When a rule is enforced, the person who is supposed to have broken it may be seen as a special kind of person, one who cannot be trusted to live by the rules agreed on by the group." That person is "regarded as an outsider" (1963, 1).

As Becker's statement suggests, labeling theorists are guided by two assumptions: (1) rules are socially constructed; that is, people make rules, and (2) rules are not enforced uniformly or consistently. Labeling theorists maintain that whether an act is deviant depends on whether people notice it and, if they do notice, whether they label it as a violation of a rule and then proceed to apply sanctions. In other words, simply violating a rule does not automatically make someone deviant. From a sociological point of view, a rule breaker is not deviant unless someone notices the violation and decides to take corrective action.

Categories of Conformists and Deviants

Labeling theorists point out that for every rule a social group creates, four categories of people exist: conformists, pure deviants, secret deviants, and the falsely accused. The category to which someone belongs depends on a combination of two factors: whether a rule has been violated and whether sanctions are applied (see Table 6.3a).

▼ Table 6.3a: Typology of Deviance Applied to Cheating

The table summarizes the four categories of people as they relate to a particular social norm or law. Think of some offending behavior, such as cheating on a test. Either a person engages in that behavior or not. Someone has to notice and apply sanctions. If no one notices and applies sanctions, a person is a secret deviant.

	Noticed/Sanctions Applied	Not Noticed, Sanctions Not Applied
Engaged in offending behavior (Cheating)	Pure deviant (Engaged in cheating; professor noticed and applied sanctions)	Secret deviant (Engaged in cheating but professor did not notice or apply sanctions)
Did not engage in offending behavior (Cheating)	Falsely accused (Did not engage in cheating; professor misperceives behavior that looks like cheating and applies sanctions)	Conformist (Did not engage in cheating and professor noticed no behavior that would suggest otherwise)

From Table 6.3a we can see that **conformists** are people who have not engaged in offending behavior and are treated accordingly. **Pure deviants** are people who have engaged in offending behavior and are caught, punished, and labeled as outsiders. **Secret deviants** are those who have engaged in the offending behavior but no one notices or, if it is noticed, no one applies sanctions. Becker maintains that we cannot really know how many secret deviants exist, but he is convinced that the number is sizable, many more than "we are apt to think" (1963, 20).

A survey of crime victims in the United States substantiates Becker's contention that the number of secret deviants is quite sizable. The U.S. Department of Justice (2011) has found that each year about 4.2 million violent victimizations and 14.8 million property victimizations are committed in the United States. Only 50 percent of victims of violent crimes and 40 percent of property crimes are reported to police. From the perspective of law enforcement, at least 50 percent of violent crimes and 40 percent of property crimes are committed by people who remain secret deviants. (Of course, simply reporting a crime does not mean that the perpetrator will be caught.)

THE FALSELY ACCUSED. Like secret deviants, we can also never know the number of people who are **falsely accused**, those who have not engaged in offending behavior but are treated as if they have. The ranks of the falsely accused include victims of eyewitness errors and police cover-ups; they also include innocent suspects who make false confessions under the pressure of interrogation. For their book *In Spite of Innocence*, sociologist Michael L. Radelet and colleagues (1994) reviewed more than 400 cases of innocent people convicted of capital crimes and found that 56 had made false confessions.

Apparently, some innocent suspects admitted guilt, even to heinous crimes, to escape the stress of interrogation (Jerome 1995). As with secret deviance, no one knows how often false accusations occur, but it probably occurs more often than we imagine. Moreover, the taint of guilt lingers even after the falsely accused is cleared of all charges.

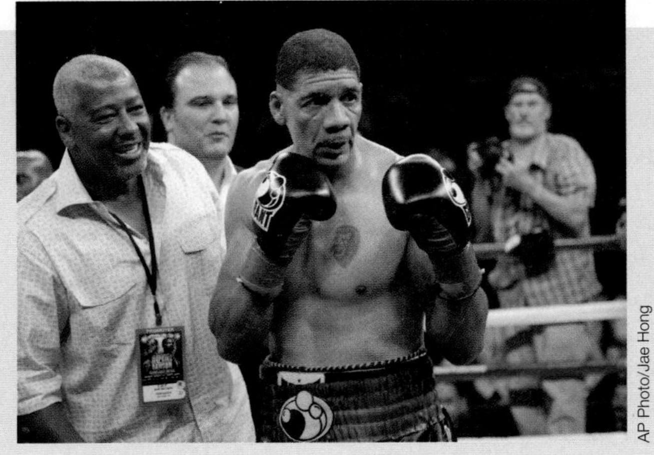

AP Photo/Jae Hong

◀ Dewey Bozella was falsely convicted and imprisoned for the murder of a 92-year-old woman. He spent 26 years in prison before his conviction was overturned in 2010 on the grounds that critical evidence, including fingerprints of the actual killer at the scene of the crime, had been suppressed. Bozella is one of hundreds of people who have been exonerated with the help of the Innocence Project, a legal initiative that works to free those falsely accused of crimes (Innocence Project 2011). After being released, Bozella, who took up boxing while in prison at age 52, fought his first and only professional fight and won.

Sociologist Kai Erikson (1966) identified the situation under which people are likely to be falsely accused of a crime—when the well-being of a country or a group is threatened. The threat can take the form of an economic crisis (such as a depression or recession), a health crisis (such as AIDS or H1N1), or a national security crisis (such as war). At times like these, people need to identify a clear source of the threat. Thus, whenever a catastrophe occurs, it is common to blame someone for it. Identifying the threat gives an illusion of control. In such crises, the person blamed is likely to be someone who is at best indirectly responsible, someone in the wrong place at the wrong time, or someone who is viewed as different. This defining activity can take the form of a witch hunt.

WITCH HUNTS. **Witch hunts** are campaigns to identify, investigate, and correct behavior that has been defined as dangerous to the larger society. In actuality, a witch hunt rarely accomplishes these goals, because the real cause of a problem is often complex, extending far beyond the behavior of a targeted category. Often, people who are identified as the cause of a problem are simply being used to make the problem appear as if it is being managed. Although the FBI did not keep statistics on the ethnicity or religious affiliation of the people questioned about the attacks, it is believed that most people interrogated were or appeared to be Muslim or Middle Eastern. Many federal agents, haunted by the September 11 attacks, acted on "information from tipsters with questionable backgrounds and motives, touching off needless scares and upending the lives of innocent suspects" (Moss 2003, A1).

The Status of Deviant

The status of deviant can be primary or secondary in character. **Primary deviants** include those people whose rule breaking is viewed as understandable, incidental, or insignificant in light of some socially approved status they hold. For example, it is unlikely that employees who use a company-provided mobile phone for personal calls or who take office supplies for personal use will be branded a thief. Similarly, a woman who kills a male partner because she fears for her physical safety has a greater chance of being labeled as suffering from battered person syndrome than being labeled a murderer. A male who kills a female partner claiming that he feared for his physical safety is likely to be labeled a cold-blooded murderer (Pfuhl and Henry 1993).

Secondary deviants include those whose rule breaking is treated as something so significant that it cannot be overlooked or explained away. Secondary deviants assume a **master status of deviant**, an identification that "proves to be more important" than most other statuses that person holds, such that he or she is identified first and foremost as a deviant (Becker 1963, 33). A master status of deviant is linked to an expected set of auxiliary statuses. An abusive spouse (master status) is expected to be male and young to middle-aged (auxiliary statuses); criminals (master status) are expected to be male and black or Hispanic, or high school dropouts (auxiliary statuses); prostitutes (master status) are expected to be young and women (auxiliary statuses). Expected auxiliary statuses are often what allow the secret deviant to remain undercover. An 80-year-old shoplifter may escape detection because store security is focused on those who possess the expected auxiliary statuses of a shoplifter (Pfuhl and Henry 1993).

(Write a Caption)

Write a caption that uses labeling theory to frame the arrest of this man.

Hints: In writing this caption

- review labeling theory,
- consider what had to take place for this man to be arrested, and
- think about whether this man is potentially a pure deviant, secret deviant, and/or falsely accused.

U.S. Army photo by Staff Sgt. Sean A. Foley/Released

Critical Thinking

Can you think of situations in your life when you assumed the status of falsely accused, secret deviant, and/or pure deviant? Explain.

Key Terms

conformists	primary deviants	secret deviants
falsely accused	pure deviants	witch hunt
master status of deviant	secondary deviants	

MODULE

6.4 Differential Association

Objective

You will learn the meaning of differential association and the role it plays in putting people in contact with deviant influences.

Missy Gish

When you were a teenager, did your parents or some other adult ever tell you that they did not like the people you were hanging out with?

The parent or other adult making this statement was thinking about a phenomenon sociologists refer to as differential association. Coined by sociologists Edwin H. Sutherland and Donald R. Cressey (1978), **differential association** states that it is exposure to criminal patterns and isolation from anticriminal influences that put people, especially juveniles, at risk of turning criminal. These criminal contacts take place within deviant subcultures.

Deviant Subcultures

Deviant subcultures are groups that are part of the larger society but whose members share norms and values favoring violation of that larger society's laws. People learn criminal behavior from closely interacting with those who engage in and approve of law-breaking activities. It is important to keep in mind, however, that contact with deviant subcultures does not by itself make criminals. Rather, there is some unspecified tipping point of exposure in which criminal influences offset exposure to law-abiding influences (Sutherland and Cressey 1978).

If we accept the premise that criminal behavior is learned, then criminals constitute a special type of conformist in that they are simply following the norms of the subculture with which they associate. The theory of differential association does not explain how people make initial contact with a deviant subculture or the exact mechanisms by which people learn criminal behavior, except that the individual learns the deviant subculture's rules the same way any behavior is learned.

Sociologist Terry Williams's (1989) research shows how teenagers can make contact with a deviant subculture. He studied a group of teenagers, some as young as 14, who sold cocaine in the Washington Heights section of New York City. These youths were recruited by major drug suppliers because, as minors, they could not be sent to prison if caught. Williams argues that the teenagers were susceptible to recruitment for two reasons: they saw little chance of finding high-paying jobs, and they perceived drug dealing as a way to earn money to escape their circumstances. Williams's findings suggest that once teenagers become involved in drug networks, they learn the skills to perform their jobs the same way everyone learns to do a job. Indeed, success in an illegal pursuit is measured in much the same way that success is measured in mainstream jobs: pleasing the boss, meeting goals, and getting along with associates.

Differential Opportunity

Williams's research suggests that criminal behavior is not simply the result of differential association with criminal ways. There are other factors at work, including what he terms **illegitimate opportunity structures**, social settings and arrangements that offer people the opportunity to commit particular types of crime. In the case of Williams's study, drug suppliers recruited 14-year-olds because as minors they would not go to prison. The larger society offers minors an opportunity structure that allows them to engage in criminal activity without risking the full penalties of the law. Opportunities to commit crimes are also shaped by the environments in which people live or work (Merton 1997). For someone to embezzle money, another person has to have entrusted a would-be embezzler with a large sum of money; the act of entrusting money has to occur before the embezzler can set money aside for some unintended purpose.

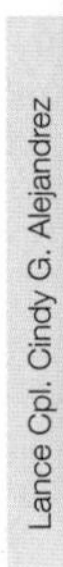

U.S. Navy Photo by Mass Comm. Spc. 3rd Class David Didier

▲ Illegitimate opportunity structures figure into the type of crimes people commit. Working as a pharmacist offers the opportunity to steal prescription drugs to sell or to feed a personal or friend's addiction. For addicts without money, one option available to them is to hold up a pharmacy.

The concept of an illegitimate opportunity structure undermines the belief that the uneducated and members of minority groups are more prone to criminal behavior than are those in other groups. In fact, crime exists in all social strata, but the type of crime, the opportunities to commit crime, the extent to which the laws are enforced, access to legal aid, and the power to sidestep laws vary by social strata (Chambliss 1974).

White-Collar and Corporate Crime

White-collar and corporate criminals also benefit from differential opportunity. **White-collar crime** consists of "crimes committed by persons of respectability and high social status in the course of their occupations" (Sutherland and Cressey 1978, 44). **Corporate crime** is committed by a corporation in the way that it does business as it competes with other companies for market share and profits. In the case of white-collar crime, offenders are part of the system: they occupy positions in the organization that permit them to carry out illegal activities discreetly. In the case of corporate crime, everyone in the organization contributes to illegal activities simply by doing their jobs. Even activities considered legal can have adverse consequences on people and the environment. Both white-collar and corporate crimes are often aimed at impersonal—and often vaguely defined—entities and involve evading taxes, polluting the environment, putting competitors out of business, and so on. They are "seldom directed against a particular person who can go to the police and report an offense" (National Council for Crime Prevention in Sweden 1985, 13).

◀ The Occupy Wall Street and the Occupy movements that it spawned have targeted Wall Street and financial institutions for the "corporate crimes" that triggered the Great Recession of 2008. Wall Street and financial institutions engaged in or benefited from risky lending practices backed by Moody's Investors Service and Standard & Poor's triple-A ratings of subprime mortgage securities that proved to be worthless when millions of borrowers defaulted on loans. To compound matters, many holders of these mortgage securities took out insurance policies that would pay off in the event of mass default. The backup protection undermined any incentive to issue loans to only those who could afford them.

White-collar and corporate crimes—such as engaging in risky lending practices, manufacturing and marketing unsafe products, unlawful disposal of hazardous waste, tax evasion, and money laundering—are usually handled not by the police but by regulatory agencies (such as the Environmental Protection Agency, the Federal Bureau of Investigation, and the Food and Drug Administration), which have minimal staff to monitor compliance. Escaping punishment is easier for white-collar and corporate criminals than for other criminals.

(Write a Caption)

Write a caption that distinguishes between white-collar and corporate crimes. One photo shows a polluted body of water; the other a surgeon operating on a patient.

Petty Officer 3rd Class Nick Ameen

Mass Comm. Spc. 3rd Class Maddelin Angebrand

Hints: In writing this caption

- review the types of white-collar and corporate crimes,
- consider how a physician might have the opportunity, given his or her position, to perform unnecessary surgeries, and
- think about who is affected by illegal dumping of waste.

Critical Thinking

Give an example of an illegitimate opportunity structure that could provide you with the means to commit a crime.

Key Terms

corporate crime

deviant subcultures

differential association

illegitimate opportunity structures

white-collar crime

MODULE

6.5 Structural Strain Theory

Objective

You will learn how structural strain generates deviant responses.

What is the American dream? How do you know when someone has achieved it? Is owning a boat evidence that the dream has been achieved?

Chris Caldeira

Based off materialistic things

The American dream is grounded in the belief that everyone has the opportunity to achieve material prosperity regardless of social position in life. What does it mean to live in a society in which people are told that they can be anything they want if they work hard enough? As we will see, sociologist Robert K. Merton's (1938, 1957b) theory of structural strain shows how belief in the American dream is connected to deviant behavior.

The Structure of Strain

Robert K. Merton's theory of structural strain takes two elements of social structure into account:

1. the *goals* a society defines as valuable (such as economic success, upward mobility, home ownership), and
2. the culturally legitimate *means* to achieve those valued goals (such as go to college, work hard), including the actual number of legitimate opportunities available to achieve valued goals (such as the number of jobs paying over $100,000).

Structural strain is a situation in which there is an imbalance between culturally valued goals and the legitimate means to obtain them. An imbalance exists when

1. opportunities for reaching the valued goals are limited or closed off to a significant portion of the population; that is, there are not enough opportunities to satisfy demand;
2. people are unsure whether following the legitimate means will lead to success; or
3. the sole emphasis is on achieving valued goals by any means necessary.

Merton (1938) argues that structural strain induces a state of cultural chaos, or **anomie**. Under such conditions people are susceptible to abandoning the legitimate means to achieve culturally valued goals and are even susceptible to abandoning those goals.

Anomie Applied to Financial Crisis

The ongoing U.S. and global financial crisis grew out of structural strain in which many in banking and financial services focused solely on a valued goal of material gain by whatever means necessary. That focus encouraged a state of anomie in which lenders abandoned the previously legitimate means of making safe loans, which required borrowers to have a good credit history and equity to secure loans. Instead, many lenders arranged terms that exceeded borrowers' actual capacity to pay.

Lisa Southwick

➤ In the years leading up to the financial crisis, lenders encouraged and even pushed borrowers to buy homes and cars that they could not afford or, in the case of homeowners, to take out equity loans to advance their standard of living. Many used credit to live beyond their means and as a quick path to visible economic success.

Two unregulated financial strategies—securitization and credit default swaps—fueled and encouraged this high-risk lending. *Securitization* involves lenders bundling hundreds, even thousands, of loans and selling them to investors. Upon selling the loans, the lender makes an immediate profit and leaves investors to assume the risk of borrowers defaulting. *Credit default swaps*, an insurance-like system in which investors pay the equivalent of premiums to shift the risk of defaults onto yet another party, allow investors to feel they can't lose even if borrowers default. As one example, American International Group, Inc. (AIG) was a world leader in insurance and financial services with operations in at least 130 countries. AIG collapsed in 2008, however, because of securitization and credit default swaps, when millions of borrowers defaulted on home and other loans.

Securitization and credit default, by fueling easy access to loans, created anomie in 2008 as illustrated by these facts:

- Homeowners were accepting loans at terms they could not afford, thereby achieving the American dream by whatever means necessary.
- Loan officers made risky loans to meet performance goals; they fabricated borrowers' worth, ignored poor credit histories, and made adjustable rate loans so that clients could afford more expensive homes.
- Homebuilders overbuilt to meet market demand for high-priced housing. But because they did not build affordable homes to coincide with buyers' real incomes, as a result, housing prices fell and a major source of equity and credit collapsed.

The financial crisis must be considered in the context of a society that places great emphasis on economic success and upward social mobility regardless of the circumstances. Such a viewpoint suggests that those who fail only have themselves to blame. Merton argues that "Americans are bombarded on every side" with the message that this goal is achievable (1957b, 137). Furthermore, when people do achieve monetary success, at no point can they say they feel secure. "At each income level . . . Americans want just about 25 percent more" and this "'just a bit more' continues to operate even after it is obtained" (136).

Responses to Structural Strain

Merton identified five ways that people respond to structural strain. The responses involve some combination of acceptance and rejection of the valued goals and means.

➤ Table 6.5a: Typology of Responses to Structural Strain

The table summarizes the five responses to structural strain. Which response involves acceptance of culturally valued goals and the rejection of legitimate means to achieve them? Which responses involve rejecting (or abandoning) culturally valued goals?

Mode of Adaptation	Goals	Means
Conformity	+	+
Innovation	+	–
Ritualism	–	+
Retreatism	–	–
Rebellion	+/–	+/–

+ Acceptance/achievement of valued goals or means
– Rejection of/failure to achieve valued goals or means
Source: Adapted from Merton (1957b, "A Typology of Modes of Individual Adaptations," 140)

Conformity is the acceptance of cultural goals and the pursuit of those goals through legitimate means. This category includes people who did not take out credit to live beyond their means, bank executives who stuck with safe lending practices, and those who invested responsibly during the so-called housing boom years.

Innovation is the acceptance of cultural goals but the rejection of legitimate means to achieve them. For the innovator, success is equated with winning the game rather than with playing by the rules; that is, the innovator seeks to achieve a desired end by whatever means necessary. With regard to the housing crisis, the great innovators were the Wall Street firms that created the investment strategies (credit default swaps) that supposedly reduced the risk of making bad loans and the loan officers who set up risky loan schemes and persuaded clients to accept them.

Ritualism involves the rejection of the cultural goals but a rigid adherence to legitimate means society has in place to achieve them. This response is the opposite of innovation: the game is played according to the rules despite defeat.

Merton (1938) maintains that this response can be a reaction to the status anxiety that accompanies the ceaseless competitive struggle to stay on top or to get ahead. For ritualists, the culturally valued goals, even though valued, are defined as being beyond their reach. Instead of taking out subprime and larger than necessary loans during the time when such loans were easy to secure, ritualists lived within their means even though others may have viewed them as unsuccessful, or at the very least of modest means.

Retreatism involves the rejection of both culturally valued goals and the legitimate means of achieving them. Retreatism is a response of those who have internalized the valued cultural goals but no longer have access to the legitimate means promising success. According to Merton, retreatists face a mental conflict in that it is against their moral principles to use illegitimate means, yet the legitimate means have failed them.

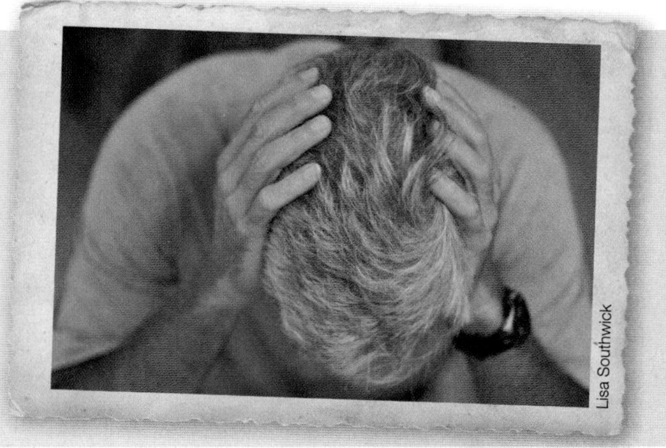
Lisa Southwick

➤ Frustrated and handicapped because they believed they played by the rules yet the system failed them, retreatists find that they cannot cope and decide to drop out. Examples include prominent high-status professionals who thought they did "everything they were supposed to do in life" but whose investments have been wiped out (Kotbi 2009).

Rebellion also involves the rejection of both the valued goals and the legitimate means of attaining them. Whereas retreatists simply give up, rebels seek a new set of goals and means of obtaining them. While this response is often confined to a small segment of society, it provides the potential for the emergence of subgroups as diverse as street gangs and the Old Order Amish. When rebellion is the response of a large number of people who wish to reshape the entire structure of society, a great potential for revolution exists.

Robert K. Wallace

➤ Notice this woman, who is part of an Occupy Cincinnati protest, is wearing a shirt that proclaims "Stop bitching and start a revolution." That advice captures the spirit of rebellion as a response to structural strain. The Occupy movement is seeking to make revolutionary changes to the financial system that, from the demonstrators' point of view, benefits Wall Street and big investors.

(Write a Caption)

Write a caption that relates a culturally valued goal in the United States that is conveyed to any child who lives there. Consider also how the culturally valued means might undermine access to that valued goal.

Lance Cpl. Jimmy Serena Jr.

Hints: In writing this caption

- identify a culturally valued goal in the United States that applies to all Americans, and
- think about how the culturally valued means of achieving that goal might not support it.

Critical Thinking

Use Merton's typology to explain how someone you know responded during the housing boom or the banking and financial crisis.

Key Terms

anomie
conformity
innovation
rebellion
retreatism
ritualism
structural strain

MODULE 6.6

Medicalization of Deviance

Objective

You will learn the process by which a deviant behavior or appearance becomes defined as a medical disorder.

When you hear someone has an addiction to food, sex, or alcohol or some other drug, do you trace the cause to a medical problem, to a lack of discipline, or to something else?

Tony Rotundo

Medicalization

If you define the cause of an alcohol addiction in medical terms, then you have medicalized the problem. **Medicalization** is the process of defining a behavior as an illness or medical disorder and then treating it with medical intervention. In this module we are particularly interested in the behaviors and appearances that first have been defined as deviant and then later medicalized. It does not matter whether the behavior or appearance is one that actually needs medical attention; what matters is that a behavior or appearance once considered deviant came to be defined as a medical condition. In this regard sociologists have written about the medicalization of alcoholism, violence, drug addiction, behavior problems (ADHD), aging, erectile dysfunction, sadness, baldness, obesity, and short stature. Sociologist Peter Conrad (1975, 2005) has laid out a framework for thinking about the medicalization process that includes the agents, stages, and consequences of medicalization.

Agents of Medicalization

Conrad (2005) identified at least five interdependent stakeholders who can act as agents in support of medicalization: (1) the medical profession, (2) grassroots social movements, (3) the pharmaceutical and biotechnology industries, (4) consumers, and (5) the managed care industry.

THE MEDICAL PROFESSION. The medical profession includes physicians and other professionals, such as nurses and technicians, but it is arguably physicians who have the most power and authority. Sociologists study how the medical profession expands its jurisdiction and control over the human body through medicalization. The rise of developmental-behavioral pediatrics is just one example of a way in which the medical profession has extended its reach. That field emerged in an official sense in 2002, the year the Academy of Pediatrics certified the first physicians in this specialty. The establishment of this specialty expanded physicians' control over areas that were once the domain of social workers, the family, educators, friends, and other caregivers.

Tony Rotundo

◀ At the time of this writing, weight loss and weight management was not yet a medical specialty certified by the American Board of Medical Specialties. But some physicians are pushing for such a specialty. This would entail certifying doctors to treat obesity using techniques such as stomach surgery or prescribing medications that target hormones or reduce cravings (Fudge 2010).

GRASSROOTS SOCIAL MOVEMENTS. Grassroots, or from-the-ground-up, social movements occur when enough people organize to push for recognition and change. Such movements often evolve from difficult circumstances that participants seek to address or alleviate. A case in point is removal of homosexuality as a mental illness from the *Diagnostic and Statistical Manual of Mental Disorders* (DSM) published by the American Psychiatric Association (APA). The decision to drop homosexuality as a mental illness was certainly influenced by a grassroots movement led by the Gay Liberation Front and other anti-psychiatry activists that culminated in a protest at the APA meetings in San Francisco in 1971. In DSM-V (forthcoming 2013), ego-dystonic sexual orientation is listed as a mental illness. This new grassroots-influenced diagnosis does not challenge any sexual orientation; rather it targets those who are unhappy with their perceived sexual orientation (whatever that orientation).

PHARMACEUTICAL AND BIOTECHNOLOGY INDUSTRIES. The pharmaceutical industry promotes pills for a variety of conditions—Ritalin (for hyperactivity), Viagra (for erectile dysfunction), and Paxil (for depression and anxiety), to name but a few. This industry advertises in medical journals, on the Internet, and on television; sends sales representatives to doctors' offices; provides free samples for doctors to distribute to patients; and sponsors medical events such as annual meetings and health fairs. Among other things, the pharmaceutical industry develops and markets to physicians what Conrad (2005) calls "biomedical enhancement products," be they to enhance height, memory, speed, or stamina.

CONSUMERS. Those who choose to use and purchase health care services and products are also part of the medicalization process. Among other things, consumers choose a health insurance plan, health care providers, and a hospital. In addition, consumers even engage in self-diagnosis and will buy over-the-counter medicines to treat a condition they believe they have. Sometimes consumers (after listening to advertisements or browsing the Internet) present to their doctors what they believe their medical condition to be and even insist on a specific treatment.

Consumer demand for cosmetic procedures and surgeries, from Botox treatments to liposuction, is another example of medicalization. "The body has become a project, from extreme makeover to minor touch ups, and medicine has become the vehicle for improvement. In a sense, the whole body has become medicalized piece by piece" (Conrad 2005, 8).

THE MANAGED CARE INDUSTRY. Over the past 20 years, managed care organizations have come to play increasingly important roles in the kinds of care patients receive. Managed care organizations require doctors to gain preapproval before embarking on major medical treatments; they also set limits on the number of visits and the kinds of drugs covered. The managed care industry both constrains and promotes certain kinds of medical care. For example, many managed care organizations cover gastric bypass surgery for what they refer to as the morbidly obese with the hope that, in the long run, it will cost less to pay for that surgery than for treatment of obesity-induced illnesses such as diabetes, stroke, heart conditions, and musculoskeletal-related problems. Managed care organizations have significantly limited the number of psychotherapy sessions between doctor and patient, and instead support treating mental and emotional problems with medications.

Stages of Medicalization

Peter Conrad and Joseph Schneider (1992) proposed a five-stage model by which a behavior or appearance considered deviant becomes medicalized. In stage 1 (Definition of Behavior as Deviant), the behavior or appearance is defined as deviant (a deviant sexual orientation, a deviant behavior, sagging body parts) before a medical name or definition is applied or before any medical writings on the condition appear. A medical explanation eventually follows that validates and gives scientific support to the existing label of deviant.

Stage 2 (Prospecting) occurs when a condition defined as deviant is presented as needing medical attention. Generally, that condition is announced in some formal way through an article or opinion piece published in a medical journal or at a medical meeting or conference. The announcement may include the name of the new disorder, a description of its proposed cause (etiology), and/or recommendations for treating it. Conrad cautions that such announcements do not guarantee that the new diagnosis will become recognized and accepted.

In stage 3 (Claims Making), those with an interest in medicalizing a behavior considered deviant lobby to have it added to the list of conditions that the medical profession treats. A doctor may set up an institute or sponsor a workshop. A pharmaceutical company may engage in a publicity campaign to raise awareness. Whatever the strategy or source, the new medical disorder needs some governmental body to recognize it as something that needs treated.

In stage 4 (Legitimacy), the government gives the medical profession its mandate to treat the condition. It can hold hearings; issue grants and contracts for medical research and care; and approve drugs (such as Botox), treatments (laser), and medical devices (breast implants).

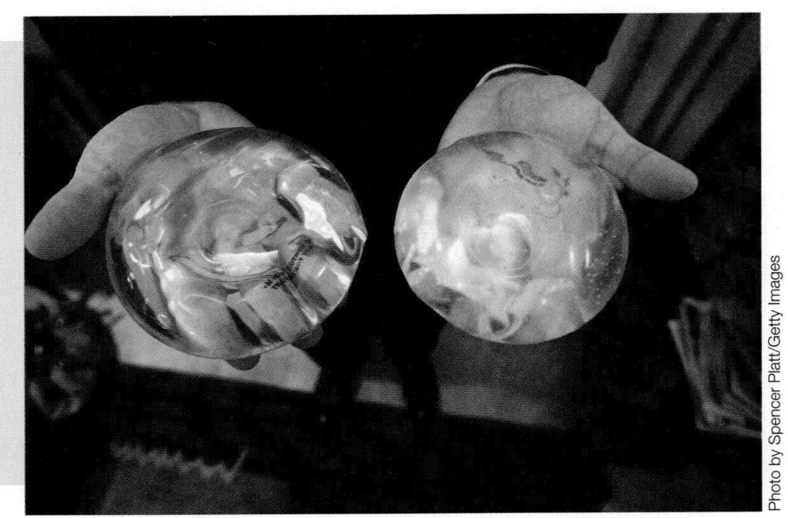

Photo by Spencer Platt/Getty Images

➤ The U.S. Food and Drug Administration (2009) has approved breast implants as well as Botox and wrinkle filler products, giving legitimacy to products that medicalize breasts and wrinkles as things to be treated.

In stage 5 (Institutionalization), the medical diagnosis and methods of treatment become accepted. The name of the medical condition, a description of symptoms, and recommended treatments assume a place in medical textbooks and manuals. Institutionalization also involves the creation of organizations that support and conduct medical research related to the condition with goals of improving the lives of people affected, preventing new cases, and eradicating or curing the condition.

Consequences of Medicalization

When a problem is medicalized, it changes the way we think about behavior once defined as deviant. Specifically, the behavior is less likely to be considered a "sin or even a moral weakness, it is now a disease" (Conrad 1975, 18). A child described as hyperactive, for example, now has a medical reason for being disruptive and/or disobedient. Conrad argues that medicalization places a humanitarian angle on a situation that was once condemned. The humanitarian angle notwithstanding, there are also some real drawbacks to medicalization.

First, each instance of medicalization represents an advance in medicine's control over the human body, and by extension human behavior. It places the "deviant" behavior or appearance squarely in the medical domain, which in theory offers a "technical solution administered by disinterested politically and morally neutral experts" (19) and moves it away from ordinary people who previously managed the condition before it became known by a medicalized name.

Second, medical treatment is not just a "cure"; it can also be a form of social control that restrains individuals from behaving or appearing in ways that disrupt, challenge, or offend a dominant value system. Using surgery, psychoactive drugs, drug testing, or other medical interventions to solve the problems assumes that something is wrong with the individual and not the system against which he or she is rebelling.

Third, the medicalization of deviant behavior with its emphasis on fixing the person reinforces what Conrad calls the **individuation of social problems**, a point of view whereby people tend to view the "problem individuals" as the cause and "fixing" them as the solution rather than looking at the social system in which so-called deviant behavior or appearances are embedded.

➤ With medicalization of complex social problems, we direct our energy and resources to changing the victims of that system with pills, surgery, and other medical treatments rather than the society in which they live. Medical intervention essentially supports the existing ways of doing things rather than seeing the deviant "behavior as symptomatic of some disorder in the social structure" (19).

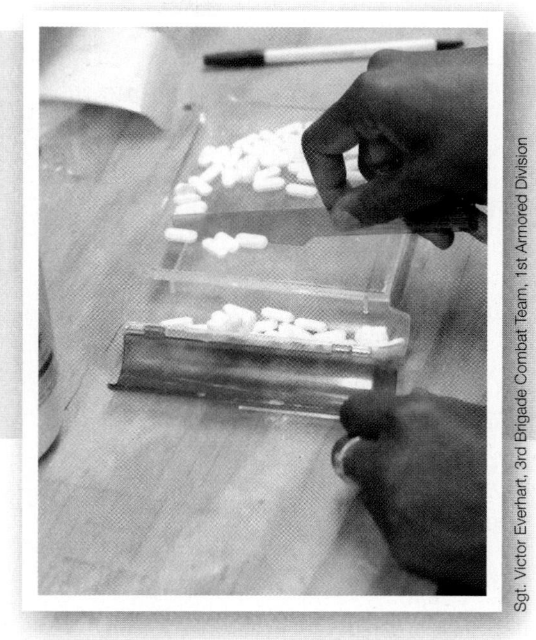

Sgt. Victor Everhart, 3rd Brigade Combat Team, 1st Armored Division

(Write a Caption)

Write a caption that illustrates how illegal drug testing via urine samples extends the reach of medicine.

Hints: In writing this caption

- think about the consequences of a positive drug test for a student, an employee, or an athlete;
- consider the options available to a person who has tested positive with regard to restoring his or her standing (via some medical treatment); and
- think about the role medical authorities may play in restoring a person's standing.

U.S. Navy photo by Photographer's Mate 2nd Class Jim Watson. (RELEASED)

Critical Thinking

Give an example of a time when you used the Internet, a home health testing product, or a pharmaceutical advertisement to make a self-diagnosis. Did you share your diagnosis with a physician?

Key Terms

individuation of social problems

medicalization

MODULE

6.7 Crime

Objective

You will learn the meaning of crime and the process by which people are punished for crimes.

Which country has the highest incarceration rate in the world?

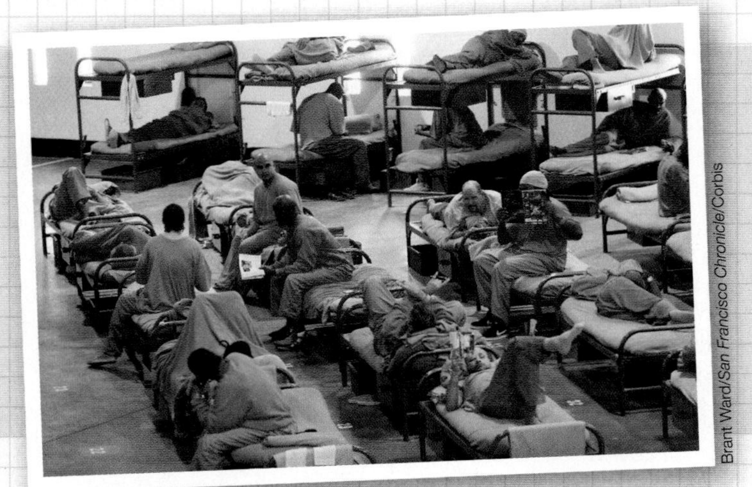

Brant Ward/San Francisco Chronicle/Corbis

Worldwide, an estimated 7.6 million people are in prison and jails. The incarceration rate varies by country with the United States having the highest rate at 730 prisoners per 100,000. After the United States are the Republic of Georgia (536) and the Russian Federation (525) (International Centre for Prison Studies 2012).

It is important to remember that high rates are not a measure of a criminal justice system's effectiveness. It could be that countries with the highest rates have the most serious crime problems, are more effective at bringing to justice those who commit crimes, or simply can pay for a large prison system (Walmsley 2002). However, it is still important to emphasize that the United States has 4.6 percent of the world's population but holds 29 percent of all incarcerated around the world (International Centre for Prison Studies 2012).

A **crime** is an act that breaks a law. A **law** is a rule governing conduct created by those in positions of power and enforced by entities given the authority to do so, such as police. A law specifies the prohibited behavior, the categories of people to whom the law applies, and the punishment to be applied to violators. Laws apply to virtually every area of life. When sociologists study crime they ask, Who avoids detection, who gets caught, and who goes to prison?

Who Goes to Prison?

Going to prison is a multistage process. In studying this process, one thing is clear: only a small portion of those who commit crimes are in prison. Why? First, someone must know about the crime and then decide to report it (or self-report it). U.S. Department of Justice (2011) victimization studies estimate that only about 40–50 percent of known crimes are reported to police. Among those reported, only 44 percent lead to an arrest (the equivalent of 16 percent of all crimes). Of those arrested, some portion are prosecuted, but not everyone prosecuted is convicted. And not everyone convicted is imprisoned (see Figure 6.7a).

Figure 6.7a: The Road to Prison

This flowchart shows the process by which people are arrested and then make their way through the criminal justice system. The prison population includes those who could not avoid one or more of the following: (1) detection, (2) someone reporting their offense, (3) being arrested, (4) arraignment, (5) prison or jail.

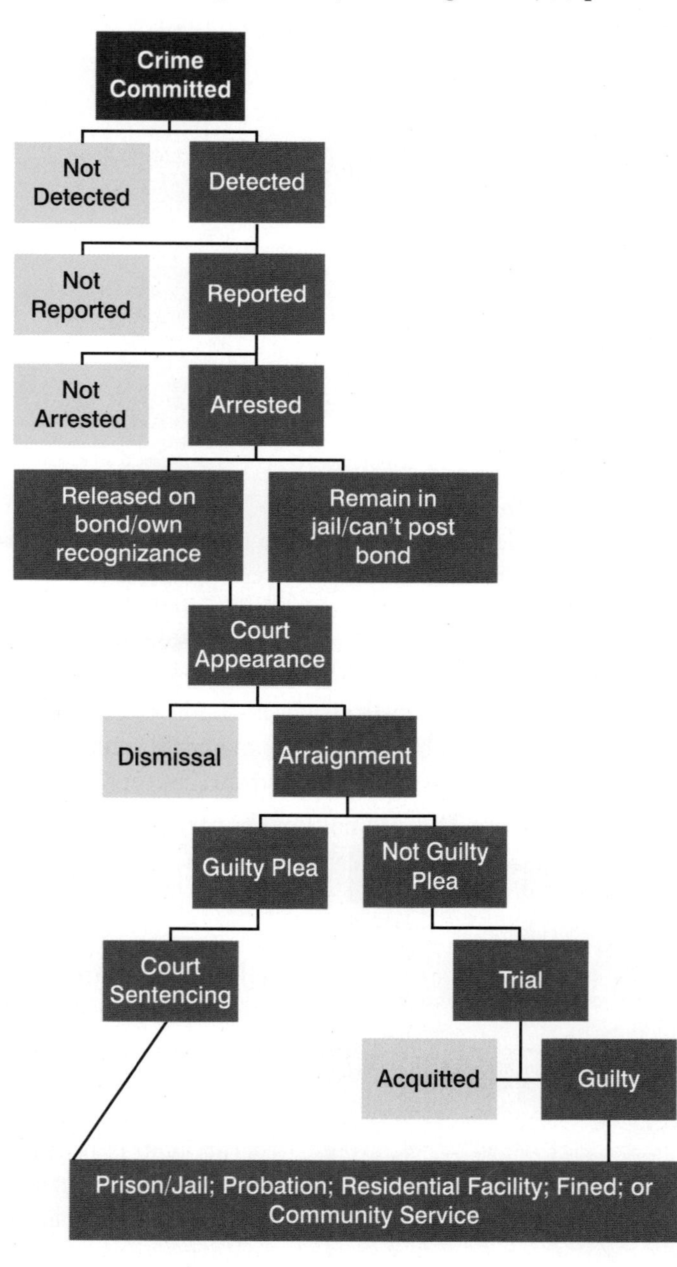

When we look at the characteristics of the prison population in the United States, we can gain some clues about which race/ethnic groups seem least able to avoid prison sentences (see Table 6.7a). According to the U.S. Department of Justice (U.S. Department of Criminal Justice Statistics 2010), 1 in 15 people in the United States will end up in prison at some point in their lifetime. But that rate is particularly high for men classified as black and Hispanic. Over the course of a lifetime, an estimated 32 percent of all black males and 17 percent of all Hispanic males will spend some time in prison. For white males that percentage is 5.9 percent. If we ask who is under some form of correctional control—on probation, home incarceration, in jail, in a residential facility, performing community service—we see that at any one time 1 in every 32 adults (or about 6.1 million) is involved in some way with the criminal justice system.

➤ **Table 6.7a: Sex and Race Characteristics of People Involved with U.S. Correctional System (in prison, jail, on probation or parole), 2010**

Characteristics	Number Involved in U.S. Correctional System
Blacks	1 in every 11
Men	1 in every 18
Hispanics	1 in every 27
All	1 in every 32
White	1 in every 45
Women	1 in every 89

No data available on Native Americans or Asian populations.
Source of data: U.S. Department of Justice (U.S. Department of Criminal Justice Statistics 2010)

Reasons for High Incarceration Rates

There are many possible explanations for the high rate of incarceration in the United States, including the propensity of elected officials to pass strict crime legislation so they can present themselves as tough on crime. This propensity explains, in part, the long mandated prison sentences handed out to those convicted of drug possession and other nonviolent crimes. In addition, critics claim the United States has a high rate of repeat offenders in prisons because the system places little emphasis on rehabilitating prisoners. Finally, the **prison-industrial complex**—the corporations and agencies with an economic stake in building and supplying correctional facilities and in providing services—fuels an ongoing "need" for prisoners so that companies can maintain or increase profit margins. It should come as no surprise that these private corporations represent a significant lobbying force shaping legislation and correctional policy.

Don B. Stevenson/Alamy

◀ Another incentive for maintaining a large prison population is that local, county, state, and federal governments have come to rely on prison labor. This dependency has increased the incentive to keep the prison population large. Many correctional institutions have short- and long-term contracts with the various government agencies to do roadwork and other routine maintenance (North Carolina Department of Correction 2009).

(Write a Caption)

Write a caption that places this prisoner's situation in the context of the process by which people end up in prison.

LCpl. Alfredo V. Ferrer

Hints: In writing this caption

- notice that this mug shot is of someone who is classified as black,
- assume that this man is guilty of a crime (not falsely accused),
- think about the percentage of people who commit crimes that are caught and eventually end up in prison, and
- draw some conclusion about his situation as it relates to his ability to avoid going to prison.

Critical Thinking

How does understanding the process by which people become imprisoned affect your understanding of the prison population?

Key Terms

crime law prison-industrial complex

MODULE 6.8

Applying Theory: Post-Structural Theories

Objective

You will learn how being watched—including the possibility that someone is watching—shapes behavior.

How often are you aware that your activities are being monitored?

Lisa Southwick

Most of us know that we are being watched at some point over the course of a typical day, if only by a store or ATM surveillance camera. What does it mean to know that behavior is being monitored at any time? How does that possibility shape behavior? These are the questions that interested French philosopher and social theorist Michel Foucault (1977). Foucault's work falls under the theoretical category known as post-structuralism, a term we will consider at the end of this module.

Foucault (1977) sought to identify the turning points that make the society we live in today fundamentally different in structure from the society that preceded it. In this regard Foucault identified a historical shift or turning point in the way society punishes people from what he called a culture of spectacle to a carceral culture.

> A **culture of spectacle** is a social arrangement by which punishment for crimes—torture, disfigurement, dismemberment, and execution—is delivered in public settings for all to see.

Prints & Photographs Division, Library of Congress, LC-USZ62-64582

This very public way of punishing began to change in Europe and in the United States as part of prison reform movements (1775–1889), an era that ushered in what Foucault called a **carceral culture**, a social arrangement under which the society largely abandons physical and public punishment and replaces it with surveillance as the method of controlling people's activities and thoughts. Foucault attributes this change not to a rise in humanitarian concern but to a transformation in the technologies available to control others.

To understand the magnitude of this shift, consider that before what we call the prison reform movement in the United States and Europe, the death penalty was applied to any number of crimes, including murder, denying the existence of God, and homosexuality. Severe physical punishments were issued to those who committed less serious crimes. Prisons existed, but they were used to hold those awaiting trial, debtors, and sometimes even eyewitnesses to crime. Everyone, regardless of age, race, or sex, was crowded together in holding pens (Johnston 2009). From Foucault's point of view, prison reforms—and by extension reforms related to punishment—were connected to the need to establish discipline and order, not to achieve some moral objective.

Prints & Photographs Division, Library of Congress, LC-USZ62-137901

‹ Crowded prison quarters facilitated the spread of disease and increased opportunities for prisoners to conspire against the guards and plan an escape. This kind of holding pen also made it difficult to monitor relationships among prisoners.

The Panopticon

The prison reform movement coincides with the period of history in which "a whole set of techniques and institutions emerged for measuring, supervising and correcting" those considered abnormal, including criminals (Johnston 2009, 401). The **panopticon** was designed by British philosopher Jeremy Bentham in 1785. *Pan* means a complete view and *optic* means seeing. The design represented his effort to create the most efficient and rational prison—the perfect prison. Foucault (1977) used it as a metaphor for the mentality driving 19th-century society.

Lee Lockwood/Time Life Pictures/Getty Images

› The architectural plan of the panopticon had the following features: the guard tower was positioned in the center of the facility inside a circular gallery of cells. The front of each cell was barred and the guard standing in the tower could see into each cell. The side and back walls of each cell were solid so that the prisoners could not see or interact with each other. Under this arrangement one guard could watch the inmates housed in hundreds of cells. Since the inmates could not see the guard in the central tower, they were never sure when or if they were being watched. The threat of surveillance pushed inmates to discipline themselves (Foucault 1977).

The panopticon is a metaphor for what Foucault calls the **disciplinary society**, a social arrangement that normalizes surveillance, making it expected and routine. This disciplinary society that Foucault wrote about in the mid-1970s has been further magnified by new technologies. Those technologies can monitor offenders, keep an eye on frail elderly people in their homes, follow teens as they drive, supervise workers, track Internet use, and survey public spaces (Felluga 2009).

Post-Structuralism

Foucault is considered a post-structuralist, adhering to a theoretical position associated with many influential French thinkers writing in the second half of the 20th century (Appelrouth and Edles 2007). Post-structuralism is a response and challenge to **structuralism**, a framework that portrays social structures as having a coercive and constraining power over people's thought and behavior. **Post-structuralists** contend that the constraining/coercive powers of social structures are exaggerated (Appelrouth and Edles 2007). Rather post-structuralists emphasize that it is people who create the social structures in the first place. Foucault writes about the disciplinary society, an aspect of social structure that institutionalizes surveillance and fuels an anxiety about being watched—an anxiety that promotes good behavior. In this sense the disciplinary society can be thought of as a coercive social structure. On the other hand, it is people who enact surveillance and who also seek and find creative ways to circumvent or expand its possibilities and uses.

Critical Thinking

Have you ever censored or disciplined yourself because of the possibility that someone was watching?

Key Terms

carceral culture	disciplinary society	post-structuralists
culture of spectacle	panopticon	structuralism

CHAPTER 6

Summary: Putting It All Together

Deviance is any behavior or physical appearance that is socially challenged and/or condemned because it departs from the norms and expectations of a group. The critical factor in defining deviance is not the behavior per se, but whether someone takes notice and, if so, whether the offender is "punished."

Ideally, from society's point of view, people should want to conform. When conformity cannot be achieved voluntarily, other mechanisms of social control are employed to enforce norms. Methods of social control include positive and negative sanctions, censorship, surveillance, authority, and group pressure.

The sociological contribution to understanding deviance lies not with studying deviant individuals per se, but with studying the context under which something is deemed deviant. To understand the context, sociologists employ labeling theory, differential association, and structural strain theories.

Medicalization is the process of defining a behavior as an illness or medical disorder and then treating it with a medical intervention. Major stakeholders in the medicalization process are (1) the medical profession, (2) grassroots social movements, (3) the pharmaceutical and biotechnology industries, (4) consumers, and (5) the managed care industry. Medicalization is a five-stage process: (stage 1) a behavior or appearance is defined as deviant, (stage 2) a formal announcement is made, (stage 3) interested parties lobby to make the behavior a disorder, (stage 4) a government gives the medical profession its mandate to treat it, and (stage 5) organizations emerge to conduct medical research and support those with the condition.

A crime is a subcategory of deviance that involves breaking a law. When sociologists study crimes they ask, Who avoids detection? Who gets caught? Who goes to prison? From a global perspective, the United States sends the largest proportion of residents to prisons. Still, its prison population includes only a small portion of those people who have actually committed crimes. Prisons include those who did not have the resources, connections, or luck to avoid being detected, noticed, arrested, arraigned, convicted, and sentenced to prison. Possible explanations for high rate of imprisonment include the propensity of elected officials to pass tough crime legislation, laws that mandated long prison sentences for drug possession and other nonviolent crimes, the prison-industrial complex, and the fact that many communities have come to depend on prisoners who do work at a low cost.

Go to cengagebrain.com to link to Aplia and CourseMate for the chapter quiz and other activities.

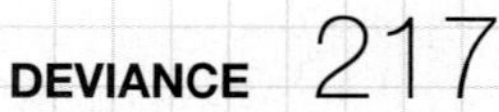

Missy Gish

Missy Gish

CHAPTER 7
SOCIAL STRATIFICATION

When sociologists examine social stratification, they look at how wealth, income, and other valued resources are distributed unequally among people. The two photos give visual representation to the concept of unequal distribution. The brown bags contain the freshly laundered clothing of people without homes. The bags—each of which represents a closet of sorts—sit inside a nonprofit organization that offers the homeless a place to shower and to leave clothing to be washed. The laundered clothes are ready the next time they drop in to shower. Contrast this grocery bag "closet" with that of a person who can afford a closet the size of a bedroom. Sociologists seek to explain how wealth, income, and other resources are distributed and the effects that inequality has on life chances, the probability that an individual's life will follow a certain path and turn out a certain way. Life chances apply to virtually every aspect of life, including the chances that someone will be without a home, have a room-sized clothes closet, go to college, live a long life, and much more.

Go to Sociology CourseMate on cengagebrain.com to watch a video that profiles the financial struggles of two young women—one black, the other white—in Greenville, MS, where the poverty rate is far above the national average.

MODULE 7.1

Conceptualizing Social Stratification

Objective

You will learn about the processes by which societies categorize and rank people on a scale of social worth.

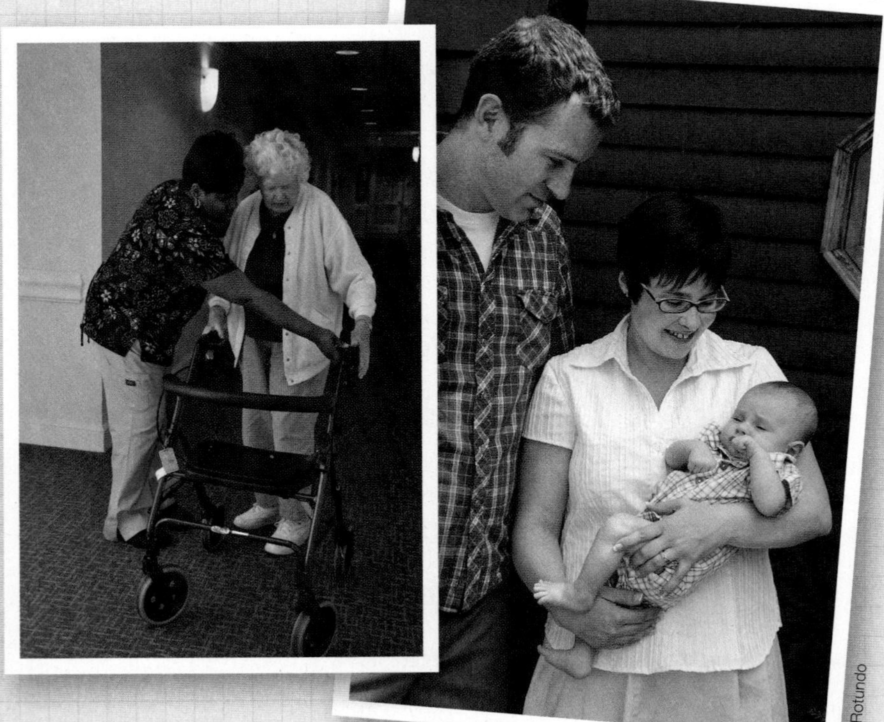

Lisa Southwick

Tony Rotundo

Look at the people in these photographs.
Imagine that you had to assign a dollar value between $250,000 (least valued) and $7.1 million (most valued) to each person's life. Could you do it?

This is what the people administering a Victim's Compensation Fund did—they assigned a monetary value to each of the 2,800 people who died as a result of the September 11, 2001, attacks on the United States. What criteria do you think the administrators used to award money? Those criteria speak to the principles underlying social stratification.

Social Stratification

Social stratification is the systematic process of categorizing and ranking people on a scale of social worth such that one's ranking affects life chances in unequal ways. Sociologists define **life chances** as the probability that a person's life will follow a certain path and turn out a certain way. As noted in the introduction, life chances apply to virtually every aspect of life—the chances that someone will survive the first year after birth, complete high school, see a dentist twice a year, work while going to school, travel abroad, major in elementary education, own 50 or more pairs of shoes, or live a long life.

Every society in the world stratifies its people. Almost any criterion—eye color, hair texture, age, sexual preference, weight, occupation, income, test scores—can be used (and probably has been used at one time or another) to assign people to categories that are unequal in status. People's status in society can be ascribed or achieved. **Ascribed statuses** are social positions assigned on the basis of attributes people possess through no fault of their own—those attributes may be inherited (such as skin shade, sex, or hair color), develop as a result of time (such as height or wrinkles), or be otherwise possessed through no personal effort (inherited wealth). **Achieved statuses** are attained through some combination of choice, effort, and ability. That is, people must act in some way to acquire an achieved status. Achieved statuses include earned wealth, income, occupation, and educational attainment.

The various achieved and ascribed statuses hold **social prestige**, a level of respect or admiration for a status apart from any person who happens to occupy it.

Cherie Cullen

➤ The achieved status of a college graduate brings with it a level of social prestige not connected with any person who happens to occupy that status. The level of prestige associated with that degree varies depending on the reputation of the university from which a person earned his or her degree.

Sociologists are especially interested in the ascribed and achieved characteristics that take on **status value**, a situation in which people who possess one characteristic (white skin versus brown skin, blond hair versus dark hair, high income versus low income, married versus single) are regarded and treated as more valuable or worthy than people who possess other characteristics (Ridgeway 1991). The compensation guidelines for the September 11, 2001, attacks on the United States show how many factors—age, annual income, occupation, marital status, potential earnings, and family status (with children versus childless)—affected victims' "worth" or status value and the money awarded to survivors. The actual awards ranged from $250,000 (least valued life) to $7.1 million (most valued life). One of the "least valued" categories was single, childless persons age 65 and older with an annual income of $10,000. Under the guidelines, their survivors were to be awarded a one-time payment of $250,000. One of the "most valued" categories was married persons age 30 and younger with two children and an annual income of $225,000. Their survivors were to be awarded a one-time payment of $3,805,087 (Chen 2004; September 11 Victim Compensation Fund 2001).

Stratification systems fall somewhere on a continuum between two extremes: a **caste system,** in which people are ranked according to ascribed statuses, and a **class system,** in which people are ranked on the basis of their achievements related to merit, talent, ability, or past performance.

Caste Systems

Sociologists use the term *caste* to refer to any form of stratification in which people are categorized and primarily ranked by characteristics over which they have no control (ascribed characteristics) and that they usually cannot change. In a caste system there is a clear association between caste rank and life chances. People in lower castes are seen and portrayed as innately inferior in intelligence, morality, ambition, and many other traits. Conversely, people in higher castes consider themselves to be superior in such traits. Moreover, a person's caste rank is treated as if it is absolute and unalterable. Finally, there are heavy restrictions on interactions between people in higher and lower castes. For example, marriage between people of different castes is forbidden.

U.S. Navy photo by Mass Comm. Spc. 1st Class Molly A. Burgess

◀ In 1967 (*Loving v. Virginia*), the U.S. Supreme Court struck down what was known as the Virginia Racial Integrity Act of 1924, a law making marriage between whites and nonwhites a felony in Virginia. That law supported a caste system of stratification intended to prohibit marriage, and by extension procreation, between whites and nonwhites. Accordingly, the law stated, it was "unlawful for any white person in this State to marry any save a white person, or a person with no other admixture of blood than white and American Indian. For the purpose of this act, the term 'white person' shall apply only to the person who has no trace whatsoever of any blood other than Caucasian; but persons who have one-sixteenth or less of the blood of the American Indian and have no other non-Caucasian blood shall be deemed to be white persons" (Virginia Racial Integrity Act of 1924).

Class Systems

While class systems contain inequality, it is based on differences in talent, ability, and past performance, not on attributes over which people have no control, such as skin shade or sex. In pure class systems, people assume that they can achieve a desired education, income, or other outcome through personal effort. Furthermore, people can raise their class position during their own lifetime, and their children's class position can differ from (and ideally be higher than) their own. Movement from one social class to another is termed **social mobility**.

➤ *American Idol* winner (season 4) Carrie Underwood represents the ideals of a class system. She experienced *inter*generational upward (vertical) mobility as her father worked at a sawmill and her mother taught at an elementary school. She also experienced *intra*generational upward mobility. Before winning *Idol*, Underwood worked at odd jobs and was three credit hours from completing a college degree (which she eventually completed in 2007). As winner of *American Idol*, Underwood received a million-dollar recording contract, a car, and use of a private plane (Associated Press 2006). According to *Forbes* (2011), Underwood earned $20 million between May 2010 and May 2011.

Staff Sgt. Patrick N. Moes

Many kinds of social mobility exist in class systems of social stratification. Mobility can be vertical or horizontal—**vertical mobility** involves a gain or loss in a person's social status, most notably economic status, relative to an earlier point in life or relative to his or her parents' or grandparents' status at a similar point in the life cycle. Mobility is considered **horizontal mobility** when there is no change in a person's social status (see Table 7.1a).

Table 7.1a: Types of Social Mobility

The hallmark of class systems is social mobility. In theory, people in class systems "rise and fall on the strength of their abilities" (Yeutter 1992, A13). This chart lists the various forms of mobility.

Type of Social Mobility	Definition	Example
Upward Vertical	A gain in social status	A medical student moves up in rank and becomes a physician
Downward Vertical	A loss in social status	A wage earner loses a job, goes on unemployment
Intragenerational	A loss or gain in social status over the course of a person's lifetime	A laid-off factory worker takes a job with a lower salary (cashier at a fast-food establishment) or a higher status (police officer)
Intergenerational	A loss or gain in social status relative to a previous generation	A son or daughter goes into an occupation that is higher or lower in rank and prestige than a parent's

Is the United States a Class System?

Neither the caste nor class systems exist in a pure form. Even in the most rigid caste system, some people manage to change their social status. Likewise, in class systems people are denied opportunities based on ascribed statuses such as race, gender, age, and disability. Most Americans—92 percent—believe that people are rewarded for hard work. Only 16 percent of Americans believe that a person's race or gender is a significant factor in explaining the economic mobility (Corak 2009). The evidence, however, shows that although there is mobility in the United States, the ascribed characteristics of race and gender do shape life chances, sometimes in very dramatic ways (see Chapters 8 and 9).

To what extent is there intergenerational mobility in the United States? While data to answer this question is hard to come by, the best and most recent data comes from the Panel Study of Income Dynamics (PSID), a nationally representative longitudinal household survey following 18,000 Americans from 5,000 families since 1968. Researchers have used this data to compare the family income of native-born adults who were in their late 30s and early 40s (1995–2002) with that of their parents when they were in that same age range. After adjusting for inflation, the findings indicate that

- two out of three adult children have higher family incomes than their parents,
- adult children who lived in households in the bottom fifth quintile were the most likely to surpass their parents' income, and
- the higher the parents' family income, the higher the family income of their adult children.

Table 7.1b shows the chances that adult children's economic status increases, decreases, or remains the same as their parents' family income.

Table 7.1b: Percentage of Adult Children with Family Income That Exceeds, Falls Below, or Stays Even with Their Parents' Family Income

The table divides household income into five categories, or quintiles, and allows us to compare the family income of adult children with that of their parents when their parents were in the same stage of the life cycle (in this case late 30s and early 40s). The table shows that 42 percent of children who grew up in households with incomes in the bottom fifth quintile now live in households as adults that are in the bottom fifth. Nine percent of children who grew up in homes in the bottom fifth quintile now live in households with incomes in the top fifth. By contrast, 39 percent of the children who grew up in households with incomes in the top fifth also live in households as adults that are in the top fifth. The bold numbers along the diagonal represent the percentage of adult children who remained in the same income grouping as their parents.

	Quintile of Adult Children's Family Income (%)				
Quintile of Parents' Family Income	**Bottom Fifth**	**Second Fifth**	**Third Fifth**	**Fourth Fifth**	**Top Fifth**
Bottom Fifth	**42**	25	17	8	9
Second Fifth	23	**23**	24	15	15
Middle Fifth	19	24	**23**	23	14
Fourth Fifth	11	18	27	**24**	23
Top Fifth	6	10	19	26	**39**

Source of data: Isaacs (2011)

While it is clear from Table 7.1b that there is a relationship between parents' and their children's economic status as adults, it is clear that parents do not simply pass on their economic advantages or disadvantages to their children. If that were the case, 100 percent of children from each income group would live in households of the same income group as their parents. Still, 64 percent of those born in households where family income was in the top quintile have family incomes as adults that place them in the top two quintiles. Conversely, 66 percent of those born in households where family income was in the bottom quintile have family incomes as adults that place them in the bottom two quintiles.

(Write a Caption)

Write a caption that illustrates how the prestige of the achieved status "basketball player" in the United States is complicated by ascribed statuses.

Hints: In writing this caption

- review the concept of prestige,
- consider how being female complicates the status of basketball player, and
- consider what measure you might use to objectively determine prestige (e.g., salary of professional basketball players, size of fan base).

Mike Kaplan

Critical Thinking

Assign a monetary value to your life. Explain your rationale. (Do not try to deflect this question by saying it's impossible to put a price tag on anyone's life.)

Key Terms

- achieved statuses
- ascribed statuses
- caste system
- class system
- horizontal mobility
- life chances
- social mobility
- social prestige
- social stratification
- status value
- vertical mobility

MODULE

7.2 Social Class

Objective

You will learn about the theoretical traditions in sociology that give meaning and significance to the concept of social class.

Tony Rotundo

Tony Rotundo

Can you tell by looking into these cars the social class of their owners? What clues do you look for to determine that?

The cars are **status symbols**, visible markers of economic and social position and rank. Sociologists use the term **social class** to designate a person's overall economic and social status in a system of social stratification. They see class as an important factor in determining life chances. In our exploration of the concept of class, we begin with the writings of Karl Marx and Max Weber, who represent the "two most important traditions of class analysis in sociological theory" (Wright 2004, 1).

Karl Marx and Social Class

Karl Marx believed that the most important engine of change is class struggle. In the *Communist Manifesto*, written with Friedrich Engels, Marx (1848) describes how conflict between two distinct social classes propelled society from one historical epoch to another. He observed that the rise of factories and mechanization created two fundamental class divisions: the owners of the means of production (the bourgeoisie) and the workers (the proletariat), who must sell their labor to the bourgeoisie. The struggle between the two derives from the bourgeoisie's relentless drive to lower wages and the proletariat's efforts to resist wage cuts and increase pay. For Marx, then, the key variable in determining social class is source of income.

Although Marx pointed to the broad class struggles between the bourgeoisie and the proletariat, he identified other class categories. In the *Class Struggles of France, 1848–1850*, Marx named another class, the finance aristocracy, who lived in obvious luxury among masses of starving, low-paid, and unemployed workers. The finance aristocracy includes bankers and stockholders, seemingly detached from the world of work, who speculate in search of wealth. Marx described their income as "created from nothing—without labor and without creating a product or service to sell in exchange for [that] wealth" (Marx 1856).

Chris Caldeira

◀ If Marx were alive today, he would likely support the Occupy Wall Street movement. Marx (1856) labeled the financial institutions of Wall Street "temples of speculation." He viewed this speculation as "the plague of society and the states." Marx believed that the finance aristocracy put the welfare of the stock exchanges ahead of everything.

Max Weber and Social Class

Karl Marx clearly states that social class is based on people's sources of income (labor versus profiting from others' labor, for example). For Max Weber (1947), the basis of social stratification extends beyond source of income. He defined a social class as composed of those who hold similar life chances, determined not just by their sources of income but also by their marketable abilities (work experience and qualifications), by their access to consumer goods and services, and by their ability to generate investment income.

Tony Rotundo

▶ According to Weber, people completely lacking in skills, property, or employment, or who depend on seasonal or sporadic employment, constitute the very bottom of the class system. They form the **"negatively privileged" property class**.

Individuals at the very top—the **"positively privileged" property class**—monopolize the purchase of the highest priced consumer goods, have access to the most socially advantageous kinds of education, control the highest executive positions, own the means of production, and live on income from property and other investments. Between the top and the bottom of this social-status ladder is a series of rungs. Weber (1948) argued that a "uniform class situation prevails only among the negatively privileged property class." We cannot speak of a uniform situation regarding the other classes, because people's class standing is complicated by status groups and power. He defines a **status group** as an amorphous group of persons held together by virtue of a lifestyle that has come to be "expected of all those who wish to belong to the circle" and by the level of social esteem and honor others accord them (Weber 1948, 187). Weber's definition suggests that people can possess equivalent amounts of income and wealth and yet have very different lifestyles and be accorded different levels of social esteem. While accorded social esteem can be the result of income and wealth, it can also derive from physical attractiveness, a particular talent, race, sex, and so on.

Mr. Justin Matthew Ward (USACE)

◂ Status groups are held together by a lifestyle that is expected of all who belong. What lifestyle characteristics might bind together the people who live in the housing complex pictured?

In addition to class and social status, power is the third dimension of social stratification. Weber defined **power** as the probability that one can exercise his or her will in the face of resistance. Power derives from social class and social status but also from the **political parties**, organizations that try to acquire power to influence social action (Weber 1982). Parties are organized to represent people of a certain class or social status. The means of obtaining power can include engaging in violence, canvassing for votes, bribery, donations, and the force of speech. Examples of political parties are endless. Some examples within the United States include NOW (National Organization for Women), the NRA (National Rifle Association), and AARP (American Association of Retired Persons). Weber also recognized that power derives from occupying a position in a bureaucracy that bestows on someone tools and resources to rule and influence others.

Determining Social Class

Sociologist Erik Orin Wright (2009), who has spent his professional career working to clarify the concept of social class and its historical, structural, and personal significance, offers a series of questions to think about class.

IS THERE AN OBJECTIVE WAY TO ASSESS THE DISTRIBUTION OF MATERIAL INEQUALITY? Once sociologists settle on the number of class categories relevant to their analysis, they work to identify objective criteria by which to classify people into those categories. If they decide, for example, to categorize class in terms of upper, middle, and lower, the next step is to decide how to objectively measure social class: should accumulated wealth and/or annual income be used to determine one's social class?

Wealth refers to the combined value of a person's income *and* other material assets such as stocks, real estate, and savings minus debt. If we divide U.S. households into five wealth classes, we see that the top 20 percent of households control 87.2 percent of the wealth (the equivalent of $43.6 trillion). The bottom 80 percent of households control 12.8 percent of the wealth. The top 1 percent of all households control 35.6 percent of wealth.

Income refers to the money a person earns, usually on an annual basis through salary or wages. Within the United States, the average after-tax income of the richest 20 percent is $198,300, or 11.2 times that of the poorest 20 percent, who average $17,700. That is, for every $1,000 of taxed income earned by the poorest one-fifth, the top one-fifth earns $11,200. When we compare the after-tax income of the top 1 percent with that of the bottom 20 percent, the inequality is even greater. That 1 percent's after-tax income is $1.32 million, or 74.5 times greater than the bottom 20 percent (see Table 7.2a).

Table 7.2a: Average After-Tax Income in the United States by Income Group, 1979 and 2007 (in 2007 dollars)

The table compares after-tax income in the United States across six income groups at two points in time—1979 and 2007. Perhaps most striking is that the greatest financial gains during this time period were made by those in the top fifth, particularly the top 1 percent. The financial gain made by those in the top 1 percent is 405 times greater than that of the lowest one-fifth. (That 405 figure is obtained by dividing the dollar change of the top 1 percent [$973,100] by the dollar change of the lowest fifth [$2,400]).

Income Category	1979 Average After-Tax Income	2007 Average After-Tax Income	Percent Change	Dollar Change
Lowest Fifth	$15,300	$17,700	15.6%	$2,400
Second Fifth	$31,000	$38,000	22.5%	$7,000
Middle Fifth	$44,000	$55,300	25.6%	$11,300
Fourth Fifth	$57,700	$77,700	34.6%	$20,000
Top Fifth	$101,700	$198,300	94.9%	$96,600
Top 1 Percent	$346,600	$1,319,700	280%	$973,100

Source: Congressional Budget Office (2010)

HOW MUCH INCOME OR WEALTH QUALIFIES SOMEONE TO BE UPPER, MIDDLE, OR LOWER CLASS? There is no agreement on the amount of income or wealth that distinguishes the social classes. The Pew Research Center (2008) defines people as middle-income class in the United States if they live in a household with an annual income from 75 to 150 percent of the annual median household income. A person with a household income above that range is upper-income class; a person whose household income is below that range is low-income class. In 2010 the median annual household income was $50,046 (U.S. Census Bureau 2011c). Seventy-five percent of that number is $37,434; 150 percent is $75,069. So, a middle-class household is one with an annual income between $37,434 and $75,069. Of course, this represents just one way to conceptualize social class. Another estimate of what constitutes middle-class household income is based on census data, which places 60 percent of households as earning between $28,636 and $79,040 (U.S. Census Bureau 2011c). In evaluating these estimates, household composition matters—obviously a four-person household earning $28,636 is not middle class; but a one-person household might be considered as such.

HOW DO PEOPLE SEE THE CLASS STRUCTURE AND WHERE DO THEY LOCATE THEMSELVES AND OTHERS? Objective measures of social class do not usually correspond with people's self-assessment of their class location. For example, the Pew Research Center (2008) found that an objective measure of social class—household income between $37,434 and $75,069—would put about 65 percent of households in that class. A subjective measure asking people to classify their household as upper, middle, or lower resulted in 41 percent of Americans with family incomes less than $20,000 declaring themselves middle class.

Pew researchers found that the greater a person's income, the higher the estimate of the amount of money it takes to be middle class. On average, those with household incomes between $100,000 and $150,000 a year believe that it takes an $80,000 annual income to live a middle-class lifestyle. Conversely, those from households earning less than $30,000 a year maintain that it takes about $50,000 a year (Pew Research Center 2008, 15).

(Write a Caption)

Write a caption that uses one or more of Weber's concepts to describe the 99% this sign references versus the 1%.

Hints: In writing this caption

- review the various concepts associated with Max Weber,
- think about how the concepts apply to the top 1%, especially the concept of positively privileged, and
- think about why the Occupy Wall Street movement is seeking to define the 99% as a status group.

Critical Thinking

Define your social class using concepts from this module.

Key Terms

income

negatively privileged property class

political parties

positively privileged property class

power

social class

status group

status symbols

wealth

Global Inequality

MODULE 7.3

Objective

You will learn that income, wealth, and other valued resources are distributed unequally across and within countries.

To what extent do you think your life has been shaped by the country in which you were born?

Sgt. Michael Conner

Norris Jones, CIV, USACE

Clearly, we have no control over which of the world's 200-plus countries we are born in, yet where people are born affects their life chances in dramatic ways. The chances that young girls will spend a large portion of their day carrying water from a community water well to their home, which may be miles away, is quite high in Iraq. The chances that young girls will spend time using personal technologies is quite high in the United States. In assessing global inequality, sociologists look to see how wealth, income, and other valued resources—whether they be electronic technologies, water, or even toys—are distributed among the 7.0 billion people living in the 240 or so countries on the planet. That distribution is marked by extremes of poverty and wealth.

Poverty and Wealth

It is difficult to define what it means to live in poverty, except to say that it is a situation in which people have great hardship meeting basic needs for food, shelter, and clothing. Poverty can be thought of in absolute or relative terms. **Absolute poverty** is a situation in which people lack the resources to satisfy the basic needs no person should be without. Absolute poverty is usually expressed as a living condition that falls below a certain threshold or minimum. For example, the World Bank (2010) has set that threshold at US$1.25 per day and estimates that 1.4 billion people live in a state of absolute poverty.

Relative poverty is a situation in which a person is disadvantaged when compared with a person in an average or more advantaged situation. Thus, one thinks not just about survival needs but also about a lack of access to goods and services that have become defined as essential.

Missy Gish

➤ Although people do not "need" the latest mobile phone to survive, many today in the United States can feel they are at a disadvantage if they possess a model (right) that does not support a touch screen (left), for example, when those around them have the latest technologies.

In contrast to poverty, **extreme wealth** is the most excessive form of wealth, in which a very small proportion of people in the world have money, material possessions, and other assets (minus liabilities) in such abundance that a small fraction of it, if spent appropriately, could provide adequate food, safe water, sanitation, and basic health care for the 1 billion poorest people on the planet (United Nations 2006). Table 7.3a presents estimates of how many people in the world are extremely wealthy.

▼ Table 7.3a: The Distribution of Global Household Wealth

The table shows various income categories and the estimated number of adults in each category. To be classified as "extremely wealthy," a person must have a minimum wealth (assets minus liabilities) of $1 billion. Worldwide, only about 800 adults are categorized as such. To be among the richest 1 percent, a person must have between $500,000 and $1 million (excluding the value of his or her home) in wealth. The richest 1 percent holds 40 percent of the world's wealth, which is estimated to be $125 trillion. Note that the poorest 50 percent of the world's adult population—1.9 billion people—possess less than 1 percent of the world's wealth.

Category	Estimated Number of Adults in Category	Minimum Wealth* (assets – liabilities) to Qualify for Income Category	Percent of Total Household Wealth	Estimated Combined Wealth
Extremely Wealthy	800	$1 billion	Not known	Not known
Ultra Rich	85,400	$30 million	Not known	Not known
Richest 0.025%	8.3 million	$1 million	Not known	Not known
Richest 1%	37 million	$500,000	40%	$49.6 trillion
Richest 10%	370 million	$61,000	85%	$105 trillion
Middle 40%	1.4 billion	$2,200	14%	$17 trillion
Poorest 50%	1.9 billion	Under $2,200	Less than 1%	$1.2 trillion

Source: United Nations University (2006)
*Does not include the value of the house in which the person resides

INEQUALITIES ACROSS COUNTRIES. When studying inequalities across countries, sociologists identify a valued resource and estimate the chances of achieving or acquiring it by country. For example, parents everywhere want their children to survive childbirth and beyond, so sociologists compare the chances that a baby will survive the first year of life by country (see Table 7.3b). Infant mortality is considered a key indicator of the overall well-being of a nation, as it reflects maternal health, socioeconomic conditions, and quality of and access to medical care (Centers for Disease Control 2011a).

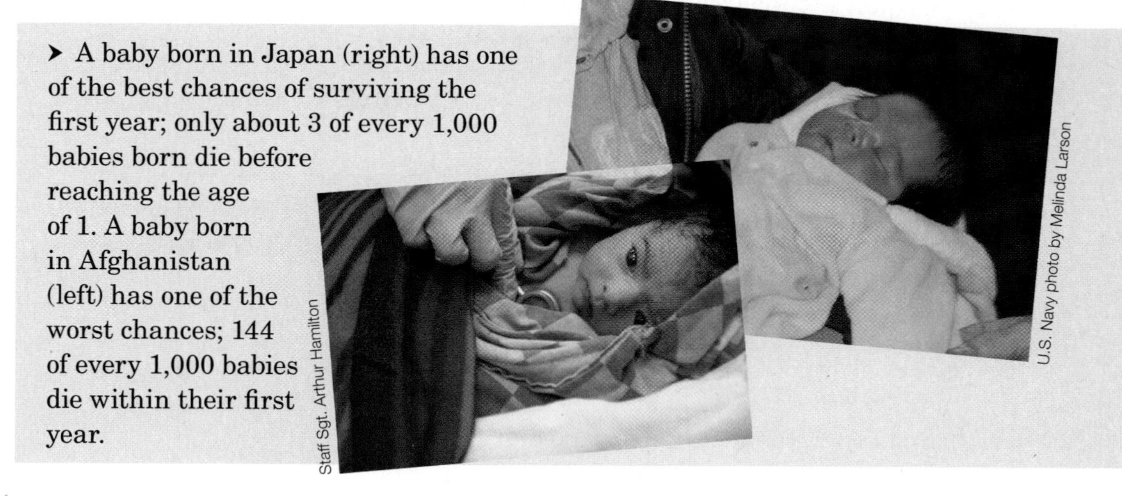

➤ A baby born in Japan (right) has one of the best chances of surviving the first year; only about 3 of every 1,000 babies born die before reaching the age of 1. A baby born in Afghanistan (left) has one of the worst chances; 144 of every 1,000 babies die within their first year.

▼ **Table 7.3b: Countries with the Best and Worst Chances of Infants Surviving the First Year of Life (infant mortality per 1,000 live births)**

The chances of staying alive during the first year of life vary from country to country. Infant mortality is the number of deaths in the first year of life for every 1,000 live births.

Worst Chances (deaths per 1,000 live births)		Best Chances (deaths per 1,000 live births)	
Afghanistan	144.2	Singapore	2.6
Sierra Leone	134.3	Iceland	2.9
Democratic Republic of the Congo	121.4	Japan	3.1
Liberia	119.6	Sweden	3.2
Angola	119.4	Finland	3.3

Source: Data from United Nations (2011b)

Other startling examples of how life chances vary across countries emerge when we compare the consumption patterns of people living in the highest-income countries against those living in low-income countries. One example is meat consumption.

▲ The chances of consuming large quantities of meat on a regular basis are high in the United States—the per capita (per person per year) consumption of meat (beef, veal, chicken, and pork) is 262.21 pounds (577 kilograms). The United States, which has 4.6 percent of the world's population, consumes 21 percent of the meat consumed in the world. In a country like Laos, the annual per capita consumption of meat is 15 pounds (32.2 kilograms), and more cuts of meat (such as feet, heart, and intestines) are consumed (U.S. Census Bureau 2012f).

INEQUALITIES WITHIN COUNTRIES. Sociologists look within countries to consider how wealth, income, and other valued resources are distributed. They ask questions like, How does the income of the richest 10 percent compare with that of the poorest 10 percent? The country in which the gap between the richest 10 percent and bottom 10 percent is greatest is Sierra Leone. There, the richest 10 percent earn 87.2 times that of the poorest 10 percent. To put it another way, for every $1 earned by the poorest 10 percent, the richest 10 percent earns $87.20. The country with the smallest gap between the richest and poorest 10 percent is Japan. In Japan the richest 10 percent earn 4.5 times that of the poorest 10 percent. Thus, for every $1.00 earned by the poorest 10 percent, the richest 10 percent earns $4.50. In the United States, the richest 10 percent earn 16 times the poorest 10 percent (World Bank 2011a).

Given that Japan is considered the most equal society on this measure, it might come as a surprise to learn that many Japanese see this gap between the top and bottom 10 percent as still too wide and express concern about it. In general, when compared with the way wages are distributed elsewhere, Japan has wage equality. One notable example involves the pay of CEOs who head Japan's largest 100 corporations. The average salary is $1.6 million compared with their CEO counterparts in the United States and Europe, who earn average salaries of $13.3 and $6.6 million, respectively (Hall 2009).

Sierra Leone's income inequality reflects the contradiction of a mineral-rich country in which a substantial portion of people live in poverty. Its mineral resources include diamonds, titanium, bauxite, gold, and rutile. A former colony of Great Britain, Sierra Leone gained independence in 1961. Control of the country's resources fueled a decade-long civil war from 1991 until 2001, in which virtually the entire population was displaced. The atrocities of this civil war have been documented and include amputations and systematic abuses of women too horrific to mention here (Ben-Ari and Harsch 2005).

Mass Comm. Spc. First Class Darryl Wood

◀ These children, who were just given soccer balls by the U.S. military, live at the Christian Mission Foster Homes for Orphans (CMFO), a nongovernmental organization (NGO). The CMFO was established in April 1999 to rescue children affected by 11 years of civil war but especially those children who became orphans as a result of a particularly brutal phase on January 6, 1999, at which point world leaders intervened. For Sierra Leone, that date has the significance that September 11, 2001, has for Americans (Christian Mission Foster Homes for Orphans 2012).

There are essentially two views on why some countries remain poor or underdeveloped. One view—modernization theory—is that poor countries are poor because they have yet to develop into modern economies and that their failure to do so is largely the result of internal factors. Another view—dependency theory—holds that for the most part poor countries are poor because they are products of a colonial past.

▼ **Figure 7.3a: The World's 25 Most Advantaged and Disadvantaged Populations** The map highlights the 25 countries that scored highest and the 25 countries that scored lowest on the UN Human Development Index (HDI) that takes into account three areas: "a long and healthy life (health), access to knowledge (education) and a decent standard of living (income)." The country that ranked highest in 2011 was Norway, and the country that ranked lowest was Democratic Republic of the Congo.

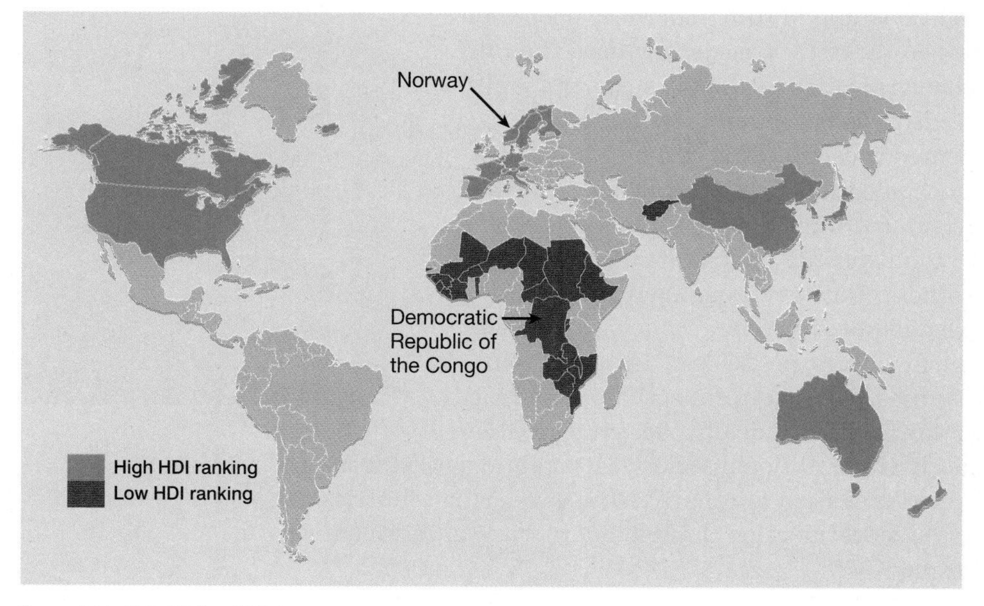

Source of data: United Nations (2011a)

Modernization Theory

Modernization is a process of economic, social, and cultural transformation in which a country "evolves" from an underdeveloped to a modern society. A country is considered modern when

1. a high proportion of the population lives in and around cities;
2. the energy to produce food, make goods, and provide services does not come primarily from physical exertion (human and animal muscle) but from machines powered by fossil fuels;
3. goods and services are widely available;
4. people have a voice in economic and political affairs;
5. literacy is widespread and there is a scientific orientation to solving problems;
6. a system of mass media and communication is in place that offsets the influence of the family and local cultures as the source of information;
7. large-scale, impersonally administered organizations, such as government, businesses, schools, and hospitals, reduce dependence on family for child care and education and serve as a safety net; and
8. people feel a sense of loyalty to a country rather than to an extended family and/or tribe (Naofusa 1999).

➤ This village school in Afghanistan enrolls over 600 students. It has no modern conveniences, and the children may attend on the condition that their fathers allow them. Modernization theorists would argue that a country will not modernize until such decisions are government-mandated and not left in the hands of a male parent.

Sgt. Robert M. Storm

Modernization theorists seek to identify the conditions that launch underdeveloped countries on the path to modernization. The road to modernization is a multi-stage and multi-decade process beginning with a tradition-oriented way of life and ending with a level of technological and economic maturity that can support high mass-consumption. Western countries can jump-start modernization in other countries through foreign aid and investments, including technology transfers (e.g., fertilizers, pesticides), contraception programs, loans, cultural exchange, and medical interventions. Ideally, these interventions "shock" the traditional society, setting into motion ideas and sentiments that support a modern alternative. The developing countries can also take steps to hasten modernization through appropriate government reforms and policies (Rostow 1960).

According to modernization theorists, modernization involves a transformation of cultural values emphasizing fatalism and group orientation to ones that emphasize a work ethic, deferred gratification, future orientation, and individualism (attitudes and traits believed to be essential to the development of a free-market economy). As the country modernizes, "the idea spreads, not merely that economic progress is possible, but that economic progress is a necessary condition" (Rostow 1960).

Dependency Theory

Dependency theorists reject the basic tenet of modernization theory—that poor countries fail to modernize because they lack the cultural values that drive economic prosperity. Rather, dependency theorists argue that poor countries are poor because they have been, and continue to be, exploited by the world's wealthiest countries and by global and multinational corporations. This exploitation is an extension of **colonialism**, a form of domination in which a foreign power uses superior military force to impose its political, economic, social, and cultural institutions on an indigenous population so it can control their resources, labor, and markets (Marger 1991).

Colonization reached its peak in the early 1900s, and it was not until around the midpoint of the 20th century and beyond that 130 countries and territories gained political independence from their colonial masters in a process known as **decolonization**. Once independence is achieved, civil war between rival factions often takes place as each seeks to secure the power once wielded by the colonizer. Some scholars argue that the Americas (which include the United States, Canada, and Central and South America) are technically still colonized because the indigenous peoples did not revolt and declare independence; rather, it was the European colonists and/or their descendants who did so. Once independence was gained, those in power simply continued colonizing and exploiting the land and resources that once belonged to indigenous peoples (Cook-Lynn 2008; Mihesuah 2008).

Gaining political independence does not mean, however, that a former colony no longer depends on the powers that once controlled them. The countries of the African continent—90.4 percent of which were once controlled by colonial powers—still extract, process, and produce primary products for the West (Rodney 1973). This continuing economic dependence on former colonial powers is known as **neocolonialism**. In other words, neocolonialism is a new form of colonialism under which foreign governments and foreign-owned businesses continue to exploit the resources and labor of the postcolonial peoples.

Some critics would argue that the U.S. military presence in and around the continent of Africa is a form of neocolonialism. In 2007 the U.S. Department of Defense put the continent of Africa under one military command known as AFRICOM. Many critics question AFRICOM, calling it a U.S. grab for African resources. The official U.S. position is that AFRICOM "allows the U.S. military to help the Africans help themselves, provide security, and to support the far larger U.S. civilian agency programs on the continent" (United States African Command 2008).

(Write a Caption)

Write a caption that relates this image to a characteristic of modernization.

Hints: In writing this caption

- review the characteristics of modernization, and
- consider which characteristic of modernization is not realized.

Chris Caldeira

Critical Thinking

Study Table 7.3a, which shows the distribution of global household wealth. Where does your household fall? Does this surprise you?

Key Terms

absolute poverty	extreme wealth	relative poverty
colonialism	modernization	
decolonization	neocolonialism	

MODULE

7.4 Responses to Global Inequality

Objective

You will learn about a plan to reduce global inequality by 2015.

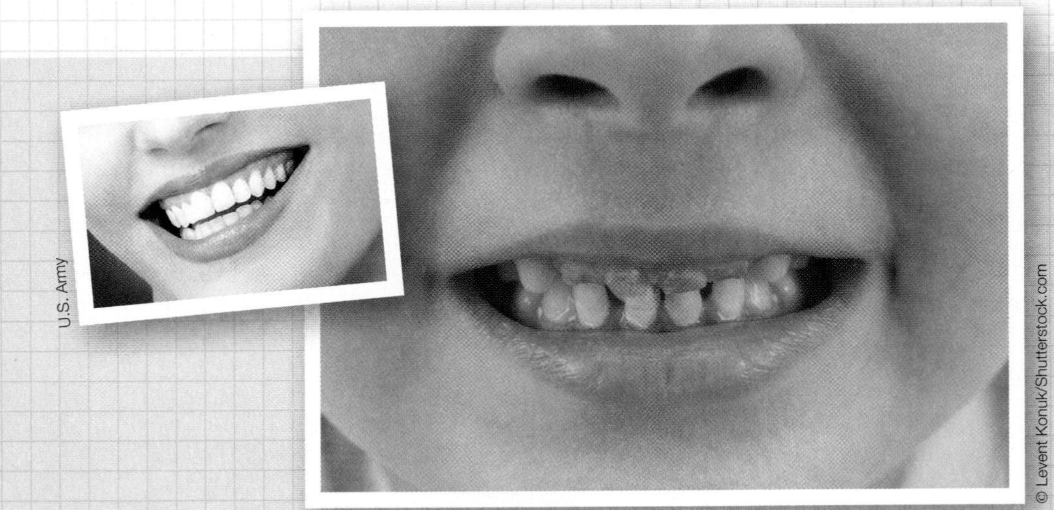

U.S. Army

© Levent Konuk/Shutterstock.com

Why is there such dramatic inequality in the world? Should we take from the rich and give to the poor?

One obvious way to reduce global inequality is to redistribute wealth by transferring some of it from the wealthiest to the poorest. While there is no plan in the works to take from rich people per se, since 2000 a plan known as the Millennium Declaration has been in place to redistribute wealth from the richest countries to the poorest ones. The question is whether that plan can be realized.

Millennium Development Project

The United Nations Millennium Development Project (United Nations 2010b) sets 18 targets to be reached by 2015, including these three:

- halve the proportion of people whose income is less than one dollar a day,
- halve the proportion of people who suffer from hunger, and
- reduce by three-quarters the maternal mortality rate.

Clearly these are ambitious targets. Their success hinges on at least two major commitments from the world's 22 richest countries:

1. increase current levels of foreign aid by investing seven-tenths of 1 percent of GDP, and
2. develop an open, nondiscriminatory trading system.

INCREASE FOREIGN AID. Under this plan the United States, the richest country as measured by its gross domestic product (GDP) of $14.6 trillion, must increase its current level of foreign aid from $34.5 billion to $102 billion. The 22 richest countries, with a combined GDP of $36.7 trillion, should increase foreign aid from $102.6 to $256 billion (U.S. Central Intelligence Agency 2012; United Nations Millennium Development Project 2000). To date the richest countries have not met this commitment.

While it is true that the United States and other wealthy countries offer assistance in other, less officially recognized forms, including private donations, wage remittances sent home from immigrants, and trade investments, we will focus on how its foreign aid is invested. Jeffrey Sachs (2010), director of the UN Millennium Project, argues that USAID, an agency funded by the State Department, disproportionately allocates funds to countries where the U.S. is at war "as an adjunct to U.S. military and foreign policy."

➤ USAID provides economic and humanitarian assistance worldwide. That agency pays for emergency food shipments but about half of its budget goes to companies and organizations that transport the food. In addition, while USAID also supports road construction and clean water projects, a high percentage of that budget goes to paying salaries of American consultants on these projects.

MC2 Chris Lussier

In fact a review of the $34.5 billion in foreign aid that the U.S. Department of State allocates each year shows that the bulk of the assistance goes not toward development but toward crisis intervention, such as disaster and famine relief and refugee programs, military training, security, and narcotics control (U.S. Department of State 2012). For the UN Millennium Development plan to work, countries must not only increase their foreign aid but also change the type of foreign aid that they give.

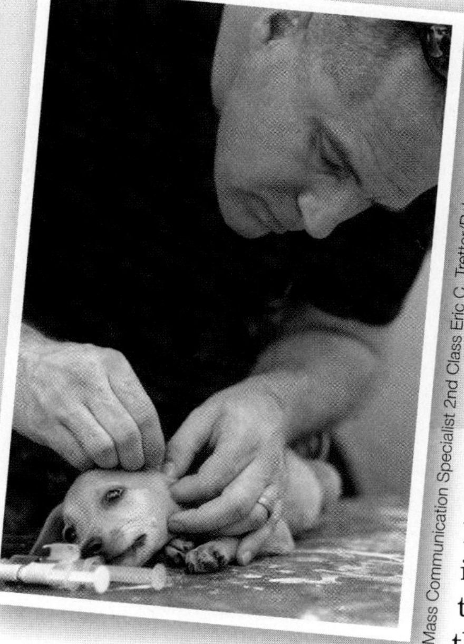

➤ There is no question that the United States often plays an important humanitarian role helping people (and animals) in resource-poor countries. Here we see a Navy veterinarian removing ticks from a puppy at a temporary animal clinic in Port-au-Prince, Haiti. The service was part of the 5-month humanitarian mission called Continuing Promise to the Caribbean, Central and South America. While this type of assistance is valued, it does not contribute to lasting development.

Mass Communication Specialist 2nd Class Eric C. Tretter/Released

END SUBSIDIES, TARIFFS, AND QUOTAS. Even though the wealthiest countries have agreed in principle to eliminate tariffs, subsidies, and quotas on products imported from the poorest countries, the wealthiest have resisted dismantling a system of trade structured to their advantage (Bradsher 2006). In particular, the United States, Japan, the European Union, and other high-income countries continue to subsidize agriculture and other sectors, such as steel, so that producers in these countries are paid more than world market value for their products. Considerable attention has been given to agricultural subsidies, which give farmers in wealthy economies an estimated $376 billion in support (United Nations 2010b). It is well documented that those subsidized are the large agricultural corporations, not small farmers. For example, Riceland Foods (the top recipient) received $554.3 million in subsidies between 1995 and 2010 (Environmental Working Group 2012).

In addition, the wealthiest countries apply tariffs and quotas to many imported items, thereby increasing their cost to consumers. Consider sugar. The U.S. government sets quotas limiting the amount of raw sugar imported from Brazil and about 40 other countries that sells for about 12.2 cents per pound on the world market. The U.S. government also adds a tariff to the sugar that is imported. These policies limit competition with U.S. sugar growers who sell their raw sugar to processors in the United States for 20.8 cents per pound. Of course this means that sugar prices in the United States are artificially high (Tampa Bay Times 2012).

Such policies affect not just workers in the poorest countries but also workers in wealthy countries. For example, Brach's and Kraft Foods closed their U.S.-based candy plants and outsourced more than 1,000 jobs to Argentina and Canada, where sugar can be purchased at lower, world-market prices (Kher 2002).

Of course there are often good reasons that the U.S. government applies subsidies and tariffs. For example, the United States imposed a 35 percent tariff on tires from China after a surge in tire imports from that country lowered tire production in the United States, from 218.4 million to 160.3 million tires per year. The U.S. government argued that China was subsidizing its domestic tire production by undervaluing its currency. Its low labor costs also gave China an unfair competitive advantage (Chan 2010). In this situation, one would hardly blame the United States for imposing tariffs. Despite these complications, as of 2010, 81 percent of the goods imported from the poorest countries by the richest countries now enter duty-free. The UN goal is 97 percent (United Nations 2010b).

Chris Caldeira

➤ The wealthiest countries are not the only ones imposing tariffs on imports. Anyone who visits Vietnam, especially its urban areas, quickly observes that it is a land of motorbikes. The widespread use of motorbikes can be explained in part by the tariffs Vietnam places on imported cars and small trucks, which was once 90 percent and is now being reduced to between 47 and 70 percent by 2017. The high tariffs make it expensive to purchase trucks, so people turn motorbikes into "trucks" (*Thanh Nien News* 2006; Business-in-Asia.com 2012).

Criticism of the Millennium Declaration

Critics of the UN plan argue that there are other factors that make it difficult for poor economies to compete in the global economy. As a case in point, multinational and global corporations take advantage of those who will work for less. The 159 million Apple products sold in 2011—iPhones, iPads, and other electronic products—were manufactured overseas, primarily in China and other Asian countries where employees live in company dorms, work 60 hours or more each week, and earn $17 or less per day. Millions compete for these contract jobs that support the operation known as Apple that employs 63,000 workers and earns $400,000 in profit per employee (Duhigg and Bradsher 2012).

A second reason why critics say the UN plan is not sufficient relates to **brain drain**, the emigration from a country of the most educated and most talented people, including actual or potential hospital managers, nurses, accountants, teachers, engineers, political reformers, and other institution builders (Dugger 2005). The rich economies have facilitated brain drain by implementing immigration policies that give preference to educated, skilled foreigners such as nurses.

Photographer's Mate 2nd Class Timothy Smith

◂ The concept of brain drain is particularly relevant to health care workers. In fact, the British Medical Association (Mayor 2005) has grown so concerned about the shortage of health care workers around the world and the migration of such workers from poor to rich countries that it has called for efforts to reduce this trend. Special immigration policies lure these Indonesian and other nurses away from their home countries to work in the richest countries.

Finally, critics argue that more focus should be placed not on the UN Millennium Project but on the hundreds of creative and successful efforts to reduce poverty. One is the Grameen Bank microlending project, which was first piloted in 1976 in Bangladesh, a country whose economy is among the poorest. The goal was to examine the possibility of extending tiny loans to the poorest of rural poor women. The anticipated outcome was to eliminate high interest rates moneylenders charge the poor and to create opportunities for self-employment among the unemployed living in rural Bangladesh. Today, the bank that piloted microlending has 1,175 branches, with an estimated 2.4 million borrowers living in 41,000 villages (about 60 percent of all villages in Bangladesh). An estimated 90 percent of borrowers repay the loans (Grameen 2005).

(Write a Caption)

Write a caption that contrasts the development value of the two forms of foreign aid—one symbolized by a woman carrying off a box of donated bottled water, the other by a child drinking clean water from a new water well constructed by the U.S. military.

Chief Mass Comm. Spc. James G. Pinsky

U.S. Air Force photo by Staff Sgt. Joseph L. Swafford Jr./Released

Hints: In writing this caption

- review the United Nations Millennium Development Project plan for eliminating poverty, and
- consider the goals for foreign aid and what type of foreign aid has lasting effects.

Critical Thinking

Use the Internet to find a nongovernmental organization (NGO) with programs that reduce poverty in some of the world's poorest countries.

Key Term

brain drain

MODULE 7.5

Why Inequality?

Objective

You will learn three sociological perspectives on social inequality.

Lisa Southwick

Cpl. Paula M. Fitzgerald

Which occupation is more important to society—physician or sanitation worker?

Social inequality is the unequal access to and distribution of income, wealth, and other valued resources. Sociologists draw upon three perspectives to explain inequality and to understand how it is manifested in daily life. Those perspectives are functionalism, conflict, and symbolic interaction.

Functionalist Perspective

Sociologists Kingsley Davis and Wilbert Moore (1945) argue that social inequality is the device by which societies ensure that the most functionally important occupations are filled by the best-qualified people. That is the reason sanitation workers in the United States earn an average of $33,660 per year and physicians earn an average of $180,000 per year. The $146,340 difference in average salary represents the greater functional importance of the physician relative to the sanitation worker (U.S. Bureau of Labor Statistics 2011d, 2011e).

Davis and Moore argue that society does not have to offer extra incentives to attract people to the occupation of collecting garbage because most can do that job with little training. Society does have to offer extra incentives to attract the most talented people to occupations such as physician that require long, arduous training and a high level of skill. Davis and Moore concede that the stratification system's ability to attract the most talented and qualified people is weakened when

- capable people are overlooked,

- some categories of people are deemed ineligible to apply;
- factors other than qualifications are involved in filling positions.

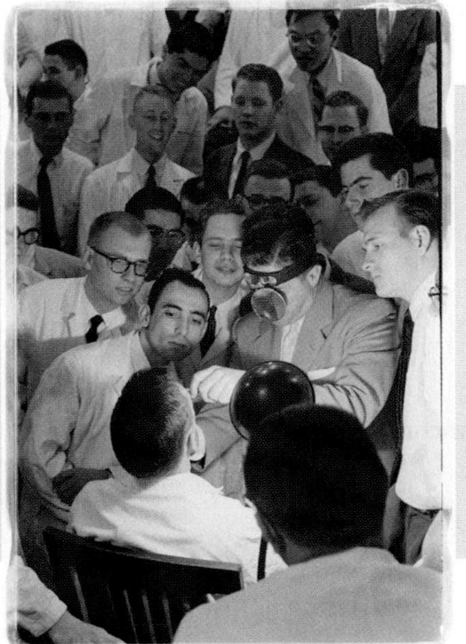

➤ This 1958 photo of all-male medical students at George Washington University Medical School shows that, for the most part, capable women and nonwhites were denied access to medical schools at that time. As a result, few women and minorities applied for medical school and those who did very likely faced rejection. According to Davis and Moore's theory, that practice weakened the stratification system's ability to attract the most talented and qualified people.

Prints & Photographs Division, Library of Congress, LC-DIG-ppmsca-03097

Davis and Moore maintain, however, that society eventually corrects the problem, as evidenced by the fact that schools eventually began admitting people from groups once denied access.

Conflict Perspective

Melvin M. Tumin (1953) and Richard L. Simpson (1956) challenge the functionalist assumption that social inequality is a necessary device societies use to attract the best-qualified people for functionally important occupations. These sociologists point out that some positions command large salaries even though their contributions to society are questionable. Consider the salaries of Division 1 college basketball coaches; the median salary is slightly under $1 million. The pay of the 68 coaches whose teams qualified for the NCAA tournament in March 2010 ranged from a high of $7.5 million to a low of $424,000 (with a potential to earn $257,000 in bonuses). We might argue that Division 1 basketball coaches deserve large, in some cases enormous, salaries because they generate income for the university and surrounding communities (Tyler 2011). But are their contributions more essential than that of the players, most of whom are compensated only with scholarships? Should we assume that coaches require large salaries without which they would refuse to coach? Moreover, a close look at the data shows that only about 10 percent of Division I schools generate enough revenue to cover expenses (Wilson et al. 2011).

In critiquing the functionalist perspective on inequality, Tumin and Simpson also ask why some workers receive a lower salary than other workers for doing the same job, just because they are of a certain race, age, sex, or national origin. After all, the workers are performing the same job, so functional importance is not the issue. For example, why do women working full time as registered nurses in the United States earn a median weekly wage of $1,039 while their male counterparts earn $1,201? In fact, U.S. Bureau of Labor Statistics (2011c) data shows only 4 of 300 occupational categories in which median weekly earnings for females exceeds those of their male counterparts: food preparation and serving workers ($388 versus $346); counselors ($818 versus $780); bill and account collectors ($634 versus $579); and stock clerks and order fillers ($495 versus $471).

In addition to the issue of pay equity, we must raise the question of comparable worth. Assuming comparable worth, why should full-time workers at a child day care center who are overwhelmingly female earn a median weekly salary of $398 while auto mechanics who are overwhelmingly male earn $683 (U.S. Bureau of Labor Statistics 2011c)?

Conflict theorists also ask if dramatic differences in pay are really necessary to make sure that someone takes the job of CEO over, say, the job of a factory worker. Probably not. In the United States, the median base salary of the CEO of the largest corporations is $9.0 million. This median base salary (excluding bonuses, stock options, and other benefits) is 250 times the median household income. Consider that the CEO of Verizon Communications was awarded a $17.9 million compensation package in 2010. Aside from the CEO's skills, the company's success can be partly attributed to the fact that its smart phones are made by workers in China and other Asian countries who earn $17.00 per day. Assuming a six-day work week, workers earn the equivalent of $5,304 a year. The Verizon CEO earns 3,374 times more than the factory worker (Krantz and Hansen 2011; Duhigg and Bradsher 2012).

Finally, Tumin (1953) and Simpson (1956) further criticize the functionalist position by arguing that specialization and interdependence make every position necessary. Thus, to judge that physicians are functionally more important than sanitation workers fails to consider the historical importance of sanitation relative to medicine. Contrary to popular belief, advances in medical technology had little influence on death rates until the turn of the 20th century—well after improvements in nutrition and sanitation had caused dramatic decreases in deaths due to infectious diseases.

Symbolic Interactionist Perspective

When symbolic interactionists study social inequality, they seek to understand the experience of social inequality; specifically, they seek to understand how social inequality shapes interactions and people's experiences.

In the tradition of symbolic interaction, journalist Barbara Ehrenreich (2001) studied inequality in everyday life as it is experienced by those working in jobs that paid $8.00 or less per hour. Ehrenreich, a "white woman with unaccented English" and a professional writer with a Ph.D. in biology, decided to visit a world that many others, as many as 30 percent of the workforce, "inhabit full-time, often for most of their lives" (6). Her aim was just to see if she could "match income to expenses, as the truly poor attempt to do every day" (6). In the process, Ehrenreich worked as a "waitress, a cleaning person, a nursing home aid, or a retail clerk" (9). Ehrenreich learned that "low-wage workers are no more homogeneous in personality or ability than people who write for a living, and no less funny or bright. Anyone in the educated classes who thinks otherwise ought to broaden their circle of friends" (8).

Ehrenreich's on-the-job observations reveal the many ways inequality is enacted. Ehrenreich tells of a colleague who becomes "frantic about a painfully impacted wisdom tooth and keeps making calls from our houses (we are cleaning) to try and locate a source of free dental care" (80). She tells of a colleague who would like to change jobs but the act of changing jobs means "a week or possibly more without a paycheck" (136); and then there is the colleague making $7.00 per hour at K-Mart thinking about trying for a $9.00-per-hour job at a plastics factory (79).

➤ Among the low-wage workers to which Ehrenreich's research applies are the hundreds of thousands of workers who assemble sandwiches and other quick service foods.

Lisa Southwick

(Write a Caption)

Write a caption using the conflict perspective to critique the functionalist assumptions about purpose of unequal pay for mechanics and day care providers.

Emily Brainard

Ignacio Rubalcava (USAG Baumholder)

Hints: In writing this caption

- review the functionalist perspective on inequality, and
- consider the conflict theorist critique.

Critical Thinking

Which view of social inequality do you find most compelling? Why?

Key Term

social inequality

Poverty and Its Functions

MODULE 7.6

Objective

You will learn what poverty is and who benefits from its existence.

An estimated 63 million computers are deemed obsolete each year in the United States. Where do they go?

© David Maska /Shutterstock.com

You might be surprised to learn that most of the obsolete computers are "donated" to poor countries. Each month Nigeria alone receives 400,000 such computers, 75 percent of which are nonfunctional and end up in landfills, where toxic materials pollute the surrounding environment (Dugger 2005). This dumping can be considered a function of poverty.

The Functions of Poverty

In his classic essay, "The Positive Functions of Poverty," sociologist Herbert Gans (1972) asked, "Why does poverty exist?" He answered that poverty performs at least 15 functions, several of which are described here.

FILL UNSKILLED AND DANGEROUS OCCUPATIONS. The poor have no choice except to take on the unskilled, dangerous, temporary, dead-end, undignified, menial work of society at low pay. Obviously, the lower the wage, the lower the labor cost needed to make goods and services, and the greater the employer's profits. Hospitals, hotels, restaurants, factories, and farms draw employees from a large pool of workers who are forced to work, who are willing to work, or who act willing to work at minimum wage or below. In the United States, for example, there are about 751,000 full-time maids and housekeeping cleaners who are overwhelmingly female (88 percent) and Latino (44 percent) whose median wages are $399 per week (U.S. Department of Labor 2012b).

PROVIDE LOW-COST LABOR FOR MANY INDUSTRIES. Many industries, including health care, fast food, hotels, and agriculture, depend on low-wage labor to ensure a profit.

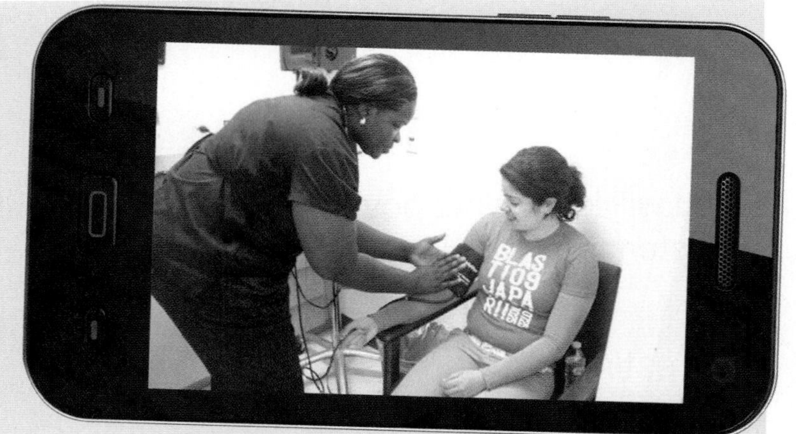

Lisa R. Rhodes

➤ When we think of low-wage laborers we often think of those who harvest fruits and vegetables, work as nannies, change beds and launder sheets and towels in hotels, and change diapers and bedpans of those dependent on care (Murphy 2004). Many who work in health care support also earn low wages. The median weekly earnings of nurses' aides is $387 per week (Bureau of Labor Statistics 2012).

SERVE THE AFFLUENT. Affluent (or relatively affluent) persons contract out and pay low wages for many time-consuming activities, such as housecleaning, yard work, and child care. On a global scale, millions of poor women work outside their home countries as maids in middle- and upper-class homes. Consider that an estimated 460,000 Indonesian women from poor villages work as maids in Malaysia and Saudi Arabia (Perlez 2004). Even the U.S. military depends on low-wage workers from the Philippines. While only 51 Filipino troops served as part of the U.S.-led coalition in the Iraq war, more than 4,000 Filipinos prepared food, cleaned toilets, and performed other support-related tasks for the American forces (Kirka 2004).

VOLUNTEER FOR DRUG TRIAL TESTS. The poor often volunteer for over-the-counter and prescription drug tests. Most new drugs, ranging from AIDS vaccines to allergy medicines, must eventually be tried on healthy human subjects to determine their potential side effects (such as rashes, headaches, vomiting, constipation, or drowsiness) and appropriate dosages. Money motivates people to volunteer as subjects for these clinical trials. Because payment is relatively low, however, the tests attract a disproportionate share of low-income, unemployed, or underemployed people (Morrow 1996).

SUSTAIN ORGANIZATIONS AND EMPLOYMENT. Many businesses, governmental agencies, and nonprofit organizations exist to serve poor people or to monitor their behavior, and of course the employees of these entities draw salaries for performing such work. In the United States there are 235,000 nonprofits with a combined budget of $230 billion serving the disadvantaged in mental health/crisis intervention, employment, nutrition, housing, and community improvement (National Center for Charitable Statistics 2009).

PURCHASE OR USE PRODUCTS THAT WOULD OTHERWISE BE DISCARDED. Poor people use goods and services that would otherwise go unused and be discarded. Day-old bread, used cars, and secondhand clothes are purchased by or donated to the poor.

Missy Gish

➤ These pants hang in the basement of a nonprofit organization that serves the homeless. Homeless clients can shower at the facility and secure clean clothes. If people did not donate clothes that they no longer wear to charity, the estimated 28 billion tons of clothes that end up in landfills each year would be even greater (Reuters 2009). The poor recycle clothes by wearing donated items; by doing so they lessen the burden on landfills.

Gans (1972) outlined the functions of poverty to show how a part of society that everyone agrees is problematic and should be eliminated remains intact: it contributes to the supposed stability of the overall system. The point is not to justify the exploitation of those who are poor but to show that the poor may not be the drain on society that many present them to be.

(Write a Caption)

Write a caption that relates this image to a function of poverty.

Lance Cpl. Andrew D. Young

Hints: In writing this caption

- review the functions of poverty, and
- consider which function this photo best represents.

Critical Thinking

Can you think of some ways those who earn poverty-level wages (minimum wage or less) provide services and goods upon which you depend?

MODULE 7.7

The Truly Disadvantaged in the United States

Objective

You will learn about the most disadvantaged groups who live in the United States.

Chris Caldeira

Rebecca Cook/Reuters/Corbis

What happens to people when a factory or other business closes?

This abandoned Detroit automobile plant (right), built in 1907, used to employ 40,000 people, enough employees to sustain a department store, two schools, and a grocery store on the premises. What happened to those who worked in the stores and schools when the plant closed in 1950? Of course, there were other waves of plant closings, restructuring, and layoffs, as Detroit automakers steadily automated factories, moved jobs to lower-wage locations within the United States and overseas, and downsized.

After 40 years, Borders closed the doors of 399 stores in July 2011. The 10,700 people who still worked for Borders lost their jobs. At its peak, Borders (and Waldenbooks) operated 1,249 stores, employed 35,000, and was known for its mega stores that housed thousands of books (Blaine 2011; CBS News 2011). The company could not compete against online bookseller Amazon or against Barnes & Noble.

Inner-City Poor

In a series of important studies, sociologist William Julius Wilson (1983, 1987, 1991, 1993) describes how structural changes in the U.S. economy helped create what he termed the ghetto poor, now known as the urban poor. Since the 1970s a number of economic transformations have taken place, including:

- the restructuring of the American economy from a manufacturing base to a service and information base,
- the rise of a labor surplus marked by the entry of women and the large baby boom segment of the population into the labor market,
- a massive exodus of jobs from the cities to the suburbs, and
- the transfer of millions of manufacturing jobs out of the United States.

These changes, combined with other historical factors, are major forces behind the emergence of the **urban poor**—diverse groups of families and individuals residing in the inner city who are on the fringes of the American occupational system and as a result are in the most disadvantaged position of the economic hierarchy (Wilson 1983). Wilson (in collaboration with sociologist Loïc J.D. Wacquant) studied Chicago, but the plight of that city applies to every large city in the United States. In 1954 Chicago was at the height of its industrial power. Between 1954 and 1982, however, the number of manufacturing establishments within the city limits dropped from more than 10,000 to 5,000, and the number of jobs declined from 616,000 to 277,000.

Chris Caldeira

◀ The loss of manufacturing jobs in the inner cities, accompanied by an exodus of working- and middle-class families to suburbs, resulted in the closing of hundreds of local businesses, including drugstores and other service-oriented establishments. These changes profoundly affected the daily lives of the people left behind.

The single most significant consequence of these economic transformations was the "disruption of the networks of occupational contacts that are so crucial in moving individuals into and up job chains." The urban poor lack connections to the stably employed who can offer knowledge about possible job openings and provide contact to employers who are hiring (Wacquant 1989, 515–516).

Poverty in the United States

One in eight people—35 million or 12.5 percent of the population—is officially classified as living in poverty (see Figure 7.7a). How does the government determine the point at which someone lives in poverty? The U.S. Census Bureau sets a dollar value threshold that varies depending on household size and age (under 65 and 65 and over). If the total household income is less than the specified dollar value, then that household is considered as living in poverty. The 2012 poverty threshold for a four-person household with three children is $23,050 per year (U.S. Department of Health and Human Services 2012).

Figure 7.7a: Percentage of Population with Income 50 Percent Below Poverty Threshold by Selected Characteristics, 2010

Poverty rates vary according to race, ethnicity, age, and sex. Those with incomes 50 percent below poverty can be considered among the "truly disadvantaged." What percentage of non-Hispanic whites falls 50 percent below the poverty threshold? What percentage of households with children under age 18 live on incomes 50 percent below poverty?

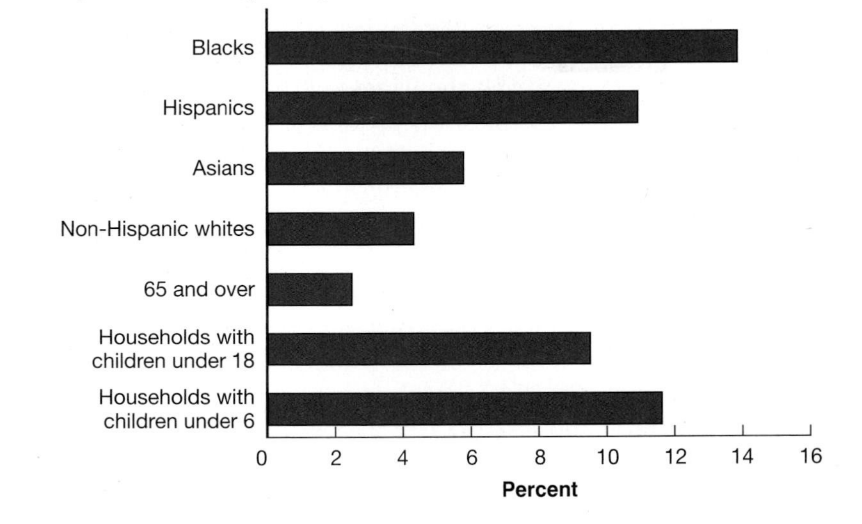

Source of data: U.S. Census Bureau (2010a)

Because the urban poor are the most visible and most publicized underclass, many Americans associate poverty with urban areas. In fact, the suburban poor outnumber the urban poor. Many of the suburban poor were pushed out of the city when factories and other businesses closed; they headed to the suburbs in search of jobs and low-cost housing (Jones 2006). The rural poor, which includes an estimated 2.6 million children, is another segment of the population that remains largely invisible. More than two-thirds of the less visible rural poor are white.

The rural poor also have felt the effects of economic restructuring, including the decline of the farming, mining, and timber industries and the transfer of routine manufacturing out of the United States (see Figure 7.7b).

Lisa Southwick

➤ Figure 7.7b: Persistent Poverty Counties in the United States

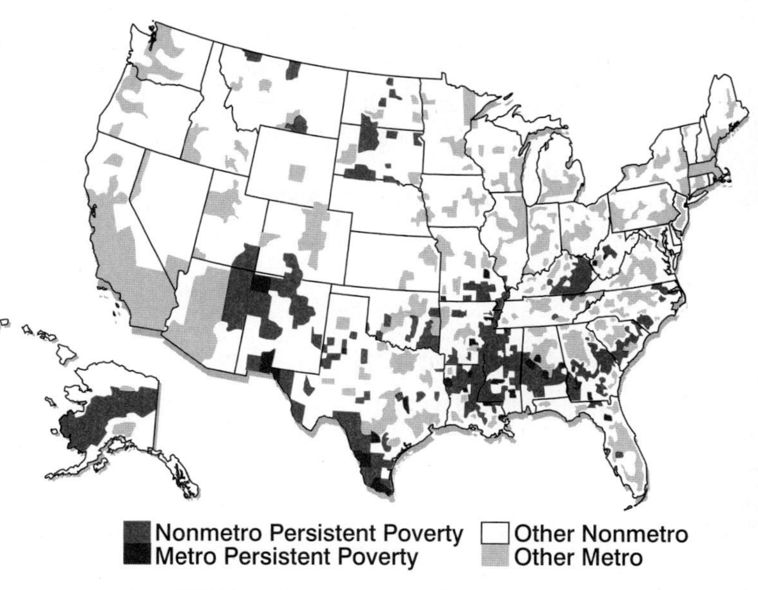

The map shows counties in the United States where the poverty rate has held at 20 percent or more for over 40 years, suggesting that it is chronic and intergenerational. According to this measure, there are 500 persistently poor counties (16 percent of all counties). Almost 90 percent of these 500 counties ($n = 450$) are considered rural or not part of a metropolitan area. This map is based on 2000 Census data and has not been updated but given decades of persistent poverty it is highly likely that poverty status has not changed.

(Write a Caption)

Write a caption connecting this closed sandwich shop in downtown Raleigh, North Carolina, to issues of economic transformation.

Hints: In writing this caption

- review William Julius Wilson's work on the economic transformations that created the truly disadvantaged,
- think about what happens to small businesses that once supported manufacturing companies that relocate, and
- consider how massive loss of jobs in inner cities affects chances of obtaining a job.

Critical Thinking

Do you know someone who has been affected by an economic transformation, such as outsourcing? Describe his or her situation.

Key Term

urban poor

MODULE

7.8 Applying Theory: Critical Social Theory

Objective

You will learn the concept of intersectionality and its importance for understanding life chances.

Jeremiah Evans

Tim Hipps

How do race and gender affect the chances of being wealthy or poor and the experiences of being poor or wealthy?

Sociologists look at how the various categories people occupy affect their chances of living in poverty or acquiring wealth. They also examine how these categories affect the experience of wealth, poverty, and statuses in between. In this regard the concept of intersectionality is particularly useful.

Intersectionality

Over her career as a sociologist, Patricia Hill Collins has made a consistent effort to theorize about **intersectionality**, the interconnections among the various categories people occupy, including race, class, gender, sexual orientation, religion, ethnicity, age (generation), nationality, and disability status. Collins maintains that sociologists must recognize these categories are interlocking categories of analysis that when taken together "cultivate profound differences in our personal biography" and the structure of our relationships (2000a, 460).

Collins (2006) places her theorizing about intersectionality within the tradition of **critical social theory**, a sociological perspective that examines

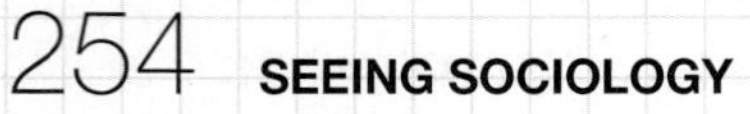

human-constructed categories and institutionalized practices to shed light on the central issues, situations, and experiences of groups differently located within "specific political, social, and historic contexts characterized by injustice" (15). In her research and writings, Collins applies intersectionality to understanding the experiences of black Americans, a constructed racial category that profoundly shapes the personal biography of those who find themselves in that category. Obviously the concept can be applied to any racial or other ascribed category. Collins emphasizes that the category "black" does not stand alone; it intersects with other socially significant and constructed categories, including social class, age, religious affiliation, sexual orientation, ethnicity, disability status, sex, gender, and nation.

Tony Rotundo

Lisa Southwick

Pete Souza

Photo courtesy of Phil Patterson

U.S. Navy photo by Mass Comm. Spc. 3rd Class William Selby

▲ To grasp the significance of intersectionality, think about the categories by which others know those pictured. While they are all classified as black and share the experience of being black in the United States, that experience is complicated by other statuses each person holds related to gender, age, occupation, disability, and social class.

The concept of intersectionality helps us to see that (1) no social category is homogeneous, (2) the multiple categories by which people are known place them in a complex system of domination and subordination, and (3) the effects of each of the multiple categories a person occupies cannot simply be added together to obtain some grand effect.

Penalties and Privileges

Collins maintains that each of us derives varying amounts of penalty and privilege from the multiple systems of oppression that frame our lives. **Penalties** include constraints on a person's opportunities and choices, as well as the price paid for engaging in certain activities, appearances, or choices deemed inappropriate of someone in a particular category. That price may be rejection, ridicule, or even death. A **privilege** is a special, often unearned, advantage or opportunity. We know some categories of people have the privilege of "not being a suspect" when in fact they have committed a crime. A 75-year-old woman who shoplifts is less likely to be noticed because her advanced age (and to some extent her sex) makes her largely invisible to store security.

A **system of oppression** is one that empowers and privileges some categories of people while disempowering other categories of people. The act of disempowering includes marginalizing, silencing, or subordinating another. We gain insights about these interlocking systems of oppression by asking these questions:

- Can you think of times that you have felt constrained by one or more categories to which you belong or have been assigned?
- Can you think of times you felt empowered by one or more of those categories?
- Have you ever resisted being labeled as belonging to one or more of those categories?
- Have you ever taken pride in a category to which you belong or felt superior to someone in another category?
- How are each of your categories represented in the media? Do those representations make you proud or defensive?

Collins (2000b) argues that while most of us have "little difficulty assessing our own victimization within some major system of oppression . . . we typically fail to see how our thoughts and actions uphold someone else's subordination" (459). Many white feminists, for example, "routinely point with confidence to their oppression as women but resist seeing how much their white skin privileges them. African Americans who possess eloquent analyses of racism often persist in viewing poor white women as symbols of white power. . . . In essence, each group identifies the type of oppression with which it feels most comfortable as being fundamental and classifies all other types as being of lesser importance" (459).

Collins maintains that each of us carries around the cumulative effects of these multiple structures of oppression and that we enforce that system in the personal choices we make about who to include or exclude in our lives and in the errors in judgment that we make about others who are not like us.

Critical Thinking

Give an example of how systems of oppression (institutionalized, symbolic, or personal) have affected your chances to connect with people in other categories.

Key Terms

critical social theory

intersectionality

penalties

privilege

system of oppression

CHAPTER 7

Summary: Putting It All Together

Social stratification is the systematic process of categorizing and ranking people on a scale of social worth where the various rankings affect life chances in unequal ways. Every society in the world stratifies its people according to ascribed and achieved statuses. Associated with each status are varying levels of prestige and esteem. Stratification systems fall on a continuum between two extremes: caste and class systems, neither of which exists in a pure form.

Sociologists use the term *social class* to designate a person's overall status in a system of social stratification. The views of Karl Marx and Max Weber are central to class analysis. For Marx, the key variable in determining social class (and life chances) is source of income. Weber defined social classes as composed of those who hold similar life chances determined not just by their sources of income but also by their marketable abilities, access to goods and services, and ability to generate investment income. Social class is complicated by status groups (lifestyle) and power. Sociologists often use income and wealth as a measure of social class, but they disagree over the amount of wealth and income associated with various class rankings. They do agree, however, that there are dramatic differences in the ways wealth and income are distributed, with the most dramatic differences being between highest and lowest strata.

When sociologists examine social stratification on a global scale, they look to see how wealth, income, and other valued resources are distributed unequally among the 7.0 billion people living in 243 or so countries. The contrasts in wealth, income, and life chances are startling. There are two views on why some countries remain poor or underdeveloped. One view—modernization theory—holds that poor countries are poor because they have yet to develop into modern economies and that their failure to do so is largely the result of internal cultural factors. Another view—dependency theory—holds that, for the most part, poor countries are poor because they are products of a colonial past. In 2000, United Nations member countries endorsed the Millennium Declaration, an ambitious plan to significantly reduce global inequality by 2015.

Sociologists draw upon three perspectives to explain inequality and to understand how it is manifested in daily life. Those perspectives are functionalism, conflict, and symbolic interaction. Sociologists also look to structural changes in the economy to understand how they disrupt networks of occupational contacts, so that large segments of society such as the urban poor cannot connect with those who are steadily employed.

Finally, to understand how stratification systems intersect to shape life chances, sociologists apply the concept of intersectionality, a term that captures the interconnections among race, class, gender, sexual orientation, religion, ethnicity, age (generation), nationality, and disability status. Sociologists recognize these traits as interlocking categories of analysis that when taken together profoundly shape how we see ourselves, how others see us, and the structure of our relationships.

Go to cengagebrain.com to link to Aplia and CourseMate for the chapter quiz and other activities.

Photo by Chris Caldeira; Mural by Paul Ygartua/www.ygartua.com

CHAPTER 8

RACE AND ETHNICITY

Each of the children depicted in this mural painted by Paul Ygartua is meant to represent one of the five "universal" races. Do you know the names of those five races? In the United States, there are five officially recognized races—"Asian," "Black or African American," "White," "American Indian and Alaskan Native," and "Native Hawaiian and other Pacific Islander." Can you match the face of each person on the mural with just one of these five races? Your ability to do so (or not) speaks volumes about the idea of race and its usefulness for classifying humanity. You might be surprised to learn that race, and by extension racial categories, makes no logical sense. Still, race has assumed great significance in structuring human affairs. This contradiction is what makes race an especially fascinating topic for sociological analysis.

Go to Sociology CourseMate on cengagebrain.com to watch a video about the Implicit Association Test (IAT), which helps researchers understand that conscious and unconscious attitudes regarding race are often not aligned.

MODULE

8.1 Race

Objective

You will learn why sociologists define race as a social construction.

Eboni Myart

What race is Barack Obama?

Obama is considered the first black president of the United States. Do you think that odd considering he described his father as a Kenyan immigrant who was "black as pitch" and his Kansas-born mother as "white as milk" (Obama 2004)? What race do you consider his wife to be? Does it matter that Michelle Obama carries "within her the blood of slaves and slave owners" (Obama 2009)? The Obamas' complex ancestries help us to understand why most biologists and social scientists have come to agree that race is not a biological fact. The reason is that parents who are considered different racial categories can produce offspring. Each offspring, by definition, is a blend of the two races and cannot, in a biological sense, belong to just one racial category.

Defining Race

Sociologists define **race** as human-created or -constructed categories that have come to assume great social importance. Although on some level we can say that race has something to do with skin shade, hair texture, eye shape, and geographical origins of ancestors, it is so much more than that. When sociologists study race, they study its social importance—the meanings assigned to physical traits, the rules for placing people into racial categories, and the effect race has on opportunities in life. We know that race is human-created if only because racial categories vary across time and place. This variation suggests that it is people who "determine what the categories will be, fill them up with human beings and attach consequences to membership" (Cornell and Hartmann 2007, 26).

U.S. Air Force photo by Staff Sgt. Lanie McNeal

➤ What race is the baby in this photo? Because of his appearance, this baby is seen as belonging to the race of just his mother. In the United States people look to skin shade and hair texture to determine a person's race, but those physical markers hide the connections this baby has with the history and ancestors associated with his father's "race."

Race as Illusion

The race we have assigned President Obama requires that we forget about his mother and his connections to her ancestors. While on the surface, it may seem natural to divide people into racial categories, upon close analysis it is illogical. First, there are no sharp dividing lines specifying the physical boundaries that distinguish one race from another.

Courtesy of Jason Eric Dustin

➤ These four brothers are offspring of the same parents, yet they do not appear to belong to the same race. In fact, it is difficult to specify the exact point at which hair texture or skin shade marks one brother as white and another as black. That is because no clear line separates so-called black from white skin or tightly curled hair from wavy or straight hair.

Second, just like the four brothers pictured above, millions of people in the world are products of sexual unions between people of different races. Obviously, the offspring of such unions cannot be one biological race. Even if we are able to classify each child as a single race, the biological reality does not support a clear-cut classification.

Third, the diversity of people within any one racial category is so great that knowing someone's race tells us little about him or her. For example, in the United States, people expected to identify as Asian include those who have roots in very different places: Cambodia, India, Japan, Korea, Malaysia, Pakistan,

Siberia, the Philippines, Thailand, Vietnam, and dozens of other Far Eastern or Southeast Asian countries. Similar diversity exists within populations labeled as "Black or African American," "White," or "American Indian and Alaskan Native."

➤ According to U.S. definitions of race, the people pictured here who live in Vietnam, Thailand, India, and Kazakhstan are Asian. What characteristics do they share? They do not share a language, a country of origin, or a culture. The only trait those pictured have in common is that each has at least one blood relative or ancestor considered as being from a country designated as Asian.

Finally, the meaning of race varies by time and place. In Brazil almost everyone thinks of themselves as multiracial. Americans think of race in categorical terms and apply the label "biracial" or "multiracial" only to those who appear "almost white."

◀ Moreover, most Americans would not think of the boy on the right as multiracial. They would be more likely to label the boy on the left as multiracial. Since these two brothers are offspring of the same mother and father, they are both multiracial.

Racial Formation Theory

With regard to race, sociologists are interested in how something that cannot be supported by logic has come to assume such great importance. They are interested in the strategies people use to make people like the two brothers pictured fit in existing racial categories. One answer is that they make rules that support those categories. At various times in the United States, rules have been employed to make the categories work. For example, children of parents who were considered different races have been classified, at various times, as "race of mother," "race of father," and "the race named first" (e.g., if a person said "I am Asian and white" that person was classified as Asian, but if the person said "I am white and Asian," that person was classified as white).

Sociologists Michael Omi and Howard Winant (1986) offer **racial formation** theory as a way to understand the social importance of race. Race is clearly not a concrete biological category; in fact, race is a product of the system of racial classification. An honest look at history reveals that race was a European creation used to justify and explain away grave inequalities and discrimination (see Module 8.6). It is also important to realize that the number of racial categories and the names assigned to racial categories have been challenged and transformed through political struggle.

Omi and Winant argue that for race to exist in its present form, anyone who lives in the United States must learn to see race—that is, they must learn to give arbitrary biological features, such as skin color and hair texture, social significance. Moreover, they must develop what the two sociologists call **racial common sense**, ideas people hold in common about race that are believed to be so obvious or natural that they need not be questioned. Racial common sense, then, relates to unquestioned assumptions held about anyone who possesses certain physical traits that associate with a specific racial category. Most people are reluctant to discuss commonsense thinking about race for fear that it will open a minefield of misunderstanding (Norris 2002).

In an interview about his book *How to Be Black*, Baratunde Thurston (2012) offers racial common sense as it relates to the role of "black friend" to people considered white.

> A lot of white people like black people. . . . Having a black friend is a mark of progressive success as a white person. And the black person is usually seen as their asset. It's like: I'm cooler by proxy. . . .

The racial common sense that underscores black–white friendships is mirrored on network television in which the 26 new shows that aired in 2011 all had white main lead actors, and many had a "cool" black best friend (Deggans 2011). Omi and Winant (2002) maintain that racial common sense informs our expectations and interactions that involve race and that this commonsense understanding of race persists even when it is challenged.

Race as a Social Construction

Sociologists see race as socially constructed. This means that the characteristics we have come to believe define race are products of social beliefs and values imposed by those who had (or have) the power to create the labels and categories. Once those labels and categories were put in place, it became easy to reify them. "Reify" means to treat them as if they are real and meaningful and to forget that they are made up. When we reify categories, we act as if people are those categories. When people do things or appear in ways that don't fit their assigned category, we act as if something is wrong with them or as if they are exceptions to the rule rather than questioning the category scheme.

➤ These children, all cousins, show there is a continuum of skin shades and not everyone thought to belong to the same race shares one skin shade. However, people have managed to use skin shade as an indicator of race, thereby reifying racial categories and making the experience of race real.

Missy Gish

(Write a Caption)

Write a caption relating Omi and Winant's racial formation theory to the situation of this father and daughter.

DoD photo by Petty Officer 2nd Class Stephanie Tigner, U.S. Navy

Hints: In writing this caption

- think about how we have been taught to see this little girl and her father in a racial sense,
- consider what physical features we draw upon and ignore to classify the child and her father, and
- ask yourself whether it is likely that the two see themselves (or at least know that others see them) as belonging to different racial categories.

Critical Thinking

Think of your living and deceased relatives on both sides of the family. Can you identify anyone who is considered a different race from you? Explain.

Key Terms

race   racial common sense   racial formation

Ethnic Groups

MODULE 8.2

Objective

You will learn the processes by which ethnicity assumes social and personal significance.

Can you tell this woman's ethnicity by looking at her? If yes, what clues did you use to make that determination?

Mr. Charles Melton (IMCOM)

If you guessed that this woman is Samoan, then you have correctly named the ethnicity with which she identifies. One way this woman announces her ethnicity to all is by performing a Samoan dance at a cultural celebration. Sociologists are interested in the processes by which ethnicity takes on personal and social significance.

Defining an Ethnic Group

An **ethnic group** consists of people who share, believe they share, or are believed by others to share a national origin; a common ancestry; a place of birth; or distinctive social traits (such as religion, style of dress, or language) that set them apart from other ethnic groups. Distinguishing between race and ethnicity is complicated because racial and ethnic identities are intertwined. In the United States, for example, Samoans are considered to belong to the racial category "Native Hawaiians and Pacific Islanders."

It is impossible to list all the ethnic groups that exist in the United States and elsewhere in the world. It is also difficult to specify the unique social characteristics and markers that place people into a particular ethnic group. For example, does performing a Samoan dance make someone Samoan? Or is it the ability to speak Samoan? Because language and cultural traditions are imprecise markers of ethnicity, one way to determine someone's ethnicity is to simply ask, "What is your ethnicity?" Self-identification is problematic, too, because people's sense of ethnicity can range in intensity from nonexistent to all encompassing (Verkuyten 2005).

In claiming ethnicity, people may point to ancestors they have never met (e.g., my great-grandfather was from Ireland) or to a distinct community to which they feel a deep connection (e.g., growing up I lived in American Samoa). For some, ethnicity is based on a sentimental connection that may manifest itself in rooting for a particular soccer team (i.e., the Italian soccer team). For others, ethnicity is a complete lifestyle that involves being born in a particular place, speaking the language, dressing a particular way, and interacting primarily with others in that ethnic group.

Ethnic identity is also affected by **selective forgetting**, a process by which people forget, dismiss, or fail to pass on a connection to one or more ethnicities. This decision is affected by larger societal forces. For example, in the United States, some races have more freedom than others in claiming an ethnic identity. Americans classified as white have a great deal of freedom to claim an ethnic identity as long as it is associated with the category of white. But it is difficult for people who appear white to claim an ethnicity associated with races considered nonwhite (e.g., Kenyan). Americans classified in racial terms as black have less choice; they are expected to identify as simply "black" or as of African descent, even though they know they have ancestors of specific African and other ethnicities or feel a special connection to ancestors from countries other than Africa (Waters 1994).

Spc. Hannah Frenchick (20th Public Affairs Detachment)

➤ We think of St. Patrick's Day as a holiday that Irish (who were not considered "white" when they first immigrated to the United States) and other "whites" celebrate. But what are some ways people who appear Asian might have a connection to an Irish heritage or ancestry?

Involuntary Ethnicity and Ethnic Renewal

Then there is the phenomenon of **involuntary ethnicity**. In this situation, a government or other dominant group creates an umbrella ethnic category and assigns people from many different cultures and countries to it. The category becomes the label by which these diverse peoples are known and with which they are forced to identify. In the United States, "Hispanic" (and non-Hispanic) are the only officially recognized ethnic categories. This Hispanic category, created in 1970, includes anyone who has roots in a Spanish-speaking country, of which there are at least 19. In contrast to the United States, China recognizes 56 different ethnic groups, including the Han (which encompasses 92 percent of the population), Tibetans, Uighur, and Yoa (Lilly 2009).

Finally, people's sense of ethnicity, even when they are free to define it, can shift through a process known as **ethnic renewal**. This occurs when someone discovers an ethnic identity, as when an adopted child learns about and identifies with newly found biological relatives or a person learns about and revives lost traditions. Ethnic renewal includes the process by which people take it upon themselves to find, learn about, and claim an ethnic heritage.

In light of the information presented thus far, is it any wonder sociologist Max Weber (1922a) argued that "the whole conception of ethnic groups is so complex and so vague that it might be good to abandon it altogether"? So why study ethnicity? One answer is that sociologists are interested in studying the processes by which people make ethnicity important (or not important).

➤ A person may seek to connect to his or her Samoan heritage by getting a tattoo or performing a traditional Samoan dance. In performing this dance for the Asian-Pacific Heritage festival, this man works to keep alive an ethnic culture, history, and tradition.

Catherine Johnson, MVIC

By Lance Cpl. Vanessa M. American Horse

Dominant Group Ethnic Identity

Sociologist Ashley W. Doane (1997) defines a **dominant ethnic group** as the most advantaged ethnic group in a society; it is the ethnic group that possesses the greatest access to valued resources, including the power to create and maintain the system that gives it these advantages. Dominant status is achieved over a long history that includes events such as conquest, colonialism, and forced and invited labor migrations that establish patterns of domination and subordination. Those who are part of the dominant ethnic group, however, tend to dismiss that history as irrelevant to any of its advantages. In the United States, the dominant ethnic group began as Anglo Americans, then expanded to encompass Protestant European Americans and eventually to encompass European Americans. In comparison to other ethnic groups, Americans from Western European ancestries several generations removed are the least likely to recall incidents in which they have personally faced prejudice, discrimination, or disadvantage because of ethnicity. In other words, because ethnicity is viewed as largely insignificant to their lives, they think of themselves as not having an ethnicity. It is important to clarify that everyone who is part of the dominant ethnic group does not hold a powerful or advantaged status. The term *dominant* refers to the fact that European Americans are overrepresented among those holding advantaged statuses.

Doane (1997) refers to the dominant ethnic group as possessing a **hidden ethnicity**, a sense of self that is based on no awareness of an ethnic identity. This is because the dominant group's culture is considered normal, normative, or mainstream. In the case of the United States, Western European standards dominate the form and content of its educational and judicial systems and permeate its mass media and political institutions. The normalization of the dominant culture promotes the belief that European Americans are cultureless; they are simply being American. Normalization creates a framework that makes it difficult for those in subordinate ethnic groups to challenge the existing system. Specifically, those who work to hold onto their ethnicity are portrayed as unwilling to give up a "foreign" culture and be "Americans." This failure to let go is labeled as undesirable. Those—even Native Americans—who seek to hold on to their cultural identity, advocate for cultural pluralism, or ask that their cultural experiences be recognized as significant to American history are viewed as divisive, guilty of political correctness, or asking for special treatment.

(Write a Caption)

Write a caption that relates these flags of 19 countries considered Hispanic to the concept of involuntary ethnicity.

Army Sgt. Nina Ramon, 345th Public Affairs Detachment

Hints: In writing this caption

- consider in what year the category Hispanic was first applied in the United States,
- think about who qualifies as Hispanic according to the U.S. definition, and
- consider that people with "Hispanic" roots must learn to see themselves as such.

Critical Thinking

Is there an ethnicity with which you identify? Explain. How do you express your ethnicity?

Key Terms

dominant ethnic group	ethnic renewal	involuntary ethnicity
ethnic group	hidden ethnicity	selective forgetting

Race and Ethnicity in Brazil and the United States

MODULE 8.3

Objective

You will learn that a society's system of racial/ethnic classification is the lens through which we view our own and others' race/ethnicity.

Rich McFadden

What race is Dora the Explorer and the child touching her?

You likely classified Dora as Hispanic/Latino, which, as we will learn, is not considered an official racial category in the United States. It is also likely that you classified the little girl as black or African American. But what if you lived in Brazil? In Brazil, Dora and this little girl would likely be seen as brown, or even white. Both most certainly would be viewed as multiracial.

When sociologists study race, they study its social importance—the meanings assigned to physical traits such as skin color and hair texture, the rules for assigning people to racial categories, and the effect race has on human relationships and life chances. The differences in the way people in Brazil and the United States view race force us to think about **systems of racial classification**, the processes by which people are divided into racial categories that are implicitly or explicitly ranked on a scale of social worth.

The U.S. System of Racial Classification

The U.S. government officially recognizes five racial categories plus a sixth, "some other race" (a category of last resort for those who resist identifying, or cannot identify, with one of the five official categories):

- American Indian or Alaskan Native—a person having origins in any of the original peoples of North, Central, or South America and who maintains tribal affiliation or community attachment;
- Asian—a person having origins in any of the original peoples of the Far East, Southeast Asia, or the Indian subcontinent;
- Black or African American—a person having origins in any of the Black racial groups of Africa;
- Native Hawaiian or Other Pacific Islander—a person having origins in any of the original peoples of Hawaii, Guam, Samoa, or other Pacific Islands; or
- White—a person having origins in any of the original peoples of Europe, the Middle East, or North Africa (U.S. Census Bureau 2011g).

It is significant that "Black or African American" is the only racial category that does not refer to original peoples. Notice that the words *original peoples* appear in every other definition. If the words *original peoples* were included in the definition of Black or African American (rather than *Black racial groups of Africa*), everyone in the United States would have to claim "black" as their race. We know from existing archaeological evidence that all humans evolved from a common African ancestor.

➤ **Table 8.1a: Percent of Total U.S. Population by Official Racial Category**

As of 2000, the United States officially recognized five racial categories plus "some other race." About 79 percent of the 313 million people in the United States self-classify as white. If people identify with a race other than the recognized five, they can check "other" on the census form. About 5 percent of the American population identify as such.

Official Racial Category	Percentage of Population
White	79.1%
Black/African American	12.3%
Asian	3.6%
American Indian/Native Alaskan	0.9%
Native Hawaiian/Pacific Islander	0.1%
Other	5.5%

Source: U.S. Census Bureau (2011g)

➤ Knowing that Latino/Hispanic is not considered an official racial category in the United States, to which of the racial categories would you assign these college students, who were born in Guatemala (left), Mexico (two females center), and Puerto Rico (right) and identify as Latino/Hispanic? Racially these students identify as "other" or "white." Can you see why Hispanics/Latinos also happen to be the group most likely to check "other race" as their racial identity?

Brazil's System of Racial Classification

The U.S. system of racial classification may seem a natural way to divide humanity until we contrast with another system such as that of Brazil. The Brazilian idea of race holds that Africans, indigenous peoples, and Europeans had mixed to the point that race was no longer important. For administrative purposes, the Brazilian government uses three broad categories (more like segments on a continuum)—*branco* (white), *pardo* (brown), and *preto* (black), with *branco* and *preto* considered ends on that continuum (see Figure 8.3a). The three categories apply to 99 percent of the country's population. Two other categories that apply to the remaining 1 percent are *amarelo* (yellow) and *indigena* (indigenous).

Chris Caldeira

◀ Although the Brazilian system recognizes a continuum of hair textures and skin shades, it does not specify a place on the continuum that marks the point at which the white (*branco*) segment gives way to the brown (*pardo*) and brown gives way to the black (*preto*) segment.

▼ **Figure 8.3a: Brazil: Race as a Continuum of "Categories"**
Most Brazilians do not see themselves as a particular race per se; rather, they see themselves on a continuum of color with *branco* (white) and *preto* (entirely African/Negro) as endpoints. That is why even those Brazilians who define themselves as white or black are likely to view themselves as multiracial. In Brazil, theoretically any drop of "white" blood makes a person "not Negro" or "not entirely African."

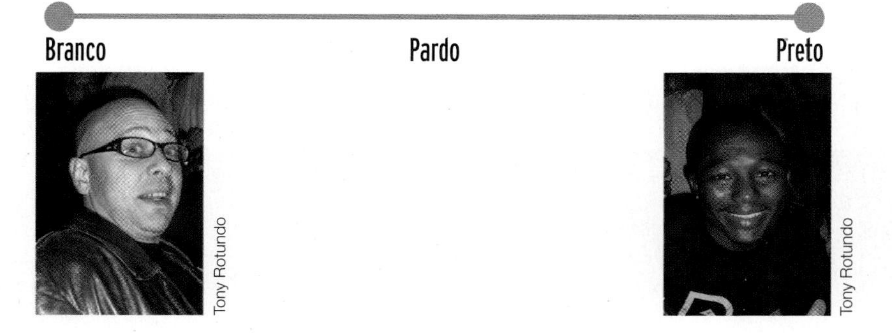

The continuum-like view of race is reflected in the popular language of Brazil. Sociologist Edward E. Telles (2004) found that when presented with an open-ended question asking people their race, Brazilians answered with 135 distinct terms, 45 of which were named only once or twice. Ninety-eight percent of respondents used one of 6 popular terms to describe their color: *branco*/white (42 percent), *moreno*/no clear race (32 percent), *pardo*/brown (7 percent), *moreno claro*/light of no clear race (6 percent), *preto*/black (5 percent), Negro (3 percent), and *claro*/light (3 percent). Telles was intrigued by the fact that more than one-third (38 percent) used the term *moreno* or *moreno claro*. *Moreno*, a term not used by the Brazilian census, means a "colored person" of ambiguous or no clear race. *Moreno claro* means a person of light color of ambiguous race.

Changing Ideas of Race

Until the 2000 census, the U.S. system of racial classification required that an individual identify with only one of its official racial categories. While Americans have always acknowledged racial mixture unofficially by using (often derogatory) words like *mixed*, *mulatto*, *half-breed*, *mongrel*, and *biracial*, the government still forced anyone who has more than one racial background to choose only one racial category. This practice changed with the 2000 Census, when for the first time in its history, the United States allowed people to identify with two or more of its official racial categories. This change is monumental because until the 2000 Census, the United States had never officially recognized intermixture. While the U.S. government now allows people to identify with more than one official race, it has yet to decide what to call people who do so. One thing is clear, to date the government has stated that it will not classify them as multiracial (Schmid 1997). It also appears that, even when given the choice to identify with more than one racial category, almost 98 percent of Americans still identify with only one. Because the U.S. government now allows people to identify with more than one race, there are officially 63 race categories, including the 6 official racial categories and 57 combinations of those official categories (e.g., "Black-Asian" or "White-Black-Asian").

Like the United States, Brazil's system of racial classification is undergoing change after supporting multiracial identities for hundreds of years. In an effort to acknowledge and remedy discrimination against those with the darkest skin shades, Brazilian public universities have instituted affirmative action policies that now require applicants to identify with one of two racial categories—white or black ("Negro"). This system of racial classification is a two-category scheme advanced by those in Brazil's black consciousness movement—someone is either *preto* (Negro) or *branco*. That movement seeks to dismantle ideas of race as a continuum, to destigmatize "blackness," and to challenge the unspoken assumption that brown (which is treated as "not black") is superior to black. Black consciousness movement activists argue that the emphasis on multiracialism has discouraged browns and blacks from mobilizing to fight well-documented discrimination and prejudices. Thus, the movement encourages all people who see themselves as *moreno* and *pardo* to identify as Negro (Telles 2004; Bailey 2008).

Lance Cpl. Cory D. Polom

▲ Study the faces of the people in this photograph. Using Brazil's new two-category system, decide who is *preto* (Negro) and *branco*. Imagine that you had grown up thinking that any amount of "white blood," no matter how dark one's appearance, would make someone "not black." With what category would you identify—*preto* or *branco*?

Ethnic Categories in the United States

In the mid-1970s, the U.S. government chose to count two official ethnicities: (1) Hispanic or Latino, and (2) Not Hispanic or Latino. The U.S. government defines Hispanic/Latino as "a person of Mexican, Puerto Rican, Cuban, Central or South American culture or other Spanish culture or origin, regardless of race" (U.S. Office of Management and Budget 1997). Note that a person from Brazil is not considered Hispanic or Latino because that country is a former Portuguese, not Spanish, colony. When the Census Bureau first classified the Hispanic population in the 1970s, it represented less than 5 percent of the total population. Today, that ethnic category includes 16.3 percent of the population and is expected to account for 24.4 percent of the total U.S. population by 2050. Approximately 50.4 million Hispanics live in the United States.

▾ Figure 8.3b: Hispanic or Latino Population as a Percent of Total Population by County, 2010

About half (52 percent) of the Hispanic population identifies as white; 36.7 percent identify as "other." Just 2.5 percent identify as black (Humes, Jones, Ramirez 2011). Notice that the Hispanic population is concentrated in the western United States in the area that was once part of Mexico.

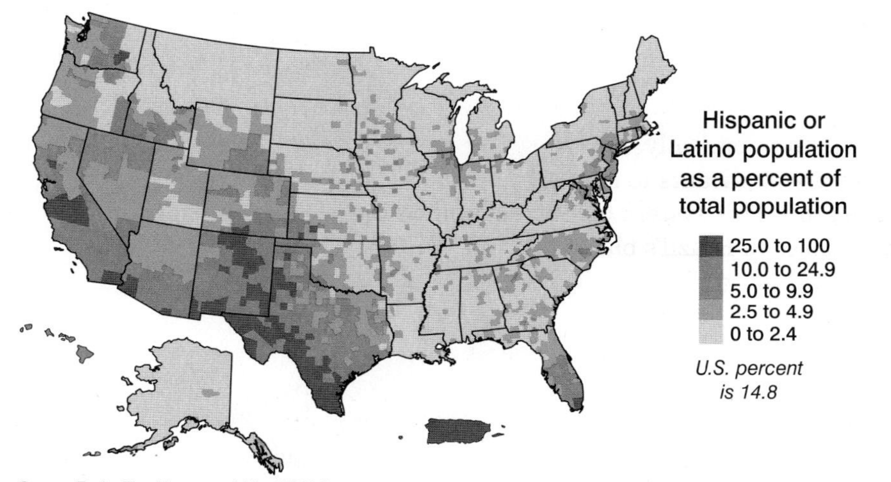

Source: Ennis, Rios-Vargas, and Albert (2011).

➤ The U.S. government asks people to identify with only one of two ethnicities: Hispanic or non-Hispanic. In other words, those with one Hispanic and one non-Hispanic parent cannot claim both ethnicities. This little girl's father identifies as Dominican, which puts him in the ethnic category Hispanic. The girl's mother is white, non-Hispanic. What ethnicity is the child? When people have a Hispanic and a non-Hispanic parent and cannot choose one ethnicity, the U.S. government considers them to be the mother's ethnicity. If the mother happens to be both Hispanic and non-Hispanic, the child is assigned to the first ethnic category named when describing the mother (U.S. Census Bureau 2012c).

Billy Santos

The Term *Hispanic*

Most people we think of as Hispanics have to learn to define themselves as such (Novas 1994). In addition, many so-called Hispanics reject the label because it forces them to identify with conquistadors and settlers from Spain, who imposed their culture, language, and religion on them. "For Latin Americans, who, like North Americans, fought hard to win their independence from European rule, identity is derived from their native lands and from the heterogeneous cultures that thrive within their borders" (Novas 1994, 2).

Panethnicity is a process by which various people with distinct histories, cultures, languages, and identities are lumped together and viewed as belonging to one catchall category, such as Hispanic/Latino.

Hispanic / Latino is a panethnic category because it applies to people from, or with ancestors from, 19 Central and South American countries (excluding Brazil) that were once under Spanish control. To complicate matters even further, the history of these 19 countries is intertwined with that of Asia, Europe, the Middle East, and Africa. As a result, each country consists of many ethnic and racial groups with multiple histories, cultures, and languages (Toro 1995; see map below).

The U.S. government also groups people with many ethnicities into a single racial category. As one example, it classifies 2,000 tribes, each with distinct cultures, as Native American. The United States has always grouped diverse peoples into single racial and ethnic categories for administrative, statistical, surveillance, and other purposes. The people classified as such often do not see the grouping as natural and must learn to see themselves as belonging to a government-imposed category. Sometimes, as a result of widespread discrimination in the society, a significant number of members of this imposed category identify with the label, transcend differences, and organize as a political force to advocate for political inclusion, needed services, and basic rights.

(Write a Caption)

Write a caption that addresses how these two men (who live in Cuba and identify as such) might be viewed if they lived in the United States versus Brazil.

Chris Caldeira

Hints: In writing this caption

- review the official racial and ethnic categories of the United States and Brazil, and
- think about how these men may view themselves if they lived in the United States or Brazil.

Critical Thinking

How do you think the systems of racial classification in the United States and Brazil have shaped family relationships in the two countries?

Key Terms

panethnicity

system of racial classification

MODULE

8.4 Minority Groups

Objective

You will learn the criteria sociologists use to determine minority group status.

Can you identify the minority in this photograph? What criteria did you use to make your selection?

Missy Gish

Minority groups are populations within a society that are regarded and treated as inherently different from those in the mainstream. They are systematically excluded (whether consciously or unconsciously) from full participation in society and are denied equal access to power, prestige, and wealth.

Characteristics of Minority Groups

Many groups can be classified as minorities, including some racial, ethnic, and religious groups; women; those in the LGBT community; the very old and the very young; and the physically impaired. Although we focus on ethnic and racial minorities, the concepts described here can apply to any minority. Sociologist Louis Wirth (1945) identified a number of essential traits that are characteristic of minority groups: involuntary status, lack of control of valued resources, exclusion from societal advantages, isolation from dominant group, and minority group status overshadowing individual characteristics.

First, according to Wirth, membership in minority groups is involuntary. In fact, people are generally born into them. He argues that as long as people are free to join or leave a group, they do not constitute a minority. The idea that one cannot be free to join or leave a group to qualify for membership in a minority group raises questions about the meaning of being free to join or leave.

> ➤ For example, one could argue that Muslim women in the United States who wear headscarves are not a minority because they can remove them. The problem is that, for many, removing the scarf also means violating some deeply held religious convictions.

Polish Warrant Officer 2nd Adam Roik

Second, minority status is not based on size. A minority may be a numerical majority in a society. In the United States, legally recognized racial minorities are the ethnic category Hispanic and the four nonwhite racial groups—American Indian/Alaskan Native, Asian, Black/African American, and Native Hawaiian/Other Pacific Islander. All five are numerical minorities. However, there are geographic areas of the United States—counties, cities, neighborhoods—where a nonwhite group is the numerical majority.

Third, minority groups are excluded from full participation in the larger society. That is, as a group minorities do not enjoy the advantages that members of a dominant group take for granted. **Advantaged** refers to symbolic support (positive images) and access to valued resources that give one group more opportunities relative to another, including the opportunity to live a long life, to make a good income, to survive the first year of life, and much more (see Table 8.4a).

▼ Table 8.4a: Differences in Life Chances by Race and Sex (United States)

Life chances are a critical set of potential opportunities and advantages, including the chance to survive the first year of life, to grow to a certain height, to receive medical and dental care, to avoid a prison sentence, to graduate from high school, to live a long life, and so on (Gerth and Mills 1954). Why do you think Asian females live almost 17 years longer on average than black males? This dramatic difference speaks to the way in which race affects life chances. Which racial category has the highest chance of dying before age one? The lowest chance? Why might this be the case?

Chance of . . .	Highest Chance	Lowest Chance	Difference
Living a long life (average life expectancy)	Asian female 86.7 years	Black male 70.2 years	**16.5 years**
Graduating from high school on time*	Asian 91.4%	Black 61.5%	**29.9%**
Going to prison	Black male 32.5%	White female 1%	**31.5%**
Earning a high median weekly income (average salary working full time)	Asian male $936	Hispanic female $508	**$428 per week**
Dying before reaching one year of age	Black 1,100 per 100,000	Asian 389 per 100,000	**711 babies per 100,000**
Living in poverty*	Native American/Black 25.8%	White (non-Hispanic) 12.3%	**13.5%**
Having no health insurance	Native American 35%	White 11.9%	**23.1%**

Sources of data: U.S. Bureau of Labor Statistics (2012b); Stillwell (2010); U.S. Department of Justice (2007); U.S. Census Bureau (2011j); Murphy, Xu, Kochanek (2012).
*Data not available for sex categories.

In addition, minorities do not enjoy the freedom or the privilege to move within the society the way those in the dominant group do. Peggy McIntosh (1992), professor of women's studies, has identified a number of such **privileges**, special taken-for-granted advantages and immunities or benefits enjoyed by a dominant group relative to minority groups, including the following:

- Most of the time—when I am at school, work, or just out walking—I am in the company of people of my race or ethnic group.
- I feel confident that I can rent or purchase housing in any area in which I can afford to live.
- I can go shopping and not worry that I am being followed or targeted for surveillance by store detectives.
- I can do poorly on a test without worrying that my classmates or professor will attribute it to my race or ethnicity.
- Whether I use checks, credit cards, or cash, I can count on my skin color not to work against the perception that I am financially reliable.
- If I perform well at some activity that is associated with my racial or ethnic group, people will still recognize my hard work and not attribute my success to natural abilities. (McIntosh 1992, 73–75)

Brian Prechtel

Phil Sussman

⋏ If you look at these photographs and think that blacks are naturally good singers and Asians are naturally good scientists, then you fail to acknowledge the hard work that goes into being a singer or a scientist. You also fail to acknowledge the larger social forces that push people to "choose" careers that are more open to and expected of those in specific racial categories.

Wirth's fourth essential trait characteristic of minorities' disadvantaged status is that minorities are socially and spatially isolated from those in the dominant group. This social and spatial isolation manifests itself in

- segregated residential arrangements (ethnic neighborhoods and/or gated communities),

- the portrayal of specific minorities as particularly dangerous,
- laws prohibiting certain racial/ethnic groups from marrying those in dominant groups (miscegenation),
- acts of profiling that target minority groups for special surveillance, and
- underrepresentation of minorities in key political and economic positions (Woodell and Henry 2005).

➤ Clearly, Muslims are not represented in key political positions in the United States. Keith Ellison made history in 2007 when he became the first Muslim congressman in U.S. history. Blacks also are underrepresented in Congress. Ellison is one of 40 blacks in the House of Representatives. This minority group constitutes 12.3 percent of the population but accounts for 9 percent of all representatives. In the history of the Senate, only 6 black senators have served. There are no black senators at the time of this textbook's printing.

Michaela McNichol/Library of Congress

The fifth essential trait is that those who belong to minority groups are "treated as members of a category, irrespective of their individual merits" (Wirth 1945, 349). In other words, people focus on those physical and cultural characteristics that identify someone as belonging to a minority. Those characteristics are considered so important to the individual's identity that they overshadow other characteristics that person might possess.

The five characteristics of minority group status apply especially to the situation of **involuntary minorities**, those who did not choose to be a part of a country (nor did their ancestors); rather, they were forced to become part of it through enslavement, conquest, or colonization. Those of Native American, African, Mexican, and Hawaiian descent are examples. There are also some groups considered white today (most notably the Irish) that came to the United States involuntarily as indentured servants (Painter 2010). Unlike peoples who voluntarily immigrated to the United States expecting to improve their way of life, those who were forced to become part of the United States had no such expectations. In fact, their initial forced incorporation involved a loss of freedom and status (Ogbu 1990).

Native Hawaiians represent a minority group that was pulled into the United States against its will. In 1993 Congress issued the Apology Resolution in which it acknowledged that "the overthrow of the Kingdom of Hawaii occurred with the active participation of agents and citizens of the United States" and that "the Native Hawaiian people never directly relinquished to the United States their sovereignty as a people over their national lands." The Queen of Hawaii in 1893 at the time of the overthrow was Liliuokalani.

AP Photo/Victor Jose Cobo

◂ The Hawaiian Sovereignty Movement, shown demonstrating, considers the 1993 Apology Resolution an important step toward eventual self-determination. However, the resolution did not change the legal status of Hawaii, nor did it place any obligation on the United States to make reparations or provide other kinds of special assistance to native Hawaiians.

(Write a Caption)

Write a caption that connects this image to racial/ethnic minority group status.

Lisa Southwick

Hints: In writing this caption

- review the characteristics of minority groups,
- think about which applies to the simple act of shopping, and
- consider whether this woman is a minority in the ethnic/racial sense of the word.

Critical Thinking

Describe a specific context in which you were a minority (not necessarily a racial/ethnic minority). Use Wirth's characteristic traits to inform your description of the experience.

Key Terms

advantaged
life chances
privileges
involuntary minorities
minority groups

MODULE 8.5

Racism

Objective

You will learn that racism relies on flawed logic to explain and justify differences and inequalities.

Do you view one of these players as possessing more natural talent for the game of basketball?

Tony Rotundo

Is your prediction based on a belief that blacks are superior athletes by nature and that whites are just not as good at basketball? Such a prediction is the result of racist thinking.

Racism is a set of beliefs that uses biological factors to explain and justify inequalities between racial and ethnic groups. Generally people who hold racist beliefs treat them as accurate explanations of the existing state of affairs. On closer analysis, however, race-based explanations fall apart. Racism is structured around two notions:

- people can be classified into racial categories according to physical traits such as skin color, hair texture, and eye shape; and
- these physical attributes are deemed so significant that they are used to explain and determine behavior and inequalities. There is no other possible explanation for these inequalities.

Origins of Racism

Racism, or some variation on racism, as a way of explaining differences between groups of people has probably always existed. Modern racism, however, emerged as a way to justify European exploitation of people and resources in Africa, Asia, and the Americas. Between 1492 and 1800, Europeans learned of, conquered, and colonized much of North America, South America, Asia, and Africa and set the tone for international relations for centuries to come. Among other things, this exploitation took the form of enslavement and **colonialism**, which occurs when a foreign power uses superior military force to impose its political, economic, social, and cultural institutions on an indigenous population in order to control its resources, labor, and markets (Marger 1991).

When Europeans' labor demands could not be met by native-born populations of places they colonized in the Western hemisphere (the United States, Brazil, and the Caribbean, for example), they imported slaves from Africa or indentured workers from Asia and Europe. In fact, an estimated 11.7 million enslaved Africans survived their journey to the Americas between the mid-fifteenth century and 1870 (Chaliand and Rageau 1995). After slavery ended, the Europeans colonized the African continent. By 1914, nearly all of Africa had been divided into European colonies. The Europeans forced local populations in Africa to cultivate and harvest crops and to extract minerals and other raw materials for export to the colonists' home countries and beyond.

Racism helped justify this exploitation of nonwhite peoples and their resources by pointing to the so-called superiority of the white race. More precisely, the exploitation was justified by **scientific racism**, the use of faulty science to support systems of racial rankings and theories of social and cultural progress that placed whites in the most advanced ranks and stage of human evolution. These racial rankings shaped how the people, not just in the United States and Europe, saw themselves. In Brazil, for example, many of the political elite embraced the doctrine of white superiority and instituted policies to "whiten" the populations, arguing that in the end miscegenation would eliminate, or at least dilute, the black population.

Flaws in Racist Arguments

Anyone who takes the time to look will find that race-based explanations of differences among people fall short. To take one example, people who argue that blacks are naturally superior athletes usually point to physiological differences that give black athletes an advantage over other athletes. As evidence that such differences exist, they point to the disproportionately high number of blacks participating in the money sports—the most visible, best-paid, and most televised sports (basketball, football, and boxing). Such evidence does not convince sociologists that blacks are naturally superior athletes.

In evaluating the argument that blacks are superior athletes, we must ask why black athletes from African countries do not dominate international competitions, such as the Olympic Games. In fact, athletes from western and west-central African countries have competed in the Olympics for more than 50 years, and during that time, they have won a total of 22 Olympic medals. American athletes classified as black earn more than twice that number during each Summer Olympics alone (Darmoni 2005).

Sociologists look to the social processes that channel athletes of a certain racial classification into a sport considered the domain of that race. With regard to black athletes, note that athletes and entertainers have always been the most highly publicized black achievers, and they have arguably been just as influential in shaping black identity as Martin Luther King Jr., Malcolm X, and other black leaders. That high visibility has surely played some role in channeling black athletes' talent toward the money sports (Hatfield 2006).

Sociologists would also point to other factors that channel black, white, and other athletic talents toward particular sports and away from others. Those factors include financial resources to pay for equipment, lessons, and playing time; encouragement from parents and peers; perceptions that a sport "belongs" to a particular race; and geographic location related to warm- and cold-weather sports.

Lance Cpl. Corey A. Blodgett

Tony Rotundo

◂ Finally, can we make the case that black basketball players are really better athletes than white water polo players or Koreans who excel in martial arts? Are speed, strength, and quick reflexes qualities that water polo and martial arts athletes also possess? What about hockey, gymnastics, swimming, soccer, cycling, sailing, rowing, archery, volleyball, skiing, and other less lucrative sports dominated by athletes of other races? Can we explain white dominance in water polo or Asian dominance in martial arts only as products of athletes' race?

(Write a Caption)

Write a caption that relates this 1870 illustration to the concept of racism.

Hints: In writing this caption

- review the definition of racism and its two components,
- consider that each woman represents a particular racial category (Irish in top left corner; black in top right corner),
- think about which image is depicted as the ideal "goddess liberty," and
- consider whether there is an implied ranking among the women.

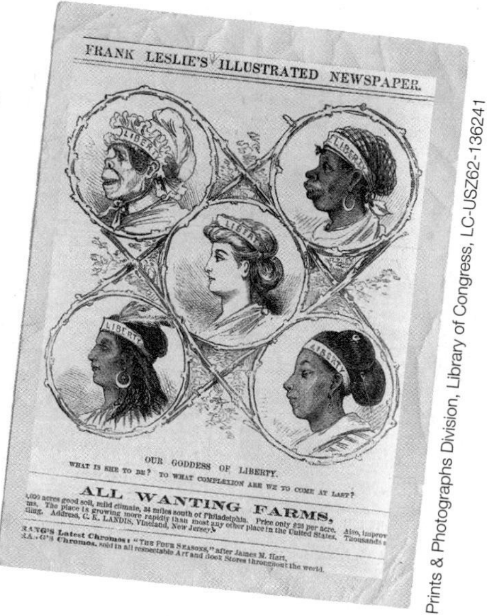

Prints & Photographs Division, Library of Congress, LC-USZ62-136241

Critical Thinking

Has your understanding of what constitutes racist thinking changed as a result of reading this module? Explain.

Key Terms

colonialism　　racism　　scientific racism

MODULE

8.6 Prejudice and Discrimination

Objective

You will learn about prejudice and discrimination and their consequences.

Lisa Southwick

Imagine that you are standing alone late at night waiting for the elevator in an underground parking garage. Are you thinking about who might be on the elevator when the doors open? Are you hoping a person of a particular sex, race, age, social class, or nationality is on the elevator? Or is not?

Prejudice

If you are hoping a person of a particular sex, race, age, social class, or nationality is on the elevator (or not), you likely hold prejudice toward those groups. A **prejudice** is a rigid and, more often than not, unfavorable judgment about a category of people that is applied to anyone who belongs to that category. Prejudices are supported by **stereotypes**—generalizations about people who belong to a particular category that do not change even in the face of contradictory evidence. Stereotypes give holders an illusion that they know the other group and that they possess the right to control images of the other group (Crapanzano 1985).

Stereotypes are supported and reinforced in a number of ways. First, in **selective perception**, prejudiced persons notice only the behaviors that support their stereotypes. These people use stereotypes as facts supported by their own observations (Merton 1957a). Many people stereotype Asians as being naturally good at math and sciences and think that their natural ability, in turn, explains why

Asians dominate engineering in the United States. Yet, these same people do not point to natural ability to explain why almost all airline pilots in the United States are white.

Second, stereotypes persist in another way: when a prejudiced person encounters a person in the stereotyped category who contradicts associated stereotypes, the former sees the latter as the exception to the rule, rather than questioning the merits of the stereotype.

Third, prejudiced individuals keep stereotypes alive when they evaluate the same behavior differently depending on the race or ethnicity of those involved (Merton 1957a). For example, incompetent behavior of racial and ethnic minority members is often attributed to innate flaws in their biological makeup; in contrast, incompetence exhibited by someone from the advantaged group is almost always treated as a personal shortcoming.

➤ Look at the photos of these men listed on the "Crime Alert" page. Do you find yourself explaining terrorist charges by noting an assumed Muslim connection or explaining drug charges by noting a Hispanic background? Do you apply that same standard to the white men wanted for kidnapping and child sexual assault? After all, a disproportionate number of those wanted by the FBI (2009) for crimes against children appear white.

Federal Bureau of Investigation

Finally, stereotypes flourish through a process known as the **self-fulfilling prophecy** (Thomas and Thomas 1928, 572). A self-fulfilling prophecy begins with a false definition of a situation that is assumed to be accurate. People behave as if it were true. In the end, the misguided behavior produces responses that confirm the false definition (Merton 1957a). With regard to race, a self-fulfilling prophecy occurs when teachers, coaches, and parents channel a child's interests in areas that they believe are appropriate to that child's race. Over time, real differences in quantity, quality, and content of instruction create seemingly race-based differences in talent. The cycle of self-fulfilling prophecies can be broken only by questioning the original assumption and redefining the situation.

Discrimination

In contrast to prejudice, which is an attitude, discrimination is a behavior. **Discrimination**, intentional or unintentional, is the unequal treatment of racial or ethnic groups without considering merit, ability, or past performance. Discrimination blocks access to valued experiences, goods, and services. Sociologist Robert K. Merton argues that knowing whether people are prejudiced does not help predict whether they will discriminate. This is because prejudiced people do not always discriminate and unprejudiced people sometimes do. To illustrate this point, Merton describes four types of people.

1. **Nonprejudiced nondiscriminators** (all-weather liberals) accept the creed of equal opportunity, and their conduct conforms to that creed. They represent a "reservoir of culturally legitimized goodwill" because they not only believe in equal opportunity but also take action against discrimination (Merton 1976, 193).
2. **Nonprejudiced discriminators** (fair-weather liberals) accept the creed of equal opportunity but discriminate because they simply fail to consider discriminatory consequences or because discriminating gives them some advantage. For example, whites decide to move out of their neighborhood after a black family moves in—not because they are prejudiced against blacks per se, but because they are afraid of declining property values.
3. **Prejudiced nondiscriminators** (timid bigots) reject the creed of equal opportunity but refrain from discrimination, primarily because they fear possible sanctions or being labeled as racists. Timid bigots rarely express their true opinions about racial and ethnic groups and often use code words such as "inner city" or "those people" to camouflage their true attitudes.
4. **Prejudiced discriminators** (active bigots) reject the notion of equal opportunity and profess a moral right, even a duty, to discriminate. They derive significant social and psychological gains from the conviction that anyone from their racial or ethnic group is superior to other such groups (Merton 1976).

Bill Pugliano/Getty Images

➤ Prejudiced discriminators are the most likely to initiate hate crimes, actions aimed at humiliating someone in the target group or destroying their property or lives. Nonprejudiced discriminators are vulnerable to going along with this suggestion.

Sociologists distinguish between individual discrimination and institutionalized discrimination. **Individual discrimination** occurs when a person acts to block another's opportunities or does harm to life or property. **Institutionalized discrimination** is the established, customary way of doing things in society—the unchallenged laws, rules, policies, and day-to-day practices established by a dominant group that keep minority groups in disadvantaged positions (F. Davis 1978). Such discrimination is difficult to identify and rectify because the discrimination results from simply following established practices that seem on the surface to be impersonal and fair or part of the standard operating procedures.

Lt. Col. David Konop

◀ Imagine these men are standing on a street corner, simply waiting for a bus or a ride. A person reports them to the police as acting in suspicious ways, even though they are simply waiting. Are these men experiencing institutional or individual discrimination?

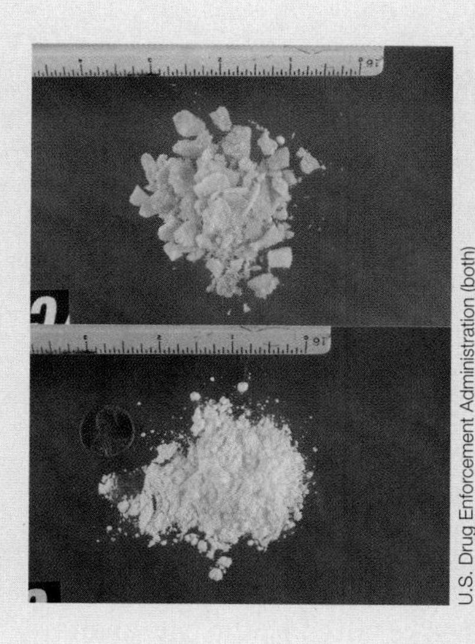

U.S. Drug Enforcement Administration (both)

◀ One example of institutional discrimination relates to federal sentences issued to those convicted of trafficking in crack cocaine (top) versus powder cocaine (bottom). Crack and powder cocaine have the same physiological and psychotropic effects but are handled very differently for sentencing purposes. On average, sentences for crack offenses are three to six times longer than those for offenses involving equal amounts of powder. Approximately 85 percent of defendants convicted of crack offenses in federal court are black, whereas 78 percent of defendants in powder cocaine cases are white; thus, the severe sentences are imposed "primarily upon black offenders" (*Kimbrough v. United States* 2007).

Redlining

Redlining refers to institutionalized practices that deny, limit, or increase the cost of services to neighborhoods because residents are low-income and/or minority. Redlining can affect access to financial services (loans, checking accounts, credit cards, mortgages), insurance, health care, and grocery stores. The term *redlining* refers to a 1960s practice when banks actually marked red lines on maps, highlighting the communities in which they would not invest. The term was later applied to systematic discrimination against a geographically based population because as a group it possesses a characteristic labeled as not good for business. A 2008 study found that lenders were more likely to issue higher-interest loans to African Americans and Hispanics than to whites with comparable credit histories (see Table 8.6a).

- In 71 percent of the 165 metro areas studied, middle- and upper-income blacks were at least twice as likely as their white counterparts to receive high-interest loans (such as 9.25 percent versus a low-interest rate of 6.25 percent).
- In 22 percent of the 165 metro areas studied, middle- and upper-income Hispanics were at least twice as likely as their white counterparts to receive high-interest loans.
- In 47.3 percent of the 165 metro areas studied, lower-income blacks were at least twice as likely as their white counterparts to receive high-interest loans.

▼ Table 8.6a: Monthly and Total Home Mortgage Payments by Interest Rates

This table shows the economic consequences of discriminatory lending patterns. A $140,000, 30-year mortgage financed at 6.25 percent translates to an $862 monthly payment, or $310,000 over the life of the loan. By contrast, a 30-year loan at 9.25 percent translates to a $1,152 monthly payment and a total of $415,000 over the life of the loan.

Cost of House	Interest Rate	Monthly Payments	Total Payments After 30 Years
$140,000	6.25%	$862	$310,000
$140,000	8.25%	$1,052	$379,000
$140,000	9.25%	$1,152	$415,000

Source: National Community Reinvestment Coalition (2008)

Segregation

Racial and ethnic **segregation** is the physical and/or social separation of people by race or ethnicity. Segregation may be legally enforced (de jure) or socially enforced without the support of laws (de facto). The segregation may be spatial or hierarchical. Spatial segregation occurs when racial or ethnic groups attend different schools, live in different neighborhoods, and use different public facilities, such as restaurants and even drinking fountains. It also occurs when people of different racial or ethnic groups are in the same buildings but sit in different places for lunch or work on different floors or rooms (in the kitchen versus dining area of a restaurant). Segregation is hierarchical when people in advantaged categories occupy the most prestigious positions while those in the disadvantaged categories are concentrated in the least prestigious positions, such as servants, maintenance workers, and laborers.

In the United States, the Jim Crow laws enforced racial segregation between whites and nonwhites from 1880 to 1964. These laws resulted in the establishment of separate but unequal race-specific bathrooms, recreational facilities, hospitals, and drinking fountains.

Tony Rotundo

◀ While racial segregation in public spaces is no longer supported by law in the United States, we must acknowledge that segregation exists in fact. For most Americans, all the important and meaningful events in our lives—weddings, funerals, graduation parties, holiday gatherings—are experienced with people who are considered the same racial/ethnic group (M. Gordon 1978, 204).

Ethnic Cleansing

Ethnic cleansing is an extreme form of forced segregation. It is a process by which a dominant group uses force and intimidation to remove people of a targeted racial or ethnic group from a geographic area, leaving it ethnically pure, or at least free of the targeted group. Ethnic cleansing also involves the destruction of cultural artifacts associated with the targeted groups, such as monuments, cemeteries, and churches. One example of ethnic cleansing was the forced removal of Native Americans from their ancestral lands.

Genocide is the calculated and systematic large-scale destruction of a targeted racial or ethnic group. The destruction can take the form of killing an ethnic group en masse, inflicting serious bodily or psychological harm, creating

intolerable living conditions, preventing births, "diluting" racial or ethnic lines through rape and forced births, or forcibly removing children to live with another group (United Nations 1948).

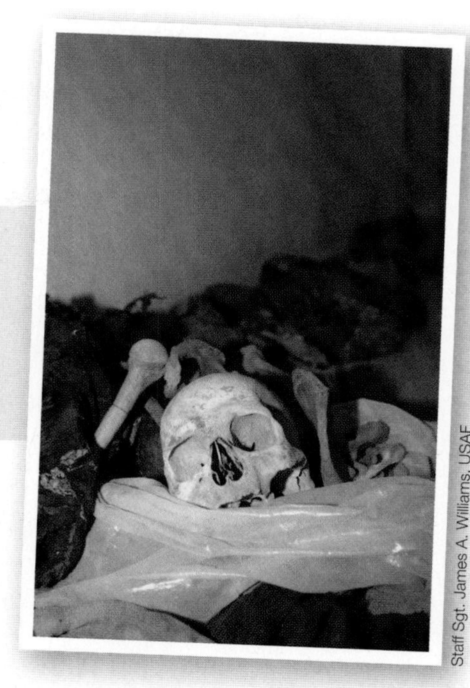

Staff Sgt. James A. Williams, USAF

➤ By one estimate, 38 million people around the world died as a result of genocide in the 20th century (Oberschall 2000). Genocide is more often than not state-sponsored, in that a dominant group uses state apparatus (police, military, surveillance) to eliminate those targeted.

(Write a Caption)

Write a caption that relates this 1939 scene (during the Jim Crow era) to the concept of segregation. Notice the sign on the drinking fountain and signs directing people to public bathrooms.

Hints: In writing this caption

- review the concept of segregation,
- consider whether the segregation is de facto or de jure, and
- decide which type of segregation is most emphasized in this image.

Prints & Photographs Division, Library of Congress, LC-DIG-fsa-8a26761

Critical Thinking

Consider sociologist Robert K. Merton's four categories relating prejudice and discrimination. Can you think of an example when you fit one these four types? Explain.

Key Terms

discrimination
ethnic cleansing
genocide
individual discrimination
institutionalized discrimination
nonprejudiced discriminators
nonprejudiced nondiscriminators
prejudice
prejudiced discriminators
prejudiced nondiscriminators
redlining
segregation
selective perception
self-fulfilling prophecy
stereotypes

MODULE 8.7

Assimilation, Integration, and Pluralism

Objective

You will learn about the process by which racial and ethnic distinctions disappear, blend, or coexist.

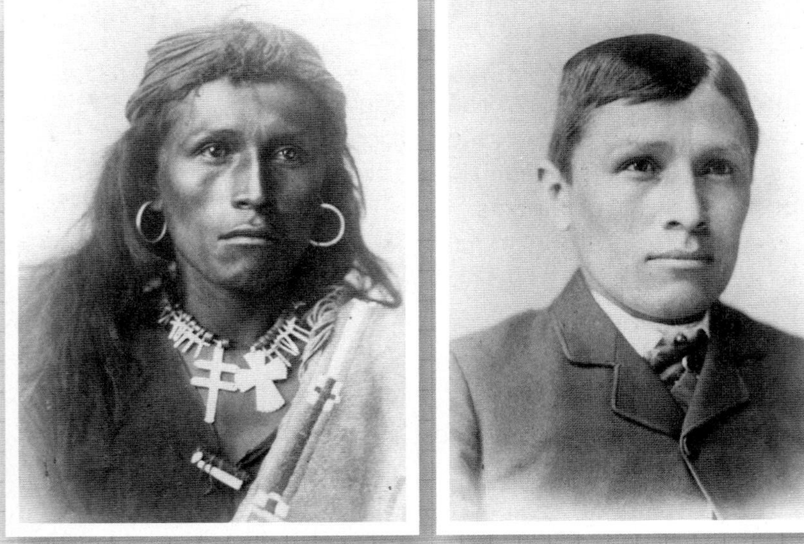

Before   After

Study these before-and-after photographs. Do you see the changes as merely physical or do you think this man has changed as a person?

These before-and-after photographs were taken in the mid- to late 1800s to showcase how Carlisle and other boarding schools "effectively" changed Native Americans from an uncivilized to civilized state of being. They capture what sociologists call **assimilation**, a process by which ethnic, racial, and/or cultural distinctions between groups disappear because one group is absorbed, sometimes by force, into another group's culture or because two cultures blend to form a new culture. Two main types of assimilation exist: absorption assimilation and melting pot assimilation.

Types of Assimilation

In **absorption assimilation**, members of a subordinate ethnic, racial, and/or cultural group adapt to the ways of the dominant group, which sets the standards to which they must adjust (M. Gordon 1978). According to sociologist Milton Gordon, absorption assimilation has at least seven levels in which a subordinate group

1. abandons (by force or voluntarily) its culture for that of the dominant group,
2. enters into the dominant group's social networks and institutions,
3. intermarries and procreates with the dominant group,
4. identifies with the dominant group,
5. encounters no widespread prejudice by those in the dominant group,
6. encounters no widespread discrimination by those in the dominant group, and
7. has no value conflicts with those in the dominant group.

A subordinate group is completely absorbed into the dominant group once all seven phases are achieved. Gordon maintains that level 1 assimilation is likely to take place before the other six are achieved. He also states that even when a group "abandons" its culture (level 1), it does not always lead to the other levels of assimilation.

Prints & Photographs Division, Library of Congress, LC-USZ62-107203

➤ In the mid-to-late 1800s and well into the next century, there were schools established to Americanize (and Christianize) Native American peoples including the Potawatomi, Winnebago, Chippewa, and Mesquakie, shown here, who attended the Mesquakie Day School, near Toledo, Iowa. These schools sought to achieve level 1 assimilation—that is, to force students to abandon their cultures by cutting their hair, assigning them "white" names, and requiring them to speak only English.

Gordon proposes that if those in the subordinate group are able to join the advantaged groups' social circles on a large enough scale (level 2 assimilation), a substantial number of interracial or interethnic marriages are bound to occur (level 3 assimilation) because of social interactions between the groups (M. Gordon 1978). Of the seven levels of assimilation, Gordon believes that gaining access to the advantaged racial group's social networks and institutions is the most important; if that occurs, the other levels of assimilation inevitably follow. Yet, in practice, gaining such access is very difficult.

Assimilation need not be a one-sided process in which a minority group is absorbed into the dominant group. Ethnic and racial distinctions can also disappear through amalgamation or **melting pot assimilation** (M. Gordon 1978). In this process, previously separate groups accept many new behaviors and values from one another, intermarry, procreate, and identify with a blended culture. The term *Blackanese* is used in reference to those who seek to identify as both black and Japanese. Those who see themselves as Blackanese seek to embrace both cultures and maintain that each culture has equal influence in their lives.

Integration

Assimilation, by definition, involves some level of integration. The term *integration* is often used in conjunction with the legal term **desegregation**, the process of ending legally sanctioned racial separation and discrimination. Desegregation often involves removing legal barriers to interaction and offering legal guarantees of protection and equal opportunity. **Integration** occurs when two or more racial groups interact in a previously segregated setting. That integration may be court-ordered, legally mandated, or the natural outcome of people crossing the "color line" once legal barriers have been removed.

> When people celebrate the entry of a minority into social circles previously closed off to them, they are celebrating a first and necessary step toward integration (e.g., first Muslim elected to the House of Representatives, first black chosen Miss America, first Asian selected as a first-round draft pick in the NFL, first white to attend a historically black college). Shown here is Dr. Mae Jemison, the first African American woman to travel into space. Jemison, born in 1956, was admitted into the astronaut training program in 1987 and eventually traveled into space in 1992 to conduct experiments on weightlessness and motion sickness.

U.S. Air Force

When evaluating the extent of integration in society, it is important to ask this question: With whom do you live, learn, pray, celebrate, and mourn? If the answer involves only people of a single race or ethnicity, then one must conclude that, in fact, he or she lives a segregated life. Many Americans believe that racial and ethnic integration has been realized and that the United States is a color-blind society. This belief may be the result of a phenomenon known as **virtual integration**, in which simply seeing other racial groups on television and in advertisements gives "the sensation of having meaningful, repeated contact with other racial groups without actually having it" (Lynch 2007).

Pluralism is a situation in which different racial and ethnic groups coexist in harmony; have equal social standing; maintain their unique cultural ties, communities, and identities; and participate in the economic and political life of the larger society. These groups also possess an allegiance to the country in which they live and its way of life. In a pluralistic society, there is no one race or ethnic group considered as the standard to which other races should aspire. Rather, cultural differences are respected and valued.

Although it is difficult to find an example of a country that practices pluralism in every way, it is possible to find descriptions of that ideal. The United States presents its ideal of pluralism as a melting pot in which the country welcomes immigrants from all over the world who bring a vitality and energy to the country's way of life.

➤ Arguably the best example of pluralism in action is the United States military, which recruits people from every racial and ethnic group into its ranks.

Sgt. Jasmine Chopra

The melting pot analogy overlooks the complex history of the United States. That history involved the European conquest of Native American peoples; the enslavement of African peoples; the annexation of Mexican territory, along with many of its inhabitants (who lived in what is now New Mexico, Utah, Nevada, Arizona, California, and Texas); and an influx of voluntary and involuntary immigrants from practically every part of the world. In addition, Puerto Ricans, American Samoans, Hawaiians, and other peoples all became part of the United States through a form of domination known as conquest or colonialism. The most celebrated group is voluntary immigrants—the millions of people who more or less chose to move to the United States.

One of the most interesting, significant, and long-lasting aspects of this global story is the U.S. government's establishment of a racial and ethnic classification scheme that applied to all who lived in and immigrated to the United States. The categories to which people were assigned reflected and reinforced the prejudices and discrimination of the times and set the tone for race relations then and far into the future. Perhaps as many as 2,000 distinct groups of Native American peoples, speaking seven different language families, were placed in a single category: "Indian." The millions of voluntary and involuntary immigrants from Europe eventually became "white." The peoples from all of Latin America became "Hispanic." Those from the Far East, Southeast Asia, and the Indian subcontinent were lumped into the category "Asian." The peoples from Hawaii and other Pacific islands (such as American Samoa and Guam) were eventually lumped into the category of "Native Hawaiian and Pacific Islander." Those of African descent became "black."

The Civil Rights Movement

Shortly after slavery was abolished in the United States, a state-sanctioned system of racial discrimination, known as Jim Crow (1877), was put into place. Under Jim Crow, blacks (and other minorities) were denied the right to vote and sit on juries; subjected to racial segregation (separate and unequal facilities); disadvantaged with regard to employment opportunities; and subjected to widespread, systematic discrimination, including violence against person and property. After decades of struggle and resistance, that system was overturned with the ratification of the Civil Rights Act of 1964, the Voting Rights Act of 1965, and the Fair Housing Act of 1968. The civil rights movement was a response to that systematic discrimination, not just in the South but across the nation. The civil rights movement reached its most organized phase in the late 1950s and

1960s. It encompassed other related movements as well, including the American Indian Movement, La Raza Unida (the Unified Race), the antiwar (Vietnam) movement, and the women's movement.

In the popular imagination, the civil rights movement involved confronting blatant white supremacists such as the Ku Klux Klan. However, activists also confronted institutional discrimination as embodied in local, state, and federal agencies; judicial systems; and legislative bodies, including police, the National Guard, judges, and all-white citizens' town/city councils. Most notably, police departments, especially in the South, arrested civil rights activists on false or trumped-up charges, and all-white juries found whites who murdered blacks not guilty. Some officials, such as Alabama Governor George Wallace, used the National Guard and state police to prevent school integration.

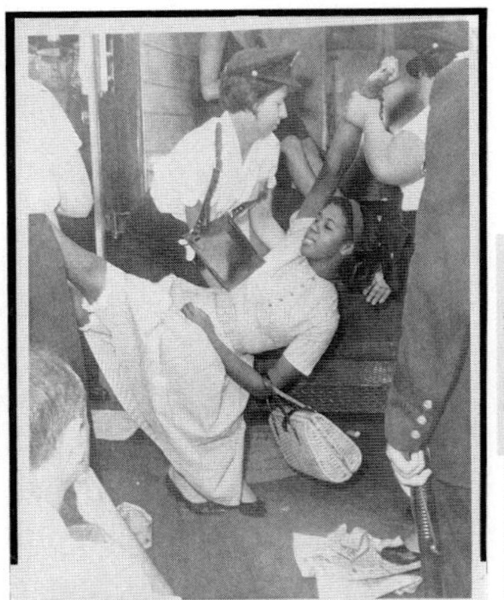

Prints & Photographs Division, Library of Congress, LC-USZ62-134715

➤ College and high school students played key roles in the civil rights movement. They participated in bus boycotts, sit-ins, freedom rides, and school integration.

The black churches played a key role in the civil rights movement as well. In an atmosphere of profound discrimination and inequality, the black church had become not just a place to worship but also served as a community clearinghouse, a credit union, a support group, and a center of political activism (National Park Service 2009). Churches were the context for the emergence of key civil rights leaders such as Reverends Martin Luther King Jr., Ralph Abernathy, Bernard Lee, and Fred Shuttlesworth.

Church and student groups founded organizations to coordinate their efforts. They included the Southern Christian Leadership Conference (SCLC), the Student Nonviolent Coordinating Committee (SNCC), Congress of Racial Equality, the National Association for the Advancement of Colored People (NAACP), and the National Urban League.

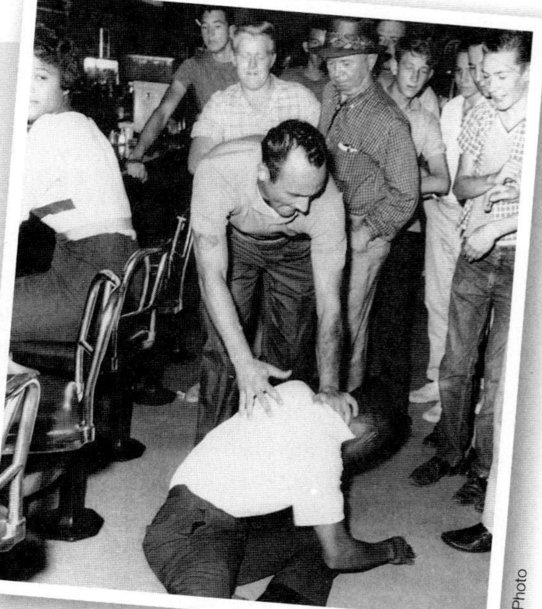

AP Photo

➤ This photograph was taken in 1960, when four black students sat down at Woolworth's department store lunch counter to eat. They were refused service but continued to sit at the lunch counter while white onlookers poured sugar, mustard, and ketchup over their heads and pulled them out of their chairs.

The Selma-to-Montgomery March was an especially significant moment for the civil rights movement, because ABC television broadcasted it, making it the first event of the activist movement to be televised. What Americans saw created enough outrage to lend national support to the movement. It began on what is known as Bloody Sunday, March 7, 1965, when state and local law enforcement agents stopped 600 demonstrators six blocks into the march and attacked them with clubs and tear gas. Civil rights leaders sought protection from the courts to march and it was granted. The escalating intensity of this movement pushed the federal government to become involved. Most notably, President John F. Kennedy used the power of his office to enforce desegregation in schools and public facilities. Attorney General Robert Kennedy filed suits against at least four states to secure the right to vote for blacks. When Congress passed the Civil Rights Act of 1964 and the Voting Rights Act of 1965, President Lyndon Johnson signed them into law knowing that it might cost him the next presidential election and severely weaken the Democratic Party's chance of winning in the next election cycle. Of course, federal and Supreme Court judges also played key roles in ruling against segregation and discrimination (National Park Service 2009).

(Write a Caption)

Write a caption that connects this photograph of children at a Fort Bliss Child Development Center with key concepts covered in this module to account for the racial diversity you see.

Lacey Justinger, USAG Fort Bliss, Texas

Hints: In writing this caption

- review the key concepts covered in this chapter, and
- consider that the children are sons and daughters of U.S. soldiers.

Critical Thinking

Assess the degree to which your life is integrated with people of other races and ethnicities by asking: With whom do I live, learn, pray, celebrate, and mourn? Does your life experience best identify with assimilation, integration, or pluralism?

Key Terms

absorption assimilation
assimilation
desegregation
integration
melting pot assimilation
pluralism
virtual integration

MODULE 8.8

Applying Theory: The Three Perspectives

Objective

You will learn how the three theoretical perspectives help us think about racial classification.

With what race do you think this person identifies—American Indian/Native Alaskan, Asian, Black/African American, Native Hawaiian/Pacific Islander, White, or some other race?

Tony Rotundo

Every 10 years the U.S. Census Bureau requires that someone in each household across the United States complete a survey asking the race of every person living in that household. Why does the United States do this? Each of the three theoretical perspectives—functionalist, conflict, and symbolic interaction—offers a framework for thinking about this U.S. policy. Keep in mind that while the United States now allows people to identify with more than one race, less than 3 percent of the population does so (U.S. Census Bureau 2011g).

Functionalist Perspective

Functionalists ask, What are the anticipated and unanticipated effects of a part on order and stability? They apply the concepts of manifest and latent functions and dysfunctions to answer that question. In this case, the part we are analyzing is the U.S. system of racial classification. One anticipated or manifest function is that racial classification is a tool the U.S. government uses to count and manage its population. The government's official reason for classifying people and collecting data on race is that various federal programs require such data. Those programs include the Federal Affirmative Action Plans, Home Mortgage Disclosure Act, Community Reinvestment Act, and Public Health Service Act. The federal government uses this data to assess successes or failures at combating discrimination against all racial groups and their communities.

An unanticipated or latent function of the U.S. system of racial classification is that asking people to declare their race actually reinforces the racial reality the government has constructed. Recall that almost 98 percent of the population sees itself as belonging to just one racial category. Despite evidence to the contrary, the system stays in place at least in part because people continue to apply it to themselves and others.

Missy Gish

➤ One manifest dysfunction of the U.S. system of racial classification is that the act of declaring a race separates people into racial categories, even people to whom they are biologically related, such as these cousins. Most Americans would likely never consider that these two children are biologically related. The two must establish a relationship in the context of a society that treats them as belonging to distinct racial categories.

Finally, a latent dysfunction of the U.S. system of racial classification is that it offers people a vocabulary and a set of assumptions for thinking about the self and others. This U.S. system of racial classification was built on the illusion that there is no overlap between racial categories. When it divides the 313 million residents of the United States into five racial categories, it reinforces difference and competing interests. The small percentage of residents who identify with more than one race often feel pressure to declare loyalty to or sympathy for one race over others.

Conflict Perspective

Conflict theorists seek to identify advantaged and disadvantaged groups, document their unequal access to scarce and valued resources, and identify the practices that advantaged groups establish to promote and protect their interests. Conflict theorists ask, Who benefits from classifying people by race and at whose expense? Conflict theorists also identify the ways that dominant groups disguise and justify a system that gives them an advantage.

Conflict theorists would point out that today's system of racial classification cannot be separated from its historical roots, which can be traced to exploitation of people and resources in Africa, Asia, and the Americas. Racism and the racial categories that form the cornerstone of racist thinking helped to justify this exploitation of nonwhite peoples and their resources. Therefore, conflict theorists would not believe that the primary reason we classify people by race is to check for patterns of discrimination and to monitor progress toward equal opportunity.

➤ Conflict theorists maintain that racial classification is used to divide nonwhite peoples into competing racial categories, thereby diluting their numbers and power in society. For example, conflict theorists question why the U.S. government maintains that Hispanics can be of any race and why Hispanic is the only recognized ethnic category in the United States. If, for example, Hispanics were classified as black, then together the two would represent 28.6 percent of the population. Separately, however, they represent 16.3 and 12.3 percent, respectively.

Chris Caldeira

Symbolic Interactionist Perspective

Symbolic interactionists ask, How do people experience, interpret, influence, and respond to what they and others are saying and doing? Symbolic interactionists ask how race shapes these processes. The system of racial classification gives people racial identities. During social encounters that involve people of different races, each party imagines how their race figures into the way others are viewing and evaluating them as persons and interpreting their words, actions, and motives. Symbolic interactionists study social interaction with particular emphasis on the interpretive and negotiating processes.

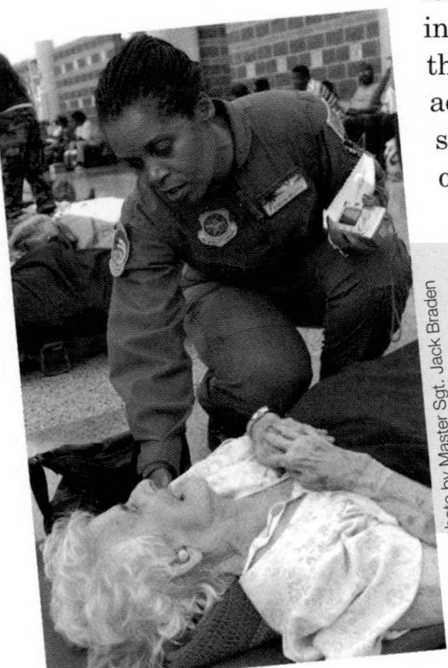
U.S. Air Force photo by Master Sgt. Jack Braden

➤ Symbolic interactionists are particularly interested in the ways in which people's race affects the course and outcome of social interactions. Specifically, the U.S. system of racial classification gives people categories in which to place themselves and those with whom they are interacting.

Critical Thinking

Which of the three theoretical perspectives do you find most compelling for explaining why the United States classifies people by race? Explain.

Summary: Putting It All Together

CHAPTER 8

Sociologists make a distinction between the concepts of race and ethnicity. Race is a physiologically based classification scheme in which racial categories are typically associated with selected physical traits, such as hair texture and skin shade, that have been assigned extraordinary social and political significance. The fact that Brazil and the United States have very different ideas about race tells us that race is human created. That is, people make the categories, attach meanings to them, and give them social significance with real consequences.

An ethnic group consists of people who share, believe they share, or are believed by others to share a national origin; a common ancestry; a place of birth; and/or distinctive social traits that set them apart from other ethnic groups. Ethnic identification is not just a matter of individual choice. People selectively remember and forget some ethnicities that make up their heritage. Some racial groups have more freedom than others in claiming or disclaiming an ethnic identity. Sometimes ethnicity is imposed, and sometimes it is revived and renewed. The United States recognizes six official racial categories and two ethnicities. Like all such classification schemes, the U.S. system is characterized by fatal flaws in logic.

Disadvantaged racial and ethnic groups are considered minority groups. Among other things, minorities do not choose that status. Disadvantaged status is justified and perpetuated by racism, prejudice, and discrimination. Prejudice, discrimination, and racism are the cornerstones of segregation, ethnic cleansing, and genocide.

Assimilation is a process by which ethnic and racial distinctions between groups disappear because one group is absorbed into another group's culture (absorption assimilation) or because two cultures blend to form a new culture (melting pot assimilation). Assimilation, which involves some level of integration, is facilitated by desegregation. Pluralism is a situation in which different racial and ethnic groups coexist in harmony; have equal social standing; maintain their unique cultural ties, communities, and identities; and participate in the economic and political life of the larger society.

Go to cengagebrain.com to link to Aplia and CourseMate for the chapter quiz and other activities.

Courtesy of Asia and Pacific Transgender Network Development

CHAPTER 9

GENDER AND SEXUALITIES

This photograph is striking in that it shows people, once considered biologically male, who transitioned to females. These women are part of the Asia Pacific Transgender Network (APTN), an organization devoted to saying "no" to discrimination and marginalization and to championing transgender women's and men's health, legal, and social rights. These women are seeking to expand our understanding of gender as a socially created and learned distinction specifying the physical, behavioral, mental, and emotional traits believed to make one masculine or feminine. While there is no fixed line distinguishing masculine from feminine traits, for the most part people act as if there is. Sociologists seek to uncover the ways in which ideas of gender shape life chances and are interested in why some people embrace these socially created distinctions that treat gender as binary (just two categories) while others resist and even challenge such notions.

Go to Sociology CourseMate on cengagebrain.com to watch a video about "stereotype threat" and how it impacts a young female mathematician and a black college student.

MODULE

9.1 Sex and Gender

Objective

You will learn that sex is an anatomical distinction and that gender is a social construction.

Missy Gish

Missy Gish

This is the same baby dressed in pink and blue.
Can you guess the baby's sex?

The decision to dress the baby boy in pink or blue relates to the concept of gender, a society's beliefs about what boys, girls, women, and men should be. While this baby does not seem to care whether it is dressed in pink or blue, color does matter to most parents. The baby will eventually learn the social importance of the two colors.

Sex

A person's **sex** is based on **primary sex characteristics**, the anatomical traits essential to reproduction. Most cultures classify people into two sex categories—male and female—based on what are considered to be clear anatomical distinctions. Biological sex, however, is not clear-cut, if only because some babies are born intersexed. The medical profession uses the broad term **intersexed** to classify people with some mixture of male and female primary sex characteristics. Although we do not know how many intersexed babies are born each year, one physician who treats intersexed children estimates that number to be one in every thousand (Dreifus 2005).

If some babies are born intersexed, why does society not legally recognize an intersexed category? In the United States and most other countries, no such category exists because such children are typically treated with surgery and/or hormonal therapy. The rationale underlying medical intervention is the belief that the condition "is a tragic event" about which something must be done (Dewhurst and Gordon 1993, A15). Consider those who are transgender as further evidence that no clear line separates male from female.

Why does no clear line exist to separate every newborn into one of two categories, male or female? One answer lies in the biological sequences of events that occur in the first weeks after conception, at which point the human embryo develops the potential to form either ovaries or testes. Approximately eight weeks into development, the ovaries *or* the testes disintegrate. About a week later the outer appearance we come to associate with male or female begins to develop. This complex chain of events does not always occur as theorized (Lehrman 1997, 49).

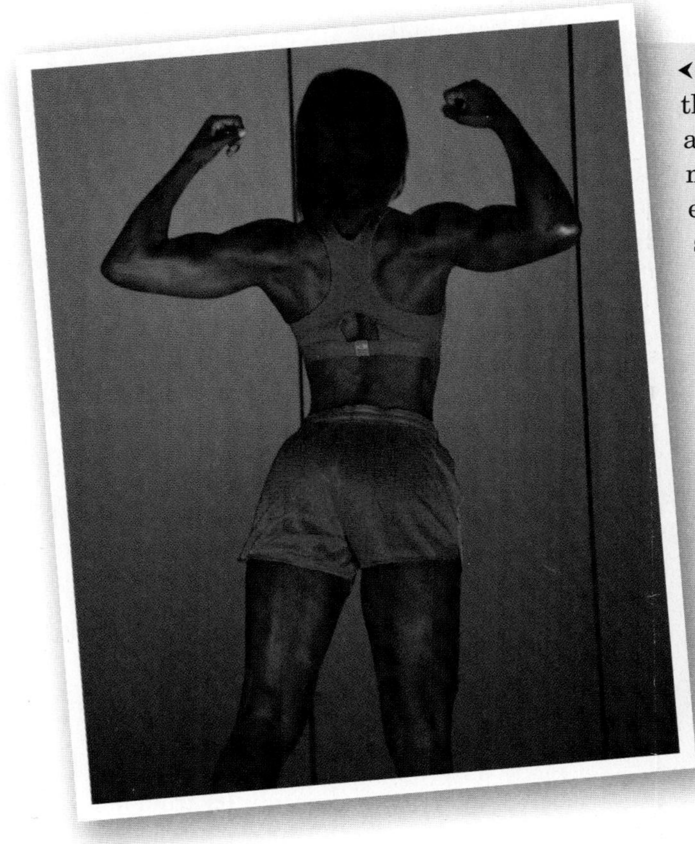

◂ Do you question whether this athlete is genetically a woman? While one might think it would be easy to determine, the so-called gender test female athletes face if a competitor challenges their sex can take several weeks to complete. That test involves a medical examination and reports from a gynecologist, an endocrinologist, a psychologist, an internal medicine specialist, and an expert on gender (Dreger 2009).

U.S. Army

In addition to using primary sex characteristics to distinguish one sex from another, we use **secondary sex characteristics**, physical traits not essential to reproduction, such as breast development, quality of voice, and distribution of facial and body hair, that supposedly result from the action of so-called male (androgen) and female (estrogen) hormones. We use the term *so-called* because all people produce androgen and estrogen (Garb 1991). Like primary sex characteristics, there is no clear line to mark any secondary sex characteristic as distinctly male or female. For example, there is no demarcation that separates a male voice from that of a female or a female pattern of body hair distribution from that of a male.

Gender

Gender is the socially created and learned distinctions that specify the physical, behavioral, and mental and emotional traits characteristic of males and females. The terms **masculinity** and **femininity** refer to traits believed to be characteristic of males and females. Masculinity and femininity are concepts that are taught, learned, emulated, and enforced (Lorber 2005). Ideas of what constitutes masculinity and femininity are often expressed as **gender ideals**. A gender ideal is at best a caricature, in that it exaggerates the characteristics believed to make someone the so-called perfect male or female. In fact, some gender ideals may not exist in reality. Consider that few, if any, women have 4–6 inch long feet. Yet at one time that was considered the *ideal* foot size for women in China—an impossible standard that no female could achieve without enduring foot binding as a young girl.

Prints & Photographs Division, Library of Congress, LC-USZ62-101143

➤ Likewise, few grown women have 18-inch waists, and yet at one time in the United States, that was the ideal. Women in the United States and elsewhere have worn corsets and even removed lower ribs to achieve that impossible standard.

Master Sgt. Scott Wagers

◂ Ideals, then, are socially created standards we use to judge ourselves and others. Just as women strive to meet gender ideals, so do men. As this little boy flexes his biceps, it is likely that he has some ideal in mind, perhaps the flexed biceps of a wrestler.

Tony Rotundo

We must note that no fixed line separates masculinity from femininity. Often we mistakenly attribute certain differences between men and women to nature, when in fact those differences are socially created. In the United States, for example, norms specify that females present themselves as having no facial or body hair. It is deemed acceptable for women to have eyelashes and well-shaped eyebrows but certainly not to have hair above their lips, under their arms, on their inner thighs, or on their chin, shoulders, back, chest, breasts, abdomen, legs, or toes. We fail to consider how hard women must work to achieve these cultural standards. Women's compliance makes males and females appear more physically distinct in terms of hair distribution than they are in reality.

To this point, we have shown that gender is a social construction and that gender ideals are often impossible to realize. Moreover, not everyone fits into one of two categories—male and female. Some refuse to do the work required to comply with ideas of masculinity and femininity; others challenge existing ideals and seek to create alternative gender categories with which they feel more comfortable. It should not surprise us, then, to learn that some societies recognize (although they may not completely accept) a third gender.

A Third Gender

In her research on *fa'afafine*, Jeannette Mageo (1992) describes the guests attending a wedding shower in Samoa. Of approximately 40 women, 6 were *fa'afafine*—people who are not biologically female but who have taken on the "way of women" in dress, mannerism, appearance, and role. During that wedding shower, the *fa'afafine* staged a beauty contest in which each sang and danced a love song. Such beauty contests are well-known in Samoa, and the winner "is sometimes the 'girl' who gives the most stunningly accurate imitation of real girls, such that even Samoans would be at a loss to tell the difference; sometimes the winner is the most brilliant comic" (Mageo 1998, 213). Often *fa'afafine* imitate popular foreign female vocalists, such as Britney Spears, Madonna, or Kelly Clarkson.

Mageo (1998) argues that *fa'afafine* could not have become commonplace in Samoa unless something about that society supported gender blurring. She notes that Samoans do not have a tradition of making sharp distinctions "between men and women, boys and girls." For example, "boys and girls take equal pride in their skills in fights . . . personal names are often not marked for gender, and outside school little boys and girls still wear much the same clothing" (1998, 451). Another factor relates to a decline in male status; men were once treated as the "strength of the village," acting as the village police force or army reserve (Mageo 1992, 444; M. Mead 1928). The introduction of mass education, a shift away from an agriculture-based economy to a wage-based one, and the introduction of new labor-saving technology eventually reduced the community's reliance on male strength. Men's status was further undermined by high unemployment rates, sometimes exceeding 20 percent, and by employment opportunities limited to working in tuna canneries or for the Samoan government (U.S. Central Intelligence Agency 2012). These factors left many males uncertain about their place in Samoan society. For some men in American Samoa, becoming a *fa'afafine* allowed them to vicariously experience the status of well-known female impersonators.

Another popular option for men in Samoa is to play football for the U.S., Canadian, or European leagues. In fact, 30 players considered to be of Samoan descent are listed on NFL team rosters. An estimated 15 percent of all male high school graduates in Samoa leave their country to play college or junior college football in the United States (Uperesa, 2010; Saslow 2007).

Christina Douglas, 9th Mission Support Command

◀ Another option available to Samoan men is to join the U.S. military. These Samoan reservists are being honored for tsunami relief efforts.

The point of focusing on the situation of men in Samoa is to reinforce the idea that gender is a social construction. Specifically there are many social forces that shape beliefs about what males and females should be and that channel behavior in what are considered to be gender-appropriate directions. Even when people decide not to conform, gender expectations still retain their importance because people still hold them up as something to be resisted.

(Write a Caption)

Write a caption that connects the concept of gender ideals to this X-ray showing feet disfigured from the practice of foot binding to meet an ideal of 4-inch-long feet.

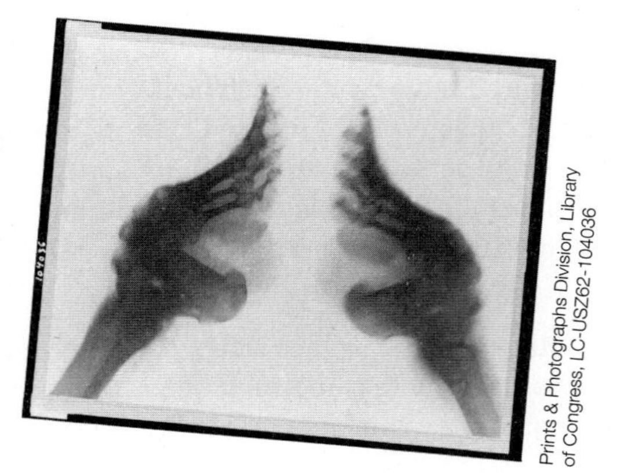

Prints & Photographs Division, Library of Congress, LC-USZ62-104036

Hints: In writing this caption

- review the concept of gender ideals,
- consider whether 4-inch-long feet (among adults) exist in reality, and
- think about the effort expended to achieve this ideal and what it suggests.

Critical Thinking

Can you think of a time in your life when you worked particularly hard to achieve a gender ideal or to resist a gender ideal?

Key Terms

femininity	intersexed	secondary sex characteristics
gender	masculinity	sex
gender ideals	primary sex characteristics	

Life Chances and Structural Constraints

MODULE 9.2

Objective

You will learn the extent to which a person's gender determines his or her life chances.

U.S. Air Force photo by Airman 1st Class Cassandra J. Locke

U.S. Army photo by 2nd Lt. Christian Venhuizen/Released

How much time do you spend on a typical day working to present yourself as a particular gender?

Life Chances

Sociologists define **life chances** as the probability that an individual's life will turn out a certain way. Life chances apply to virtually every aspect of life—the chances that a person will become an airline pilot, play T-ball, major in elementary education, spend an hour or more getting ready for work or school, or live a long life.

Sociologists are interested in the processes by which being male or female increases the probability that a person's life will be a certain way. Ideas about what men and women should be shape every aspect of life, including how people dress, the time they wake up in the morning, what they do after they wake up, the social roles they take on, the things they worry about, and even ways of expressing emotion and experiencing sexual attraction (Bem 1993).

To understand the power of gender in shaping life chances, consider the now-classic research by Alice Baumgartner-Papageorgiou (1982) on elementary and high school students. She asked them how their lives would be different if they were the other gender. Their responses reflected culturally conceived and learned ideas about sex-appropriate behaviors and appearances and about the imagined and real advantages and disadvantages of being male or female (Vann 1995). The boys generally believed that their life chances would change in negative ways if they became girls. Among other things, they would become less active and more restricted in what they could do. In addition, they would become more conscious about tending to their appearance, finding a husband, and being alone and unprotected in the face of a violent attack—"I'd use a lot of makeup and look good and beautiful" and "I would not be able to help my dad fix the car and his two motorcycles" (2–9).

The girls, on the other hand, believed that if they became boys they would be less emotional, their lives would be more active and less restrictive, they would be closer to their fathers, and they would be treated as more than "sex objects"—"My father would be closer, because I'd be the son he always wanted" and "People would take my decisions and beliefs more seriously" (5–13).

Baumgartner-Papageorgiou's findings about how one's life is shaped by gender seem to hold up across time. When I asked my students how their lives would change if they were the other gender, their responses were remarkably similar to those described above. Decisions about how early to get up in the morning, which subjects to study, whether to show emotion, how to sit, and what sports to play are influenced by society's ideals of masculinity and femininity rather than by criteria such as self-fulfillment, interest, ability, or personal comfort.

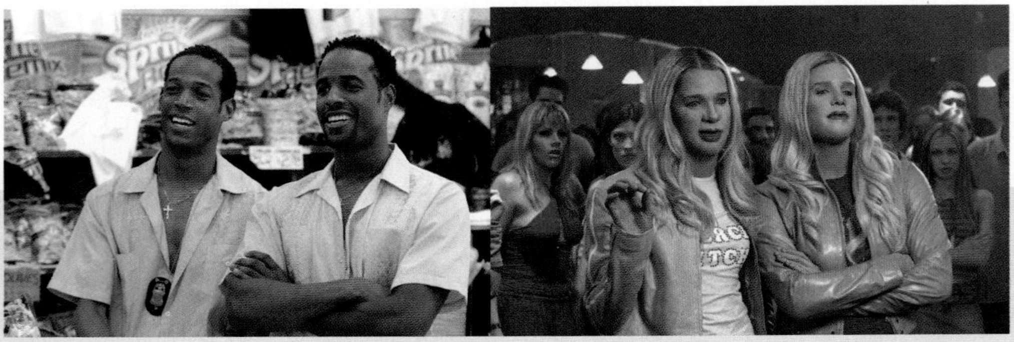

▲ While the movie *White Chicks*, starring Shawn Wayans and Marlon Wayans, is a comedy, it allows us to consider how two men with a change of clothes, wigs, makeup, and body language can present themselves as females. These photos offer clues to how people's lives change as a result of the gender they present to the world.

When selecting a college major, many students consider, even if subconsciously, the "sex" of the major: if a major matches their sex, they consider the major to be a viable option, and if it does not match, they may reject the major outright (Bem 1993). Note that about 85 percent of bachelor's degrees in engineering and in computer-information sciences are awarded to males, whereas 94 percent of bachelor's degrees in library sciences are awarded to females. Other majors dominated by women include education, health professions, and public administration/social services. Approximately 80 percent of all bachelor's degrees awarded in these fields go to women (National Center for Education Statistics 2011a).

Life chances include not just the probability of choosing a college major based on whether it is viewed as sex-appropriate but also the probability of wearing uncomfortable clothing. One study by the Society of Chiropodists and Podiatrists found that 40 percent of women admit to buying shoes that they know do not fit—that is, they are too narrow (by design) or too small by one or two sizes (Harris 2003). This finding suggests that many women decide to wear uncomfortable shoes so they can appear more fashionable. Apparently, 17 percent of men buy shoes they know don't fit (BBC 2009). It is not known whether men buy shoes that are larger or smaller than needed.

Structural Constraints

How do we explain that, for the most part, nurses are females and carpenters are males? Sociologists believe that one answer to this question lies with **structural constraints**, the established and customary rules, policies, and day-to-day practices that channel behavior in a certain direction and that shape a person's life chances. One example relates to the structural constraints that push many men and women to careers that correspond with society's ideals about sex-appropriate work. Women are more likely to be "pushed" into work roles that emphasize personal relationships and nurturing skills or that pertain to products and services labeled as family-oriented and feminine. Men are more likely to be pushed into work roles that emphasize decision-making and control and that pertain to machines, products, and services considered masculine.

Sociologists argue that we must also consider how the jobs men and women "choose" channel behavior in stereotypically male and female directions. The point is that it is not the day care worker per se that is feminine; it is the skills needed to do the job well that makes the day care worker behave in ways we associate with femininity. Presumably, anyone holding the job of day care worker will display those "feminine" characteristics (Anspach 1987).

➤ To be successful at the job of drill instructor, a person must be aggressive, relentlessly critical, and forceful—qualities we associate with masculinity. All people who take on this job—male or female—will find themselves displaying "masculine" characteristics.

Sgt. Jose Nava

The concept of structural constraints offers insights as to how institutions are gendered. Sociologists look for ways in which gender is embodied and embedded in institutional arrangements. Institutions are **gendered** when there is an established pattern of segregating the sexes into different workspaces or jobs, of disproportionately assigning one sex to positions of power, and of otherwise disadvantaging one sex relative to the other (Britton 2000; P. Martin 2004). To determine whether an institution is gendered, sociologists ask the following kinds of questions: Do work institutions hold expectations about whether a male or female is best suited for a particular job? Are some occupations disproportionately occupied by females (administrative assistant) and others by males (vice president)? Are women the social studies teachers and men the science teachers? Do employers offer female employees maternity leave but do not offer male employees paternity leave? If the answer to any of these questions is "yes," then the institution is gendered.

Staff Sgt. David Chapman, 5th MPAD

➤ Have you ever played a sport? Were your coaches male or female? No matter your own gender, it is likely that your coach was male. For the most part it is commonplace for men to coach females, but not for females to coach men. Simply consider that 58 percent of women's college teams have a head coach who is male; less than 2 percent of men's teams have a female head coach. This state of affairs qualifies as a gendered pattern (Rhode and Walker 2008).

Molly Hayden, U.S. Army Garrison—Hawaii Public Affairs

(Write a Caption)

Write a caption that relates the concept of gendered institutions to team mascots.

Hints: In writing this caption

- review the concept of gendered institutions,
- identify the institutions of which mascots are typically a part, and
- think about those mascots in terms of their purpose and presumed sex.

Critical Thinking

How do you think your life would change if you presented yourself as another gender?

Key Terms

gendered life chances structural constraints

Gender Stratification

MODULE 9.3

Objective

You will learn that sociologists seek to understand situations that put one sex at a disadvantage relative to the other.

Mass Comm. Spc. 2nd Class Erick S. Holmes

Mass Comm. Spc. 2nd Class Ronald Gutridge

Is there any country in the world where women are considered equal to men?

Each year the World Economic Forum publishes a report on the global **gender gap**, defined as the disparity in opportunities available for men and women. The report considers the situation of women relative to men in 135 countries with regard to four areas: economic participation and opportunity, health and survival, educational attainment, and political empowerment. In doing so, the World Economic Forum is considering **gender stratification**, the extent to which opportunities and resources are unequally distributed between men and women.

According to this report, there is no country in which women have more *overall* opportunities than their male counterparts, but there are countries in which men and women share more equally in the available resources and opportunities. According to the measures used, Iceland ranks first as the country with the least inequality; the United States ranks 17th, and Yemen ranks 134 or last with greatest inequality (see Table 9.3a).

Table 9.3a: Indicators Used to Rank Countries on Gender Equality/Inequality

Notice that female life expectancy exceeds that of males in all three countries. With regard to political power, the data show that women in Yemen have virtually no political power. Compare Iceland and the United States on all indicators. On which indicators is Iceland ahead of the United States? On which indicators, if any, are Iceland and the United States similar? Why do you think Iceland is ranked ahead of the United States with regard to gender equality?

	Iceland		United States		Yemen	
	Females	Males	Females	Males	Females	Males
Indicator	Economic Opportunity					
% in Labor Force	81	90	68	80	21	74
Median Income	$27,675	$40,000	$35,436	$40,000	$857	$4,046
% Professionals/ Workers	56	44	55	45	2	98
% of Legislators, Senior Managers/ Officials	33	67	43	57	4	96
	Educational Attainment					
% Literacy Rate	99	99	99	99	45	80
	Health and Survival					
Life Expectancy	75	73	72	68	51	48
	Political Empowerment					
% in Parliament/ Congress	43	57	17	83	0	100
Number of Years (of last 50) with Female or Male Head of State	18	32	0	50	0	50

Source of Data: World Economic Forum (2011)

Explaining the Gender Gap

Using a complex formula, the World Economic Forum estimates that in Iceland, 85 percent of the overall gender gap has been closed. In the United States, 74 percent of the gap has been closed, and in Yemen, 49 percent has been closed. Sociologists seek to identify the social factors that put one sex at a disadvantage relative to the other. Inequality exists when one sex relative to the other

1. faces greater risks to physical and emotional well-being,
2. possesses a disproportionate share of income and other valued resources, and/or
3. is accorded more opportunities to succeed.

YEMEN. In Yemen there are many customs and traditions that work to keep women in positions inferior to men. In particular, Yemen is a **patriarchy**, an arrangement in which men have systematic power over women in public and private (family) life. Male power is supported by law, and in the case of Yemen, those laws are Islamic (*Shari'a*) and govern all aspects of life. In addition, "women are prohibited from interpreting the religious texts that define Islamic laws, and they cannot serve as family court judges. One can speculate that if women had this right, they would interpret Islamic texts differently than men, who for the most part have defined a woman's duty as obeying her husband" (Freedom House 2008). Because 55 percent of Yemenese women are not literate, most are unable to read Islamic texts. Thus, the low literacy rate of women (45 percent) *relative* to men (80 percent) affects their ability to gain independence from men (World Economic Forum 2011).

➤ These Yemenese women are participating in a U.S. military-sponsored animal husbandry training program with the goals of improving women's position in society and the country's overall livestock health and productivity. Here a U.S. Staff Sergeant is showing a thermometer and explaining its uses.

Tech. Sgt. Carrie Bernard

ICELAND. Iceland has one of the highest per capita incomes in the world. In 2011 the country ranked number one in the world with regard to gender equality. Gender discrimination is prohibited in Iceland, and laws mandating gender equality in schools and education have been in place since 1976. In 2008 a new legal mandate—the Act on the Equal Status and the Equal Rights of Women and Men—was passed. The act promotes gender equality in all spheres of society. Among other things, the act "stipulates that equal participation of women and men shall be promoted in committees, boards and councils under the auspices of the government and local authorities, the gender proportion being not less than 40 percent where there are more than three members." In addition to an administrative function, The Centre for Gender Equality is the national bureau that provides counseling and other support education to governments, corporations, and nonprofits (Guđmundsson 2008). Among other things, Iceland is known for its family leave policy that guarantees both parents the opportunity to care for newborns and for its extensive system of day care and child development centers.

THE UNITED STATES. The United States ranks 17th of 134 countries studied in closing the gender gap. In comparison to women in Iceland, American women earn significantly more income relative to their male counterparts. In addition, the United States does rank number one with regard to equality of educational opportunity. The United States falls short in the areas of political empowerment (ranks 39th in the world) and female labor force participation. Unlike Iceland, the United States does not mandate equal representation in government and other public organizations. As a result, only 17 percent of the U.S. Congress consists of women (see Figure 9.3a).

Figure 9.3a: The Gender Gap: 10 Most Equal and 10 Least Equal Countries in the World

The map shows the 10 countries in which women's overall opportunities relative to the men in that country are most and least equal. The 10 countries with greatest equality for women are Norway, Finland, Sweden, Iceland, New Zealand, the Philippines, Denmark, Ireland, the Netherlands, and Latvia. The 10 least equal are Nepal, Oman, Benin, Morocco, Cote d'Ivoire, Saudi Arabia, Mali, Pakistan, Chad, and Yemen.

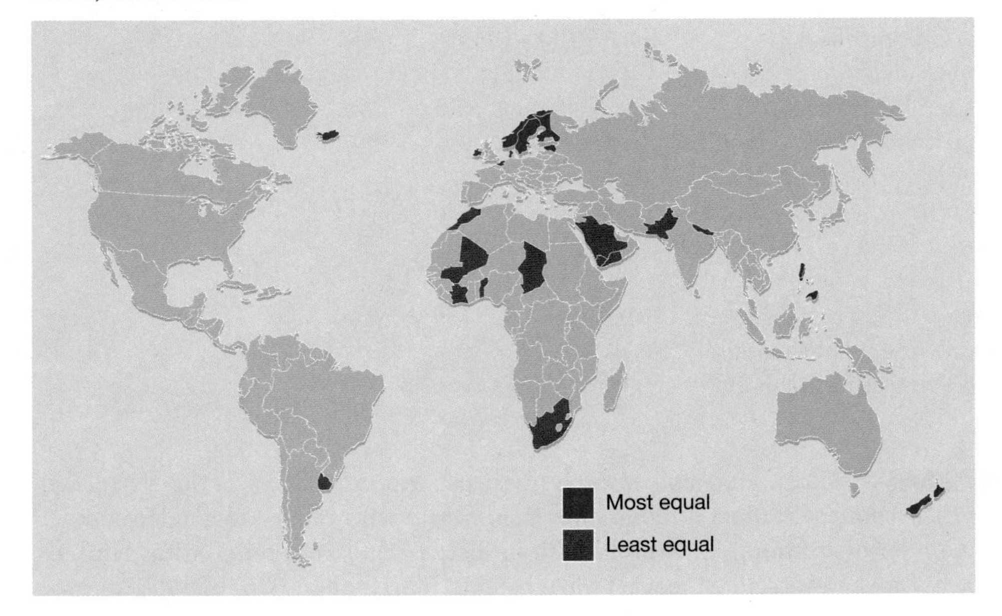

Source of Data: World Economic Forum (2011)

While the income gap between men and women is significantly smaller in the United States than in Iceland, there is still income inequality between men and women. Figure 9.3b offers a graphic depiction of gender inequality in pay as it relates to full-time wage and salary workers.

Figure 9.3b: Women's Earnings as a Percent of Men's Full-Time Wage and Salary, 1979–2010

In 1979 women working full-time earned about 63 cents for every dollar earned by their male counterparts. In 2010 that gap had decreased by 29 cents on the dollar, so women earned about 81 cents for every dollar earned by men.

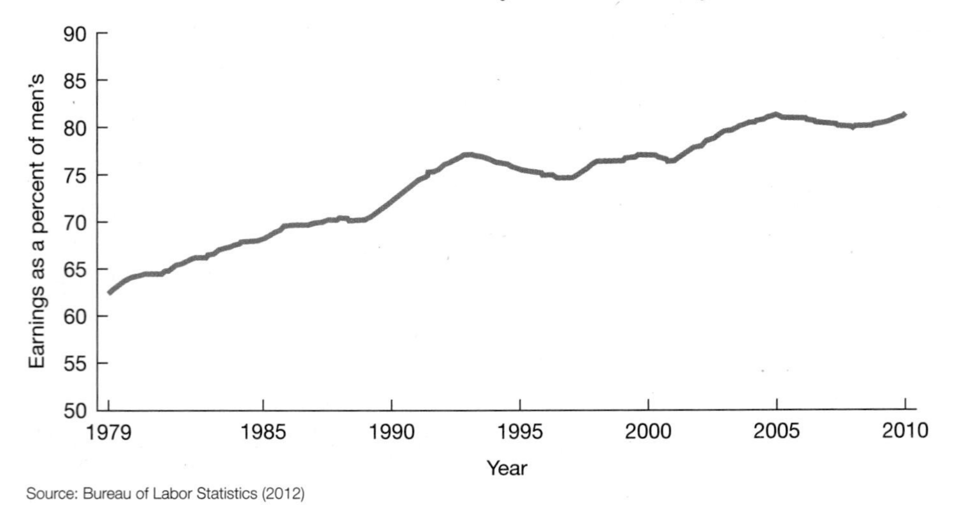

Source: Bureau of Labor Statistics (2012)

Male–female income differences also vary by age group, with the greatest inequality between men and women ages 45–54 and the least inequality between men and women ages 16–34. Women ages 16–34 earned 90 cents for every dollar earned by men, and women 45–54 and over earned 76 cents for every dollar earned by men (U.S. Department of Labor, Bureau of Labor Statistics 2012d).

If we take a broader view and examine total earnings of men and women over an extended period of time, such as a 15-year period, we find that the average woman earned about 38 percent of what the average man did (Madrick 2004). Economists Stephen J. Rose and Heidi Hartman (2004) compared men's and women's total earnings between 1983 and 1998 and found that the average woman earned $273,592 while the average man earned $722,693.

Overall we can say that men earn more income than women. But it is important to point out that while women are gaining ground, some of those gains are achieved because men's wages have declined (see Chart 9.3a).

Chart 9.3a: Percent Change in Weekly Median Income (in constant dollars) for Full-Time Wage and Salary Workers, Men and Women, Compared by Level of Education, 1979–2010

While men as a group earn more than women, this chart helps us see that, except for those men with a bachelor's degree or higher, the value of men's labor has declined dramatically over the past three decades or so. While it is worth noting that the gap between male and female wages has declined significantly, as noted in Figure 9.3b, women's gains have been accompanied by men's falling wages. Men's falling wages are a result of structural transformations in the economy (decline of manufacturing and increase in service, management, and information jobs), the mortgage crisis that particularly impacted construction-related jobs, and technologies that automated or computerized many tasks.

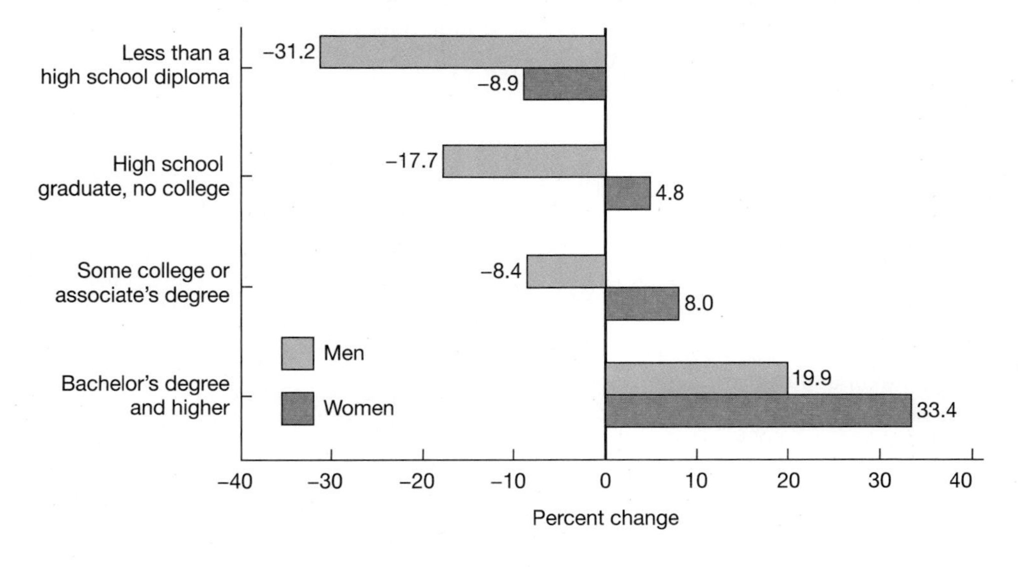

Source: U.S. Bureau of Labor Statistics (2012b)

Explaining the Income Gap

There are many possible explanations for the overall male-female income gap and the gender gap in political participation.

- Women are disproportionately employed in lower-paying, lower-status occupations. Specifically, they choose or are forced into lower-paying positions that are considered sex-appropriate, such as teacher, secretary, and caregiver. In addition, women are channeled into positions that offer fewer and more flexible hours to meet caregiving responsibilities. Women choose or are forced into occupations that will not require them to relocate, work in unpleasant environments (such as mines), or take on hazardous assignments—three activities associated with higher pay (Sahadi 2006).

- Some employers underinvest in the careers of childbearing-age women because they assume the women will eventually leave the workplace to raise children. Because society still expects women to take on primary caregiving responsibilities, many do leave the labor market to take care of children and elderly parents and then re-enter it later. The associated time away from the workplace puts them behind with regard to wages and promotion.
- Employers often view women's salary needs as less important than men's and pay women accordingly. Unfortunately, many still consider women's earnings as supplemental to a presumed male partner—earnings that can be used to buy "extras"—when in reality many women are heads of households. When negotiating for salaries, women underestimate their worth to employers and ask for less than their male counterparts. Some employers steer males and females into different gender-appropriate assignments (such as sales clerks in baby clothes departments rather than hardware) and offer them different training opportunities and chances to move into better-paying jobs (Love 2007).
- Women encounter a **glass ceiling**, a term used to describe a barrier that prevents women from rising past a certain level in an organization, especially for women who work in male-dominated workplaces and occupations. The term applies to women who have the ability and qualifications to advance but who are not well-connected to those who are in a position to advocate for or mentor them. With regard to men who work in female-dominated professions, they encounter the **glass escalator**, a term that applies to the invisible upward movement that puts men in positions of power, even within female-dominated occupations. In this case, management singles out men for special attention and advancement such that men are encouraged to move from school teacher to assistant principal to principal or from social worker to program director.

➤ This woman represents one of the 90.5 percent of all registered nurses who are female. The male represents one of the 9.5 percent of all nurses who are male. The median weekly income for female nurses is $1,039. The median weekly income for male nurses is $1,201, $162 per week more than female counterparts (U.S. Department of Labor 2012).

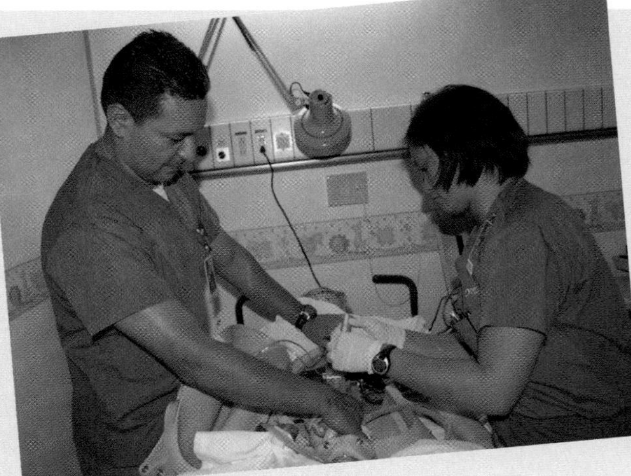

U.S. Navy photo by Mass Comm. Spc. 2nd Class Chantel M. Clayton

Workplace Fatalities and Injuries

There are many areas in which men are disadvantaged relative to women. Men work 60 percent of all hours worked but suffer 92 percent of workplace fatalities. (In 2010 there were 4,547 fatal workplace injuries in the United States.) Men suffer disproportionately more workplace fatalities than women because they are employed in the most dangerous industries, including construction, mining, and logging (see Figure 9.3c). There is also evidence that males (relative to their female counterparts) fail to take preventive steps that would protect them from fatalities.

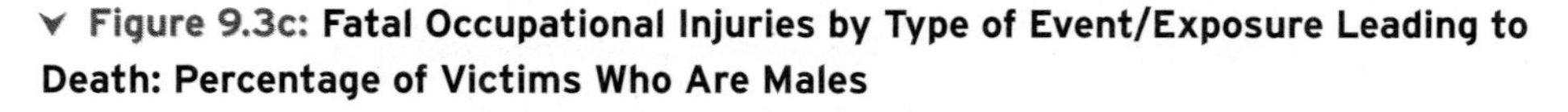

Figure 9.3c: Fatal Occupational Injuries by Type of Event/Exposure Leading to Death: Percentage of Victims Who Are Males

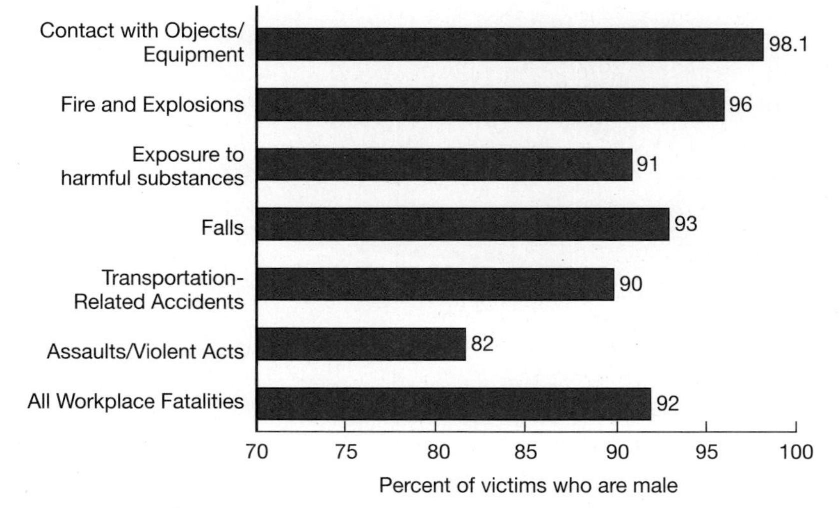

Source of data: U.S. Bureau of Labor Statistics (2010)

Men also experience more nonfatal workplace injuries than females. There are approximately 1.2 million nonfatal job-related injuries at work that result in a median of 8 days lost. For every 10,000 employed males, 128 are injured on the job each year. For every 10,000 employed females, 106 are injured on the job each year. Women's injuries are likely to be sprains and strains from lifting and otherwise overexerting muscles (lifting patients, picking up children, lifting mattresses). Men, too, suffer from sprains and bruises but are much more likely than women to suffer injuries related to falls, chemical and heat burns, severed limbs, and exposure to harmful chemicals (U.S. Department of Labor, Bureau of Labor Statistics 2011c).

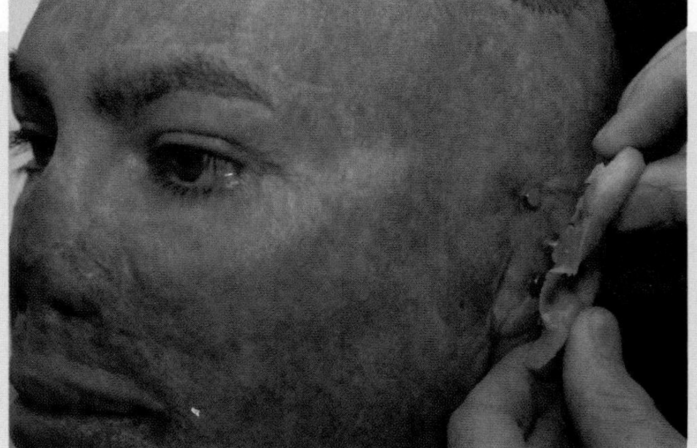

➤ This man who suffered severe burns while serving his country is being fitted for a prosthetic ear. He is one of almost 100,000 men wounded in Iraq and Afghanistan compared to 865 women. The lower numbers of women relative to men can be explained by the fact that women make up 15 percent of soldiers and are largely barred from combat career positions (Tilghman 2012).

Staff Sgt. Robert Barnett

(Write a Caption)

Write a caption that applies the concept of gender gap to describe the disadvantages of occupying the position of firefighter (99 percent of whom are male) and other dangerous occupations.

Pfc. Daniel Boothe

Hints: In writing this caption

- review the concept of gender gap,
- think about the disadvantages that come with the job of firefighter, and
- consider why it is that mostly men come to occupy this position.

Critical Thinking

Is there an area of your life where you feel advantaged relative to the other sex?

Key Terms

gender gap

gender stratification

glass ceiling

glass escalator

patriarchy

MODULE 9.4

Gender Socialization

Objective

You will consider the process by which people learn to be male or female.

How much do you think the pan of charcoal weighs? Do you know any men who could do this? Does this pregnant woman's ability to balance and carry this load on her head challenge assumptions that women are the weaker sex?

U.S. Navy photo by Chief Mass Comm. Spc. Robert Fluegel

While you may not know any men (or women) who could match this woman's physical feat, there is no doubt that, given the "right" environment, a person, male or female, could learn to do it. Sociologist Cynthia Fuchs Epstein points out that "throughout the world, where water is a scarce commodity it is women who carry heavy buckets and vessels of water, usually on foot and over long distances, because this has been designated as a woman's job and men regard it as a disgrace to help them" (2006, 10).

What is considered masculine and feminine is not something that is innate or natural. People learn to be masculine or feminine. Once a baby is labeled male or female, everyone who comes in contact with the child begins to treat him or her as such. With encouragement from others, children learn to talk, walk, and move in gendered ways (Lorber 2005). They also learn **gender roles**, the behavior and activities expected of someone who is male or female. These expectations channel male and female energies in different gender-appropriate directions. As children learn their society's expectations about how boys and girls should look and behave, most will, by their own behavior and appearance, reproduce and perpetuate those expectations. When children fail to behave in gender-appropriate ways, their character becomes suspect (Lorber 2005). At the minimum, people call girls who violate the rules tomboys and boys who do so sissies.

The gender socialization process may be direct or indirect. It is indirect when children learn gender expectations by observing others' words and behavior, such as the jokes, comments, and stories they hear about men and women or portrayals of men and women they see in magazines, books, and on television (Raag and Rackliff 1998). Socialization is direct when significant others intentionally convey the societal expectations to children.

Lisa Southwick

Cpl. Im Jin-min

▲ Socialization theorists argue that an undetermined yet significant portion of male–female differences are products of the ways in which males and females are socialized. These little boys and girls are too young to choose on their own whether they want to be football players or dancers, but their parents have invested money and time in activities on the basis of what they consider gender-appropriate for their children.

Agents of Socialization

Agents of socialization are the significant people, groups, and institutions that act to shape our gender identity—whether we identify as male, female, or something in between. Agents of socialization include family, classmates, peers, teachers, religious leaders, popular culture, and mass media. Child development specialist Beverly Fagot and her colleagues (1985) observed how preschool teachers shape gender identity. Specifically, the researchers focused on how toddlers, ages 12 and 24 months, in a play group interacted and communicated with one another and how teachers responded to the children's attempts to communicate. Fagot found no differences in the interaction styles of 12-month-old boys and girls: All of the children communicated by gestures, gentle touches, whining, crying, and screaming. The teachers, however, interacted with them in gender-specific ways. They were more likely to respond to girls when they communicated in gentle, "feminine" ways and to boys when they communicated in assertive, "masculine" ways. That is, the teachers tended to ignore girls' assertive acts but respond to boys' assertive acts. Thus, by the time these toddlers were two, they communicated in very different ways.

Fagot's research was conducted more than 25 years ago. A more recent study found that early childhood teachers are more accepting of girls' cross-gender behaviors and explorations than they are of boys'. According to this research, teachers believe that boys who behave like "sissies" are at greater risk of growing up to be homosexual and psychologically ill-adjusted than are girls who behave like "tomboys." This finding suggests that while American society has expanded the range of behaviors and appearances deemed acceptable for girls, it has not extended the range for boys in the same way (Cahill and Adams 1997).

Children's toys and celebrated images of males and females figure prominently in the socialization process. Barbie® dolls, for example, have been marketed since 1959 with the purpose of inspiring little girls "to think about what they wanted to be when they grew up." The dolls are available in 67 countries. An estimated 95 percent of girls between ages 3 and 11 in the United States have Barbie® dolls, which come in several skin colors and 45 nationalities (Mattel 2010).

Lisa Southwick

➤ For boys, G.I. Joe was the first action figure toy on the market, launched in 1964, and it was followed by a long line of action figures, including Transformers™, Micronauts™, Star Wars™, Power Rangers™, X-Men™, Street Fighter™, Bronze Bombers™, and Mortal Kombat™. The popularity of these toys is boosted by comic books, motion pictures, and cartoons, and they appear on school supplies, video games, card games, lunch boxes, posters, and party supplies (Hasbro Toys 2010; Son 1998).

Norms Governing Body Language

Learning to be male or female involves learning norms governing the way males and females present themselves. That includes learning the sex-appropriate norms governing body language. Norms governing male body language suggest power, dominance, and high status, whereas norms governing female body language suggest submissiveness, subordination, vulnerability, and low status. These norms are learned, and people give them little thought until someone breaks them, at which point everyone focuses on the rule breaker.

Lisa Southwick

➤ "In this typical office scene, the man in the photo holds power with an authoritative stance—one hand in pocket and the other at mid-chest, straight posture, and head high; the woman is submissive with smile, arms and hands close to her body. Note the man's wide, stable stance and the woman's unstable stance. Many women tend to slip into a posture similar to that shown when talking to a shorter male authority figure.

Lisa Southwick

➤ In [this] photo, the man defers to authority by assuming a feminine, subordinate posture—with scrunched-up spine, constricted placement of arms and legs, canted head, and smiling attentiveness" (J. Mills 1985, 8).

Such norms governing appropriate body language for males and females can prevent women from conveying a sense of security and control when they are in positions that demand these qualities, such as a lawyer, politician, or physician. In this regard women face a dilemma: To be perceived as feminine and nurturing, a woman needs to appear "passive, accommodating, affiliative, subordinate, submissive, and vulnerable." To be perceived as a competent manager, a woman needs to appear "active, dominant, aggressive, confident, competent, and tough" (J. Mills 1985, 9).

(Write a Caption)

Cpl. Cristina Noelia Gil

Write a caption that relates the concepts of gender role and socialization to this boy being handed a Big Wheel truck.

Hints: In writing this caption

- review the concept of gender roles and its connection to gender socialization,
- consider how children's toys celebrate images of masculinity and femininity,
- think about the messages about masculinity conveyed through giving this particular toy, and
- comment on how those observing this handoff show their approval.

Critical Thinking

Have you ever tried to engage in an activity that was not considered sex-appropriate? What strategies, if any, did others use to discourage you from pursuing that activity (or encourage you to pursue it)?

Key Terms

agents of socialization   gender roles

MODULE 9.5

Sexualities and Sexual Orientations

Objective

You will learn the meaning of sexuality and sexual orientation.

Do you think your parents wondered before you were born what your sexual orientation might be?

Chris Caldeira

We are bombarded daily with messages about sexuality. They may come from sex education classes that warn of the health dangers of unprotected sexual activities; from fairy tales; from song lyrics; or from commercials, movies, and news events. Other messages come from those close to us: friends who come out to us or parents who kid us about having a boyfriend or girlfriend. Finally, messages come from observing the treatment of people around us: we notice uncomfortable reactions toward women who breast-feed their babies in public and toward a man who appears feminine or a woman who appears masculine; we take notice of the boy and girl everyone wants to date or not date; we take note of reactions and facial expressions when someone says he or she is from San Francisco.

Sexuality

Sexuality encompasses all the ways people experience and express themselves as sexual beings. The study of sexuality considers the range of social activities, behaviors, and thoughts that generate sexual sensations and experiences and that allow for sexual expression. Sexuality is not an easy subject to present for several reasons.

First, for most of us sex/sexuality education focused on the dangerous consequences of sexuality (sexually transmitted diseases) uninformed by any discussions of what to make of sexual excitement, sexual attraction, or the relentless messages regarding sexuality all around us. Second, people who have had difficult sexual experiences—those molested or raped as children, men who cannot achieve erections or orgasms, and women and men who have been sexually assaulted—may be uncomfortable with the topic (N. Davis 2005). Third, it is very difficult to discuss human sexuality in all its dimensions when heterosexuality and all that it entails is presented as normal and legitimate and any sexuality outside that norm is considered deviant and in need of fixing.

Sexual Orientation

Sexual orientation is an expression of sexuality. While sociologists are interested in the topic of sexual orientation, the American Sociological Association does not issue a statement about what sexual orientation means. According to the American Psychiatric Association (2009), **sexual orientation** refers to "an enduring pattern of emotional, romantic, and/or sexual attractions to men, women, or both sexes. Sexual orientation also refers to a person's sense of identity based on those attractions, related behaviors, and membership in a community of others who share those attractions." The word *enduring* suggests that one encounter does not make someone gay or lesbian. This caveat speaks to the fact that many people have experienced at least one same-sex sexual encounter at some point in their lives. Results from the most recent survey conducted by the Centers for Disease Control (2011b) found that 1 in 8 women and 1 in 16 men ages 15–44 have had a sexual experience with someone of the same sex.

Sexual orientation falls along a continuum, with its endpoints being exclusive attraction to the other sex and exclusive attraction to the same sex. In the United States, we tend to think of sexual orientation as falling into three distinct categories: heterosexual (attractions to those of the other sex), gay/lesbian (attractions to those of one's own sex), and bisexual (attractions to both men and women). It is important to realize that there are other labels that cultures apply to expressions of human sexualities (APA 2009).

Sexual orientation should not be confused with other related and intertwining terms that shape the experiences of sexuality and sexual orientation, including

- **biological sex**—the physiological, including genetic, characteristics associated with being male or female;
- **gender role**—the cultural norms that guide people in enacting what is considered to be feminine and masculine behavior;
- **gender identity**—the awareness of being a man or woman, of being neither, or something in between (gender identity also involves the ways one chooses to hide or express that identity); and
- **transgender**—the label applied to those who feel that their inner sense of being a man or woman does not match their anatomical sex, so they have to undergo medical procedures and behave and/or dress in ways that actualize their gender identity.

People enact sexual orientation in relationships with others. Thus, according to the APA (2009), "sexual orientation is closely tied to the intimate personal relationships that meet deeply felt needs for love, attachment, and intimacy."

Based on what we know to date, the core attractions that emerge in middle childhood through early adolescence prior to sexual experiences are the foundation of adult sexual orientation. The experiences of coming to terms with sexual orientation vary. People can be aware of their sexual orientation even if they are celibate or have yet to engage in sexual activity. Others come to label their orientation after a sexual experience with a same-sex and/or other-sex partner. Still others ignore, suppress, or resist pulls toward those of the same sex because of widespread social disapproval (APA 2009).

Courtesy of Serina Beauparlant

➤ In addition to sexual behaviors, sexual orientation includes "nonsexual physical affection between partners, shared goals and values, mutual support, and ongoing commitment. . . . One's sexual orientation defines the group of people in which one is likely to find satisfying and fulfilling romantic relationships that are an essential component of personal identity for many people" (APA 2009).

Every group has established **sexual scripts**, responses and behaviors that people learn, in much the same way that actors learn lines for a play, to guide them in sexual activities and encounters. These scripts are gendered in that males and females learn different scripts about the sex-appropriate responses and behavioral choices open to them in specific situations (Stein 1989). Even if they resist following the script, people know the script they are expected to follow and must come to terms with accepting or rejecting that script. The sexual scripts of the dominant culture call for behaviors and responses that support its definitions of what it means to be heterosexual. Other sexual scripts constructed by those in lesbian, gay, bisexual, and transgender (LGBT) communities and other cultures are dismissed as deviant.

How Is Sexuality Commodified?

Commodification occurs when economic value is assigned to things not previously thought of in economic terms such as an idea, a natural resource (water, a view of nature), or a state of being (youth, sexuality). The **commodification of sexuality** occurs when companies create products for people to buy so that they can express themselves as sexual beings or elicit a sexual response from others.

Sociologists refer to this as the **commercialization of sexual ideals**—the process of introducing products into the market using advertising campaigns that promise consumers they will achieve a sexual ideal if they buy and use the products. The list of available products is endless, especially for women. Relatively new products on the market for men include the erectile dysfunction drugs that are being advertised even to men who do not have the medical condition for which these drugs were designed. Millions of men have tried these drugs to date, and the manufacturers hope to attract millions more "by suggesting that if men cannot have an erection 'on demand,' if they 'fail' even once, they are candidates" (Tuller 2004).

Lisa Southwick

◀ The trend to market cosmetics to preteens has led critics to claim that young girls are being hypersexualized and in the process are joining the ranks of older women, many of whom have been socialized to believe that they fall short in their natural state (Wiseman 2003).

Social Movements

If we simply think about the men and women we encounter every day, we quickly realize that many people cannot, do not, or outright refuse to express their sexuality in idealized ways. Social movements occur when enough people organize to make a change, resist a change, or undo a change in some area of society.

It is not easy to piece together the complete history of LGBT revolutionary social movements and the reactionary movements that emerged to oppose any gains made. In the United States, LGBT movements can be dated to the establishment of the first gay rights organization in Chicago in 1924 (the Society for Human Rights). Over time the movements have involved various players and assumed various names, including gay liberation, lesbian feminism, the queer movement, and the transgender movement (Bernstein 2002).

Sociologist Mary Bernstein describes the movements' goals as both cultural and political: the "cultural goals include (but are not limited to) challenging dominant constructions of masculinity and femininity, homophobia, and the primacy of the gendered heterosexual nuclear family (**heteronormativity**). Political goals include changing laws and policies in order to gain new rights, benefits, and protections from harm" (2002, 536). Strategies to achieve these goals include building communities, lobbying legislators, voting for politicians sympathetic to LGBT issues, holding street marches of celebration and protest, and promoting LGBT culture through international, national, and community events, magazines, films, literature, and academic research.

Chris Caldeira

◀ Demonstrations, such as this gathering in San Francisco to oppose a proposed constitutional amendment banning same-sex marriage, represent another strategy to achieve political goals.

The relative successes of the LGBT movements and by extension opposition movements can be gauged by examining each of the 50 states' position on same-sex marriage and unions. Voters in 37 states have approved measures amending their constitutions to ban same-sex marriage. On the other hand, nine states allow same-sex marriages or unions.

AP Photo/Nick Ut

◀ The various LGBT movements have met resistance from social movements broadly labeled the Religious Right and from state, local, and federal politicians who have sought to preserve, and have succeeded in preserving, the gendered heterosexual nuclear family and reserving marriage as a right granted only to a man and a woman.

Cpl. Patrick Fleischman

(Write a Caption)

Write a caption that places this image in the overall context of the meaning of sexuality and sexual orientation.

Hints: In writing this caption

- review the concepts of sexuality and sexual orientation,
- think about the sexuality that this couple represents, and
- consider the extent to which this image reflects dominant norms of sexuality.

Critical Thinking

Have you had the chance to vote on a state constitutional amendment to ban same-sex marriage? If so, how did you vote? If not, imagine how you would vote. Explain.

Key Terms

biological sex
commercialization of sexual ideals
commodification
commodification of sexuality
gender identity
gender role
heteronormativity
sexual orientation
sexual scripts
sexuality
transgender

MODULE

9.6 Sexism and Feminism

Objective

You will learn about the meaning of sexism and feminism.

Do you think any of the following words—*sexist*, *misogynist*, *homophobic*, *hypermasculinized*—apply to this photograph of the hip-hop recording artist "Mario"?

Ray Tamarra/Getty Images

Sexism

Sexism is the belief that one sex—and by extension, one gender—is innately superior to another, justifying unequal treatment of the sexes. Sexism revolves around three notions:

1. People can be placed into two categories—male and female.
2. A close correspondence exists between a person's reproductive organs (e.g., primary sex characteristics) and other characteristics such as emotional state, body language, personality, intelligence, the expression of sexual desire, and athletic capability.
3. Primary sex characteristics are viewed as so significant that they explain and determine behavior and the inequalities that exist between the sexes.

Sexism rationalizes unequal treatment of men, women, and the transgendered as natural and ignores individual aspirations, ability, or capability.

Sexism can be so extreme that it can involve a hatred for one sex. That hatred is called **misogyny** when it is directed at women; it is called **misandry** when it is directed at men. Arguably one of the most publicized charges of misogyny is leveled against those hip-hop/rap artists who portray themselves as pimps and the women around them as prostitutes and sex objects who must obey them. In addition to the accusation of misogyny, some hip-hop artists have also been accused of promoting homophobia and hypermasculinity. As we will learn, hip-hop is much more complex than how it is portrayed in the popular media.

Sgt. Joel A. Chaverri

➤ Hypermasculinity involves exaggerating the traits and behaviors believed to be characteristic of males by placing excessive emphasis on strength to the point that a man's muscles and reproductive organs are presented as impossibly large. Other portrayals of hypermasculinity involve armed males surrounded by dozens of scantily clad women whose only role in life appears to be that of a sex object.

The Connection Between Homophobia and Hypermasculinity

The term **homophobia** is used in at least two ways. The word refers to an irrational fear held by some heterosexuals that any same-sex person will make a sexual advance toward them. Some people are so afraid that they engage in or support violence toward gays and others who do not conform to prevailing notions of masculinity and femininity; other discriminatory acts include banning gays from certain occupations (military service, child care).

In assessing whether the words *sexist*, *misogynist*, *homophobic*, and *hypermasculinized* apply to hip-hop, it is important to consider a number of questions:

1. Who is promoting misogyny, homophobia, and hypermasculinity? One overly simplistic answer is black and other rap artists who are considered part of hip-hop culture. The more complex answer is the $9 billion industry promoted through primarily white-owned record companies, magazines, radio stations, and retailers to consumers who are believed to be 70 percent white males (Public Broadcasting System 2008).
2. Are misogyny, homophobia, and hypermasculinity unique to hip-hop culture? No, they are not. Many advertisers portray women as sex objects to sell products. Likewise, jokes and derogatory comments about gays are commonplace in American society. In addition, hypermasculinity has deep roots in American society. "From the outlaw cowboy in American history to the hypermasculine thug of gangster rap, violent masculinity is an enduring symbol of American manhood itself" (Dyson 2008).
3. Are misogyny, homophobia, and hypermasculinity related? Yes, hypermasculinity reinforces misogyny and homophobia. Avoiding and fearing close contact with same-sex people, responding to a same-sex sexual advance with violence, and treating women as sex objects are elements of hypermasculinity. Why? From a hypermasculine point of view, any characteristic that threatens masculinity and departs from the extreme ideals about what men should be, such as homosexuality and femininity, must be demeaned.

4. Do misogyny, homophobia, and hypermasculinity capture the essence of hip-hop culture? No, for most, hip-hop is a culture or lifestyle that uses music to promote constructive political engagement and social activism.
5. Who is drawn to the commercialized hip-hop emphasis on hypermasculinity? Hypermasculinity may be most appealing to those men who lack financial status and power. Hypermasculinity sends the message to those with little power that one way men can communicate power is through their bodies and taking control over women's lives (PBS 2008).

Hypermasculinity and the Military

Hypermasculinity and all that it entails point to another dimension of sexism—the belief that people who behave in ways that depart from ideals of masculinity or femininity are considered deviant and in need of fixing and should be subject to negative sanctions. This ideology was reflected in U.S. military policy toward gay men and lesbians that until recently banned gays from openly serving in the military under "Don't Ask, Don't Tell" (DADT). The U.S. Department of Defense (1990) maintained that homosexuality was "incompatible with military service" and that the "presence of such members adversely affects the ability of the Armed Forces to maintain discipline, good order, and morale" (25). Over the course of a decade, the U.S. military discharged 12,500 servicemen and women for homosexuality (Bumiller 2009).

U.S. Navy photo by Mass Comm. Spc. 2nd Class Patrick Gordon

◂ Whether taking part in training exercises or relaxing afterward, soldiers often make physical contact with other soldiers. Would the presence of openly gay men and lesbians in the military disrupt these kinds of bonding activities? Is physical contact of any kind sexual? Might the presence of two soldiers who share a strong friendship also be disruptive?

People who opposed the presence of gay men and lesbians in the military stereotyped them as sexual predators just waiting to pounce on heterosexuals while they shower, undress, or sleep. Opponents seem to believe that any same-sex person is attractive to a gay man or lesbian. But as one gay ex-midshipman noted, heterosexuals "have an annoying habit of overestimating their own attractiveness" (Schmalz 1993, B1). In December 2010, Congress passed legislation repealing DADT, officially ending the policy on paper. U.S. servicemen and women were already serving with coalition partners who allowed gays to serve openly. In fact, openly gay soldiers serve in the British (major coalition partner) and 23 other militaries without any significant problems. The British military even recruits soldiers at gay pride events (Lyall 2007).

Feminism

When women living in the United States are asked, Do you consider yourself a feminist or not? only one in four answers yes. When asked the same question accompanied by this definition of feminist—someone who believes in social, political, and economic equality of the sexes—65 percent of women answer yes. One possible explanation for the difference is that few people consider the label of feminist a compliment (CBS News Polls 2005).

People correctly or incorrectly associate feminists with many mainstream and controversial positions, including support for equal rights, equal pay, affordable day care, abortion rights, opposition to sexual harassment, a lack of respect for "stay-at-home" moms, and a dislike or even hatred of men (Time.com 2008). **Feminism** is a perspective that seeks to understand the position of women in society relative to that of men in the context of economic, political, and cultural structures in which their lives are embedded. In addition feminists advocate for equal opportunity. Questions about what that equality looks like and how equality should be achieved distinguish feminist camps from one another.

➤ This early-20th-century political cartoon speaks to a widespread misperception that feminists as a group disdain men and see them as the weaker sex. Here four women are inspecting what appears to be a bug. It is actually someone of the "male species" on his knees begging for mercy.

Prints & Photographs Division, Library of Congress, LC-DIG-ppmsc-05887

Many feminists believe that any inequality between males and females, including that which gives females an advantage over males, needs to be addressed. The following quotations from well-known feminists demonstrate the range of concerns and positions feminists hold:

> It's important to remember that feminism is no longer a group of organizations or leaders. It's the expectations that parents have for their daughters, and their sons, too. It's the way we talk about and treat one another. —Anna Quindlen

> If divorce has increased by one thousand percent, don't blame the women's movement. Blame the obsolete sex roles on which our marriages were based. —Betty Friedan

> Women do not have to sacrifice personhood if they are mothers. They do not have to sacrifice motherhood in order to be persons. Liberation was meant to expand women's opportunities, not to limit them. The self-esteem that has been found in new pursuits can also be found in mothering. —Elaine Heffner

> No one sex can govern alone. I believe that one of the reasons why civilization has failed so lamentably is that it had one-sided government. —Nancy Astor

I myself have never been able to find out precisely what feminism is: I only know that people call me a feminist whenever I express sentiments that differentiate me from a doormat, or a prostitute. —Rebecca West

◀ One of the most memorable quotes associated with feminist thought is *"Remember, Ginger Rogers did everything Fred Astaire did, but she did it backwards and in high heels."* The person who observed this—Faith Whittlesey—was commenting on the fact that Ginger Rogers, although a great talent in her own right, was best known as Fred Astaire's dance partner.

Feminism's Activist Roots

The history of feminism cannot be separated from efforts to bring about change in the lives of women, and by extension the lives of children and men. The following very selective list highlights some major events in feminist history. They include gaining the right to vote, securing fair labor standards, and opening doors that were previously closed. These are rights that most of us—male and female—have come to take for granted.

1920 The 19th Amendment to the U.S. Constitution becomes law, guaranteeing women the right to vote.

1943 With many men fighting in World War II, over 6 million women hold factory jobs as welders, machinists, and mechanics.

1963 The Equal Pay Act, signed by President John F. Kennedy, prohibits the practice of paying women less money than men for the same job.

1964 President Lyndon B. Johnson signs the Civil Rights Act, outlawing discrimination in unions, public schools, and the workplace on the basis of race, creed, national origin, or sex.

1965 Under Title VII of the Civil Rights Act of 1964, the Equal Employment Opportunity Commission is established to prohibit discrimination in the workplace on the basis of sex, religion, race, color, national origin, age, or disability.

1966 The National Organization for Women (NOW) is founded by Betty Friedan with the purpose of challenging sex discrimination in the workplace.

1972 The U.S. Senate approves the Equal Rights Amendment, including Title IX, making sex discrimination in schools that receive federal funding illegal and requiring schools that receive such funds to give females an equal opportunity to participate in sports.

1973 The U.S. Supreme Court rules that the Texas law restricting abortion in the first trimester is unconstitutional. As a result, anti-abortion laws in nearly two-thirds of the states are declared unconstitutional, legalizing abortion nationwide.

1975 President Gerald Ford signs a defense appropriations bill to allow women to be admitted into U.S. military academies.

1993 President Bill Clinton signs the Family Medical Leave Act, allowing eligible employees to take up to 12 weeks of leave for reasons of illness, maternity, adoption, or a child's serious health condition.

1996 U.S. women's successes in the Summer Olympics (19 gold medals, 10 silver, 9 bronze) are attributed to the Title IX legislation that supported and encouraged girls' participation in sports.

1997 The Supreme Court rules that college athletic programs must actively involve men and women in numbers that reflect the proportions of male and female students.

Source: Adapted from Barbara Boxer, U.S. Senator from California, Historical Timeline for Women's History (2007).

A feminist viewpoint emphasizes the following kinds of themes:

- the right to bodily integrity and autonomy;
- access to safe contraceptives;
- the right to choose the terms of pregnancy;
- access to quality prenatal care, protection from violence inside and outside the home, and freedom from sexual harassment;
- equal pay for equal work;
- workplace rights to maternity and other caregiving leaves; and
- freedom for both men and women to make choices in life that defy gender expectations.

(Write a Caption)

Write a caption that connects the concept of hypermasculinity to football.

Hints: In writing this caption

- review the concept of hypermasculinity, and
- think about the ways in which football embodies the elements of hypermasculinity.

U.S. Navy photo by Damon J. Moritz/Released

Critical Thinking

Identify some example of popular culture—such as a movie, a song, a sport, a cartoon, a toy—that exhibits misogyny, misandry, homophobia, or hypermasculinization.

Key Terms

feminism
misandry
sexism
homophobia
misogyny

MODULE 9.7

Applying Theory: The Three Perspectives

Objective

You will learn how the theoretical perspectives help us think about sex testing.

Did your mother know from the results of an ultrasound or genetic test your sex months before you were even born?

Ms. Marie Berberea (TRADOC)

If your parents knew your sex before you were born, why do you think it was important to them to know beforehand? If your parents did not, why do you think they resisted knowing? The sociological theories—functionalist, conflict, and symbolic interaction—help us to go beyond individual cases and think about the larger consequences and issues associated with sex testing months before babies are born.

Functionalist Perspective

Functionalists focus on how parts of society contribute to the stability of an existing social order. They also pay attention to how parts contribute to disorder and instability. Functionalists ask, What are the anticipated and unanticipated effects of a part on society? In this module, the part we will analyze is the practice of testing for the sex of a baby before it is born.

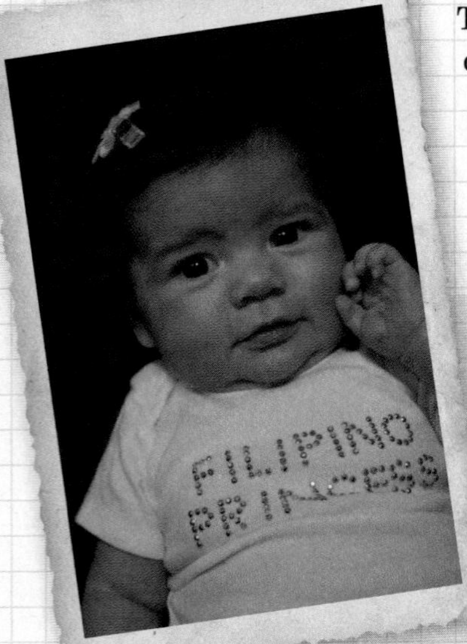

◂ One manifest, or expected, function of using an ultrasound to test for a baby's sex is that parents can prepare for a baby boy or baby girl. They can choose a name, decorate a room, and buy clothes (such as a Princess T-shirt) that correspond with the expected sex. Relatives and friends are able to buy sex-appropriate gifts.

Missy Gish

An unexpected, or latent, function of sex testing is that it allows parents who might be hoping for a baby of a particular sex time to deal with disappointment if the sex tests show that their baby is the other sex. The logic is that it is better to experience and come to terms with disappointment prior to the baby's birth.

An expected, or manifest, dysfunction of sex testing is that sometimes the tests fail to correctly predict the sex of the baby. The parents have prepared for a baby of a particular sex only to learn when the baby is born that it is not the predicted sex. Now room color has to be changed and clothes and other sex-coded items have to be exchanged.

A latent, or unexpected, dysfunction of sex testing is its worrisome connection to **female infanticide**, the targeted abortion of female fetuses because of a cultural preference for males and corresponding low status assigned to females. Note there seem to be no cases in which a society has a cultural preference for female babies. The widespread practice of female infanticide is believed to be the cause of sex ratios skewed in favor of males, especially in China and India. Another serious latent, or unexpected, dysfunction of sex determination testing and resulting imbalances in the number of females relative to males is that the imbalance increases the likelihood that one sex will have greater power over the other in heterosexual relationships. For example, the less prevalent sex (females) could use the power that comes with a numerical shortage to oppress their partner, who may accept subordination as the price of having a mate (Hollingsworth 2005).

Conflict Perspective

Conflict theorists seek to identify the advantaged and disadvantaged groups, document unequal access to scarce and valued resources, and identify the practices that advantaged groups establish to promote and protect their interests. With regard to sex testing, conflict theorists ask, Who ultimately benefits from knowing the sex of the baby in advance and at whose expense? Conflict theorists reject the argument that a good reason to sex test is that it allows parents time to plan for a boy or girl. There should be no difference in planning for the needs of a baby boy or girl, both of whom have the same needs. In reality, conflict theorists say, sex tests simply allow parents to plan out the baby's future in very gender-specific ways that thereby narrow or expand their opportunities in life.

Conflict theorists also maintain that another advantaged group is those who own/sell sex testing equipment, kits, and tools. One sex testing kit, for example, is marketed over the Internet and allows parents to test for the sex of their baby 7 weeks after conception (13 weeks before physicians typically know the sex through ultrasound technology). The cost is $380 (Medical News.net 2011). Critics argue that knowing the baby's sex that early in the pregnancy may encourage some parents, disappointed by the baby's sex, to choose abortion, something they might not have considered 20 weeks into a pregnancy. In fact, the availability of ultrasound technologies to determine a baby's sex 20 weeks after conception is believed to be responsible for sex ratio imbalance in India and other countries (especially rural areas) where there are clear cultural preferences for boys (Dhar 2012).

Symbolic Interactionist Perspective

Symbolic interactionists ask, How do people experience, interpret, influence, and respond to a social situation? Symbolic interactionists are interested in how knowing the sex of the baby after 20 or fewer weeks of pregnancy affects a parent's self-awareness or sense of self. Do parents feel different about themselves if they are having a baby girl versus a baby boy? Are parents' projections of future interactions different depending on whether the baby tests as a boy or a girl?

Symbolic interactionists are also interested in how parents and other involved parties negotiate interactions where the question of the baby's sex comes up. How do some parents, who choose not to sex test their baby, withstand questions like "Do you know the sex of the baby yet? Why not?" How do they explain their decision not to know? After the baby is born, do these parents interact in less gender-specific ways with their baby?

Table 9.7a: Summary of Three Perspectives as Applied to Sex Determination Testing

	Theory		
	Functionalist	**Conflict**	**Symbolic Interaction**
Question	What are the anticipated and unanticipated effects of testing for the sex of a baby before it is born?	Who ultimately benefits from knowing the sex of the baby in advance (and at whose expense)?	How do people experience, interpret, influence, and respond to a social situation?
Answer	**Manifest function**: parents can prepare for a baby boy or baby girl **Manifest dysfunction**: the test may fail to correctly predict the sex of the baby **Latent function**: parents hoping for a particular sex would have time to deal with disappointment if the sex test shows that their baby is not the wished-for sex **Latent dysfunction**: sex testing supports practice of infanticide	**Who benefits?** parents who plan out the baby's future in gender-specific ways; those who sell sex testing equipment, kits, and tools; those in advantaged sex group **Disadvantaged groups**: babies aborted because they are the wrong sex; babies whose opportunities are diminished by prematurely channeling their lives in gender-specific ways	Looks at how knowing sex of baby affects parents' sense of self; studies symbols parents use to convey sex of baby; considers how expectant parents explain a decision not to learn the sex of the baby

Critical Thinking

Which one of the three perspectives best captures your mother's and/or father's response to sex determination testing?

Key Term

female infanticide

Summary: Putting It All Together

CHAPTER 9

Sex is a biological distinction determined by the anatomical traits essential to reproduction. While most cultures classify people in two categories—male and female—sex should not be considered a clear-cut category. Gender is a social distinction about how males and females should be; it is something that is carefully constructed, taught, learned, and enforced. Not every society divides people into so-called opposite genders. For example, American Samoans and other Pacific Islander peoples accept a third gender known as *fa'afafine*.

Sociologists are especially interested in gender ideals. Often, gender ideals do not exist in reality, yet that does not stop people from trying to attain them. The commercialization of gender ideals is the process of introducing products to consumers through advertising campaigns that promise that those who buy and use the products will achieve masculine or feminine ideals.

Sex, and by extension gender, affects people's life chances, the *probability* that an individual's life will take a certain path or turn out a certain way. This is because we organize life in a gender-polarized way. The effect a person's gender has on his or her life becomes evident when we ask people to imagine life as another gender. Structural constraints are the established and customary rules, policies, and day-to-day practices that channel behavior in a certain direction and that shape a person's life chances. One example relates to the structural constraints that push many men and women into careers that correspond with society's ideals about sex-appropriate work. Institutions are gendered when there is a pattern to the relationships, practices, images, and belief system that supports segregating the sexes and empowering or subordinating one sex relative to the other. An institution is gendered when there is an established pattern of segregating the sexes into different workspaces or jobs, of disproportionately assigning one sex to positions of power, and of otherwise disadvantaging one sex relative to the other.

When sociologists study inequality between males and females, they seek to identify the social factors that put one sex at a disadvantage relative to the other. Inequality is justified by sexism, which can be so extreme that it takes the form of misogyny and misandry. Sexism also encompasses homophobia and hypermasculinization. Feminism, a response to sexism, is a perspective that advocates equality between men and women. Most feminists believe that any inequality between males and females, including that which gives females an advantage over males, needs to be addressed.

Go to cengagebrain.com to link to Aplia and CourseMate for the chapter quiz and other activities.

Chris Caldeira

10 CHAPTER

ECONOMICS AND POLITICS

When sociologists study economies, they focus on employment opportunities; the labor force; and how goods and services are produced, distributed, and consumed. When they study politics, they focus on who has the power over and access to scarce and valued resources and the power to make laws, policies, and decisions that affect others' lives. Sociologists also seek to understand how economy and politics are interconnected. As one example, consider that each year in the United States local, state, and federal governments enact tens of thousands of laws that affect employment, including laws setting minimum wage levels (San Francisco mandates a minimum wage above $10.00), prohibiting sale of certain products (an Oregon law prohibits the sale of shark fins), and requiring employers to determine whether a potential employee is a legal resident (Georgia corporations that employ 500 or more workers must use E-Verify to verify legal status of new hires). There are also tens of thousands of laws introduced that are never ratified. An example relates to laws prohibiting corporations from outsourcing U.S. jobs overseas. Many such laws have been introduced and debated in legislatures but have never been enacted (National Conference of State Legislatures 2012).

Go to Sociology CourseMate on cengagebrain.com to watch a video about two emerging economic giants—China and India.

MODULE

10.1 Economic Systems

Objective

You will learn about economic systems of socialism, capitalism, and the welfare state.

How much do you think this person gets paid to hold this sign? Do you think she gets benefits?

Chris Caldeira

Should people depend on their employer to provide health care, tuition reimbursement, day care, and so on? What about employers who cannot or will not provide such benefits? Would you be willing to pay more taxes if the government guaranteed free college education with a monthly stipend for 55 months, child care, health care, and other benefits from cradle to grave? You would receive these kinds of benefits if you lived in Finland. The catch is that you would live in a smaller house, pay more taxes (50 percent), and consume less. You would also see fewer people who are extremely poor or extremely wealthy.

Economic systems are social institutions that structure employment opportunities (formal and underground) and regulate the production, distribution, and consumption of products and services. We can classify the economies of the world as falling somewhere along a continuum that has capitalism and socialism as its extremes. Keep in mind, however, that no economy fully realizes capitalist or socialist principles and that, in practice, economic systems are some combination of the two.

Capitalism

Capitalism is an economic system in which the raw materials and the means of producing and distributing goods and services are privately owned. That is, individuals (rather than employees or the government) own the raw materials, land, machines, tools, trucks, buildings, and other inputs needed to produce and distribute goods and services. In theory, this economic system is profit-driven and free of government interference. Profit-driven is the most important characteristic of capitalist systems. In such systems, those who own the means of production and distribution are motivated to increase and maximize profits by seeking maximum return on investments and using labor and resources in cost-efficient ways. Theoretically, consumer demand and choice drives the production and distribution of goods and services. In addition, the best businesses survive because consumers choose from whom to buy goods and services.

Capitalist systems are governed by the laws of supply and demand; that is, as the demand for a product or service increases, its price rises. Manufacturers and service providers respond to increased demand by increasing production, which in turn "increases competition and drives the price down" (Hirsch, Kett, and Trefil 1993, 455). Although most economic systems in the world are classified as capitalist, in reality no system fully realizes capitalist principles. Simply consider that the United States government ignores capitalist principles anytime it intervenes to regulate an industry, stimulate the economy, or prevent a recession. Consider also that the U.S. Postal Service, national parks, the public school system, Medicaid, and libraries are public-owned enterprises that are, in theory, not profit-driven.

Karl Marx argued that capitalism ignores too many human needs and exploits human labor in the name of profit. Marx believed that if this economic system were in the right hands—those of socially conscious people motivated not by a desire for profit or self-interest but by an interest in the greatest good to society—public wealth would be more than abundant and could be distributed according to need.

➤ The search for the lowest cost—even free—labor underlies all forms of workplace exploitation. The bottom image at right shows the enslaved from Africa in severely crowded quarters sailing the Atlantic to the United States, Brazil, the Caribbean, and other slave-holding societies. The photo on the top shows a man from Mexico scaling a fence. The fence marks a boundary separating a society where farm workers earn $5 per day from one where workers earn $81.36 per day (U.S. Department of Agriculture 2011b).

Petty Officer 1st Class Matthew Tyson

Prints & Photographs Division, Library of Congress, LC-DIG-ppmsca-05933

Competition and the drive for profit push capitalists to cut the costs of production by introducing labor-saving machinery, laying off workers, and/or finding workers who will work for less. Marx argued that at some point the search to lower production costs would result in so many jobs lost and workers displaced that the system would collapse. For Marx, one contradiction of capitalism is that the capitalists must lay off and lower the wages of those they depend on to buy their products (Kilcullen 1996).

Socialism

The term *socialism* was first used in the early 19th century in response to excesses of capitalism, specifically the poverty and inequality that accompanied the capitalist-driven Industrial Revolution. In contrast to capitalism, **socialism** is an economic system in which raw materials and the means of producing and distributing goods and services are collectively owned. In other words, public ownership—rather than private ownership—is an essential characteristic of this system. Socialists reject the idea that what is good for the individual and for privately owned businesses is good for society as a whole. Instead, they believe the government or some worker or community organization should play the central role in regulating economic activity on behalf of the people as a whole.

Tony Rotundo

➤ This sculpture, a symbol of socialist ideals, celebrates the common worker, who ideally labors for the good of society. The following principle dominates: "From each according to his ability, to each according to his needs" (Marx 1875). Ideally, each person shall produce to the best of his or her ability and shall take in accordance with his or her needs. Living by such principles would eliminate the wide disparities in income and wealth that separate, even divide, people in the professional and working classes.

Socialists maintain that things like oil, banks, health care, transportation, and the media should be state-owned. In socialism's most extreme form, the pursuit of personal profit is forbidden. In less extreme forms, profit-making activities are permitted as long as they do not interfere with larger collective goals. As with capitalism, no economic system fully realizes socialist principles. The People's Republic of China, Cuba, North Korea, and Vietnam are all officially classified as socialist economies, but they permit varying degrees of profit-making activities that generate personal wealth.

Welfare States

The term **welfare state** applies to an economic system that is a hybrid of capitalism and socialism. In this economic model, the government (through taxes) assumes a key role in providing social and economic benefits to some or all of its citizens, including unemployment benefits, supplemental income, child care, social security, basic medical care, transportation, education (including college), or housing. Under one welfare state model followed by the United States (with the exception of Social Security and Medicare, to which everyone over age 65 who has contributed to the system is entitled), such benefits are provided to those who fall below a set minimum standard, such as a poverty line or a certain income level. Under a second welfare state model, the benefits are awarded in a more comprehensive way (e.g., all families with children, all college-age students, universal health care). Most European countries follow the second model. Finland, for example, funds all schools equally. There are no schools considered to be for the elite, working class, or low-income. In Finland day care and preschools are free, uniformly high in quality, and staffed by teachers who have received extensive training. Finland's investment in equality of educational opportunities explains in part why its students are among the best-performing in the world (Kaiser 2005).

(Write a Caption)

Write a caption about the production of biodiesel/ethanol fuels in the United States to show that economies defined as capitalistic often depart from the ideal.

Chris Caldeira

Hints: In writing this caption

- review the principles of the capitalist model,
- consider the fact that Congress has passed laws such as the 2005 Energy Policy Act (in effect until 2011) subsidizing corn production and turning corn into ethanol, and
- think about the principle of capitalism violated.

Critical Thinking

What are some examples of government-run (local, state, federal) programs in the United States that follow socialist principles?

Key Terms

capitalism	socialism
economic system	welfare state

MODULE

10.2 The U.S. Economy and Jobs

Objective

You will learn about some characteristics that define the U.S. economy.

Mass Comm. Spc. 2nd Class Eric C. Tretter/Released

Mr. Douglas Demaio (IMCOM)

Which occupation, veterinarian or postal clerk, is expected to be among the fastest growing jobs between now and 2020?

Veterinarians (along with personal care aides, biomedical engineers, and physical therapists) are projected to be among the fastest growing occupations in this decade. Postal clerks (along with switchboard operators, word processors, and typists) are projected to be among jobs in decline. What clues do these occupational patterns offer about the forces driving the U.S. and global economy? The aging of the population is one factor driving a demand for personal care aides, physical therapists, biomedical engineers (for medicines and medical technologies), and veterinarians (to care for animals who serve as companions and helpers). The rise of the Internet and voice-driven technologies are forces that are lowering demand for postal clerks and switchboard operators. In this module we examine some of the characteristics of the U.S. economy that are shaping career opportunities.

Job Growth by Sector

We can think of an economy as comprising three sectors: primary, secondary, and tertiary. The **primary sector** includes economic activities that extract raw materials from the natural environment. Mining, fishing, growing crops, raising livestock, drilling for oil, and planting and harvesting forest products are examples. The **secondary sector** consists of economic activities that transform raw materials into manufactured goods such as houses, computers, and cars.

The **tertiary sector** encompasses economic activities related to delivering services (such as health care, entertainment, sales) and to creating and distributing information. One way to identify the relative importance of each sector of an economy is by determining how much it contributes to total employment. Chart 10.2a shows the 10 broad occupational categories that employ the largest numbers of people. These 10 categories account for 80 percent of all employment in the United States.

Chart 10.2a: Number of People Employed (in millions) in 10 Largest Occupational Categories and Median Income, 2010

Four out of every five workers (80 percent) in the United States are employed in the broad occupational groups listed in the graph. If one takes a sociological view of occupational opportunities, it is clear that people's choices are limited to the kinds of jobs available. How many of the 10 occupational categories on this list are part of the tertiary sector? Which occupational categories are associated with the highest median wage? With the lowest? What percentage of U.S. workers are employed in office and administrative support?

Source: Lockard and Wolf (2012)

A Two-Tier Labor Market

Over time, a two-tier labor market has emerged in the United States—one in which those who lack formal education beyond high school or technical skills are at a severe disadvantage relative to those with strong educational and professional credentials. As a result, those without a college education are likely to hold occupations that pay lower wages and offer no health insurance coverage or other benefits (U.S. Central Intelligence Agency 2012). Examples of occupations from Chart 10.2a for which those with less than a high school education are qualified include health care support and personal care aides, transportation and material moving, and fast-food cooks. The median incomes of the millions of people who work full time in these professions fall between $18,770 and $28,400 per year. While these occupations tend to be filled by those with the least formal education, there are factors other than education that account for low wages.

In a comparative study of low-wage labor in six countries, researchers found that the United States has the highest percentage of low-wage workers, followed by Germany and the United Kingdom (see Chart 10.2b). The Netherlands, France, and Denmark have significantly smaller percentages of their workforces earning low wages.

Chart 10.2b: Percentage of Low-Wage Workers in the Workforce of Six Countries

One definition of low-wage work is a job that pays two-thirds of the median gross hourly wages. In the United States, that median wage is $18.75, so low-wage work would include those who earn $12.40 or less per hour before taxes (the equivalent of about $23,000 per year). This chart shows the percentage of the workforce earning two-thirds of the median gross hourly wages in six countries.

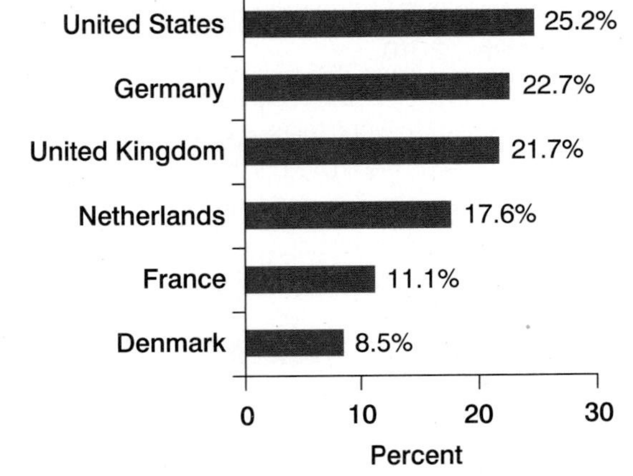

Source of data: Appelbaum and Schmitt (2009)

To understand why the percentages varied, Eileen Appelbaum and John Schmitt (2009) looked at the work situations of five low-wage occupations in each of the six countries: (1) nursing assistants, (2) cashiers and stock/salesclerks, (3) food processing operators, (4) housekeepers in hotels, and (5) incoming call operators in call centers. They found that the percentages earning two-thirds, or 66 percent, of median gross hourly wages in each occupation varied by country. As one example, 38 percent of nursing assistants in the United States are low-wage earners, compared with less than 5 percent in the Netherlands. The researchers discovered that each of the six countries have different levels of tolerance regarding wage inequality. That tolerance is reflected in each country's employment laws and views about which occupations "deserve" low wages. As a case in point, consider the plight of nursing assistants. In the United States, almost 40 percent of all nursing assistants earn $12.40 or less per hour. In the United Kingdom, that figure is 21 percent. In Germany, France, and the Netherlands, less than 10 percent of nursing assistants earn the U.S. equivalent of $12.40 or less. Why are nursing assistants more likely to be counted among the ranks of low-wage workers in the United States than in the other five countries? The researchers found that, as a general rule, European countries place high value on quality health services and see that quality as dependent on health care workers well-trained at all levels. The United States, by contrast, seems to believe that some tasks can be done with minimal training (Appelbaum and Schmitt 2009).

Instead of a registered nurse or an LPN monitoring temperature, taking blood pressure, applying sterile dressings, and drawing blood, U.S. hospitals allocate these kinds of tasks to minimally trained nursing assistants (who receive about six weeks of training) as a way of reducing the cost of delivering health care.

In the Netherlands, a hospital nursing assistant must complete a 34-month program that includes 56 weeks of theory-oriented coursework. France has taken steps to eliminate unqualified nursing assistants by giving them a choice to leave or enroll in a strong vocational training program to upgrade their knowledge and skills (Appelbaum and Schmitt 2009).

Union Membership

Of all workers in the United States, 11.8 percent (or 14.8 million) are union members. Union affiliation varies by state, ranging from 21.5 percent of workers belonging to unions (Hawaii) to 2.9 percent (North Carolina) (U.S. Department of Labor, Bureau of Labor Statistics 2012c).

Union membership in the United States has declined from a high of 35 percent of the workforce in the 1950s to 20.1 percent in 1983 to just under 12 percent in 2012 (U.S. Department of Labor, Bureau of Labor Statistics 2012d). The steady drop in union membership, especially over the past 30 years, has been connected to many factors, including

- the declining significance of the manufacturing sector (the traditional base of union membership) in the overall economy,
- increasing percentages of females in the workforce (who have tended to work in nonunion positions), and
- increasing global competition (which has added pressure to keep wages down and minimize union influence).

In the United States, about 37 percent of public-sector employees belong to unions. Among the most unionized occupations are teachers, police officers, and firefighters. Only about 7 percent of private-sector employees are unionized, with the highest unionization rates among employees in transportation (airlines), utilities, and construction. For the most part, there are no significant differences in union membership rates for men versus women or across race. Workers ages 55 to 64 are more likely (15.7 percent) to be union members than those ages 16 to 24 (4.4 percent) (U.S. Department of Labor, Bureau of Labor Statistics 2012c; Center for Responsive Politics 2009).

National and Personal Debt

As measured by the gross domestic product (GDP), the United States has the largest economy in the world at $15.4 trillion, followed by China ($11.3 trillion) (U.S. Central Intelligence Agency 2012). In per-capita or per-person terms, however, the United States has the 11th largest GDP, averaging $48,100 per person, and China has the 117th largest at $8,040 per person. The economy with the largest per-capita GDP is Liechtenstein at $141,100 (U.S. Central Intelligence Agency 2012).

Large amounts of debt relative to GDP affect a government's ability to spend. If a large portion of tax revenues must be used to repay debt and interest on debt, then, by definition, there is less money to invest in people, infrastructure, and other programs. At the time of this writing, the U.S. national debt was $15.7 trillion, which represents more than 100 percent of the country's GDP. International investors and governments held an estimated $4.0 trillion of that debt. The foreign countries from which the United States has borrowed the most money are China (about $1.1 trillion) and Japan ($1 trillion) (U.S. Department of the Treasury 2012). About $4.7 trillion of the $15.7 trillion in debt is held by other U.S. government agencies, including the Social Security Trust Fund. Other major holders of debt include U.S. state and local governments ($856 billion) and domestic private investors ($3.8 trillion).

In addition to the national debt, total U.S. consumer debt, not counting mortgage debt, stands at $2.57 trillion, with approximately $801 billion in credit card debt (U.S. Federal Reserve 2012). If the mortgage debt of $13.6 trillion were counted, total consumer debt would be $16.7 trillion (U.S. Federal Reserve 2011).

Unemployment and Underemployment

The United States has a labor force of 153.1 million people. That represents 4.6 percent of the total global labor force of 3.3 billion. India's labor force is 480 million, and China's is 819.5 million (U.S. Central Intelligence Agency 2012). As you might imagine, one of the most pressing global problems has to do with finding work for such a large number of people. We will focus on the U.S. job market with these labor force statistics in mind. Between December 2007 and April 2009, 6.5 million net jobs were lost in the United States. This job loss, along with the lending crisis, triggered what is known as the Great Recession of 2008. The impact of this recession cannot be measured only by job loss. In January 2012, for example, 23 million Americans—15.1 percent of the workforce—were unemployed and looking for work, underemployed (working fewer hours than they would like), or so discouraged that they gave up looking for work. Putting this many people back to work will take years, even in the event of a recovery (U.S. Department of Labor, Bureau of Labor Statistics 2012a). In assessing unemployment statistics, it is important to consider which occupational categories have been most and least affected by the economic downturn. The occupational category most affected by job loss is construction, with an unemployment rate of 18.4 percent. The job category most insulated from unemployment is health care practitioner and technical occupations, with an unemployment rate of just 1.8 percent. In addition, some demographic groups, such as the young and black males, have been more affected by the downturn in the economy than others.

Outsourcing

When we think of outsourcing, we often think of China as the destination for manufacturing jobs and India for technology and service jobs. Although these two countries are the best-known destinations, manufacturing and service jobs are outsourced to just about every country in the world. Outsourcing is only one factor behind the net loss of 8 million U.S. factory jobs since mid-1979, the year the number of manufacturing jobs peaked at 19.6 million (Norris 2012). We cannot offer precise accounts of where those 8 million factory jobs went, because that job loss can be attributed to many factors. First, some factory jobs have been automated out of existence.

The outsourcing of factory jobs from the United States to elsewhere has been going on for more than 50 years. For example, RCA was the first U.S.-based corporation to outsource jobs to Mexico in 1965; today there are several thousand foreign-owned manufacturing operations in that country, 85 percent of which are U.S.-owned (Barrio 1988; Export.gov 2012). Advances in digital technologies, however, have led to a second stage of outsourcing that has put a wider range of jobs up for international competition—those jobs are in the areas of office tasks and IT. At first, the outsourced jobs involved routine office work that required little training and little direct or face-to-face contact with customers or coworkers—such as bill processing, bookkeeping, data entry, and payroll. But then, the not-so-routine, high-skilled jobs became targets for outsourcing, including work done by architects, radiologists, and lawyers.

It is impossible to know the number of office, information, and other jobs that have been outsourced to foreign countries. First, the U.S. Bureau of Labor Statistics does not collect data on jobs outsourced; second, corporations rarely share outsourcing plans with the public or government policy makers. Finally, the United States does not collect data on the impact of outsourcing, positive or negative (Hira and Hira 2008).

➤ Other jobs have been outsourced to foreign countries (e.g., Made in China), and some factory jobs—like making typewriters—just disappeared because the products manufactured went out of fashion. Apparently, there is only one company in the world that still makes typewriters; it is located in India and sold 800 last year (Huffington Post 2011).

Dependence on Oil and Mineral Imports

Despite its large size and relative stability, the United States depends on foreign sources for many critical raw materials, such as oil. According to the U.S. Department of Energy (2011), the United States consumes 19.1 million barrels of oil per day; 49 percent (or 9.4 million barrels per day) of which is imported from foreign countries. The extent of U.S. dependence on foreign oil becomes evident when we consider that it has an estimated 22.3 billion barrels of proven oil reserves—less than 3 percent of the world's total known oil reserves. At the current rate of domestic production (51 percent of need per year), these reserves will last about 6 years. If the United States could produce and refine 100 percent of its crude oil needs, it would deplete those known reserves in less than 3.2 years.

Although most Americans likely know about their country's dependence on oil, they are less likely to know of the country's dependence on imported minerals. Simply consider that an automobile contains at least 32 minerals, including aluminum, carbon, copper, silicon, lead, and zinc. Hybrid cars need about 2.2 pounds of neodymium for their motor magnets. By one estimate, Americans use 25,000 pounds of minerals per person per year to sustain their current standard of living as it relates to food production, shelter, the infrastructure, health care, and information and communication. Even a green economy depends on minerals. As one example, wind power turbines require more than 700 pounds of neodymium per megawatt of energy-generating capacity. Currently, that mineral can only be acquired from China. The United States has a 100 percent dependence on at least 18 other strategic mineral commodities, including arsenic and indium needed to make semiconductors and strontium needed to make fuel cells. The competition for these minerals is likely to dramatically increase as China's and India's economies continue to grow at record-setting paces (Industrial College of the Armed Forces 2010).

(Write a Caption)

Write a caption that relates these photographs to employment opportunities, the sector of the economy, and the two-tier labor market.

Chris Caldeira

Missy Gish

Hints: In writing this caption

- notice that one photograph is of a truck representing transportation occupations and the other is of a bank manager,
- review Chart 10.2b, which shows the broad occupational categories that employ the largest numbers of people in the United States with median salary,
- think about the level of education required for the two pictured occupations, and
- consider in which sector—primary, secondary, or tertiary—the truck driver (moving trees to make lumber) and the bank manager are involved.

Critical Thinking

With what characteristic of the U.S. economy have you had the most personal experience? Explain.

Key Terms

primary sector secondary sector tertiary sector

Multinational and Global Corporations

MODULE 10.3

Objective

You will learn a way to think about the size and power of the world's largest corporations.

What does it mean to be the largest corporation in the world and have $421.8 billion in annual revenue?

Chris Caldeira

Walmart's annual revenue of $421.8 billion in 2011 made it the 36th largest economy in the world, after South Africa, which has a $422 billion GDP.

Multinational Corporations

A **multinational corporation** is an enterprise that owns, controls, or licenses facilities in countries other than the one in which it is headquartered. It is difficult to estimate the number of multinationals in the world, because digital technologies allow as few as two people in different countries to form a corporation. The last estimate made by the United Nations was 65,000 multinationals, with 820,000 foreign affiliates (Chanda 2003). A multinational corporation can range in size from fewer than 10 to millions of employees. In fact, most multinationals employ 250 or fewer people (Gabel and Bruner 2003). Regardless of size, multinationals compete, plan, produce, sell, recruit, acquire resources, and do other activities on a multi-country scale.

Multinationals establish operations in foreign countries for many reasons, including to obtain raw materials (such as oil and diamonds), to avoid paying taxes, to employ a low-wage labor force, and to manufacture goods for consumers in a host country (as does Toyota Motor North America, Inc.). Multinationals are headquartered disproportionately in the United States, Japan, and Western Europe. These global enterprises make "the key decisions—about what people eat and drink, what they read and hear, what sort of air they breathe and the water they drink, and ultimately what societies will flourish and which city blocks will decay" (Barnet 1990, 59).

The world's largest multinational corporations are often referred to as global corporations (see Table 10.3a). Theoretically, a truly global corporation should have some kind of presence in every country in the world. Probably no corporation is yet global in that sense. Still, many corporations such as McDonald's, with a presence in 119 countries, and UPS, with a presence in 200-plus countries, are called global because of their size and reach (McDonald's 2012; UPS 2010).

Table 10.3a: The World's Largest Global Corporations, 2011

Notice that 6 of the 10 largest corporations in the world extract or refine oil. In what ways might petroleum industries slow down national efforts to use less oil?

Rank	Global Corporation	Industry	Revenues (in billions)	Profits (in billions)	Headquarters
1	Walmart	General merchandisers	$421.3	$16.4	United States
2	Royal Dutch Shell	Petroleum	$378.2	$20.2	Netherlands
3	ExxonMobil	Petroleum	$354.7	$30.5	United States
4	BP	Petroleum	$308.9	–$3.7	UK
5	SINOPEC	Petroleum	$273.4	$7.6	China
6	China National Petroleum	Petroleum	$240.2	$14.4	China
7	State Grid	Electricity (power)	$226.3	–$4.6	China
8	Toyota Motors	Automobile	$221.8	$4.8	Japan
9	Japan Post Holdings	Banking, insurance, mail delivery, and over-the-counter services	$204.0	$4.9	Japan
10	Chevron	Petroleum	$196.3	$19.0	United States

Source: Fortune Magazine (2012)

When you read the names of the world's largest global corporations, 10 of which are listed in Table 10.3a, it is difficult to imagine their size and power of influence without some basis for comparison. We can get some idea of their size by comparing the annual revenues of a corporation to a country's GDP. A corporation's annual revenue is the total amount of money it receives for goods sold or services provided over the course of a given year. Taken together, the annual revenue of the top 10 global corporations is $2.825 trillion. GDP is the total value of all goods and services produced *within* the country over a year's time. Only four countries in the world—the United States, China, Japan, and Germany—have a GDP that exceeds $2.825 trillion. The annual revenue of the world's largest corporation, Walmart, is $421.8 billion. Only 35 countries have a GDP larger than that amount; those countries include Australia, Brazil, Canada, China, India, Iran, Italy, Japan, Mexico, and the United States.

In 2008 the global economic crisis alerted us to the possibility that many of the world's largest corporations were "too big to fail." That is, if these corporations failed, the economy as a whole might fail, too. So the United States and other governments intervened to prevent key financial and automobile institutions from failing. At that time, the combined annual revenues of just four troubled financial institutions—Citigroup, Bank of America, JPMorgan, and AIG—was $504 billion (Fortune Magazine 2009).

Criticism and Support for Multinationals

Critics of multinational corporations maintain that they are engines of destruction. That is, they exploit people and natural resources to generate profits. They take advantage of desperately poor labor forces, lenient environmental regulations, and sometimes almost nonexistent worker safety standards. As a case in point, in the decade leading up to the Great Recession that started in 2008, financial institutions set terms and loaned money to people and businesses who could never hope to meet payments (e.g., subprime loans) and then passed that risk on to others.

Supporters of multinational corporations, by contrast, maintain that these companies are agents of progress. Most obviously, multinationals employ millions and distribute goods, services, technology, and capital across the globe. In addition, they praise the multinationals' ability to raise standards of living, increase employment opportunities, transcend political hostilities, transfer technology, and promote cultural understanding. As one measure of their positive contributions, consider that Royal Dutch Shell and other oil producers make it possible for billions of people to drive cars, warm their homes, and so on.

On another level, however, multinationals' operations can aggravate problems related to obesity, poverty, mass unemployment, and overall inequality. Still, one can also argue that multinationals are not responsible for creating or solving social problems, such as obesity. After all, nobody forces people to eat fast foods or choose Walmart over a locally owned store (Weiser 2003).

◀ McDonald's and other fast-food establishments sell healthy items such as real fruit smoothies. As one former McDonald's CEO pointed out, "You can get a balanced diet at McDonald's. It's a question of how you use McDonald's" (Greenberg 2001).

Moreover, corporations claim that they merely respond to consumer demand. For example, virtually all the major fast-food companies have introduced healthy items on their menus, and most have proven unpopular with consumers. The typical McDonald's restaurant, for example, sells 50 to 60 salads per day versus 300 to 400 double cheeseburgers (Warner 2006). Nevertheless, critics question whether corporations should have the right to ignore the long-term effects of their products and practices on people and the environment, even if they are responding to consumer demand. Profitable products may benefit a corporation's bottom line and shareholders, but they can also be costly for a society due to **externality costs**—hidden costs of using, making, or disposing of a product that are not figured into the price of the product or paid for by the producer. Such costs include those associated with cleaning up the environment and with treating injured and chronically ill workers, consumers, and others. These costs must eventually be paid by someone (Lepkowski 1985).

While multinationals and other corporations are very powerful, consumer advocacy organizations have demonstrated that they can hold corporations in check. One thing is clear: when corporate executives feel real pressure from consumers, they act. However, if only a small number of consumers speak out, their claims are often dismissed. Consider comments from a McDonald's CEO after as many as 2,000 protesters trashed McDonald's restaurants and other businesses during four days of protest in Seattle against the World Trade Organization. The CEO noted that while 2,000 people protested, 17.5 million other people visited a McDonald's restaurant to eat (Greenberg 2001). The point is that there is no need to be concerned about the voices of 2,000 activists when 17.5 million consumers are voting with their feet—or mouths. McDonald's has changed some of its practices in response to pressure from other organizations, however. Greenpeace is one example of an organization dedicated to mobilizing people to hold governments and corporations responsible for crimes against the environment or for failing to protect the environment. It has recorded a number of successes, including pressuring McDonald's Corporation to agree not to use chickens that have been fed on soya, a feed that is grown in the Amazon rainforest. For a list of Greenpeace victories, see http://www.greenpeace.org/international/about/victories.

(Write a Caption)

Write a caption that relates Coca-Cola to the concept of a multinational corporation.

Hints: In writing this caption

- review the concept of a multinational corporation,
- find out how many countries sell Coca-Cola products, and
- identify the geographic location of its first sale.

Robert K. Wallace

Critical Thinking

Review the most current list of the world's 500 largest corporations, which can be found online by using the search term "Global 500." Identify one that has had a positive or negative impact on your life. Explain.

Key Terms

externality costs   multinational corporation

Power and Authority

MODULE 10.4

Objective

You will learn under what circumstances people can exert their will, even in the face of opposition.

Do you recognize the couple on the television screen? How might you describe the power the two hold?

Ensign Haraz Ghanbari

The marriage ceremony of Prince William and Kate Middleton attracted an estimated 2.0 billion viewers around the world, including this audience of U.S. soldiers (Sher 2011). The two possess a kind of power over others that we will discuss in this module, a power that derives from tradition.

A society's **political system** is the institution that regulates access to and use of power to control access to scarce and valued resources and to make laws, policies, and decisions that affect others' life chances. **Power** is the probability that an individual can achieve his or her will, even against opposition (Weber 1947). That probability increases when an individual has the means to force people to obey his or her commands, but it increases even more when the individual possesses authority over others. **Authority** is legitimate power—power that people believe is deserved, just, and proper. A leader has authority to the extent that people view him or her as being legitimately entitled to it.

Types of Authority

Max Weber identified three types of authority: traditional, charismatic, and legal-rational. People can possess more than one kind of authority. **Traditional authority** is grounded in the sanctity of time-honored norms that govern how someone comes to hold a powerful position, such as chief, king, queen, or emperor. Usually, the person inherits that position by virtue of being born into a family that has held power for some time. People accept that system because it has always been like that and to abandon the past is to renounce a heritage and collective identity (Boudon and Bourricaud 1989).

Ultimately, Prince William's power is grounded in traditional authority—his father is Prince Charles of Wales, and his grandmother is Queen Elizabeth II. By tradition, after Charles, William is second in line and, until William and Kate have children, his brother, Harry, is third in line to be king of the United Kingdom and its independent states.

Charismatic authority is grounded in exceptional and exemplary personal qualities. Charismatic leaders are obeyed because their followers believe in and are attracted irresistibly to the leaders' vision. These leaders, by virtue of their special qualities, can persuade followers to behave in ways that depart from rules and traditions. Charismatic leaders often emerge during times of profound crisis, such as economic depressions and wars. During such times people are susceptible to a vision of a new order. A charismatic leader is more than popular, attractive, likable, or pleasant. A merely popular person, "even one who is continually in our thoughts," is not someone for whom we would break all previous ties and give up our possessions (Boudon and Bourricaud 1989, 70). Charismatic leaders successfully persuade their followers to make extraordinary personal sacrifices, cut themselves off from ordinary worldly connections, or devote their lives to achieving a vision that the leaders have outlined.

The source of a charismatic leader's authority, however, does not rest with the ethical quality of his or her vision. Adolf Hitler, Franklin D. Roosevelt, Mao Zedong, Winston Churchill, and the Reverend Martin Luther King Jr. were all considered charismatic leaders. Each assumed leadership during turbulent times. Likewise, each conveyed a powerful vision (right or wrong) of his country's destiny. Charismatic authority results from the intense relationships between leaders and followers. From a relational point of view, charisma is a highly unequal relationship between a guide who inspires and followers who believe wholeheartedly in that guide's promises and visions (Boudon and Bourricaud 1989).

➤ Mahatma Gandhi, born in Porbandar, India, was a key spiritual and political leader in India and world history. Gandhi developed and conceived of a strategy of nonviolent resistance known as *satyagraha*—"the force of truth" or "the firmness of truth." This philosophy and strategy guided his leadership over India's largely nonviolent independence movement against British rule. Gandhi's accomplishment was "an extraordinary feat of personal magnetism," in which he inspired the participation of the illiterate and poor (Luce 2007). Among others, Gandhi's thinking and methods had a great influence on Nelson Mandela in his struggle to end South Africa's system of apartheid and on the way Martin Luther King Jr. organized campaigns for civil rights in the United States.

Over time, charismatic leaders and their followers come to constitute an emotional community devoted to achieving a goal and sustained by a belief in the leader's special qualities. Weber argues, however, that eventually the followers must be able to return to a normal life and to develop relationships with one another based on something other than their connections to the leader. Attraction and devotion cannot sustain a community indefinitely, if only because the object of these emotions—the charismatic leader—is mortal.

Legal-rational authority derives from a system of impersonal and formal rules that specify the qualifications for occupying an administrative or judicial position; the individual holding that position has the power to command others to act in specific ways, and that power is backed by the force of law or established organizational structure. These rules also specify the scope of that power and appropriate context in which it can be exercised. In cases of legal-rational authority, people comply with commands, decisions, and directives because the power belongs to the position, and by extension, to the person occupying the position.

Courtesy of the White House Transportation Agency

◀ The White House is the residence and workplace of the person who holds the office of President of the United States. Anyone who holds the position of U.S. president possesses legal-rational authority. The Constitution of the United States specifies how the position of president is filled and the powers associated with that position.

The Power Elite

Sociologist C. Wright Mills wrote about the connection between government, industry, and the military in *The Power Elite* (1956). The **power elite** are those few people who occupy such lofty positions in the social structure of leading institutions that their decisions affect millions, even billions, of people worldwide. For the most part, the source of this power is legal-rational—residing not in the personal qualities of those in power, but rather in the positions that the power elite have come to occupy.

The power elite use their positions—and the tools of their positions in a bureaucracy—to rule over, control, and influence others. These tools include surveillance systems, communication structures, and weapons. In writing about the power elite, Mills focuses on those who occupy the highest positions in the leading U.S. institutions. Those institutions are the military, corporations (especially the 200 or so largest), and the government (Mills 1963, 27).

The origins of the power elite that Mills describes can be traced to World War II, when the political elite mobilized corporations to produce the supplies, weapons, and equipment needed to fight that war. U.S. corporations, which were left unscathed by the war, were virtually the only corporations capable of providing the services and products war-torn countries needed for rebuilding. The interests of the U.S. government, the military, and corporations became further intertwined when the political elite decided that a permanent war industry was needed to contain the spread of communism. Thus, over the past 60 years, these three institutions (and by extension the U.S. workforce) have become deeply and intricately interrelated in hundreds of ways.

U.S. Navy photo by Pat Corkery, courtesy United Launch Alliance

➤ As one measure of the ties between the military, the government, and industry, consider that in 2011 the U.S. Department of Defense alone did business with more than 17,000 contractors and coordinated 335,000 active contracts. The value of these contracts ranges from less than $25,000 to billions (Defense Contract Management Agency 2012). The photo shows the launching of a tactical satellite communications system designed by Lockheed Martin, a global security and aerospace company, to coordinate ground communications among U.S. military forces. Other defense contractors include BP, Dell, FedEx, and Krispy Kreme. In 2010 the top five military contractors were Lockheed Martin ($10.9 billion), Northrop Grumman ($8.2 billion), Boeing ($5.1 billion), General Dynamics ($4.6 billion), and Raytheon Co. ($4.1 billion) (Defense Contract Management Agency 2012).

Simply think about Lockheed Martin's dependence on government contracts. About 93 percent of its sales are to the U.S. Department of Defense (DoD), the Department of Homeland Security, and other U.S. government agencies. In 2011 the company was awarded a total of $42.4 billion in contracts (Federal Procurement Data System 2012). These government contracts support 123,000 employees working in 1,000 facilities in 500 cities and 46 states throughout the United States and in 75 other nations and territories (Lockheed Martin 2012).

Because the military, the government, and corporations are so interdependent and because decisions made by the elite of one sector affect the elite of the other two, Mills believes that everyone has a vested interest in cooperation. Shared interests cause those who occupy the highest positions in each sector to interact with one another. Out of necessity, then, a triangle of power has emerged. Mills cautions that we should not assume, however, that the alliance among the three sectors is untroubled, that the powerful share exactly the same interests, that they know the consequences of their decisions, or that they are joined in a conspiracy to shape the fate of a country or the globe. At the same time, it is clear that they know what is on each other's minds. Whether they come together casually at their clubs or hunting lodges or more formally as board members, they are definitely not isolated from each other. There is a community of interest and sentiment among the elite (Hacker 1971).

Mills gives no detailed examples of the actual decision-making process at the power elite level. Rather, he focuses on understanding the consequences of this alliance. Mills acknowledges that the power elite are not free agents but are subject to controls, such as whistleblowers, congressional investigations, and budget constraints.

Pluralist Model

A pluralist model of power views politics as an arena of compromise, alliances, and negotiation among many competing special-interest groups, and it views power as something dispersed among those groups. **Special-interest groups** consist of people who share an interest in a particular economic, political, or other social issue and who form an organization or join an existing organization to influence public opinion and government policy.

Joe Flood

Boni Li

⮝ Special-interest groups are very diverse. They include those who are part of grassroots groups that are considered socially conscious about some issue, as well as those who belong to established groups such as Boeing Corporation, which was awarded $21.6 billion in government contracts in 2011 (Federal Procurement Data System 2012).

These interest groups are often represented by **lobbyists**, people whose job it is to solicit and persuade state and federal legislators to create legislation and vote for bills that favor the interests of the group they represent. There are about 106,000 people employed as lobbyists, about one-third of whom work in Washington, DC. Lobbyists can represent states, foreign governments, industries, universities, or nonprofit agencies. Some examples of organizations that lobbied to shape the 2009 health care reform legislation and the amount each spent are the Pharmaceutical Research and Manufacturers Association ($20.1 million), Blue Cross and Blue Shield ($16.1 million), American Medical Association ($12.3 million), and American Hospital Association ($12.6 million) (Center for Responsive Politics 2009).

Some special-interest groups form **political action committees (PACs)**, which raise money to be donated to the political candidates who seem most likely to support their special interests. There are more than 4,500 registered PACs, some of which are called Super PACs.

Super PACs, the products of federal court cases (*SpeechNow.org v. Federal Election Commission* and *Citizens United v. Federal Election Commission*), are allowed to raise and spend unlimited amounts in support of or against political candidates. In addition, corporations and unions can contribute unlimited amounts of money directly or though other organizations without complete or immediate disclosure, making it difficult for the public to know the source of funding behind political messages (Center for Responsive Politics 2012a).

Prominent names of Super PACS include Restore Our Future (pro-Romney) and Priorities USA Action (pro-Obama). As of April 8, 2012 (months before the 2012 presidential campaign got into full swing), 412 groups were registered as Super PACs and reported holding $154.9 million in contributions (Center for Responsive Politics 2012a).

According to the pluralist model, no single special-interest group dominates the U.S. political system. Rather, competing groups thrive and can express their views through opinion polls, unions, protests, e-mails, and PACs. One problem with the pluralist model is that we cannot conclude that every special-interest group has enough resources to represent and defend its interests.

(Write a Caption)

Write a caption that connects one of the three types of authority with this drill instructor shouting orders at a recruit.

Sgt. Pamela Shelley

Hints: In writing this caption

- review the three types of authority,
- consider the type of authority this drill instructor possesses,
- consider what this Marine is willing to tolerate because of her authority, and
- think about under what circumstances her authority would not apply.

Critical Thinking

Think of someone who has authority over you. Which of the three types of authority—traditional, charismatic, legal-rational—does the person possess?

Key Terms

authority
charismatic authority
legal-rational authority
lobbyists
political action committees (PACs)
political system
power
power elite
special-interest groups
traditional authority

Forms of Government

MODULE

10.5

Objective

You will learn about various forms of government and how they direct and coordinate political and economic activity.

What form of government does an ink-stained finger symbolize?

Spc. Joshua W. Lowery

The Iraqi in the photograph has just voted, as indicated by the finger being dipped into ink. Voters in Iraq dip their index finger into an inkwell after voting. The stained finger marks people so they do not try to vote again. This image has come to symbolize the hoped-for change in Iraq's government as it moves from a dictatorship to a democracy.

Government is the organizational structure that directs and coordinates people's involvement in the political activities of a country or some other territory, such as a city, county, or state. It is also one mechanism through which people gain power and exercise authority over others. Governments make laws and create policies that shape every aspect of life. Governments are often divided into branches of power, such as executive, judicial, and legislative. In this module we consider five forms of government: democracy, totalitarianism, authoritarianism, monarchy, and theocracy.

Democracy

Sir Winston Churchill, prime minister of the United Kingdom during World War II, once said that "democracy is the worst form of government except for all others that have been tried" (1947, 7566). **Representative democracy** is a system of government in which power is vested in citizens who vote into office those candidates they believe can best represent their interests. Because it is not feasible for citizens to vote on every matter that affects them, they make their voices heard through those they elect. In a democracy, political candidates can be from more than one party, and those in minority parties can oppose the party holding power. In addition, when a majority of voters elect to change the party in power, the transition is orderly and peaceful. In democracies, elected representatives make laws, vote on taxes, establish budgets, and support or oppose the party in power.

Democratic forms of government extend basic rights to citizens and legal residents. These rights include freedom of speech, movement, religion, press, and assembly (that is, the right to form and belong to parties and other associations), as well as freedom from unwarranted arrest and imprisonment. Other characteristics of democracies include free and fair elections, access to a free press, an educated or informed citizenry, and a constitution that sets limits on executive and other powers (Bullock 1977).

Successful democracies depend on informed voters. Although the United States is classified as a representative democracy, not everyone who is eligible to vote does so. In U.S. presidential elections, typically 65 percent of all those eligible to vote actually vote. Thus, the winner is usually awarded victory with the support of about 33 percent of all Americans eligible to vote (U.S. Census Bureau 2009). In light of these statistics, we might question whether elected officials represent everyone or just those who vote. Furthermore, we might question whether elected officials represent those powerful constituents who donate generously to their campaigns. In assessing whether a form of government is a true democracy, it is important to consider who has the right to vote. All democracies have at one time or another excluded some from the political process based on race, sex, income, property ownership, criminal status, mental health, religion, age, or other characteristics.

Totalitarianism

Totalitarianism is a system of government characterized by (1) a single ruling party led by a dictator, (2) an unchallengeable official ideology that defines a vision of the "perfect" society and the means to achieve that vision, (3) a system of social control that suppresses dissent, and (4) centralized control over the media and the economy. Ideological goals vary but may include overthrowing capitalist and foreign influences (as in Cuba's 60-year revolution to resist U.S. capitalist influence), or creating the perfect race (as in Germany under Hitler). Whatever the government's goals, the political leaders, the military, and the secret police intimidate and mobilize the masses to help the state meet its goals.

Totalitarian governments are products of the 20th century, because by that time technologies existed that allowed a few people in power to control the behavior of the masses and the information the masses could hear. Many of the governments labeled as totalitarian have followed communist principles. Traditionally, communist governments have outlawed private ownership of property, supported the equal distribution of wealth, and offered status and power to the working class or proletariat. Communist leaders also mobilize the masses to bring about change.

➤ North Korea's government is considered totalitarian. This photograph shows North Korean soldiers celebrating the 75th anniversary of the Korean People's Army. The soldiers' precise movements and synchronized timing showcase a dedication to the revolutionary ideology of their Great Leader, Kim Il Sung (1912–1994) and symbolize the entire party, nation, and military moving as one (*DailyNK* 2008).

Authoritarianism

Under an **authoritarian** government, no separation of powers exists; a single person (a dictator), a group (a family, the military, a single party), or a social class holds all power. No official ideology projects a vision of the "perfect" society or guides the government's political or economic policies. Authoritarian leaders do not seek to mobilize the masses to help realize a vision or meet ideological goals. Instead, the government functions to serve those in power, who may or may not be interested in the general welfare of the people. Common to all authoritarian systems is a lack of checks and balances on the leader's power (Chehabi and Linz 1998).

How does a single person, a group, or a social class gain control of an entire country? Authoritarian leaders typically receive support from a foreign government that expects to benefit from their leadership (Buckley 1998). Saddam Hussein is one example of an authoritarian leader whom the U.S. government both supported and opposed. The U.S. government supported Hussein during the Iran–Iraq War (1980–1988)—even though he invaded Iran, abused the human rights of Iraqi citizens, used chemical weapons on Iranians and his own people, and sought to produce nuclear weapons. Then, the U.S. government opposed Hussein during and after the 1991 Gulf War.

◀ In 2003 the United States military invaded Iraq and deposed Saddam Hussein on the grounds that the dictator had been secretly developing and building weapons of mass destruction. When no evidence of such weapons turned up, the U.S. government justified the invasion on the grounds that Hussein had abused the human rights of Iraqi citizens and used chemical weapons against his own people. This photo shows an American soldier draping the U.S. flag around a statue of Saddam Hussein, after taking the capital city.

Monarchy

A **monarchy** is a form of government in which the power is in the hands of a leader (known as a monarch) who reigns over a state or territory, usually for life and by hereditary right. Typically, the monarch expects to pass the throne on to someone who is designated as the heir, usually a firstborn son. The monarch's power may range from absolute (total power) to nominal (in name only) (U.S. Central Intelligence Agency 2012).

Theocracy

Theocracy, which means rule by divine guidance, is a form of government in which political authority rests in the hands of religious leaders or a theologically trained elite group. Thus, there is no legal separation of church and state. Government policies and laws correspond to religious principles and laws. Contemporary examples of theocracies include the Vatican under the pope, Afghanistan when it was under the Taliban, and Iran under the ayatollah. In other forms of theocracy, power is shared by a secular ruler, such as a king, and a religious leader, such as a pope or an ayatollah, or by secular government leaders devoted to the principles of the dominant religion. At one time, England was dominated by the Anglican Church, France by the Roman Catholic Church, and Sweden by the Lutheran Church.

Master Sgt. Jerry Morrison

◄ Saudi Arabia qualifies as a hereditary monarchy. Since 1932 it has been ruled by the sons and grandsons of the first king of Saudi Arabia, Abd al-Aziz ibn Saud, shown in these portraits. Saudi Arabia has no official political parties and no national elections. The king's power is tempered by consultation with the royal family and with an appointed consultative body known as the Majlis al-Shura, which debates, rejects, and amends proposed legislation; holds oversight hearings over government ministries; and initiates legislation. While the head of state and government is a king and not a religiously trained leader, Saudi Arabia's constitution and legal system are grounded in Islamic laws and principles (U.S. Central Intelligence Agency 2012).

(Write a Caption)

Write a caption that connects the pope to the type of government of which he is the head.

Hints: In writing this caption

- review the forms of government, and
- consider the specific territory over which the pope is the head.

U.S. Air Force photo by Tech. Sgt. Craig Clapper

Critical Thinking

Have you voted in local, state, and/or national elections? What does your voting record say about democracies?

Key Terms

authoritarian	monarchy	theocracy
government	representative democracy	totalitarianism

Imperialism and Related Concepts

MODULE 10.6

Objective

You will learn about concepts sociologists use to describe forms of foreign dominance.

These specially painted F-117 Nighthawks cost billions to develop and are part of a military structure that spends $700 billion a year on defense. How would you explain this spending?

U.S. Air Force photo/Senior Master Sgt. Kim Frey

In 2011 worldwide military expenditures approached $1.7 trillion. The United States budgeted $700 billion, followed by China ($129 billion). U.S. military spending accounts for 40 percent of all military spending worldwide (SIPRI 2012). Sociologists use several terms to describe governments with the capacity to exercise their will, even force it, upon others. In this module we will discuss concepts of empire, imperialism, hegemony, and militarism. As you learn about their meanings, ask if any of these terms apply to the United States.

Concepts of Power and Dominance

An **empire** is a group of countries under the direct or indirect control of a foreign power or government that acts to shape the political, economic, and cultural life of the people over which it has power. Perhaps the most well-known empire was the British Empire, considered the largest in history. By the early 1920s, this empire ruled over a quarter of the world's population and territory (Maddison 2001).

An **imperialistic power** exerts control and influence over foreign entities either through military force or through political policies and economic pressure.

Imperialists believe that the superiority of their cultural, political, or economic system justifies control over foreign peoples and their resources. From the imperialist's point of view, such control is for the greater good of those conquered and for the planet as a whole.

Hegemony is a process by which a power maintains its dominance over foreign societies. Those in power use bureaucratic structures (such as the World Bank and the United Nations) to formalize their power and to make it seem impartial and abstract. That power also uses educational programs and various forms of media (advertisements, television, music, and movies) to influence the foreign populace.

Jim Gordon, CIV

➤ There is a close correspondence between imperialistic and militaristic powers. A **militaristic power** believes that military strength, and the willingness to use it, is the source of national and even global security. Usually a peace-through-strength doctrine—peace depends on military strength and force—is cited to justify military buildups and interventions on foreign soil (see Figure 10.6a).

▼ **Figure 10.6a: Number of Active-Duty U.S. Military Personnel by Country**
The map applies to military personnel and does not take into account the contractors, who can equal or exceed the number of military personnel.

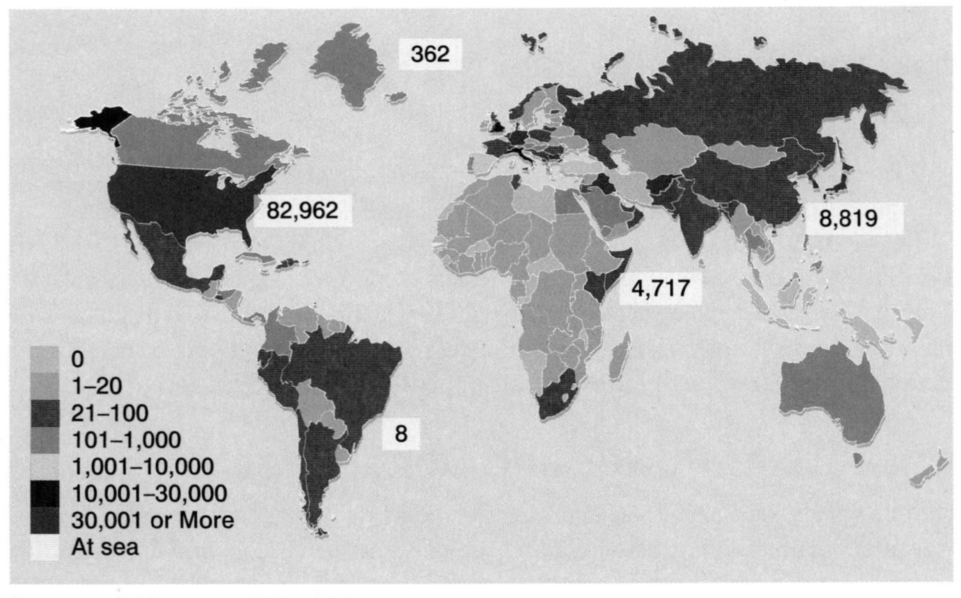

Source of data: U.S. Department of Defense (2011)

Do the terms *empire, imperialism, militarism*, and *hegemony* apply to the United States? In answering this question, our goal is not to classify the United States as an evil or a benevolent empire. Rather, we simply wish to assess the United States' world influence and power in light of these concepts. One assessment describes the United States this way: "Throughout the history of mankind certainly no country has existed that so thoroughly dominates the world with its policies, its tanks, and its products as the United States does" (*Der Spiegel* 2003).

DoD photo by Fred W. Baker III

Still, "talk of 'empire' makes Americans distinctly uneasy." Americans don't want to be part of a country that is an empire. But if it is one, they want to think of the United States as a kind and just empire (Judt 2004, 38).

Functionalist Perspective

Functionalists ask: "What are the anticipated and unanticipated consequences of the U.S. military presence throughout the world?" One manifest, or expected, function of this global-scale military presence is that it exists to safeguard U.S. interests and the American way of life. The September 11, 2001, attacks on the World Trade Center are often evoked as evidence that the United States has enemies.

An unanticipated, or latent, function of U.S. military presence is that the military often operates as a humanitarian force working to promote democracy and to improve health, education, and economic growth in some of the world's poorest countries (Bush 2007).

Mass Comm. Spc. 3rd Class Joshua Valcarcel

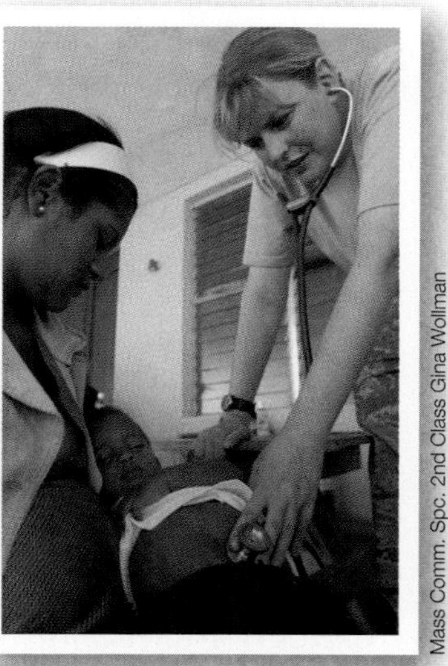
Mass Comm. Spc. 2nd Class Gina Wollman

Here we see a U.S. Navy hospital ship, one of many, that deploys to nations around the world providing medical treatment to children, expectant mothers, and others in need. The right photo shows a Navy health care worker examining a boy in a medical clinic in the Dominican Republic.

An expected, or manifest, dysfunction of this military presence around the world is that wherever there are servicemen or servicewomen, we can expect to find babies conceived with locals overseas. For example, there are significant populations of children known as Amerasians who live in Guam, Hawaii, Okinawa, Thailand, South Korea, Vietnam, and the Philippines.

A latent, or unexpected, dysfunction of U.S. global military presence is that it drives what sociologists call a **military-industrial complex**, a relationship between those who declare, fund, and manage wars (the Department of Defense, the office of the president, and Congress) and corporations that make the equipment and supplies needed to wage war. Corporations and their stockholders come to need war to maintain profits, and the millions who work for these corporations also rely on war to maintain their employment (and by extension their lifestyles). In his farewell address to the American people, President Dwight D. Eisenhower (1961) warned that this complex is "felt in every city, every state house, every office of the Federal government . . . our toil, resources and livelihood are all involved; so is the very structure of our society . . . we must guard against the acquisition of unwarranted influence, whether sought or unsought, by the military-industrial complex."

In evaluating Eisenhower's warning, keep in mind that at any one time the U.S. Department of Defense (2012) coordinates 335,000 active contracts with businesses and consultants (including universities) to supply everything from donuts to tanks. This amount of money is a powerful incentive for businesses to influence American foreign policies in ways that encourage war and other defensive measures.

Conflict Perspective

Conflict theorists would certainly focus on the corporations that benefit from military contracts and how the U.S. leaders justify this level of military presence around the world. They ask: "Who benefits from military deployments and at whose expense?" In answering this question, conflict theorists point out that the American way of life is very dependent on oil and nonfuel minerals such as coltan (needed to make cell phones and other electronic equipment). Conflict theorists would argue that it is no coincidence that the U.S. military is stationed in parts of the world that supply these needed resources. As a case in point, in 2007 the United States created a command to oversee the African continent because of Africa's increased strategic importance. Conflict theorists ask: "What factors might contribute to the rise of the continent's strategic importance?" One obvious answer is China's growing presence in Africa and interest in that continent's oil and other natural resources.

➤ Given the fact that 21 percent of U.S. oil imports come from African countries, it is no surprise that the United States has a significant military presence in that region (U.S. Department of Energy 2011). Here a U.S. navy officer advises a member of the Djiboutian navy on searching and securing vessels that transport goods in waters around the Horn of Africa.

Tech. Sgt. Ryan Labadens

Petty Officer 1st Class Sean Mulligan

➤ In spite of its size, the U.S. military, the most powerful in the world, has been battling insurgents in Iraq, Afghanistan, and other places for years. Insurgents are groups who participate in armed rebellion against an occupying or established power with the hope that those in power will retreat or pull out. From the occupier's or liberator's point of view, insurgents have no legitimate cause; from the insurgent's point of view, the authority of the occupier or liberator is illegitimate.

(Write a Caption)

Write a caption that relates this image of the Chinese military as a frame of reference for U.S. military spending.

Hints: In writing this caption

- review U.S. spending relative to world spending, and
- calculate the difference between what China spends relative to what the United States spends.

Chief Mass Comm. Spc. David Rush

Critical Thinking

Do you think any of the four terms you learned in this module—*empire*, *imperialism*, *hegemony*, and *militarism*—apply to the United States? Explain your answer.

Key Terms

empire
imperialistic power
military-industrial complex
hegemony
militaristic power

MODULE 10.7

Applying Theory: Global Society Theories

Objective

You will learn how world system theory explains the forces creating a global-scale economy.

Chris Caldeira

What does the term *global economy* mean to you? How is this woman part of the global economy? How are you part of a global economy?

The woman in the photo is part of the global economy if only because she rows products out to tourists on cruise ships docked in Vietnam ports. Among other things, her "convenience store" stocks Oreo cookies and Ritz crackers. World system theory explains how the global-scale economy came to be. This theory falls under the category of global society theories. Global society theorists focus on human activity that is embedded in a larger global context. They emphasize that seemingly local events are interconnected with events taking place in other countries and regions of the world. These interconnections are part of a phenomenon known as **global interdependence**, a situation in which human interactions and relationships transcend national borders and in which social problems within any one country—such as unemployment, water shortages, natural disasters, or drug addiction—are shaped by events taking place outside the country.

Global interdependence is part of a dynamic process known as **globalization**—the ever-increasing flow of goods, services, money, people, technology, information, and other cultural items across political borders. This flow has become more dense and quick-moving as constraints of space and time that once separated people who lived far apart geographically seemingly dissolve.

The Sociology of Immanuel Wallerstein

Immanuel Wallerstein (1984) is the sociologist most frequently associated with world system theory. He has written extensively about the ceaseless 500-year-plus expansion of a single market force—capitalism—that created the world economy. The world economy, which encompasses more than 200 countries and thousands of cultures, is interconnected by the division of labor. In this world economy, economic transactions cross national boundaries. Although each government seeks to shape the global market in ways that benefit its interests, no single political structure such as a world government holds authority over the global economy.

How has capitalism come to dominate the global network of economic relationships? One answer to this question can be found in the ways capitalists respond to changes in the economy, especially to economic stagnation. One response is to find ways to lower labor-associated production costs by moving production facilities out of high-wage zones and into lower-wage zones outside a country, introducing technologies that save labor, and forcing people to work for little to no wages (enslavement, indentured servitude).

Molly Hayden, U.S. Army Garrison-Hawaii Public Affairs

© Justin Guariglia/Corbis

▲ The label "Made in China" has come to symbolize the practice of moving production from high-wage to low-wage labor zones. The average manufacturing worker in China earns about 57 cents per hour. This labor is behind the masses of dolls and "other made-in-China bric-a-brac that the world consumes each day" (Wolf 2005).

A second response to economic stagnation is to secure at the lowest possible price the raw materials needed to make products; those products may be rubber, sugar cane, or coltan (for mobile phones). Karl Marx wrote that the drive for profit is a boundless thirst that chases the capitalist "over the whole surface of the globe" in search of not only low-cost labor but also inexpensive raw materials (Marx and Engels 1848). A third response to economic stagnation is to create new markets that expand the boundaries of the world economy. The global expansion of the Oreo cookie represents such a response (see Figure 10.7a).

Because of these responses, capitalism has spread steadily across the globe and facilitated connections among local, regional, and national economies. Wallerstein (1984) argues that the 200-plus countries that have become part of the world economy play one of three different and unequal roles in the global economy: core, peripheral, or semiperipheral.

Figure 10.7a: Largest and Fastest Growing Markets for Oreo Cookies
The case of Oreo cookies shows how expanding market share creates a global economy. The first Oreo cookie was sold in Hoboken, New Jersey, in 1912. A hundred years later the product is sold in 100 countries. The map shows the countries that count among the largest and fastest growing markets. Today Oreo cookies has a Facebook page with 23 million "friends" in more than 200 countries (Kraft Foods 2012).

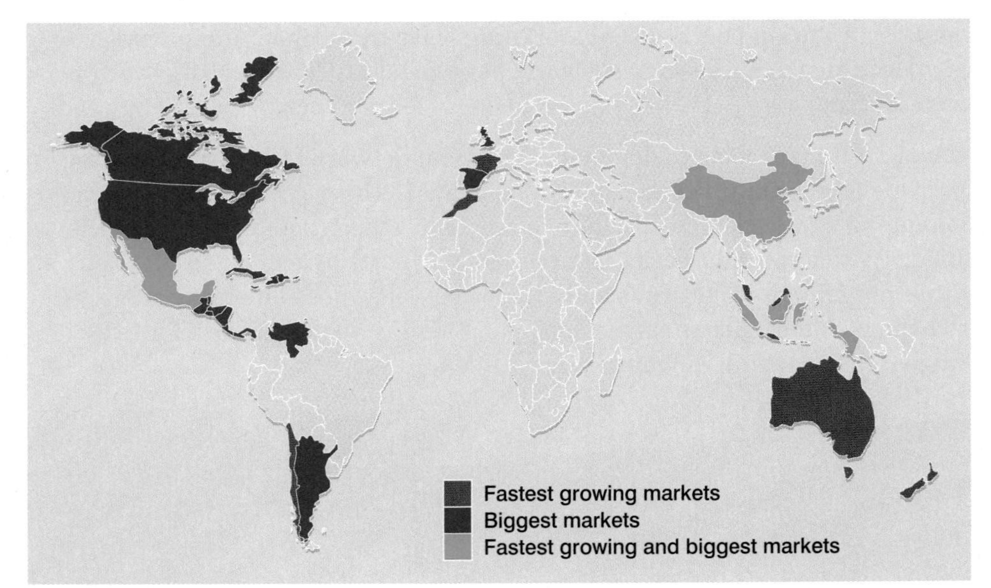

Source of data: Kraft Foods (2012)

Core Economies

Core economies include the wealthiest, most highly diversified economies with strong, stable governments. Examples of core economies include the United States, Japan, Germany, France, Canada, and the United Kingdom. The 20 percent of the world's population that constitutes the core economies accounts for 86 percent of total private consumption expenditures (World Bank 2008b). Consequently, when economic activity weakens in core economies, the economies of other countries suffer, because exports to core economies decline and prices fall. In spite of their large size and relative stability, core economies have weaknesses, one of which is their dependence on foreign sources for raw materials, such as oil needed to fuel cars and palm oil needed to make soaps.

Dimas Ardian/Getty Images

Most Americans are unaware that palm oil—produced in Indonesia, Malaysia, and Western African forests—is used to make soaps, cooking oil, and personal care products. Palm oil is also used to make biofuels. Here an Indonesian plantation worker harvests palm oil fruits. Over the past 25 years, Indonesia has lost more than 60 percent of its tropical forests in meeting the demand from core and other economies (World Watch Institute 2012).

Peripheral Economies

Peripheral economies are built on a few commodities or even a single commodity, such as coffee, peanuts, or tobacco, or on a natural resource, such as oil, tin, copper, or zinc (Van Evera 1990). As a result, commodity price fluctuations due to bad weather, a bumper crop, or lowered consumer demand have a dramatic impact on people's livelihoods and affect the incidence of poverty, as the vast majority of the poor depend on primary commodities for their livelihoods. At least 38 countries depend on one commodity for 50 percent or more of the revenues generated from exports (U.S. Central Intelligence Agency 2012). For example, coffee accounts for 50 percent of the revenue that Uganda, Ethiopia, and Burundi earn from their exports (United Nations 2010a). Peripheral economies have a dependent relationship with core economies that traces its roots to colonialism. Peripheral economies operate on the so-called fringes of the world economy. Most of the jobs that connect their workers to the world economy pay little and require few skills. Amid widespread and chronic poverty, pockets of economic activity may exist, including manufacturing zones and tourist attractions.

Extreme dependence on one or even a few commodities causes many problems. First, fluctuations in the price of that commodity can leave a government flush with money when prices rise and poor when prices fall, leading to cycles of wild government spending followed by drastic spending cuts. Second, corruption and political rivalries inevitably arise when a government or a few powerful people control a country's natural resources, the distribution of those resources, and the revenues from them (Birdsall and Subramanian 2004).

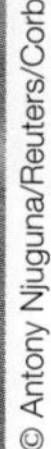

➤ The African country of Malawi represents one example of a peripheral economy as illustrated by this description in the *World Factbook*: "Landlocked Malawi ranks among the world's most densely populated and least developed countries. The economy is predominately agricultural with about 80% of the population living in rural areas. Agriculture, which has benefited from fertilizer subsidies since 2006, accounts for one-third of GDP and 90% of export revenues. The performance of the tobacco sector is key to short-term growth as tobacco accounts for more than half of exports" (U.S. Central Intelligence Agency 2012).

Semiperipheral Economies

Semiperipheral economies are characterized by moderate wealth (but extreme inequality) and moderately diverse economies. Semiperipheral economies exploit peripheral economies and are in turn exploited by core economies. According to Wallerstein (1984), semiperipheral economies play an important role in the world economy because they are politically stable enough to provide useful places for capitalist investment if employee wage and benefit demands in core economies become too great. The countries known as emerging markets—Mexico, Brazil, Argentina, South Africa, Poland, Turkey, India, Indonesia, China, and South Korea—qualify as semiperipheral.

Tony Rotundo

➤ Turkey is considered to be a rapidly emerging economy. It is located at the crossroads between East and West. One indication of Turkey's emergence as an economic power is its popularity as a tourist destination. In 2011, 31.5 million tourists visited the country, making it one of the top 10 tourist destinations in the world (Republic of Turkey 2012). Here children pose with an Anatolian Lion, a nickname for the Turkish economy.

One problem with this three-tier classification scheme—core, semiperipheral, and peripheral—is that some countries possess characteristics that apply to more than one category.

© david pearson/Alamy

Keith Farley

▲ Some agricultural regions in India revolve around human or animal muscle power, suggesting the economy is peripheral. A call center in India, on the other hand, which services customers who have purchased products from U.S. multinational corporations, indicates a semiperipheral economy.

Critical Thinking

Find an example of a country that depends on one commodity for 50 percent or more of the revenues generated from its exports.

Key Terms

core economies
global interdependence
globalization
peripheral economies
semiperipheral economies

Summary: Putting It All Together

CHAPTER 10

Economic systems coordinate human activity to produce, distribute, and consume goods and services. There are two major types of economies—capitalist and socialist. Although the two are treated as extremes along a continuum, in practice, economies are blends of these two systems. An economy is composed of three sectors: primary, secondary, and tertiary. The relative importance of each sector to a particular country's economy can be determined by examining the percentage of workers in each category.

The United States possesses the world's largest economy as measured by GDP; it is the 11th richest country when measured by GDP per capita. Among other things, the U.S. economy is characterized by high government and consumer debt and a dependence on oil and other critical minerals. The United States, along with Western European countries, China, and Japan, is headquarters to most of the world's largest multinational or global corporations. Multinational corporations, which wield considerable influence over how we live, have facilities and operations in countries outside the one in which they are headquartered.

Political systems distribute power over resources and decision making. Power is the likelihood that an individual can achieve his or her will, even against opposition. That probability increases when people possess authority or legitimate power, of which there are three types: traditional, charismatic, and legal-rational.

There are two broad models for the ways power is distributed: power elite and pluralist. The power elite model describes power as concentrated in the hands of those few people who occupy lofty positions in the social structure of the military, corporations (especially the 200 or so largest), and the government. The pluralist model views politics as an arena of compromise, alliances, and negotiation among many competing special-interest groups, and power is therefore dispersed among those groups.

There are five major forms of government: democracy, totalitarianism, authoritarianism, monarchy, and theocracy. Governments and other political bodies that have the capacity to dominate others can be described with the following terms: *empire*, *imperialism*, *hegemony*, and *militarism*.

The United States is part of a world economy, which currently encompasses more than 200 countries and thousands of cultures interconnected by the division of labor. That world economy is essentially capitalist. The force driving the creation of this global economy is economic stagnation, specifically the way capitalists respond to create growth and profit. One response is to find ways to lower labor-associated production costs by moving production facilities out of high-wage zones and into lower-wage zones. Another is to find the cheapest resources to make products, and a third is expanding markets. Over the course of 500 years these profit-generating responses have created the global economy assigning each country to core, peripheral, and semiperipheral status.

Go to cengagebrain.com to link to Aplia and CourseMate for the chapter quiz and other activities.

Courtesy of Terra Schulz

CHAPTER 11
FAMILIES

In everyday life, the family posing for this photo would be called traditional. But for this photo, the adults have switched roles with him behaving as a wife and her the husband. It is obvious, however, that the bond and emotional ties that hold this family together are unaffected by this departure from what we define as normative. Sociologists consider the *family* a social institution that binds people together through blood, marriage, law, and/or social norms. When sociologists study family, they do not have a particular family structure in mind as a standard. Instead, they consider the social forces that affect the ever-changing structure of families. Those social forces include, but are not limited to, life expectancy, women's ability to control pregnancy, and employment opportunities for males and females.

Go to Sociology CourseMate on cengagebrain.com to watch a video that profiles several politicians whose wives have stood by them after their affairs become public.

MODULE

11.1 Defining Family

Objective

You will learn about the challenges of defining family and what being part of a family means.

ABC/Photofest

Do you recognize this TV family? Does your family share any similarities?

Family is a social institution that binds people together through blood, marriage, law, and/or social norms. It is important to point out that "there is no concrete group which can be universally identified as 'the family'" (Zelditch 1964, 681).

An amazing variety of family arrangements exists in the United States and worldwide—a variety reflected in the many norms that specify how two or more people enact family life. These norms govern who can marry, the number and kind of partners one can have, the connection to paternal and maternal relatives, and the living arrangements (see Table 11.1a). In light of this variability, we should not be surprised that when people think of family, they often emphasize different dimensions, such as kinship, ideals with regard to members, or legal ties.

Table 11.1a: Norms Governing Family Household Structure and Composition

	Number of Partners/Spouses
Monogamy	One partnership/union/spouse
Serial Monogamy	Two or more successive partnerships/unions/spouses
Polygamy	Multiple partners/unions/spouses at a time
Polygyny	One husband, multiple wives at one time
Polyandry	One wife, multiple husbands at one time
Single	Person not in a committed relationship, partnership, or union
	Choice of Spouse
Arranged	Parents select/approve marriage partners for children
Romantic	Person selects partner based on love
Endogamy	Partnership/union/marriage within one's social group
Exogamy	Partnership/union/marriage outside one's social group
	Authority
Patriarchal	Male-dominated
Matriarchal	Female-dominated
Egalitarian	Equal authority between partners
	Descent
Patrilineal	Traced through male lineage
Matrilineal	Traced through female lineage
Bilateral	Traced through both male and female lineage
	Living Arrangements
Nuclear	Partner/spouse plus children
Extended	Three or more generations living in one household
Single-parent	One parent/guardian living with children
Domestic Partnership	People (with or without children) committed to each other and sharing a domestic life but not joined in marriage or civil union
Civil Union	A legally recognized partnership (with or without children) providing same-sex or transgender couples with some of the rights, benefits, and responsibilities associated with marriage
Married Partners	People joined in marriage (with or without children) with rights, benefits, and responsibilities associated with marriage

KINSHIP. Definitions of family emphasizing kinship view family as comprising members linked together by blood, marriage, or adoption. The size of any given person's family network is incalculable, because one person has an astronomical number of living and deceased kin. For this reason every society finds ways to exclude some kin from their idea of family. Some societies, for example, trace family lineage through the maternal or the paternal side only. In addition, people make conscious or unconscious decisions about which kin they will remember or "forget" to tell offspring about (Waters 1990).

MEMBERSHIP. Definitions of family emphasizing membership often focus on ideals. One ideal defines marriage, for example, as a voluntary union of a man and a woman in a lifelong covenant welcoming of children. However, in reality, few living arrangements match that so-called ideal arrangement.

Courtesy of Billy Santos

< Only about 26 percent of all households in the United States consist of a male and a female with children. When we consider just those households with children, about two-thirds consist of a mother and father. Another 35 percent include a child living with one parent, a grandparent, or a guardian. These facts do not stop people from unconditionally labeling family arrangements that do not fit the ideal as nontraditional, dysfunctional, immoral, or at-risk (Cornell 1990).

LEGAL RECOGNITION. From a legal perspective, a family consists of people whose living and procreation choices are recognized under the law as constituting a family. In the United States, federal law defines marriage as "a legal union between one man and one woman as husband and wife" (Defense of Marriage Act 1996). But some state and local governments recognize same-sex marriage and other alternatives to marriage, such as domestic partnerships and civil unions. Legal recognition means that specified benefits, responsibilities, and rights are enforced by law. As a result, those who do not meet legal definitions face obstacles and hardships. As a case in point, the U.S. General Accounting Office (2004) identified 1,138 federal statutory provisions classified to the United States Legal Code "in which marital status is a factor in determining or receiving benefits, rights, and privileges."

Functionalist View of Family Life

Because of the debate over what constitutes family, some sociologists argue that it should be defined in terms of social functions, of which there are at least four: (1) regulating sexual behavior, (2) replacing the members of society who die, (3) socializing the young, and (4) providing care and emotional support.

REGULATING SEXUAL BEHAVIOR. Marriage and family systems include norms that regulate sexual behavior. These norms may prohibit sex outside of a marriage or specify norms for *who* should engage in sexual partnerships. Such norms can take the form of laws prohibiting marriage and sexual relationships between certain blood relatives (such as first cousins), age groups (such as an adult and a minor), and racial or ethnic groups. Other laws regulate the number of people who may form a marriage.

REPLACING THE MEMBERS OF SOCIETY WHO DIE. For humans to survive as a species, society must replace those who die. Marriage and family systems provide a socially and legally sanctioned environment into which new members can be born or adopted and nurtured.

SOCIALIZING THE YOUNG. The family is the most significant agent of socialization, because it gives society's youngest members their earliest exposure to relationships and the rules of life (see Chapter 3).

PROVIDING CARE AND EMOTIONAL SUPPORT. A family is expected to care for the emotional and physical needs of its members. Without meaningful social ties, people of all ages deteriorate physically and mentally. The life cycle is such that humans experience a time of extreme dependency in infancy and early childhood, which is likely to recur at the end of life.

Conflict View of Family Life

While everyone might agree that families, no matter what they look like, should fulfill these four functions, conflict theorists point out that family members do not always care for one another and that the family also perpetuates inequalities by passing on social advantages and disadvantages to members. Moreover, marriage and family systems are structured to value productive work and devalue reproductive work, and to foster and maintain divisions and boundaries.

SOCIAL INEQUALITY. Families transfer power, wealth, property, and privilege from one generation to the next. Obviously, income affects the level of investment parents can make in their children. In the United States, 28 percent of children live in a household in which no parent has secure employment; that is, no parent worked 35 or more hours per week for at least 50 weeks in the past year (see Figure 11.1a).

Figure 11.1a: Percentage of U.S. Children with Secure Parental Employment by Living Arrangement

Under which living arrangement are children most likely to have a parent securely employed?

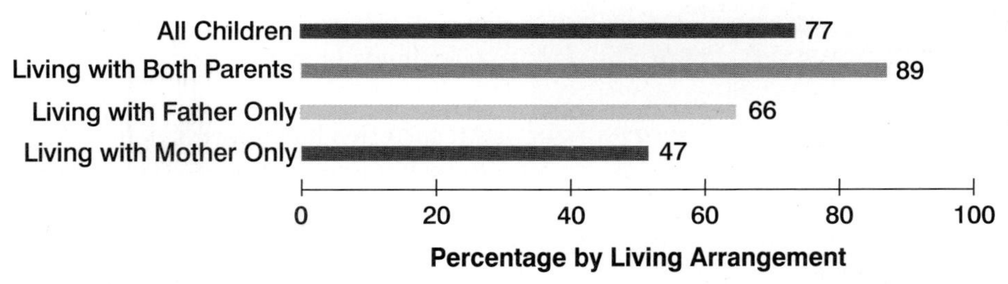

Source: ChildStats.gov (2012b)

PRODUCTIVE AND REPRODUCTIVE WORK. Friedrich Engels (1884) distinguishes between productive and reproductive work. Productive work involves the actual manufacture of food, clothing, and shelter and the tools necessary to produce them. Reproductive work involves bearing children, caregiving, managing households, and socializing children. Both are work: bearing children and caregiving activities are essential tasks if societies are to thrive. Yet, reproductive work is disproportionately performed by women, even when women are employed full time (Bianchi and Milkie 2010).

The gap between the time men and women spend on household tasks is affected by living arrangement (cohabitating couples tend to establish a more equal division of household labor) and income differences (the more equal the incomes, the less the imbalance). Using data from 17,636 respondents in 28 countries, sociologist Shannon Davis and colleagues (2007) examined the division of reproductive work within heterosexual-couple households. They found that on average men report spending about 9.4 hours per week performing household tasks; their female counterparts report an average of 21 hours.

Lisa Southwick

➤ The largest gap was found in Chile, where men reported contributing 24 percent of total hours devoted to household tasks to women's 76 percent. The smallest gap was found in Australia, where men reported contributing 39.1 percent of the total hours to women's 61.9 percent.

FOSTERING SOCIAL DIVISIONS AND BOUNDARIES. In the United States, we assume that people choose a partner or mate based on love. Upon investigating who marries, however, we find that people's choices are guided by other considerations as well: a potential partner's age, height, weight, income, education, race, sex, social class, and religion, among other things. When the conditions are right, we allow ourselves to fall in love.

Chris Caldeira

◀ The person with whom we fall in love may not be a person we can legally marry. For example, only those who meet certain legal criteria can get married at the Chapel of the Bells in Nevada or any other site that performs marriages in the United States.

All societies have norms defining who may date or marry whom. These norms may be formal (enforced by law) and/or informal (enforced by social pressure). **Exogamy** refers to the practice of choosing a partner from a social category other than one's own—for example, to choose a partner outside one's immediate family who is of the other sex or of another race. **Endogamy** refers to the practice of choosing a partner from the same social category as one's own—for example, a partner of the same race, sex, ethnicity, religion, or social class.

Tony Rotundo

◀ In the United States, there are about 58 million married couples living together who are male and female; ninety percent of all married couples in the United States are married to someone who is considered the same race. The U.S. Census Bureau counted 594,000 same-sex couples in a committed partnership, union, or marriage that live together (U.S. Census Bureau 2011i, k).

Symbolic Interactionist View of Family Life

When symbolic interactionists study family life, they seek to understand the experiences of being in a family, engaging in family-related activities, and conforming to or resisting societal ideals about what family life should be. Symbolic interactionists are interested in how established norms and ideals about family are enacted or resisted. This scenario posed in *Having It All* by Francine M. Deutsch (1999) shows how symbolic interactionists frame such an analysis. With this example, Deutsch illustrates the costs borne by men and women who decide to share in reproductive work: a husband, a wife, and their young son, Ollie, are traveling together by plane. The husband is holding Ollie, who is screaming. A male passenger sitting behind them asks, "What's wrong with the child's mother? Why doesn't the mother take care of the baby?" The husband answers, "Because I'm his father and I am perfectly capable of taking care of him." A man's decision not to hand over a crying child to a woman and a woman's decision not to grab the baby away represents a commitment by both to share in the reproductive work. The scenario Deutsch described illustrates the interaction dynamics by which even well-intentioned couples committed to equality might give in to societal pressures for women to handle the reproductive tasks (Gerson 1999).

(Write a Caption)

Chris Caldeira

Write a caption using the sociological language from this module to describe this family (a married couple with one child).

Hints: In writing this caption

- consider the type of household into which this family would be classified,
- think about whether this family meets the federal definition of marriage,
- describe how one child compares to the fertility rate of women in the United States, and
- think about whether the relationship between the couple suggests exogamy or endogamy.

Critical Thinking

Think about the household in which you grew up. How were household tasks (reproductive work) divided among those who lived in the household?

Key Terms

endogamy exogamy family

MODULE

11.2 Family Structure in Three Countries

Objective

You will learn about the ways in which the larger social context influences the structures of family.

U.S. Air Force photo by Senior Airman Ashley N. Avecilla/Released

Missy Gish

If you asked the Afghan girls and the American girls whether they would like to have children one day and if so, how many, how do you think each would answer?

If American girls answered "none" they would likely draw the response, "Why not?" If they answered "two," no one would raise an eyebrow. If they answered "eight," they would likely draw expressions of shock. But imagine this exchange taking place in Afghanistan. An answer of "none" would quite likely draw the reaction of shock. From such exchanges, children learn what their society considers an ideal number of children. They also learn that they may have to justify a future decision that departs from the ideal.

Family Systems

In this module we compare family life in the United States with that of Japan and Afghanistan. We chose Japan because it has one of the lowest fertility rates in the world; we chose Afghanistan because it has one of the highest. Both situations have prompted concern and efforts to increase low fertility and decrease high fertility. In making comparisons, we will consider three factors that shape

the experience of living in a family: (1) women's control over reproductive life, (2) the timing of death, and (3) caregiving responsibilities.

FEMALE CONTROL OVER REPRODUCTIVE LIFE. If we look at the statistics presented in Table 11.2a, we can gain some insights about the typical size of families in the three countries. We see that **total fertility**, the average number of children that a woman bears in her lifetime, is almost seven for women in Afghanistan. In the United States, total fertility is two children, and in Japan it is about one (see Table 11.2a).

▾ Table 11.2a: Fertility-Related Statistics

By age 30, 56 percent of Japanese women and almost 30 percent of American women remain childless. A woman who is childless at age 30 is a rare occurrence in Afghanistan. In which country are chances greatest that a baby will be born to a teenage mother? What percentage of women in each country report using some form of contraception?

	Japan	United States	Afghanistan
Chance of Teens Giving Birth per Year (ages 15–19)	1 in 200	1 in 20	1 in 9
Total Fertility (average number of children per woman)	1.21	2.05	5.64
Percentage of Married or In-Union Women Using Any Method of Contraception	52%	73%	10%
Percentage of All Women Who Give Birth to a Child Outside of Marriage	2%	41%	Virtually zero
Percentage of Childless Women at Age 30	56%	29.7%	1.8%

Sources: Population Reference Bureau (2009); Mason (2010); U.S. Central Intelligence Agency (2012); OECD (2011c); U.S. Census Bureau (2012b); DeParle and Tavernise (2012); Friedman (2009)

DEATH OF FAMILY MEMBERS. Death results in the loss of a family member. We can think of death in terms of the time in life at which it occurs. Specifically, sociologists ask, "What are the chances a child will survive the first year of life and beyond? What are the chances a woman will survive childbirth? And what are the chances of living beyond age 65?" The answers to these questions offer important clues about the composition of the family unit and the broad nature of family relationships.

Official U.S. Marine Corps photo by Cpl. Sarah M. Maynard

Journalist 1st Class Lynn Jenkins

U.S. Air Force photo by Master Sgt. Jim Varhegyi/Released

▴ For the most part, parents in Japan and the United States can expect their children to survive them. Look at Table 11.2b on page 386. How many babies in each country die before reaching age 5? In Afghanistan the answer is 198.6 die per 1,000 live births.

When parents can expect a baby to survive, the future of the child seems more secure and parents feel less need to have additional children. On the other hand, when children have a high mortality rate, and there is no government safety net in place to support people in old age, parents have incentive to have a large number of children to ensure at least one survives into adulthood.

Each year in the United States, 24 women die for every 100,000 births as a result of pregnancy complications. In Japan that number is 6, and in Afghanistan it is 1,400. The high fertility rate coupled with the uncertainty that women will survive childbirth, and the uncertainty about whether children will survive the first year of life and beyond, speaks to women's lack of control over their reproductive lives.

MC1 Monique Hilley

➤ Lack of access to clean water is responsible for many waterborne illnesses such as diarrhea, the number one cause of death among children under 5 years old in Afghanistan. The United Nations, the United States, and other agencies are working to find sources of clean water near where people live so that children, who are often charged with collecting water, do not have to walk far to get it. In fact, many children, especially girls, in Afghanistan and elsewhere do not go to school because they spend their days carrying water.

Life expectancy, or the average number of years after birth a person can expect to live, offers insights about the timing of death. Notice in Table 11.2b that 84 percent of people in Japan can expect to reach age 65 and to live, depending on their sex, on average another 17 or 22 years beyond that age. In the United States, 77 percent can expect to live to age 65 and then on average an additional 17 to 20 years. In Afghanistan, 56 percent of people live to age 65 and upon reaching that age can expect to live an average of 10 or 11 years longer.

▾ Table 11.2b: Selected Life Expectancy and Mortality Statistics

	Japan	United States	Afghanistan
Infant Mortality (per 1,000)	2.21	6.0	121.6
Under Age 5 Mortality (per 1,000)	3.3	7.8	198.6
Maternal Mortality (per 100,000 births)	6	24	1,400
Average Life Expectancy	83.9	78.5	49.7
Percentage of Population Expected to Reach Age 65	84.0%	77.4%	56.0%
Average Number of Years After Age 65 One Can Expect to Live	22.5 (females) 17.4 (males)	20.0 (females) 17.2 (males)	11.0 (females) 10.0 (males)

Sources: United Nations (2012); U.S. Central Intelligence Agency (2012)

How might high infant mortality and maternal mortality affect the size of families? Can you see how higher infant and maternal mortality can prompt women to have more children? Can you see how women who are confident that they will survive childbirth and that their children will also survive might decide to have fewer children? How about life expectancy—how does high life expectancy affect the structure of family with regard to the presence or absence of grandparents and great-grandparents?

CAREGIVING AND DEPENDENCY. The family plays an important role in meeting the emotional and physical needs of its members. People's emotional and/or physical needs tend to be more evident when they are very young and very old. Figure 11.2a offers broad insights about where family energies are likely to be pulled, depending on the age composition of the population.

Figure 11.2a: Age and Sex Composition of Japan and Afghanistan, 2012

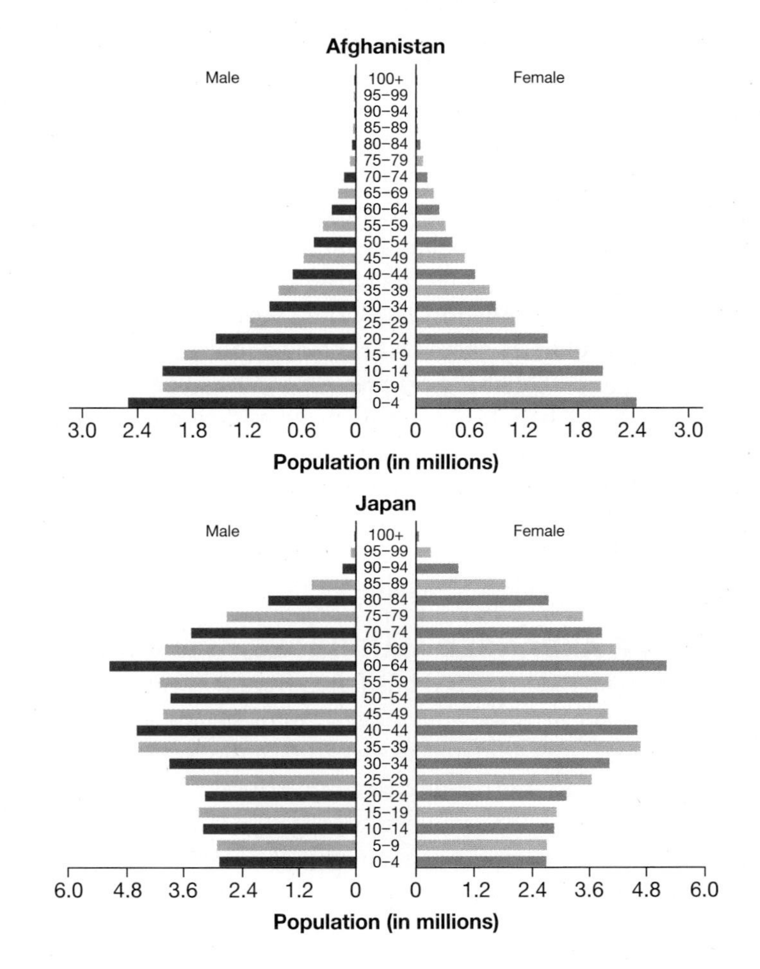

Source: U.S. Census Bureau (2012e)

The population pyramids are a snapshot of the age and sex composition of Afghanistan and Japan. The population pyramid shows that Japan has an **aging population**, one in which people tend to live longer and in which those in older age cohorts outnumber those in younger age cohorts. As a result people can expect to participate in family life for a greater number of years than ever before. Can you tell by looking at the pyramids which country has the oldest population? The youngest population? In Japan, which age group has the most men and most women? Which age group is the largest in Afghanistan? If the proportion of children under 19 years of age is considerably larger than that of adults age 65 and over, then more energy within families is directed toward caring for their young.

Japan's Family System

Japan's very low total fertility rate, combined with its long life expectancy and low immigration rate, means that the country has one of the world's oldest populations. The low total fertility rate is a major national concern, and it has prompted a variety of responses encouraging people to have children. Japan's fertility rate of 1.2 children per woman can be explained by a number of social forces including, but not limited to, the dismantling of the *ie* family structure, a dramatic decline in arranged marriages, and employment barriers.

DISMANTLING OF THE *IE* FAMILY STRUCTURE. Until Japan's defeat in World War II, everyone was officially part of a male-headed multigenerational household known as an *ie*. Relationships between family members were shaped by the Confucian values of filial piety, faith in the family, and respect for elders. Firstborn sons held privileged status, inheriting and controlling family wealth (primogeniture). Legally, women could not choose a spouse or own property. Daughters were viewed as temporary family members until marriage, at which point they moved in with their husband's family (Takahashi 1999). Brides served and obeyed the husband and his parents, including caring for them in old age. After World War II ended in 1945, the United States imposed sweeping changes on Japan's family systems granting equal rights to women, ending arranged marriage, and abolishing primogeniture. In spite of these changes to the family structure, the belief that problems should be managed within the family has persisted. This persistence is evidenced by the fact that the Japanese government provides only limited public services for the elderly; women are expected to serve as primary caregivers to the young and old. This caregiving responsibility is likely to have a depressing effect on fertility as women anticipate how to balance needs of children and aging parents.

THE DECLINE OF ARRANGED MARRIAGES. Since 1955 the proportion of all marriages considered arranged has fallen from 63 percent to 10 percent. When parents no longer arranged marriages, young people were left to their own devices. Japan has not fully adjusted to this shift, and it has yet to develop a couple's culture (Ogawa, Retherford, and Matsukura 2009). As one young Japanese woman commented, "In the U.S., you are supposed to be with your boyfriend or husband all the time. . . . In Japan, women have their ways of having fun and men have their ways. You're not expected to bring a date everywhere and you don't feel excluded if you're not involved with someone" (Orenstein 2001, 34).

In Japan, 70 percent of working females ages 20 to 39 live with their parents while contributing little to household expenses (Ogawa, Retherford, and Matsukura 2009). Sociologist Masahiro Yamada (2000) coined the phrase *parasite singles* to describe those who work but are free of the financial and emotional pressures associated with parenting and marriage. According to government surveys, among the most stressful pressures these singles wish to avoid is that of guiding children through Japan's highly competitive educational system.

Courtesy of Phillip Reakes

◂ To give their children an edge, parents enroll children in *juku* (after-school cram schools) and even hire tutors at a combined cost of $303 to $433 per month (Sato 2005). Keep in mind that scores on qualifying exams are the only factor that determines acceptance into the best high schools and colleges—GPA, letters of recommendation, athletic ability, and extracurricular activities are not taken into consideration (8Asians.com 2009). The photograph of these students at *juku* was taken at 10:00 p.m. An estimated 40 percent of all elementary school students and 75 percent of middle and high school students attend one of the 50,000 *juku* (Sato 2005).

The *juku* pressure is particularly stressful for mothers, who almost single-handedly guide children through the process—many working fathers are largely absent from the home. The workplace emphasizes work over family, as employees work long hours, take little time off for personal reasons, and accept sudden job transfers (Newport 2000).

ENTRENCHED BARRIERS TO EMPLOYMENT. Although Japan's labor laws forbid discrimination against women, it exists at every level, including recruitment, hiring, training, compensation, and promotion. For the most part, Japanese women are expected to quit working when they marry or have children. When mothers do work, they tend to accept low-paying and insecure employment in exchange for the flexibility needed to meet household and caregiving responsibilities. In addition, the tax system offers married women incentives to "choose" part-time employment (*Japan Times* 2003; Lam 2009). In sum, women must choose between work and family life. As a result, many are not willing to give up the security of employment to have children.

Afghanistan's Family System

In Afghanistan the family is *the* source of economic and social support, because there are very few public services. The typical Afghan family is extended, with members sharing a household or a valley. Extended family households span several generations, including "the male head of family and his wife, his brothers, several sons and their families, cousins within their families, as well as all unmarried and widowed females" (Blood 2001). The Afghan family is male-dominated; authority rests with male elders, and inheritance is passed through the male line. Women are expected to want children, especially sons, as infertility is "a frightening social stigma" for a woman and her family (Blood 2001).

In Afghanistan males, considered superior to females, assume control over women's lives, usually deciding whether they can pursue an education or career.

The men also approve the spouse and negotiate the price of the bride for their sons. Independent females are viewed as a threat, and families tolerating such independence are ostracized. The ideal mate for a son is a paternal first cousin. The typical Afghan bride is in her teens and is matched with a man in his mid-twenties. When price is paramount, very young girls are matched with much older men who can afford to pay the price (Blood 2001).

The average female in Afghanistan can expect to attend 4.7 years of school; the average male can expect to attend 7.6 years of school. Although there are a few exceptions, the sexes attend segregated schools for the most part (Brown 2010). Under the Taliban, who enforced a conservative interpretation of Islamic law, girls could not go to school; women could not hold public jobs and could leave their homes only under rare circumstances, and only if wearing full-body burqas. The Taliban mandated that men grow beards and banned Western-style clothing. Crimes such as theft, adultery, and drinking were subject to severe punishments, including lashings, stoning, and amputation (Shoup 2006). After 2001, the year of the U.S. military intervention, a new constitution was put into place that guaranteed equal rights for women but allowed ethnic minority groups to establish their own family law in keeping with their religious traditions. These family laws tend to place women in subordinate positions (Boone 2009).

(Write a Caption)

Write a caption that identifies larger social forces that are likely to influence these Afghan girls' reproductive experiences in the future.

Hint: In writing this caption

- think about current reproductive experiences of Afghan women and project the reproductive experiences these girls are likely to have if the situation does not change, and
- consider the larger society's view about the place of women with regard to wanting children.

Staff Sgt. Joshua T. Jasper

Critical Thinking

Use the concepts in this module to describe the family structure into which you were born and/or raised.

Key Terms

aging population   life expectancy   total fertility

Household Structures in the United States

MODULE

11.3

Objective

You will learn that household structures in the United States are not static.

Sadie Bleistein
Lisa Southwick
Courtesy of Serina Beauparlant
Courtesy of Noriko Ikarashi

Do any of these photographs represent the kind of household in which you grew up? Did your household stand out for being larger, blended, or otherwise different from what was considered the mainstream?

According to the U.S. Census Bureau (2012d), a household includes all the persons who occupy a housing unit. That unit may be a house, an apartment, a rented room, or a mobile home. The occupants may be a single family, one person living alone, two or more families living together, or any other combination of related or unrelated persons who share a housing unit. To capture the diversity of households, sociologists emphasize structural features, elements such as size (one person or more), family/nonfamily, number of children, married couple, single parent, and so on (see Table 11.3a).

Changing Household Structures in the United States

Table 11.3a shows that the distribution of different household structures in the United States is not static; over the span of 100-plus years, it has changed quite dramatically, especially with regard to single-person and single-parent households.

Table 11.3a: U.S. Households: 100-Plus Years of Change

The table is a profile of household structures for 1900, 1950, and 2011–2012. Which household characteristic has seen the greatest decline? Which household characteristic has seen the greatest increase?

	1900	1950	2011–2012
Household Characteristics			
Average Household Size (persons)	4.8	3.4	2.6
Percentage of Households with Seven or More People	20.4%	4.9%	1.2%
Percentage of Households with One Person Living Alone	5.0%	9.3%	28.0%
Same-Sex Couple Households	—	—	1.0%
Living Arrangements of *Children* by Family Status (percent of 35 million households with children)			
Two-Parent—Farm Family	41%	17%	2%
Two-Parent—Father Breadwinner, Mother Homemaker	43%	56%	23%
Two-Parent—Dual-Earner	2%	13%	40.7%
Single-Parent	9%	8%	26.5%
Children Not Living with a Parent	5%	6%	4.1%
Child Living with Step-Parent, Grandparent, or Adoptive Parent	—	—	3.7%
Same-Sex Couple Households	—	—	0.003%*

Sources: U.S. Census Bureau (2011a, 2011i, 2002); ChildStats.gov 2012a)

* For the years 1900 and 1950, the phrase "two-parent" encompassed only "opposite" sex parents because the Census Bureau likely only counted them as parents if a male and female were involved. While there were no doubt same-sex couple households in those times, they were not counted as "parents." Most sociologists object to the term "opposite sex" because it suggests that a male and a female share no common characteristics or attributes.

Sociologist Kingsley Davis (1984) points to a particularly dramatic change that affected household structure: The percentage of women age 16 and over in the *paid* labor force rose from less than 20 percent in 1900 to 60 percent in 1980. Today the percentage is 54 percent. The percentage of married women who were in the labor force rose from 15.4 percent in 1900 to 53.1 percent in 1980. Today that percentage is about 58 percent (U.S. Bureau of Labor Statistics 2011f). Davis links women's entry (and especially married women's entry) into the paid labor market to a series of interdependent social forces connected to the rise and fall of the breadwinner system.

The Rise and Fall of the Breadwinner System

Before industrialization—that is, for most of human history—the workplace was made up of the home and the surrounding land, and both men and women worked together to produce for the household. Industrialization destroyed the household economy by moving production from the home to the factory. In the process, women's contribution to production of goods declined. Davis calls this new economic arrangement the breadwinner system.

Prints & Photographs Division, Library of Congress, LC-USZC2-1242

➤ This 1887 drawing depicts an image we have come to associate with the breadwinner system. As the breadwinner, men worked not in the home or on surrounding land but with non-kin in factories, shops, and offices. The man's economic role now made him the link between the family and the wider market economy. At the same time, his personal participation in the household diminished. His wife stayed home and performed the parental and domestic duties that women had always performed. Under the breadwinner system, the woman bore and reared children, cooked meals, washed clothes, and cared for her husband's personal needs, but to an unprecedented degree her economic role changed because she could not produce what the family consumed—production had been removed from the home (K. Davis 1984).

Historically, this system was not typical. Rather, it was limited to the middle and upper predominantly white classes, and its heyday was 1860 to 1920. The breadwinner system emerged at about the same time as the average number of living children in family households reached its peak and infant and childhood mortality declined. The breadwinner system came about because many women had too many children to engage in work outside the home.

The breadwinner system did not last long because it placed too much strain on husbands and wives. The strains stemmed from several sources. Never before had

- the roles of husband and wife been so distinct,
- women not produced what their families consumed,
- men spent most of their waking hours separated from their families, and
- men assumed sole responsibility for supporting the entire family.

Davis regards these events as structural weaknesses in the breadwinner system. Given these weaknesses, the system needed strong normative controls to survive: "The husband's obligation to support his family, even after his death, had to be enforced by law and public opinion." Sexual relations considered "illegitimate," specifically when babies were born outside of marriage, "had to be condemned; divorce had to be punished, and marriage had to be encouraged by making the lot of the 'spinster' a pitiful one" (406). Davis maintains that the strains were too great and these strict normative controls eventually collapsed. Other factors contributed to the breadwinner system's collapse, including decreases in total fertility rates, increases in life expectancy, increases in divorce rates, and increases in employment opportunities for women.

DECLINES IN TOTAL FERTILITY. The decline in total fertility (number of children born over a woman's lifetime) actually began before married women entered the labor force in large numbers. Davis attributes the decline to the forces of industrialization, which changed children from economic assets to economic liabilities. Not only did women have fewer children, but the number of years separating first- and lastborn had also decreased so that women had their last child at a younger age. These changes in number and spacing of children gave women time to work outside the home, especially after their children entered school.

INCREASED LIFE EXPECTANCY. As life expectancy increases and women devote fewer years to bearing and raising children, the time devoted to child care has come to occupy a smaller proportion of her life. Because women's life expectancy exceeds that of men's and because brides tended to be younger on average than their partners, the average married woman could expect to outlive her husband by 8 to 10 years. Although few women would name increased life expectancy as a reason for working in the paid labor force, the possibility of living many years longer than her husband changed the way women looked at and planned out their lives.

INCREASED DIVORCE RATE. Davis traces the rise in the divorce rate to the breadwinner system, and specifically to the shift of economic production outside the home:

> With this shift, husband and wife, parents and children, were no longer bound together in a close face-to-face division of labor in a common enterprise. They were bound, rather, by . . . the husband's ability to draw income from the wider economy and the wife's willingness to make a home and raise children. The husband's work not only took him out of the home but also frequently put him into contact with other people, including young unmarried women who were strangers to his family. Extramarital relationships inevitably flourished. Freed from rural and small-town social controls, many husbands either sought divorce or, by their behavior, caused their wives to do so. (410–411)

Davis argues that an increase in the divorce rate preceded married women's entry into the labor market by several decades. However, once the divorce rate reached a certain threshold (a 20 percent or greater chance of divorce), more married women seriously considered seeking employment to protect themselves in case their marriages failed. When both husband and wife participate in the labor force, the chances of divorce increase even more. Working outside the home puts both in contact with others, increasing the possibility of extramarital romantic attachments.

INCREASED EMPLOYMENT OPPORTUNITIES FOR WOMEN. While women were motivated to seek work because of changes in the childbearing experience, increases in life expectancy, and the rising divorce rate, women could act on these motivations only when their opportunities for paid employment increased. As industrialization matured, it brought a corresponding increase in the kinds of jobs perceived as suitable for women. Paid employment increased their economic self-sufficiency and transformed gender roles. There is "nothing like a checking account to decrease someone's willingness to be pushed into marriage or stay in a bad one" (Kipnis 2004).

NBC-TV/The Kobal Collection

◀ The groundbreaking *Cosby Show* began airing in 1984, the same year Davis published his article describing the rise and fall of the breadwinner system. The show emphasized the dual-income arrangement and offered a model of what husbands and wives could expect of each other when the woman held a position (lawyer) essentially equal in stature and income to her husband's (physician).

Davis does not argue that the dual-career heterosexual model is a problem-free arrangement. First, it lacked "normative guidelines" such that each "couple has to work out its own arrangement" (1984, 413). Second, even in the two-income system, women tend to remain primarily responsible for domestic matters. One reason women bear this responsibility is because men and women are unequal in the labor force. [At the time Davis wrote his study, females working full time earned 66 cents for every dollar men earned. Today, they earn about 81 cents for every dollar men earn (U.S. Department of Labor 2012).] Davis believes that as long as women make an unequal contribution to the overall household income, they will do more work around the house. The problems of the post-breadwinner system are evident when we consider that a large proportion of married women with children (almost one-third) "choose" to not work outside the home (U.S. Bureau of Labor Statistics 2011f). According to Davis, that choice reflects the psychosocial costs of employment that married women face—costs that include the stress of juggling family and career, finding reliable day care, and anxiety over making those choices.

(Write a Caption)

Write a caption that highlights this woman's role in the context of the breadwinner system.

Hints: In writing this caption

- review the material on the breadwinner system,
- consider the role of men and women under this system, and
- explain how shopping represents this wife's ability (or lack of ability) to produce what the family consumes.

Prints & Photographs Division, Library of Congress, LC-USW3-003556-E

Critical Thinking

To what extent does the image of the breadwinner system and its downfall (as described by Davis) apply to the family in which you grew up?

MODULE

11.4 Economy and Family

Objective

You will learn how the economy shapes relationships among family members, but especially between men and women.

Why might someone who suffers physical and/or mental abuse at the hands of a partner not take action to change the situation?

Ingrid Barrentine

Three answers are when the abuser is physically stronger, when the person being abused does not have the financial resources to leave, and when the abused lack access to police or other sanctioned agents of violence control. With this kind of question in mind, sociologist Randall Collins (1971) proposed a theory of sexual or gender stratification, the system societies use to rank males and females on a scale of social worth such that the ranking affects life chances in unequal ways. Collins's theory is based on three assumptions. First, people tend to use their economic, political, physical, and other advantages to dominate others. Second, any change in the way resources are allocated to men and women alters the structure of domination. Third, ideology is used to support and justify one group's domination of another.

Mass Comm. Spc. 2nd Class Jason Johnston

◂ Under what conditions might women be *required* by law to cover their bodies from head to toe? Be denied access to education? Be forbidden to leave the house unless accompanied by an adult male relative? According to Collins (1971), when males hold political and economic power and when the society defines males as innately superior to females. How would things change if women were given access to education and to opportunities for earning an income independent of males?

Collins points out that in general, males tend to be physically stronger than females; thus, the *potential* for coercion by males exists in most encounters with females. He maintains that women have historically been viewed and treated as men's sexual property and that this ideology lies at the heart of gender stratification. The extent to which women are viewed as sexual property and subordinate to men depends then on whether women can call the police in the event of violence and whether their income is equal to that of men's. Collins identified four historical economic arrangements that shape relationships between men and women: (1) low-technology tribal societies, (2) fortified households, (3) private households, and (4) advanced market economies.

LOW-TECHNOLOGY TRIBAL SOCIETIES. Low-technology tribal societies include hunting and gathering societies without technologies that permit the creation of surplus wealth—that is, wealth beyond what is needed to meet basic human needs, such as food and shelter. In such societies, sex-based division of labor is minimal, because the emphasis is on collective welfare and the belief that all members must contribute to ensure the group's survival. Because almost no surplus wealth exists, marriage between men and women from different families does little to increase a family's wealth or political power. Consequently, daughters are not treated as property, in the sense that they are not used as bargaining chips to achieve such aims. Relatively speaking, women in low-technology tribal societies have greater social standing and value than women in other arrangements described in this section (Cote 1997).

FORTIFIED HOUSEHOLDS. Fortified households are preindustrial arrangements characterized by a lack of police force or other agency dedicated to social control. Rather, the household functions as an armed unit, and the head of the household acts as a military commander. All fortified households share one characteristic: the presence of a nonhouseholder class, consisting of propertyless laborers and servants. In the fortified household, "the honored male is he who is dominant over others, who protects and controls his own property, and who can conquer others' property" (Collins 1971, 12). Men treat women as sexual property in every sense: daughters are often bargaining chips for establishing economic and political alliances with other households, and male heads of household take sexual liberties with female servants. In this system, women's power depends on their relationship to the dominant men.

◀ Slaveholding households in the United States qualified as fortified households in that they included propertyless servants and enslaved females treated as sexual property. This engraving speaks to the sexual relationship between master and slaves, in which the product of their union was slaves who appeared often as white but assumed the legal status of the mother—the master's property.

HarpWeek

PRIVATE HOUSEHOLDS. Private households emerge with the establishment of a market economy; a centralized, bureaucratic state; and agencies of social control that alleviate the need for citizens to take the law into their own hands. Under the private household arrangement, men monopolize the most desirable and important economic and political positions. Men are still heads of household in that they control the property and assume the role of breadwinner; women remain responsible for housekeeping and childrearing.

The decline in the number of fortified households, the separation of the workplace from the home, smaller family size, and the existence of a police force to which women could appeal in cases of domestic violence gave rise to the notion of romantic love as an important ingredient in heterosexual marriages.

Prints & Photographs Division, Library of Congress, LC-USZ62-61085

➤ This 1916 drawing suggests one image of the private household: men offer women economic security because they dominate the important, high-paying positions. Women offer men companionship and emotional support and they strive to be attractive—that is, to achieve ideal femininity, which might include possessing an 18-inch waist or wearing high-heeled shoes. At the same time, before marriage women try to act as sexually inaccessible as possible, because theoretically they offer sexual access to men in exchange for economic security. Given the dominance of the private household, can you see how this economic arrangement might discourage same-sex partnerships? The question has relevance for understanding the forces that make same-sex partnerships acceptable in the larger society.

ADVANCED MARKET ECONOMIES. Advanced market economies offer widespread employment opportunities for women. Although women remain far from being men's economic equals, some can enter into relationships with men by offering more than an attractive appearance; they can provide an income and other personal achievements. Because they have more to offer, these women can make demands on men to be sensitive and physically attractive, to meet the standards of masculinity, and to help with reproductive work. In the United States today, about 28 percent of all full-time employed married women living in dual-income households in which their male partner is also employed full time earn more money than the man (U.S. Bureau of Labor Statistics 2011h). This situation may explain in part why increasing commercial attention has been given to male appearances and sexual performance (e.g., the erectile dysfunction drugs Viagra and Cialis).

One might argue that advanced market economies, which gave women greater opportunities in the labor market, reduced their economic and protective dependence on men and has thus reduced the incentives for women to form committed partnerships only with men. Likewise, since men could no longer count on women assuming homemaker and caretaker roles, incentives to marry for these reasons declined. It should not come as a surprise that same-sex relationships would become more common, and even accepted.

Lesbian and Gay Marriages and Partnerships

Gay partnerships, unions, and marriages (hereafter referred to as gay partnerships), like heterosexual relationships, are complex relationships that cannot be analyzed in simple ways. To what extent do same-sex or other unconventional partnerships mirror the inequalities that characterize heterosexual partnerships in general? As you might imagine, there is limited research on so-called unconventional partnerships, if only because it is difficult to identify those who are in such relationships. Most, if not all, of the existing research on unconventional partnerships focuses on gay and lesbian relationships. Those unfamiliar with same-sex commitments often frame them in terms of heterosexual ideals—ideals that assume one person enacts a traditional masculine role and the other enacts the traditional feminine role. The research, however, indicates that most childless lesbian and gay couples tend to reject the traditional heterosexual ideals associated with masculinity and femininity as a template for their relationships (Kurdek 2005). This generalization changes when same-sex couples become parents, at which point the relationship becomes unequal with regard to child care, household tasks, and employment (Biblarz and Savci 2010).

(Write a Caption)

Write a caption that identifies information you would need to know about this couple to predict the equality or lack of equality in their relationship.

Hints: In writing this caption

- review the four historical stages of sexual stratification, and
- consider which of the four stages would support an emphasis on men's sexual performance.

Missy Gish

Critical Thinking

Identify a couple in a partnership. In what ways do physical differences, occupation (paid or unpaid), and income shape their relationship?

MODULE

(11.5) Intergenerational Family Relationships

Objective

You will learn about the social forces that affect intergenerational family relationships.

Do you have a grandparent in your life? How about a great-grandparent or even a great-great-grandparent who is still alive?

Kyle Cowgill

The 4-year-old was born into a family that includes her mother (24 years old), grandmother (41 years old), great-grandmother (63 years old), and great-great-grandmother (82 years old). What does it mean that the 63-year-old grandmother still has her 82-year-old mother in her life and that a 4-year-old has three generations of grandparents?

While great-grandparents and even great-great-grandparents have always been present in some people's lives, the likelihood that most of us will know and remember these older relatives has increased substantially over the past few decades. Moreover, the likelihood that parents will be in children's lives for 70 or 80 years has increased substantially as well.

In the context of family, we define a **generation** as people in an age category passing through time who are distinguished from those in other age categories by cultural disposition (dress, language, preferences for songs, activities, entertainment), posture (walk, dance, the way the body is held), access to resources, and socially expected privileges, responsibilities, and duties (Eyerman and Turner 1998). Each generation is distinguished by titles such as baby, teenager, parent, and great-grandparent. As people age they pass into and out of generational categories. In this module we consider how three social forces have broadly affected intergenerational relationships among family members. Those forces are increased life expectancy, declines in parental authority, and change in economic status of children.

Dramatic Increases in Life Expectancy

Since 1900 the average life expectancy at birth has increased by 28 years in the world's richest economies and by 20 years (or more) in developing economies. Sociologist Holger Stub (1982) describes at least four ways in which gains in life expectancy have altered the composition of the family in the last century. First, the chance that children will lose one or both parents before they reach 16 years of age has decreased sharply. In 1900 the chance of such an occurrence was 24 percent; today it is less than 1 percent. At the same time, parents can expect their children to survive infancy and early childhood. In 1900, 250 of every 1,000 children born in the United States died before reaching age 1; 33 percent did not live to age 18. Today, 6 of every 1,000 children born die before they reach age 1; less than 5 percent die before reaching age 18. Not only can parents and children be much more secure that the other will survive, but the length of time parents, siblings, and other relatives share each others' lives has increased.

Second, the number of people surviving to old age has increased. In countries such as Japan, Italy, Germany, and the United States, where the total fertility rate is low and declining, the proportion of older people in the population is increasing.

Prints & Photographs Division, Library of Congress, LC-USZ62-104928

➤ This 1896 photographic print shows six generations of women. This portrait indicates that multigenerational and even six-generation families existed in the past. From a societal point of view, however, the number of such families was insignificant. But what happens when four-, five-, and six-generation families become commonplace? It will take some time for societies' institutions to adjust to that reality, because there is virtually no model in place for how members of families with four or more generations are expected to interact with each other.

Third, the potential length of the average marriage and other intimate partnerships has increased. Given the mortality patterns in 1900, newly married couples could expect their marriage to last 23 years before one partner died (assuming they did not divorce). Today, if they do not divorce, newly married couples can expect to be married for 53 years before one partner dies. This structural change may be one factor underlying the currently high divorce rates. When people could expect to live only a few more years in an unsatisfying relationship after retirement or children left home, they would usually resign themselves to their fate. But today, the thought of living 20, 30, or even 50 more years in an unsatisfying relationship can provoke decisive action at any age (Dychtwald and Flower 1989). Divorce dissolves today's marriages at the same rate that death did 100 years ago (Stub 1982).

Fourth, people now have more time to choose and get to know a partner, settle on an occupation, attend school, and decide whether they want children. Now, these areas of experience are occurring later in life than in past generations. Moreover, an initial decision made at any one of these stages is not final. The amount of additional living time enables individuals to change life course. In fact, one might argue that the so-called midlife crisis derives from the belief "that there yet may be time to make changes" (Stub 1982, 12).

Decline in Parental Authority

Around the beginning of the 20th century, children learned from their parents and other relatives the skills needed to make a living. As the pace of industrialization increased, jobs moved away from the home and into factories and office buildings. Parents no longer trained their children, because the skills they knew and possessed were becoming obsolete. Children came to expect that they would not make their living in the same way their parents did. In short, as the economic focus shifted from agriculture to manufacturing, the family became less involved in children's lives. Ultimately, the transfer of work away from the home and neighborhood removed opportunities for parents and children to work together.

Parental authority over adult children lost its economic support. Gradually, values and norms developed that supported privacy and intergenerational independence; for example, the ideas that elders should not interfere in the lives of adult children and that parents and adult children should reside in separate households. Popular opinion pushed governments to establish social security and health insurance programs for the retired, low-income, and disabled. Those policies further reinforced changes in family structure and values such that schooling, work, social, and leisure activities became largely age-segregated experiences, to the point that if intergenerational activities were to take place, they needed to be planned in advance.

The responsibilities once assigned to the family are now performed by organizations serving the public in general, such as day care, nursery schools, and preschools. In 1970 about 20 percent of 3- and 4-year-olds in the United States were enrolled in preschool programs. Today about half attend a full-time preschool program. The rise in such programs is connected in part to the increased number of women entering the workforce but also to the belief that early education programs are essential to educational and social success (Social Issues Reference 2009).

Andy Knell

Sr. Airman Bethann Hunt, USAF

▲ Contrast the experience of these American children (left) with that of these Afghan children from the Pashtun tribe sitting together at home in Kabul. While kindergartens exist in Afghanistan, only about 2,000 children attend them in a country where 7 million children are age 5 and under (Education Encyclopedia 2009). Can you see how parental authority is diminished when outside agencies assume responsibility for some of children's care?

The Economic Status of Children

Technological advances associated with the Industrial Revolution and the shift from an agriculture-based economy to a manufacturing-based one also changed children from economic assets to liabilities. Mechanization decreased the amount of physical effort and time needed to produce food and other commodities. Consequently, children lost the opportunity to contribute to household income and as a result lost their economic value. In agricultural and other extractive economies, children represent an important source of free unskilled labor for the family. This fact may partly explain why the highest total fertility rates in the world occur in places like Afghanistan, Somalia, the Democratic Republic of the Congo, and Niger—places where labor-intensive agriculture and extractive industries are crucial to the economy.

Demographer S. Ryan Johansson (1987) argues that in most industrialized economies, couples who choose to have children bring them into the world to provide intangible, emotional services—services such as love, companionship, an outlet for nurturing feelings, enhancement of dimensions of adult identity—rather than economic services. These desires seem to hold for all income groups and age groups (teens versus adults who delay childbearing until later in life).

Prints & Photographs Division, Library of Congress, LC-DIG-fsa-8a10151

Cpl. Im Jin-min (USAG-Yongsan)

➤ Study the two photographs of children who are about the same age. When you think of the role of children in the United States today, do you think of children performing labor or do you think of children as having talents that need to be developed? The 1938 photo of a child working in a cranberry bog depicts labor that contributes to his family's income. The child in gymnastics class is participating in an activity for which the family must pay.

The shift away from labor-intensive production stripped children of opportunities to make an economic contribution to the family and made children expensive to raise. Today in the United States, depending on income, the average yearly expenses for childrearing can range from $8,760 to $19,820 (Lino 2011). These estimates cover only the basics; when we include extras, such as summer camps, private schools, sports, and music lessons, the costs go even higher. Of course, for many parents, the cost of raising children does not stop when they turn 18. The high cost of raising children may be one reason that, in one recent survey of 2,200 adults, children were among the least cited contributors to a successful marriage. Faithfulness, a good sexual relationship, household chore-sharing, income, good housing, shared religious beliefs, and similar tastes and interests were cited more often than children (PEW Research Center 2007).

The Economy

The economy, which encompasses the labor force and existing opportunities for paid employment, certainly shapes intergenerational relationships (e.g., parent–child, child–grandparent). As one example, as the number of married and unpartnered women with children entering the workforce increased, grandparents were called on to care for children while the mother worked. The U.S. Census Bureau (2012g) shows that grandparents are primary caregivers for 23 percent of preschool-aged children. We can expect that grandparents' role in children's life increases beyond the "day care" hours to spill over into other areas of the child's life.

Another intergenerational trend that can be connected to the economy involves the increasing proportion of young adults aged 25 to 34 living with a parent (or even a grandparent). According to the U.S. Census Bureau (2011d), 19 percent of men and 10 percent of women in that age category live in the home of their parents (up from 14 and 8 percent, respectively, in 2005). Explanations include the large debt incurred from taking out college loans, a shortage of jobs that pay sufficient wages, and life expectancies that delay entry into committed partnerships, parenthood, and careers.

(Write a Caption)

These Tanzanian children are celebrating a U.S. military–sponsored project, the opening of a windmill that will pump water to seven wells in the region. Write a caption that projects how this change will affect their role in the family.

Staff Sgt. Joseph L. Swafford Jr.

Hints: In writing this caption

- note that children in Tanzania spend much of their day carrying water from distant water sources that can be hours away,
- consider what it means to have wells closer to their homes, and
- think about what children will do now that they can spend less time carrying water.

Critical Thinking

Describe a social force that affects an intergenerational relationship in your life.

Key Term

generation

Who Needs and Gives Care?

MODULE 11.6

Objective

You will learn how sociologists think about those who give and receive care.

Is there someone in your life for whom you give, have given, or expect to give care? Have you ever needed care?

Missy Gish

Obviously we needed care when we were very young and, unless we die suddenly, we will very likely need care at the end of life. In this module we explore caregiving apart from that needed to raise children.

Caregiving

We use the term **caregivers** to mean those who provide service to people who, because of physical or mental impairment, cannot do certain activities without help (Day 2009). The caregiving that family members, neighbors, and friends provide in a home setting is **informal care**. Caregiving provided, usually for a fee, by credentialed professionals (whether in the person's home or some other facility) is **formal care**. In this module we focus on the informal dimension of care, because most caregiving in the United States is informal (Day 2009). An estimated 28.8 million informal caregivers in the United States devote 30.9 billion hours of caring each year to assist a family member or friend. On average, each caregiver devotes 25.1 hours per week (Day 2009).

An estimated 36 million people age 5 and over have a physical or mental impairment. It is difficult to know how many of these 36 million people are receiving informal care in the United States. The best estimates we have come from the U.S. Census Bureau (2011h), which counts those who have serious difficulties (1) hearing and/or seeing (even when wearing glasses); (2) concentrating, remembering, and/or walking or climbing stairs; and (3) dressing, bathing, and doing errands alone, such as shopping and going to a doctor.

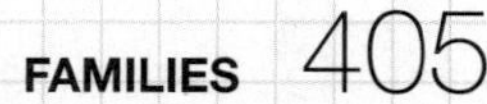

Ms. Jessica Obermeyer (3rd ID)

Five percent of noninstitutionalized persons ages 5 to 15 have a physical, mental, or sensory impairment. Generally, when one thinks of populations labeled as disabled, those age 65 and over come to mind; 14.5 million in that age group are considered disabled in the United States. But there are a significant number of children ages 5 to 15 classified as disabled. The total number counted as disabled is 2.8 million (U.S. Census Bureau 2011h).

Caregiving, especially when it is long-term or ongoing, creates strain and tension for both the provider and receiver of care. The relationship between provider and receiver is subject to any number of strains and stresses related to

- the amount of time the caregiver is able to devote and the time those receiving care need or require;
- balancing the demands of caregiving with the caregiver's other commitments, including job and family;
- the physical and emotional demands of caring and being cared for; and
- dignity and privacy (changing diapers, using the toilet, bathing) (Day 2009).

People take on caregiving responsibilities for any number of reasons, including

- the desire to pay back recipients for the sacrifices they made;
- the emotional bond between the caregiver and the recipient;
- a desire to live a life free of regret or guilt;
- a feeling of accomplishment, especially when the recipient expresses appreciation and satisfaction; and
- personal growth (Yamamoto and Wallhagen 1997).

To gain insights about caregiving, I routinely ask my students to describe anyone in their family who needs care because of a physical or mental impairment. The last time I asked my students to write on this topic, 85 students wrote about a total of 109 family members. Four examples follow:

> My uncle has a learning disability. He can barely read and write. His speech is slurred and it is very difficult to understand him. His disability does not affect him very much though; he gets around and goes to work everyday just like anyone else. Even with his learning disability he is great at working on cars and fixing miscellaneous things around the house. His disability isn't physical but whenever he has to read or sign something, my mom will have to read it and tell him what it means.

> My mother has severe arthritis. Her bones are deteriorating in her wrist and her ankles. My mother used to be a hairdresser but four years ago she was no longer able to move her hands and stand on her feet. My mom has had two wrist

> replacements and one ankle operation. She has been on disability for four years now and it's time to reapply. My mom takes many pills every morning and every night, about 20 each time. She often uses her wheelchair. She's able to take care of herself but if she pushes herself too much, she needs help and becomes stiff.

> My younger brother is 19 and was born with several problems. He is mentally retarded and has scoliosis and a club foot. He could not walk until he was 4; he cannot talk other than making simple sounds and he has had one major surgery to correct his scoliosis (he had a metal rod put into his back) and will have another next year after he graduates high school. . . . He uses a handheld computer to communicate, and he just got his first job at Kroger. Growing up with him I have come to know many disabled people. . . . It's unbelievable how many disabilities exist and the seriousness of those disabilities.

> The person in my family is me. Because I have epilepsy I cannot drive. I cannot consume certain foods that cause migraines, which trigger seizures. When I get a migraine, I have so much pressure on my head my sight blurs and it is hard to see or want to move. . . . It's scary because you don't know when you are going to have one and it's embarrassing when you do. I get sick easily. I would be stuck if it weren't for my friends and family who drive me places.

Impairment and Disability

Sociologists distinguish between impairment and disability. An **impairment** is a physical or mental condition that interferes with someone's ability to perform a major life activity that the average person can perform without assistance or without making changes to the physical environment around him or her (e.g., most people confined to a wheelchair can cook, so they do not have an impairment with regard to cooking; they do have an impairment when it comes to walking).

From a sociological point of view, a **disability** is something society has imposed on those with certain impairments because of how inventions and social activities have been organized to exclude them but to accommodate others.

➤ In this vein, one might argue that humans invented bicycles to assist those with legs in overcoming barriers to how fast they could travel by foot. Clearly, people without legs cannot pedal a traditionally constructed bike, but they can pedal with their hands and arms. This suggests that working assumptions guide the design of technologies: a design changes dramatically when the guiding assumptions change.

Cpl. Brian A. Tuthill

The point is that bicycles were designed to remove barriers for only those who can pedal with their legs. Likewise, we may treat people confined to wheelchairs as impaired with regard to cooking, but if stoves were designed to accommodate those who must sit while they cook, the so-called impairment would disappear.

Disability is imposed when the emphasis is placed on the loss of some mental or physical capacity and no consideration is given to ways of reducing barriers to full participation (L. Barton 1991).

Those with impairments and disabilities often experience what many have called the **tyranny of the normal**—a point of view that measures differences against what is thought to be normal and that assumes those with impairments fall short in other ways. Pam Evans (1991), a wheelchair-bound woman, clarifies further this tyranny by outlining common assumptions held about the life of impaired individuals:

- That our lives are a burden to us, barely worth living.
- That we crave to be normal and whole.
- That any able-bodied partner we have is doing us a favor and that we bring nothing to the relationship.
- That if we are particularly gifted, successful, or attractive before the onset of disability our fate is infinitely more tragic than if we were none of these things.
- That our need and right to privacy isn't as important. . . .

While it would be naive to believe that these kinds of thoughts never occur to those with impairments, these common assumptions capture the indirect and direct messages leveled against impaired individuals.

Caregiving for an Aging Population

Although there has always been a very small proportion of people who lived to age 80 or 90, "there is no historical precedent for the aging of our population" (Soldo and Agree 1988). The family must find ways to adapt to this situation, especially as regards caregiving. With regard to the elderly and their caregivers, we will detail the caregivers' characteristics, time spent giving care, and the **caregiver burden**, the extent to which caregivers believe that their emotional balance, physical health, social life, and financial status suffer because of their caregiver role (Zarit, Todd, and Zarit 1986).

> Keep in mind that family caregiving is a complex activity that goes beyond time commitments and perceived burdens.

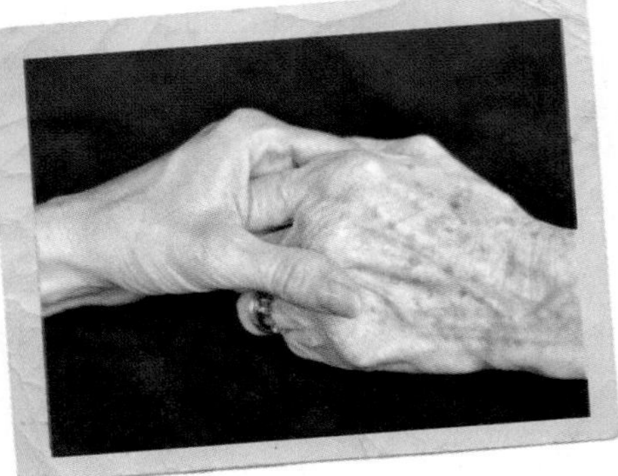
Lisa Southwick

Although most people age 65 and over in the United States do not live in nursing homes, one in four does require assistance with daily activities such as bathing, walking, dressing, and eating. An estimated 17 percent of households (18.5 million) provide some kind of informal care to a person age 50 or older (National Alliance for Caregiving and AARP 2004). That percentage varies by racial classification, with 42 percent of Asian American households and 19 percent of white households offering such informal care (AARP 2001). One might conclude from these data that nonwhites are more willing to care for older members than whites. However, older whites are the most likely to rate their health as good to excellent, which reduces their need for caregiving.

In the United States, approximately 61 percent of all caregivers are women. This percentage holds for all racial groups except Asian Americans; in that group, 54 percent of caregivers are male. About 40 percent of all people caring for seniors are also caring for children younger than age 18. The typical caregiver is a 46-year-old employed woman caring for her 76-year-old mother, who lives nearby, for an average of 18 hours per week. The average person who cares for a senior does so for 4.3 years (Jackson 2003; National Alliance for Caregiving and AARP 2004). Ironically, it seems that those who provide the most care are also the most likely to feel guilty that they are not doing enough (AARP 2001).

(Write a Caption)

Write a caption that describes the type of caregiving these children receive while living in an orphanage for handicapped children.

U.S. Navy photo by Photographer's Mate 3rd Class David J. Hewitt/Released

Hints: In writing this caption

- consider whether this caregiving arrangement is formal or informal, and
- describe what qualities make it formal or informal.

Critical Thinking

Does someone in your family have an impairment or disability? If yes, describe his or her situation.

Key Terms

caregivers	formal care	informal care
caregiver burden	impairment	tyranny of the normal
disability		

MODULE

11.7 Applying Theory: Feminist Theory

Objective

You will learn how the feminist perspective analyzes family dynamics with emphasis on dynamics surrounding caregiving.

Missy Gish

Which of the four siblings do you think is the one on call to help older relatives or otherwise respond in times of need?

If you guessed the sister, you are correct. That guess speaks to a social structure that "proclaims women as best suited and able to provide that care" and that shifts a necessary burden disproportionately onto them (Connidis and McMullin 2002, 564).

The Feminist Perspective

In studying families, feminists maintain that they cannot be studied apart from the larger economic and political structures in which they are embedded. Economic structures include the workplace and marketplace; political structures include the legal system. So, for example, caregiving and other work-family balance issues reflect the larger economic and political systems that surround the family.

Feminist Theory Applied to Caregiving

Sociologists Ingrid Arnet Connidis and Julie Ann McMullin (2002) use a feminist perspective to frame family dynamics as they relate to caregiving. The following themes guide their analysis:

- All social structures—the family, the economy, the political system—consist of social relationships involving people with varying degrees of disadvantage and privilege associated with their class, gender, age, race, ethnicity, sexual orientation, and other statuses.

- Relationships and interactions are shaped by the various social structures in which they are embedded.
- A person's investment and interest in maintaining the existing social structure depends on the extent to which that person is privileged or disadvantaged by it.
- People seek to exercise control over their lives. Human agency pushing against oppressive structures is the source of conflict and change.

➤ This sign speaks to how relationships and interactions are shaped by various social structures in which they are embedded. It is ironic that two services—divorce and health care—are advertised on this single sign. We have all heard of cases in which a spouse is contemplating divorce (or has divorced) because the partner has a chronic illness such as dementia and requires expensive 24/7 care and medical treatments. Not legally divorcing means that the spouse will live as an impoverished widow or widower. The point is that relationships between partners can be shaped by the economics of health care and the legal options available to them.

A structure is oppressive when it systematically gives certain categories of people advantages over others. In the context of caregiving, for example, men face expectations and privileges that are different from and less taxing than those faced by women. Among other things, men are handed a "legitimate excuse" to not engage in caregiving, and if they do, to assume a limited role. Excuses are considered legitimate when they are supported by longstanding beliefs and norms related to caregiving. "Hence, for men, paid employment is a legitimate excuse from caring for aging parents; for women, it is not" (Connidis and McMullin 2002, 562).

➤ When you see a group of young children around, do you expect the primary monitor of their activities to be a woman or a man? Or do you think there is an equal chance of either a man or a woman assuming such a role? Your expectations say a lot about who you view as being responsible for caregiving.

When women feel socially pressured to assume a disproportionate share of caregiving responsibilities, they can experience ambivalence toward those for whom they must care. These responsibilities and conflicting feelings can cause tensions in family, work, and other relationships. In response, a person may "choose" to quit work, to pay a third-party caretaker, to solicit others (like grandparents) to help, and so on. Of course, the response depends on the available resources from which a person can draw.

Missy Gish

➤ We know from the U.S. Census Bureau (2012g) that grandparents are primary caregivers for 23 percent of preschool-aged children. In this regard, grandparents, especially grandmothers, are acceptable substitutes for the mother.

Connidis and McMullin argue that working women have fewer options than working men (and men in general) for resisting the societal pressures to provide care. As a result, they are more likely to experience an ambivalence intensified by the fact that they are "expected not only to provide the care but also to derive satisfaction and fulfillment" from it (563).

U.S. Navy photo by Mass Comm. Spc. Jason R. Wilson

◀ In family life men face considerably less pressure to assume caregiving roles. This lack of pressure is reinforced by their lack of public visibility in such roles. For example, less than 1 percent of day care providers are male; less than 10 percent of nurses are male.

Women vary in their ability to exercise agency. Obviously, financially secure women have the option to hire others to provide some or all of the care. Women with few financial resources and employed in low-paying jobs may be forced to use vacation and personal days, work fewer hours, change shifts, or drop out of the labor force to meet caregiving obligations. Simply hiring someone to provide care may not reduce strain and accompanying ambivalence if women do not trust the providers to give their child, parent, or other relative high-quality care.

Connidis and McMullin point out that, even in light of the increasing number of family-friendly workplace policies (e.g., parental leave, job sharing, on-site day care, telecommuting), it is female employees who take the most advantage of them and who sacrifice careers to meet familial responsibilities. Even with family-friendly policies, we must remember that women are still part of a society that defines them as best able to provide care.

Mass Communication Specialist 2nd Class Devon Dow

➤ One way that men hear the message that they are not naturally suited to provide caregiving relates to the high praise they receive when they do engage in it. Who do you think gets more praise and attention for helping serve these children, the women or the man? The low expectations society places on men play a role in eliciting high praise.

Men, who face considerably less pressure to assume caregiving roles, are likely to experience a different kind of strain and ambivalence when they do so. They, too, are part of a society that tells them they are not primarily responsible for caretaking and that they are not naturally suited to provide it. When men engage in caregiving, their motives are often suspect (e.g., are they sexual predators? gerontophiliacs?). Men may question their ability to provide such care, especially with regard to intimate aspects of caregiving—bathing, changing adult diapers, and dressing. Anecdotal evidence suggests that men feel unprepared for the job and isolated from support networks (Leland 2008). One man interviewed in the *New York Times* who is a caregiver to his elderly mother shared his feelings:

> She doesn't know if I'm her husband or her boyfriend or her neighbor. She knows she trusts me. But there are times when it's very difficult. I need to keep her from embarrassing herself. She'll say things like, "I adore you." I don't know who she's loving, because she doesn't know who I am. Maybe I'm embarrassed about it—it's my mom, for Christ sakes. But it's weird how the oldest son becomes the spouse. (Nicholson 2008)

Feminist theory prompts us to explore the dynamics by which structurally generated tensions are negotiated. Connidis and McMullin point out that people employ many strategies, including humor, confrontation, substance abuse, acceptance, and abandonment. They hypothesize that the fewer perceived or real options available to people, the more likely they are to deal with oppressive social structures by resigning themselves to the situation. However, when substantial numbers of people experience tensions and accompanying ambivalence, there is a potential for fundamental change to occur.

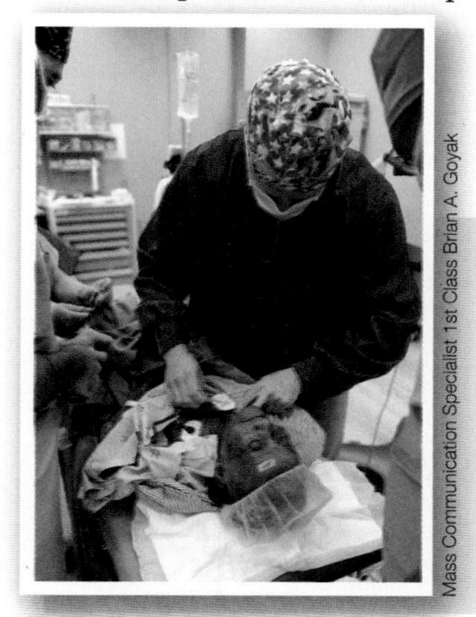
Mass Communication Specialist 1st Class Brian A. Goyak

➤ This woman, age 101, is being prepared for a routine surgery. Her needs for post-operative care are likely complicated by her advanced age. As the numbers of people aged 80 to 100+ years old increase, the pressures on younger generations to provide care are bound to increase. This demographic pressure has the potential to change current arrangements that assign primary responsibility to women, most notably daughters.

Lance Cpl. Cory Yenter, USMC

◀ One should not conclude that caregiving is only a burden. It can be a very rewarding experience. But it is also a labor-intensive experience. When social structures define only half the population, women, as naturally suited for caregiving, it eliminates an important source of support. When social structures portray men as unsuited to caregiving, those structures alienate men from establishing meaningful bonds with others.

Critical Thinking

Has someone you know taken on the role of primary caregiver? Does that person experience ambivalence? Explain.

Summary: Putting It All Together

CHAPTER 11

Family is a social institution that binds people through blood, marriage, law, and/or social norms. When sociologists study family, they do not have a particular family structure in mind as a standard. Functionalists emphasize the functions of family; conflict theorists emphasize the ways in which families perpetuate existing inequalities and social boundaries. Symbolic interactionists seek to understand the experiences of conforming to or resisting societal ideals about what family life should be.

Millions of seemingly personal decisions influence the variety of family arrangements that exist in any country—decisions about (1) whether to have children, and if so, how many to have, when to have them, and how to space them; (2) whether to marry, and if so, when; (3) whether to work for pay; and (4) whether to become a caregiver to dependent relatives. These decisions are informed, or perhaps even largely determined, by the larger social context, including life expectancy, employment opportunities, social norms about the ideal number of children, and ideas about who and when one should marry.

Sexual or gender stratification is the system by which societies rank males and females on a scale of social worth such that the ranking affects life chances in unequal ways. The more access women have to agents of violence control and the more equal women's position relative to men in the labor market, the more likely women are to have an equal relationship with men. This principle seems to apply to all types of intimate relationships, not only heterosexual ones.

A generation is a category of persons passing through time that is distinguished from other generational categories by such characteristics as its cultural disposition, posture, access to resources, and socially expected privileges, responsibilities, and duties. Industrialization and dramatic increases in life expectancy have broadly affected relationships among generations of family members in terms of the length of time the generations spend with each other, the authority parents have over children, and the status of children.

Caregiving, especially when it is informal, long-term, or ongoing, creates strain and tension for both the provider and receiver of care. The relationship between provider and receiver is subject to any number of strains and stresses related to time, conflict with other commitments, physical and emotional demands of caring and being cared for, and dignity and privacy. In family life men face considerably less pressure to assume caregiving roles than do women. This lack of pressure is reinforced by their lack of public visibility in such roles. When men do engage in caregiving, they are likely to experience a different kind of strain and ambivalence than women do. Women's strain and ambivalence derive from social pressures that define them as naturally suited for caregiving and that award them ultimate responsibility for such tasks. Men's strain derives from beliefs that they are not naturally suited to caretaking and from suspicions about why they might want to do such work.

Go to cengagebrain.com to link to Aplia and CourseMate for the chapter quiz and other activities.

Lisa Southwick

12 CHAPTER

EDUCATION AND RELIGION

Education and religion are institutions that touch the lives of just about everyone. Schools and religions seek to instill what sociologist Émile Durkheim terms a collective consciousness, "a complex of ideas and beliefs that influence ways of seeing and of feeling" (Durkheim 1961, 860). From a global perspective, almost all governments in the world mandate some amount of schooling, and every government in the world has some policy or law addressing the role of religion in education, the broader society, and everyday life. In the United States, for example, everyone is required by law to attend school beginning at age 6 until they are 16 (National Center for Education Statistics 2011c). But not every student attends a public school. About 1 in every 12 primary and secondary school students attends a faith-based school. It would be a mistake to think public schools in the United States are religious-free zones if only because public schools let students off for key religious holidays, specifically Christmas and often Easter (spring break). Among other things, public school students have the right to pray, to write about their religious beliefs in relation to class assignments, and to organize faith-based clubs. Students do not have the right to compel others to participate.

Go to Sociology CourseMate on cengagebrain.com to watch a video about teacher shortages around the country and efforts to reduce them.

MODULE

12.1 Education and Schooling

Objective

You will learn why education, and schooling in particular, is an important area for sociological study.

What is your earliest memory of being a student in school? More than likely, your memory holds important clues about the meaning of school in your life.

Spc. Christopher Wellner

Education includes any experience that trains, disciplines, and shapes the mental and physical potentials of the maturing person. While every experience has the potential to educate, sociologists who study education tend to emphasize **schooling**, a deliberate, planned effort that takes place in a brick-and-mortar or virtual classroom to impart specific skills or information. We tend to think of schooling as a liberating or positive experience, but it can be impoverishing and narrowing as well if it involves indoctrination, brainwashing, or neglect. In any case, from the point of view of those who design experiences that educate, schooling is considered a success when students acquire the skills or thoughts that those who designed the experience seek to impart.

Sociologists who study schooling seek to identify the key factors and processes—"inside and outside schools—that affect student outcomes" (Karen 2005). For insights on these factors and processes we turn to the three theoretical perspectives: symbolic interaction, functionalism, and conflict theory.

Symbolic Interaction

Schooling is a vehicle through which students are exposed to a collective consciousness, "a complex of ideas and beliefs that influence ways of seeing and of feeling (Durkheim 1961, 860). The symbolic interactionist perspective offers a framework for capturing how this collective consciousness is transmitted over the course of the school day. Since the curriculum is a key component of any school day, symbolic interactionists would certainly focus on its role in conveying meaning.

Teachers everywhere teach two curricula simultaneously: one formal and one hidden. The various academic subjects—mathematics, science, English, reading, physical education, and so on—make up the **formal curriculum**. Students do not learn in a vacuum, however. As teachers instruct students and as students learn the subject matter and complete their assignments, other activities are occurring around them. These other activities are the **hidden curriculum**, the lessons conveyed by the teaching method, type of assignments, tests (multiple choice versus essay), tone of the teacher's voice, attitudes of classmates toward school, and the frequency of teacher absences, as just some examples. The hidden curriculum sends messages to students not only about the value of the subject but also about culturally valued ways of thinking and behaving. For insights on this we consider the comparative research of Yi Che, Akiko Hayashi, and Joseph Tobin (2007), who observed preschools in the United States and China.

The three researchers filmed daily life in a Chinese and a U.S. preschool to learn how teachers in each system socialize children to participate effectively in their respective societies. Though it is impossible to make definitive generalizations about preschools in countries as large and diverse as the United States and China, we can identify some broad differences. The researchers found that, compared with U.S. preschoolers, Chinese preschoolers (four-year-olds) are taught to give constructive critiques of each other's work and to learn from those critiques. Chinese preschoolers are also taught to downplay interpersonal conflicts and play cooperatively. In contrast, American preschoolers are taught to expect praise for their work. The following scene exemplifies the extent to which Chinese preschoolers are taught to give and accept critique.

Each day in a Shanghai preschool class, 22 children engage in a storytelling activity. One child is designated the "story king" and stands in front of the class to tell a story. Upon finishing, the teacher (Mrs. Wang) makes comments, asks students questions about the story, and then calls for a vote on whether today's storyteller earned the title story king. Eighteen children vote yes. Mrs. Wang asks the four children who voted "no" for reasons: "A child remarks, 'Some words I could hear, but some I couldn't.' 'Don't think his voice was loud enough' says another." The teacher turns to the storyteller and asks if he agrees with the critique, to which he nods yes. "At that point, the teacher comments, 'Next time, he will be loud and clear'" (Che, Hayashi, and Tobin 2007, 7).

In the United States, preschoolers are not encouraged to critique each other's work, and especially not when that work is considered self-expressive or creative. Early-childhood teachers in the United States tend to believe that their job is to protect and support children's self-esteem and that it is not "developmentally appropriate" to subject a child to peer criticism (Che, Hayashi, and Tobin 2007, 7). In fact, American teachers are reluctant to correct a child's mistakes in front of other students. Rather, teachers in the United States often give empty praise—"that's wonderful"—regardless of quality.

The researchers showed an audience of U.S. teachers video clips of preschoolers in China critiquing each other and resolving conflicts. Some American teachers were bothered by the Chinese practice of allowing children to critique each other because they thought children were too young to do this kind of thing. Others were amazed at how well children gave and accepted critique. One teacher remarked, "I'm amazed how well that boy handled the criticism. I'm an adult and I think I would cry if people criticized me like that in front of a group!" (7). The researchers pointed out that perhaps four-year-olds' self-esteem may not be as fragile as Americans assume.

Nan Wylie

◀ Would you allow or encourage this child's classmates to critique her work—to identify things she might do to improve that work? If you live in the United States, it is unlikely that you would. But in Chinese preschools, such critique is encouraged and welcomed. Can you see how schooling is a vehicle through which students are exposed to a collective consciousness, "a complex of ideas and beliefs that influence ways of seeing and of feeling" (Durkheim 1961, 860)?

Functionalist Perspective

The functionalist perspective focuses on the functions of schooling. These functions include transmitting skills, facilitating personal growth, integrating diverse populations, screening and selecting the most qualified students for what are considered the most socially important careers, and solving social problems.

TRANSMITTING SKILLS. Schools exist to teach children and others the skills they need to adapt to the society and world in which they live. To ensure that this end is achieved, stakeholders in education such as employers, parents, and government officials remind teachers what they believe must be impressed on students. That might include patriotism, specific skills, and a sense of civic engagement.

FACILITATING PERSONAL GROWTH. Education can be a liberating experience that releases students from the blinders imposed by the accident of birth into a particular family, culture, religion, society, and time in history. Education can broaden students' horizons, making them aware of the conditioning influences around them and encouraging them to think independently of authority. In that sense, schools function as agents of personal change.

INTEGRATING DIVERSE POPULATIONS. Schools function to socialize (for example, to Americanize, Europeanize, "Chinesize") people of different ethnic, racial, religious, and family backgrounds. In the United States, schools play a significant role in what is known as the melting-pot process. Recall that the creation of the United States involved the conquest of the native peoples, the annexation of Mexican territory along with many of its inhabitants (who lived in what is now New Mexico, Utah, Nevada, Arizona, California, and parts of Colorado and Texas), and the voluntary and involuntary influx of millions of people from practically every country in the world (Sowell 1981). Early American school reformers—primarily those of Protestant and British backgrounds—saw public education as the vehicle for Americanizing a culturally and linguistically diverse population, instilling a sense of national unity and purpose, and training a competent workforce.

Prints & Photographs Division, Library of Congress, LC-USZ62-42810

Jorge Gomez, Fort Lee Public Affairs Office

⮝ These schoolchildren are probably too young to appreciate the meaning of the words in the Pledge of Allegiance, but the act itself reminds them that they are Americans. The fact that the pledge was written in 1892 with schoolchildren in mind speaks to the Americanizing function of schools. The photos show schoolchildren in 1942 (left) and today (right) reciting the pledge.

SELECTING. Schools use tests and grades to evaluate students and reward or punish them accordingly by conferring or withholding degrees, by rejecting or admitting students into programs of study, and giving negative or positive recommendations. Ideally, they channel the most capable and skilled students into the most desirable and important careers and the least capable and skilled into careers believed to require few, if any, special talents.

SOLVING SOCIAL PROBLEMS. Societies use education-based programs to address a variety of social problems, including parents' absence from the home, racial inequality, drug and alcohol addictions, malnutrition, teenage pregnancy, sexually transmitted diseases, and illiteracy. Although all countries likely support education-based programs that address social problems, the United States seems to place particular emphasis on education as a primary solution to many problems, such as childhood obesity, illegal drug use, poverty, hunger, and teen pregnancy. The U.S. belief that education-based programs can be used to solve social problems and address community needs manifests itself in service-learning programs aimed at addressing local, even global, problems and needs.

OTHER FUNCTIONS. Schools perform other, less obvious functions. For one, they function as reliable babysitters, especially for nursery school–, preschool–, and grade school–age children. They also function as a dating pool and marriage market, bringing together students of similar and different backgrounds and ambitions whose paths might otherwise never cross.

Conflict Perspective

Schools are not perfect: not all minds are liberated; students drop out, refuse to attend, or graduate with skill deficiencies; schools misclassify some students as slow learners when they are not; and so on. The conflict perspective draws our attention away from order and stability and toward inequalities. Conflict theorists ask questions like, Which schools have access to the most up-to-date computer or athletic facilities? Which types of students are most likely to drop out of high school? Which types of students are most likely to attend college?

Conflict theorists argue that, for the most part, schools simply perpetuate the inequalities of the larger society. This point is obvious when we consider that the poorest schools and most economically disadvantaged children usually have the highest dropout rates, lowest high school graduation rates, and lowest college enrollments.

Among other things, conflict theorists study the extent to which students are exposed to different and unequal kinds of curricula. In this regard most, if not all, educational systems engage in **tracking**, a process by which students are sorted into distinct instructional groups based on past academic performance, performance on standardized tests, or even anticipated performance. Usually these instructional groups are unequally valued, such as college or advanced placement versus remedial classes.

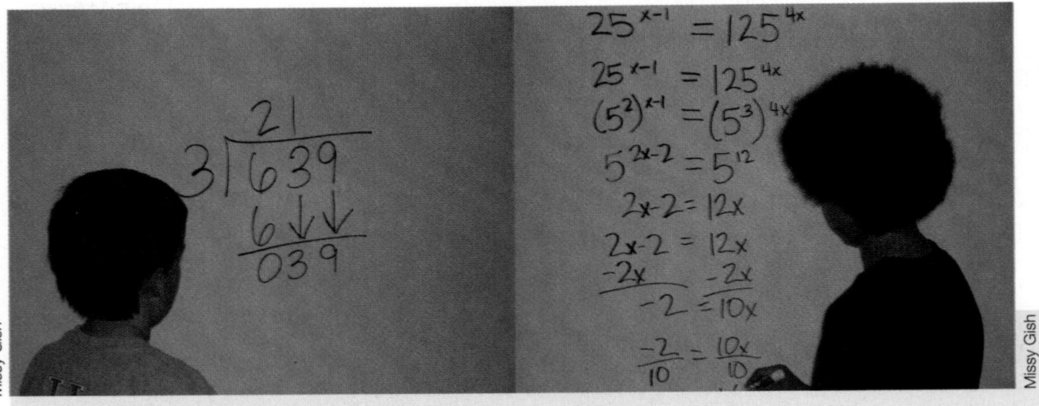

Missy Gish

Missy Gish

▲ While most countries track students according to academic abilities, they do not offer students assigned to lower tracks a very basic math, never exposing them to higher tracks such as algebra, geometry, or calculus. Everyone takes the same number and type of math courses. In the United States, it is common for students in the same grade to be learning different kinds of math.

Advocates of tracking offer the following rationales in support of the practice:

- Students learn better when they are grouped with those who learn at the same rate. The brighter students are not held back by the slower learners, and the slower learners receive the extra time and special attention needed to correct academic deficiencies.
- Slow learners develop more positive attitudes when they do not have to compete with the more academically capable.
- Groups of students with similar abilities are easier to teach than groups of students with differing abilities.

There is little evidence to indicate that placing students in remedial or basic courses contributes to intellectual growth, corrects academic deficiencies, increases interest in learning, or narrows achievement gap relative to students placed in higher tracks (Oakes 1986a, 1986b). In a now-classic study, sociologist Jeannie Oakes (1985) investigated how tracking affected the academic experiences of 13,719 middle school and high school students in 297 classrooms and 25 schools across the United States. "The schools themselves were different: some were large, some very small; some in the middle of cities; some in nearly uninhabited farm country . . . but the differences in what students experienced each day in these schools stemmed not so much from where they happened to live and which school they happened to attend but rather, from differences [related to tracking] within each school" (Oakes 1985, 2).

Oakes's findings are consistent with the findings of later studies that assess tracking:

Placement—Poor and minority students were placed disproportionately in the lower tracks.

Treatment—The different tracks were not treated as equally valued instructional groups. Clear differences existed in classroom climate and in the quality, content, and quantity of instruction. Low-track students were consistently exposed to inferior instruction—watered-down curricula and endless repetition—and to a more rigid, more emotionally strained classroom climate.

Self-image—Low-track students did not develop positive self-images because they were publicly identified and treated as educational discards, damaged merchandise, or even as unteachable. Among average- and low-track groups, tracking seemed to foster lower self-esteem and promote misbehavior, higher dropout rates, and lower academic aspirations. In contrast, placement in a college preparatory track had positive effects on academic achievement regardless of "family background and ability differences" (Hallinan 1988, 260).

Oakes argues that although many educators recognize the problems associated with tracking, efforts to undo tracking have collided with demands from politically powerful parents of high-achieving or gifted students; these parents insist that their children must get something more than the other students (Wells and Oakes 1996). As a result, tracking persists.

(Write a Caption)

Write a caption that considers the potential formal and hidden curricula in a parenting class aimed at teens.

Arthur McQueen (USAG Miami)

Hints: In writing this caption

- review the concepts of formal curriculum and hidden curriculum,
- consider what elements of the parenting class might qualify as part of the formal curriculum, and
- decide what additional "lessons" might be conveyed.

Critical Thinking

Think back to when you were in grade school, middle school, or high school. Did you experience tracking? Describe how it was done and how it shaped your educational experience and sense of self.

Key Terms

education | hidden curriculum | tracking
formal curriculum | schooling

MODULE

12.2 Learning Environments

Objective

You will learn about the many factors that affect academic achievement and other learning outcomes.

Think back on your school career. Did you go to school with students who mostly appeared to be the same race as you?

Mrs. Stephenie Tatum, Third Army Public Affairs

The classroom environment, which includes the racial composition of the student body, is one of many factors that shape the educational experience. As we will learn, the racial composition of a classroom provides important clues about the quality of educational experience.

Racial Inequalities: Then and Now

More than 45 years ago, sociologist James Coleman (1966) was the principal investigator behind the congressionally mandated report *Equality of Educational Opportunity*, popularly known as the Coleman Report. Coleman and his colleagues surveyed 570,000 students and 60,000 teachers, principals, and school superintendents in 4,000 schools across the United States. Coleman found that a decade after the 1954 Supreme Court's famous desegregation decision (*Brown v. Board of Education*), U.S. schools were still largely segregated: 80 percent of white children attended schools that were 90 to 100 percent white, and 65 percent of black students attended schools that were more than

90 percent black. Furthermore, almost all students in the South and the Southwest attended schools that were 100 percent segregated. The Coleman Report also found that white teachers taught black children, but that black teachers did not teach white children. In his study approximately 60 percent of the teachers who taught black students were black, whereas 97 percent of the teachers who taught white students were white.

Today, almost 50 years after the Coleman Report was published, the percentage of nonwhite and Hispanic students has increased from 12 to 40 percent, and school-based segregation is still a problem. According to the most recent data, 40 percent of blacks attend schools that are 90 percent black and 40 percent of Hispanics or Latinos attend schools that are 90 percent minority. Of course, the level of segregation varies by place with 92 percent of blacks in Washington, D.C., attending schools that are 90 percent minority. More than 50 percent of blacks in Illinois, Michigan, and New York attend schools that are 90 percent black (Fry 2007).

➤ *Brown v. Board of Education* declared unconstitutional state laws that established separate public schools for black and white students on the grounds that such an arrangement denied black children equal access to educational opportunities. This 1955 photograph taken shortly after the decision shows a newly integrated class at Barnard School in Washington, D.C. Does this photo suggest that the classroom is integrated?

Prints & Photographs Division, Library of Congress, LC-U9-183B-20

TEST SCORES. Coleman's study also looked at test scores and did find sharp differences across racial groups with regard to verbal ability, nonverbal ability, reading comprehension, mathematical achievement, and general knowledge as measured by the standardized tests. The white students scored highest, followed by Asian Americans, Native Americans, Hispanics (Mexican Americans and Puerto Ricans), and African Americans. For the most part, these sharp differences across racial groups still persist more than a half century later. The latest data show that, as a group, Asian students tend to score highest, followed by whites, Hispanics, blacks, and Native Americans (see Table 12.2a).

MCSN Crystal Habbershon

◄ Race's effect on academic achievement is complicated by a number of factors, including the quality of the school, the surrounding neighborhood, and parents' income and level of education.

▼ Table 12.2a: Percentage of Fourth- and Eighth-Grade Students Considered Proficient-to-Advanced in Reading by Race

What percentage of Asian students in fourth and eighth grade are considered proficient-to-advanced in reading achievement? What percentage of black students in those same grades are considered proficient-to-advanced in reading? Also note that there is still a significant percentage of students in all racial groups who are not proficient in reading.

Racial Classification	Percentage of Fourth Graders	Percentage of Eighth Graders
Asian	66%	53%
Native American	22%	24%
Black	19%	16%
Hispanic	21%	20%
White	55%	48%

Source of data: The Nation's Report Card (2011)

NEIGHBORHOOD. Coleman determined that the average minority student was likely to attend schools that served students from economically and educationally disadvantaged households located in a disadvantaged neighborhood. Thus, significantly fewer of his or her classmates complete high school, achieve high grade point averages, enroll in advanced placement classes, or are optimistic about their future. As in the 1960s, today black, Hispanic, and Native American students are significantly more likely than white and Asian students to find themselves in school environments characterized by high levels of poverty and low levels of academic achievement. The latest data show that 22 percent of all elementary students attend high-poverty schools but that more than 40 percent of black and Hispanic students attend high-poverty schools (see Chart 12.2a).

▼ Chart 12.2a: Percentage of Public Elementary School Students in High-Poverty Schools, by Race/Ethnicity: School Year 2008-2009

Public schools with more than 75 percent of the students eligible for the free/reduced-price lunch program are considered high poverty. What percentage of students classified as white attend high-poverty schools? What percentage of Hispanics attend high-poverty schools?

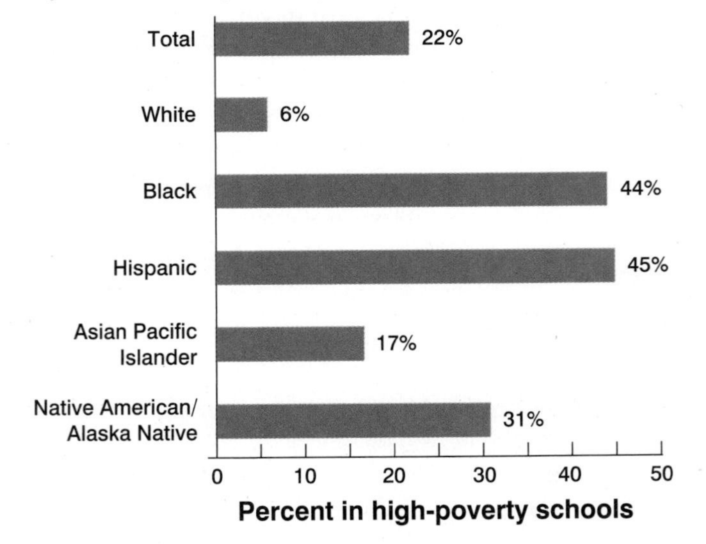

Source of data: U.S. Department of Education (2011a)

Among other things, the Coleman Report compared the test scores of disadvantaged blacks who had participated in school integration programs against disadvantaged counterparts who had not. Coleman found the test scores were higher for those who were part of integration programs. Clearly, then, the important variable in determining academic success is not race. If race were the factor, test scores would not improve when black students changed schools. Coleman attributed this newfound academic success to complex intersections among family background and neighborhood and peer group influences.

FAMILY BACKGROUND. Coleman maintained that neighborhood and peer groups are closely intertwined with family background. After all, a family's economic standing determines the housing and neighborhood in which it can afford to live. Of course schools draw students who live in surrounding neighborhoods or communities. In this sense family background was the most important factor in predicting academic success.

Many other research studies confirm the importance of family background in educational achievement (Hallinan 1988). For example, the International Association for the Evaluation of Educational Achievement (2011) routinely tests students in more than 20 countries on six subjects. Their studies find that home environment is the key factor in predicting academic achievement, interest in learning, and the number of years students will remain in school. Nevertheless, family background explains only about 30 percent of the variation in student academic performance. Clearly, as important as family background is, other factors must also be involved.

Pvt. Choi Keun-woo

➤ Another related factor Coleman identified was the influence of peer groups on academic achievement. Clearly the peer group can exert considerable influence on students' attitudes toward school, behavior in the classroom, and larger school environment.

Peer Groups

Think back to your high school days. How would you most like to be remembered by your classmates—as an athlete, as a brilliant student, as a leader in extracurricular activities, as attractive to the opposite sex, or as the most popular student? If you answered "athlete," it is likely you are male. If you answered "attractive to the opposite sex," it is likely you are female. These are the kinds of questions sociologists James Coleman and his colleagues (1961) asked in their now-classic study, *The Adolescent Society*.

The emergence of an adolescent society can be traced to industrialization. Around the turn of the 20th century—the early decades of late industrialization—fewer than 10 percent of teenagers attended high school in the United States. Young people attended elementary school to learn reading, writing, and arithmetic, but they learned the skills needed to make a living from their parents or neighbors. As the pace of industrialization increased, jobs began to move away from the home and the neighborhood and into factories and office buildings. Parents no longer trained their children because the skills they knew were becoming obsolete. Children came to expect that they would *not* make a living in the same way as their parents. As a result of this shift, opportunities for parents and children to work together disappeared, and the family became less involved in children's lives.

➤ This 1910 photograph shows a five-year-old working with parents in a field. Imagine the extent to which the family influenced day-to-day activities. How might the family's influence be diminished if the children were in school for six or seven hours each day?

Prints & Photographs Division, Library of Congress, LC-DIG-nclc-00040

This shift in training from the family to the school cut adolescents off from the rest of society so that they came "to constitute a small society, one that has most of its important interactions within itself, and maintains only a few threads of connection with the outside adult society" (Coleman, Johnstone, and Jonassohn 1961, 3).

Coleman surveyed students from 10 high schools in the Midwest to learn about the **adolescent status system**, a classification system in which participation in some activities results in popularity, respect, acceptance, and praise, and participation in other activities results in isolation, ridicule, exclusion, disdain, and disrespect. He selected schools representative of a wide range of environments: five schools were in small towns, one in a working-class suburb, one in a well-to-do suburb, and three in cities of varying sizes. Also, one was an all-male Roman Catholic school. Coleman asked students questions similar to the following:

- How would you like to be remembered—as an athlete, as a brilliant student, as a leader in extracurricular activities, or as the most popular student?
- Who is the best athlete? The best student? The most popular? The boy the girls go for most? The girl the boys go for most?
- Which person in the school would you most like to date? To have as a friend? What does it take to get in with the leading crowd in this school?

➤ Based on the answers to these questions, Coleman found that a popular boy could be a good student or someone who dressed well or had enough money to meet social expenses, but to be truly admired he also had to be a good athlete. Girls gave the highest value to social success with boys, which could be achieved through good looks and/or becoming a cheerleader.

Air Force photo by Mike Kaplan

Coleman wrote about an adolescent society that seemed to penalize academic achievement when it was not paired with some other socially valued trait or talent like good looks or athletic ability. Coleman argued that in the United States athletics is one of the major avenues open to adolescents, especially males, in which participation is considered important to the status of both the school and the surrounding community. Athletic competition between schools generates more internal cohesion among students and the surrounding community than other school-sponsored events can. It is for this reason that the male athlete is conferred so much status and women gain status when they date an athlete or assume supporting roles to athletes such as a cheerleader.

Chris Caldeira

◄ While Coleman alerted us to the adolescent status system as described above, we can find many examples of schools that develop and celebrate other talents. What do the state championships listed on these signs say about what is valued at the school? Is your high school known for a nonsport activity?

(Write a Caption)

Write a caption that addresses some of the factors that detract from students' chances of academic success.

Hint: In writing this caption

- review the Coleman studies to determine factors that have the potential to detract from any child's educational success.

Kris Gonzalez

Critical Thinking

Imagine the Coleman study on adolescent society had been done at your high school at the time you were a student. Would his findings apply?

Key Term

adolescent status system

MODULE 12.3

Rewards and Costs of Higher Education

Objective

You will learn about the rewards and costs associated with higher education.

What percentage of students in your high school class graduated in four years? Of those who graduated, what percentage went directly to college? How did they pay for it?

Chris Caldeira

Rewards of Higher Education

Most Americans are taught to equate more education with increased job opportunities and higher salaries. This belief has a basis in reality. Figure 12.3a shows that income rises as the level of education increases. On average, both males and females with four-year college degrees earn more than their same-sex counterparts who have less education. On the other hand, earnings are still affected by factors such as gender.

➤ While college graduates possess an earnings advantage over those who have obtained less education, it is important to realize that some without a college degree earn more than some who have a degree. For example, while the median weekly income of males with no high school diploma is $488, those in the bottom 10 percent of wages earn a median income of $302 per week while those in the top 10 percent earn a median income of $927. Compare that salary range with the range for males with a college degree, where the lowest 10 percent earn a median income of $617 and the highest 10 percent earn $2,894 (U.S. Bureau of Labor Statistics 2012a). The point is that educational attainment is only one factor in determining income.

U.S. Navy Petty Officer First Class Chad J. McNeeley

▼ Figure 12.3a: Median Weekly Earnings of Full-Time and Salaried Workers Age 25 Years and Over by Sex and Educational Attainment, 2011
Which education level is associated with the highest median weekly wages for males? For females? For which education level do men most outearn women? Do income differences between men and women increase or decrease as level of education increases?

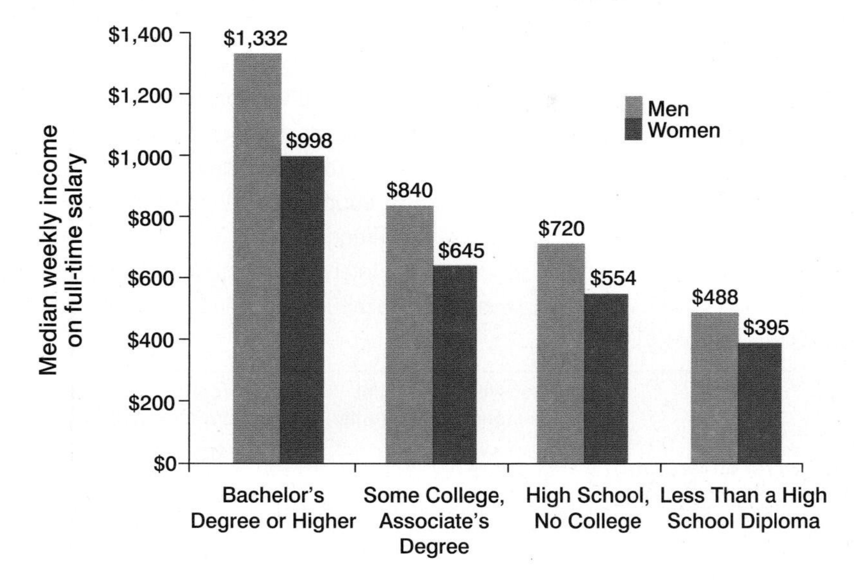

Source: U.S. Department of Labor, Bureau of Labor Statistics (2012d)

Who Goes to College?

We can see that income does rise as level of education increases. So it is important to examine who completes high school and goes on to college. In the United States, the on-time or four-year high school graduation rate is 74.5 percent. This means that 74.5 of ninth-graders graduate from high school four years later. Of course, many who do not graduate on time do go on to earn a high school diploma a year or more later, and others eventually earn a general equivalency diploma (GED). The on-time graduation rate varies by school; it can be as low as 5 percent and as high as 100 percent (*The Oregonian* 2011). It also varies for students classified as Asian (92 percent), white (82 percent), Native American (57 percent), black (63.5 percent), and Hispanic (64.9 percent) (Layton 2012; U.S. Department of Education 2010).

Among those who do graduate from high school, almost 70 percent enroll in college the following year. The probability of enrolling for the fall semester after high school varies by race/ethnicity with 84 percent of Asian, 68.6 percent of white, 61.4 percent of black, and 59.6 percent of Hispanic students going on to college (U.S. Department of Labor, Bureau of Labor Statistics 2011a).

One seemingly distinctive feature of U.S. higher education is that academically unprepared and unqualified students can attend a college, even those who did not graduate from high school or did not earn a GED. A U.S. Department of Education study found that 400,000 students—2 percent (1 in 50) of all college students—did not have a high school diploma or GED (Arenson 2006).

Graduating from high school does not necessarily mean that a student is prepared for college. About 70 percent of colleges and universities offer noncredit-bearing remedial courses, such as college algebra, for students who lack the skills needed to do college-level work. An estimated 36 percent of entering freshmen take one or more remedial courses in reading, writing, or mathematics (Aud, Hussar, Kena, Bianco, et al. 2011).

Funding Higher Education

Many countries have programs in place that pay for or offset the cost of going to college. Chart 12.3a shows four such models and lists two countries that exemplify each.

Chart 12.3a: Funding Models for Public Colleges and Universities

The dollar figure in parentheses after each country represents the average annual cost of educating a college student who attends a public university. The chart shows that Model 3 applies to the United States in that there is generous financial support, but students also pay the highest out-of-pocket tuition/fees in the world—on average, $6,312 each academic year (see footnote). This means that on average $23,598 of the total cost per student is supported by government and private sources (e.g., employer tuition reimbursement, federal grants, state grants, and private scholarships).

	Generous Financial Assistance to Students	Less Generous Financial Assistance to Students
Students Pay No or Very Low Tuition/Fees (avg. cost paid by student: $0*–$1,850 per year)	**MODEL 1**	**MODEL 2**
	Iceland ($10,429)	France ($14,945)
	Sweden ($20,864)	Ireland ($16,248)
Students Pay High Tuition/Fees (avg. cost paid by student: $3,000–$6,312 per year)*	**MODEL 3**	**MODEL 4**
	United Kingdom ($15,314)	Japan ($16,533)
	United States ($29,910)*	Korea ($10,109)

* Countries in green: students pay no fees or tuition; note that U.S. students on average pay the highest out-of-pocket tuition and fees ($6,312).
** Note that there are no countries where the average cost per student is between $1,850 and $2,999.
Source of Data: OECD (2011a)

We might speculate that students who live in countries where college tuition is free or offered at very low cost are part of a society where education is viewed as a basic right and where people believe a government's role is to ensure access to that right. Note that in the Model 1 countries, students pay no tuition or fees, and there is a generous system of support in place to help students offset costs of living while in college. Students in Model 2 countries also pay low tuition/fees relative to total cost, but assistance programs are less generous.

➤ Keep in mind that college students take out loans, even when they live in countries where tuition is free or very low, to offset living expenses. How might the financial burden be reduced if students have access to reliable and efficient systems of public transportation? For the most part, U.S. students who live off campus rely on automobiles to get to school. U.S. students must also find ways to pay for health care, which is considerably more expensive in the United States than elsewhere.

Chris Caldeira

Student Debt After College

College loan programs exist in at least 70 countries. Researchers Hua Shen and Adrian Ziderman (2009) analyzed 44 such programs in 39 countries with two questions in mind: What portion of the loan are students required to repay? And what percentage of money loaned is paid back or recovered?

The researchers found that many loan programs have "hidden grants" that forgive some of the amount owed depending on actions the borrower takes (working in public service) or on the income borrowers earn after graduation (a graduate's loan payment cannot exceed 10 percent of income). In addition, some loan programs require repayment within as few as 5 years and others as many as 20 years. The researchers also found that overall, a high percentage of students around the world who borrow money for college default on those loans but that the default rate varies by country. In response, some governments have instituted programs to decrease the rate of default including publishing the names of defaulters, barring defaulters from other credit sources, and deducting money owed from social security payments, tax refunds, and disability payments. Other countries have no such measures in place. In theory, if borrowers pay back loans plus interest, loan programs can sustain themselves. Of the 44 loan programs Shen and Ziderman (2009) studied, the Czech Republic has the best loan repayment ratio at 108 percent. This high return is possible because the Czech Republic charges at least 12 percent interest on loans so even if some students default, this high rate offsets money lost from default. Nigeria has the worst loan repayment ratio at 10.9 percent. Note that about two-thirds of college graduates report having taken out loans while in college. The average graduate who takes out a loan leaves college with $25,250 in debt (Ellis 2011). The U.S. loan repayment ratio is 73 to 78 percent, depending on source of loan.

The Credential Society

If we look at the proportion of the population age 25 and over with at least a bachelor's degree, we can see that the percentage has risen from 4.6 percent (about 1 in 20) in 1940 to 30 percent (1 in 3) in 2010 (see Chart 12.3b). What accounts for this increase? Sociologist Randall Collins is associated with the classic statement outlining the reason for the ever increasing percentages of people with college degrees.

▾ Chart 12.3b: Percentage of U.S. Population 25 Years and Over with a Bachelor's or Advanced Degree by Decade

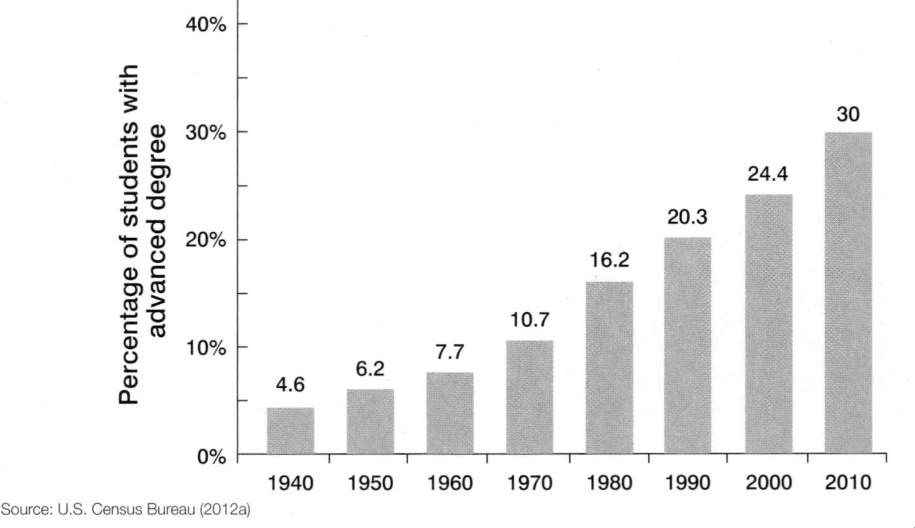

Source: U.S. Census Bureau (2012a)

Collins begins his argument by pointing out that there have been few, if any, systematic studies examining the role education plays in occupational success. In addition, it is the rare study that examines "what is actually learned in school and how long it is retained" (1961, 39). Regardless, over the decades there has been a steady increase in employer demands for job applicants with a college degree. This demand has created what Collins calls a **credential society**, a situation in which employers use educational credentials as screening devices for sorting through a pool of largely anonymous applicants. Specifically, many employers use the college degree as an indicator that an applicant is responsible, can set goals and follow through, and possesses basic skills. In addition, employers often require a college degree for promotion and advancement, even for employees who have an excellent work record and have demonstrated a high level of competence. In identifying the historical factors behind the emergence of a credential society in the United States, Collins eliminated the ever increasing need for an educated workforce as a factor, because most jobs created as a result of industrialization did not require advanced technical knowledge beyond an eighth-grade education. Rather, Collins traced the emergence of a credential society to two factors:

1. A long-standing belief going back to the colonial era that associated high economic status with educational attainment. Collins argues that simply "the existence of a relatively small group of experts in high status positions fostered the demand that opportunities to acquire such positions be available on a large scale" (37).
2. The United States federal government has always left decisions about what to teach to state and local communities. In addition, it maintains a separation between church and state (no national church). These two characteristics set the stage for various religious and other special interest groups to establish their own schools. Collins argues that it was this rivalry among interest groups to educate and socialize the young and the large number of schools and colleges it spawned that helps explain, in part, the emergence of the credential society. More specifically, the religious rivalries helped produce the Catholic and Lutheran school systems and even the public school system. Collins maintains that white Anglo-Saxon Protestant (WASP) elites founded the public school system in response to large-scale Catholic immigration from Europe.

◀ This anti-Catholic political cartoon from the 1880s, titled "Wolf at the door, gaunt and hungry—Don't let him in," shows children trying to keep the wolf, symbolizing the threat of Catholic education, from bursting through the door.

Of course, once the system of public mass education was established, the elite found private schools for their children as a "means of maintaining cohesion of the elite culture itself" (Collins 1961, 38). These rivalries set the stage for religious and elite groups to establish universities as well. As a result, opportunities for education at all levels expanded faster in America than anywhere else in the world, so that today there are an estimated 4,495 institutions of higher learning (National Center for Education Statistics 2011b).

Collins argues that this large number of schools and colleges in the United States helped explain the emergence of the credential society. The employer's demand for credentials, in turn, has fueled and perpetuated a widely held belief that a person must go to college to be successful. The ever-increasing supply of college-educated persons has made a college degree a requirement for many jobs. This may explain why, when surveyed, almost all parents in the United States (94 percent) say they expect their child to attend college (PEW Research Center 2011).

(Write a Caption)

Write a caption that projects the median income of those who graduate from college and seek employment, the size of the college-educated population of which they will become a part, the percentage who will hold debt from college loans, and the proportion who will default on that debt.

U.S. Air Force Master Sgt. Jerry Morrison

Hints: In writing this caption

- review the median income of full-time workers with a college education for men and women,
- determine the percentage of the U.S. population ages 25–34 with a college degree, and
- review the percentages of those who take on loans, the average debt, and the percentage expected to default.

Critical Thinking

How are you paying for college? What do you project your debt from college loans to total?

Key Term

credential society

MODULE

12.4 Religion

Objective

You will learn how sociologists approach the study of religion.

Can you identify the religious figure pictured? Do you think it is important to have a framework by which to think about your own and another's religious experiences?

Tony Rotundo

When sociologists study religion, they do not investigate whether God or some other supernatural force exists, whether certain religious beliefs are valid, or whether one religion is better than another. Sociologists investigate the social aspects of religion, which include the ways in which people have used religion to justify the most constructive and destructive behaviors, the ways religion shapes people's behavior and their understanding of the world, and the ways religion is intertwined with social, economic, and political issues.

Defining Religion

Max Weber (1922b) believed that no one definition could capture the varieties and essence of religion. For him, religion encompasses those human responses to the ultimate and inescapable problems of existence—birth, death, illness, aging, injustice, tragedy, and suffering (Abercrombie, Hill, and Turner 1988). To Weber, the hundreds of thousands of religions, past and present, represent a rich and seemingly endless variety of responses to these problems.

Like Weber, Émile Durkheim (1915) believed that religion is difficult to define. He cautioned that when studying religions, sociologists must assume that "there are no religions which are false" (3). Durkheim maintained that all religions are true in their own fashion; all address the problems of human existence, albeit in different ways. Consequently, he said, those who study religion must first rid themselves of all preconceived notions of what religion should be. We cannot study religion using standards that reflect our own personal experiences and preferences. Preconceived notions and uninformed opinions about the meaning of religious symbols and practices can close people off to a wide range of religious beliefs and experiences.

U.S. Army photo by Spc. Karah Cohen/Released

Missy Gish

▲ It is a challenge to see through one's own preconceptions about what is "right" in everyday life. A Western woman may look on the traditional Muslim female head covering, the *hijab*, as a sign of oppression. On the other hand, a Muslim woman may look on Western female makeup and presentation of self as a sign of oppression because many women, even young girls, seem to feel a need to present themselves as sex objects.

In formulating his ideas about religion, Durkheim remained open to the many varieties of religious experiences throughout the world. He identified three essential features that he believed were common to all religions, past and present: (1) beliefs about the sacred and the profane, (2) rituals, and (3) a community of worshipers. Thus, Durkheim defined **religion** as a system of shared rituals and beliefs about the sacred and the profane that bind together a community of worshipers.

BELIEFS ABOUT WHAT IS SACRED AND PROFANE. The **sacred** includes everything that is regarded as extraordinary and that inspires in believers deep and absorbing sentiments of awe, respect, mystery, and reverence. Sacredness stems from the symbolic power people confer on objects (such as chalices), living creatures (such as cows), elements of nature (such as mountains), places (such as mosques), states of consciousness (oneness with nature), holy days, ceremonies (such as baptism), and other activities (pilgrimages).

The **profane** encompasses everything that is not considered sacred, including things opposed to the sacred (such as the unholy, the irreverent, and the blasphemous) and things that stand apart from the sacred (such as the ordinary, the commonplace, the unconsecrated, and the bodily) (Ebersole 1967). Believers often view contact between the sacred and the profane as being dangerous and sacrilegious, as threatening the very existence of the sacred, and as endangering the fate of the person who makes or allows such contact. Consequently, people take action to safeguard sacred things. For example, some people refrain from speaking the name of God when they feel frustrated; others believe a man must remove his hat during worship.

Barbara Houghton

Tony Rotundo

⮝ Some religions define certain animals as sacred. Perhaps the most well known is the Hindu belief that cows are sacred. Mahatma Gandhi once wrote that "[i]f someone were to ask me what the most important outward manifestation of Hinduism was, I would suggest that it was the idea of cow protection" (*Nature* 2007). In some Asian countries such as Thailand, "white" elephants are considered sacred and are revered as symbols of power and good fortune.

RITUALS. In the religious sense, **rituals** are rules that govern how people behave in the presence of the sacred. These rules may take the form of instructions detailing the appropriate day(s) and occasions for worship, acceptable dress, and wording of chants, songs, and prayers. Participants follow rituals to achieve a specific goal, whether it be to purify the body or soul (as through immersion in water, fasting, or seclusion), to commemorate an important person or event (as by celebrating Passover), or to transform profane items into sacred items (for example, changing water to holy water and bones to sacred relics) (Smart 1976). Although rituals are often performed in sacred places, some rituals are codes of conduct aimed at governing the performance of everyday activities such as sleeping, walking, eating, defecating, washing, and relating to the other sex.

Lance Cpl. Lukas J. Blom

⮜ Buddhist monks of the Shingon sect located in Daishoin, Japan, participate in a ritual that involves walking over burning coals as a way of honoring the "Three Awesome Forest Deities" believed to inhabit Mount Misen. These coals were lit by a fire that is believed to have been burning continuously for the past 1,200 years.

According to Durkheim, the manner in which a ritual is performed is relatively insignificant. What is important is that the ritual is shared by a community of worshipers and that in performing it, people feel they are part of something larger.

COMMUNITY OF WORSHIPERS. Durkheim (1915) uses the word **church** to designate those who hold the same beliefs about what is sacred and profane, who share rituals, and who gather in body or spirit at agreed-on times to reaffirm their commitment to those beliefs and rituals. The act of gathering and sharing creates a moral community and cultivates a common identity.

Durkheim (1915) argued that some form of religion appears to have existed for as long as humans have lived (at least two million years). In view of this fact, one might argue that religion must serve some vital function for society. In this regard, one can argue that all religions strive to raise individuals above themselves—to help them achieve a better life than they would lead if left to their own impulses. Religion offers people a code of conduct that can evoke guilt and remorse when violated. Such feelings, in turn, motivate people to make amends.

Chris Caldeira

Chris Caldeira

Religion functions as a stabilizing force in times of severe social disturbance and abrupt change. During such times, many norms that guide behavior may break down. In the absence of such forces, people are more likely to turn to religion in search of a force that will bind them to a group.

➤ In addition, Durkheim observed that whenever any group of people has a strong conviction, that conviction almost always takes on a religious character. Religious gatherings become vehicles for affirming convictions and mobilizing the group to uphold them, especially when those convictions are threatened. People on both sides of controversial issues such as same-sex marriage evoke "God" or a higher power as supportive of their convictions.

The fact that the variety of religions is endless and that religion functions to meet societal needs led Durkheim to reach a controversial but thought-provoking conclusion: the "something out there" that people worship is actually society. In reaching this conclusion, Durkheim asked, How else might we explain the endless variety of religious response? The answer is that people create everything encompassed by religion—images of gods, rites, sacred objects. That is, people play a fundamental role in determining what is sacred and how to act in the presence of the sacred. Consequently, at some level, people worship what they (or their ancestors) have created. This point led Durkheim to conclude that the real object of worship is society itself—a conclusion that many critics cannot accept (Nottingham 1971).

AP Photo/Ed Reinke

➤ Artist Stephen Sawyer is shown sitting in front of his artwork titled "Call to Repentance." Sawyer, a self-described Christian, says that he paints these images of Christ on his own terms. Given that Jesus was born in Bethlehem (according to the Christian Bible), a town in the Middle East, is the image an accurate representation of Jesus's physical appearance? Archaeological evidence suggests that the average man at the time of Jesus was 5 feet, 3 inches tall and weighed approximately 110 pounds (Gibson 2004). Sawyer's presentation of Jesus Christ lends support to Durkheim's idea.

The Opiate of the People

Conflict theorists focus on ways in which people use religion to repress, constrain, and exploit others and how religion turns people's attention away from injustice and inequality. This perspective draws inspiration from the work of Karl Marx (1843), who believed that religion was the most humane feature of an inhumane world and that it arose in response to the tragedies and injustices of human experience.

Molly Hayden, U.S. Army Garrison-Hawaii Public Affairs

➤ Marx described religion as the "sigh of the oppressed creature, the heart of a heartless world, and the soul of soulless conditions. It is the opiate of the people." According to Marx, people need the comfort of religion to make the world bearable and to justify their existence. In this sense, he said, religion is analogous to a sedative.

Even though Marx acknowledged the comforting role of religion, he focused on its repressive, constraining, and exploitative qualities. In particular, he conceptualized religion as an ideology that justifies the status quo. That is, religion is used to rationalize existing inequities or downplay their importance. This aspect of religion is especially relevant with regard to the politically and economically disadvantaged. For them, Marx said, religion serves as a source of **false consciousness**. That is, religious teachings encourage the oppressed to accept existing economic, political, and social arrangements that limit their opportunities in this life because they are promised compensation for their suffering in the next world. This thinking inhibits protest and revolutionary change. Marx went so far as to claim that religion would be unnecessary in a truly classless society—one without material inequality—because by definition exploitation and injustice would not exist.

Critics of the conflict perspective argue that it underestimates the power of religion to inspire people to confront inequalities and injustices. Sometimes religion is a catalyst that inspires change and justice.

Prints & Photographs Division, Library of Congress, LC-U9-30656B-10

◂ Historically, African American churches have reached out to millions who have felt excluded from the U.S. political and economic system. For example, African American churches contributed greatly to the overall successes of the civil rights movement. Indeed, some observers argue that the movement would have been impossible if the churches had not been involved (Lincoln and Mamiya 1990).

The Protestant Work Ethic

Max Weber wanted to understand the role of religious beliefs in the origins and development of modern capitalism—an economic system that involves the careful calculation of costs of production relative to profits, the borrowing and lending of money, the accumulation of capital, and the drawing of labor from around the world to make products (R. Robertson 1987). In his book *The Protestant Ethic and the Spirit of Capitalism*, Weber (1958) asked why modern capitalism emerged and flourished in Europe rather than in China or India. He also asked why business leaders and capitalists in Europe and the United States were overwhelmingly Protestant.

To answer these questions, Weber focused on how different religious traditions supported different economic orientations and motivations. Based on his comparisons, Weber concluded that a branch of Protestant tradition—Calvinism—supplied a "spirit" or an ethic that supported capitalism. In particular, Calvinism emphasized **this-worldly asceticism**—a belief that people are instruments of divine will and that God determines and directs their activities. Calvinists glorified God when they accepted a task assigned to them, carried it out in an exemplary and disciplined fashion, and did not indulge in the fruits of their labor (that is, they did not use money to eat, drink, or otherwise relax in excess). In addition, Calvinists conceptualized God as all-powerful and all-knowing; they also emphasized **predestination**—the belief that God has foreordained all things, including the salvation or damnation of individual souls. According to this doctrine, relatively few people were destined to attain salvation, and people could do nothing to change their fate.

Weber maintained that beliefs in this-worldly asceticism and predestination created a crisis of meaning among adherents. This crisis led them to look for concrete signs that they were among God's chosen people, destined for salvation. Consequently, accumulated wealth became an important indicator of whether one was among the chosen. At the same time, this-worldly asceticism "acted powerfully against the spontaneous enjoyment of possessions; it restricted consumption, especially of luxuries" (Weber 1958, 171). Frugal behavior supported investment and the accumulation of wealth—important actions for the success of capitalism.

Prints & Photographs Division, Library of Congress, LC-DIG-ppmsca-05929

◀ The print features men who are considered empire builders in U.S. history, including Andrew Carnegie, Cornelius Vanderbilt, and John D. Rockefeller. The caption (not shown here) reads: "Those Christian men to whom God in his infinite wisdom has given control of the property interests of the country." These words speak to a belief that people are instruments of divine will.

Do not misread the role that Weber attributed to the Protestant ethic in the rise of a capitalist economy. According to Weber, that ethic was a significant ideological force; it was not the sole cause of capitalism but "one of the causes of certain aspects of capitalism" (Aron 1969, 204). Unfortunately, many people who encounter Weber's ideas draw a conclusion that Weber himself never reached: the reason that some groups and societies are disadvantaged is simply that they lack this Protestant work ethic. Finally, note that Weber was writing about the origins of industrial capitalism, not about the form of capitalism that exists today, which places a heavy emphasis on consumption and self-indulgence. He maintained that once established, capitalism would generate its own norms and become a self-sustaining force. In such circumstances, religion becomes an increasingly insignificant factor in maintaining the capitalist system (Aron 1969).

(Write a Caption)

Write a caption that explains why there is so much variety across religion as to what and who is considered sacred.

Hint: In writing this caption

- determine which of the three major theorists—Weber, Durkheim, or Marx—best helps to explain religious variety.

Tony Rotundo

Tony Rotundo

Critical Thinking

Whose ideas about religion—Max Weber, Émile Durkheim, or Karl Marx—do you find most compelling? Why?

Key Terms

church	profane	sacred
false consciousness	religion	this-worldly asceticism
predestination	rituals	

Civil Religion and Fundamentalism

MODULE

(12.5)

Objective

You will learn that civil religion and fundamentalism share some essential characteristics.

Do you recognize the figure looming over actor Laurence Fishburne, who is speaking to an audience on Memorial Day, a national holiday in the United States? Is there a religious quality about this scene?

DoD photo by Mass Comm. Spc. First Class Chad J. McNeeley, U.S.

The figure behind Fishburne is Abraham Lincoln, the 16th president of the United States, who has been memorialized in countless ways including through a national monument (Lincoln Memorial), postage stamps, coins, paper money, and the cities, towns, and streets named after him. In addition, Lincoln is memorialized through often-repeated phrases attributed to him ("government of the people, by the people, for the people, shall not perish from the earth" and "the proposition that all men are created equal").

Sociologist Émile Durkheim defined religion as a system of shared rituals and beliefs about what is sacred and what is profane that bind together a community of worshipers. This combination of characteristics applies not just to religion but to religious-like events such as national holiday celebrations. Thus this combination of characteristics also applies to **civil religion**, an institutionalized set of beliefs about a nation's past, present, and future and a corresponding set of rituals. These beliefs and rituals can take on a sacred quality and elicit feelings of patriotism.

Civil Religion

A nation's values, such as individual freedom and equal opportunity, and its rituals (parades, fireworks, singing the national anthem, 21-gun salutes), often assume a sacred quality. Even in the face of internal divisions centered on race, religion, or gender, national rituals and shared values can inspire awe, respect, and reverence for the country and what it stands for. These sentiments are most evident during times of crisis and war, on national holidays (Veterans Day), and in the presence of national monuments or symbols (the Vietnam Memorial, the flag).

David McNally (USAG-Yongsan)

Chris Caldeira

➤ In many societies leaders defined as great are memorialized and the things they stood for are treated as sacred to the nation. Koreans celebrate King Sejong (left), who is credited with inventing an alphabet known as Hangeul with clear rules governing pronunciation of characters. Sejong created this alphabet with the goal of encouraging literacy among all people, not just the elite. Hangeul Day is celebrated on October 9. Che Guevara (right) is a revolutionary leader celebrated in Cuba for his role as physician to rebel forces and key strategist in the 26th of July 1959 Movement after which Cuba was declared a socialist country. Guevara was also active in other "third world" revolutionary movements. He was killed at age 39 while fighting the Bolivian army.

Sociologist Roberta Cole (2002) argues that America's civil religion found its voice in a 19th-century political doctrine known as manifest destiny, a longstanding ideology that the United States, by virtue of its moral superiority, was destined to expand across the North American continent to the Pacific Ocean and beyond (Chance 2002). Manifest destiny included the beliefs that the United States had a divine mission to serve as a democratic model to the rest of the world, that the country was a redeemer exerting its good influence upon other nations, and that it represented hope to the rest of the world (Cole 2002).

Especially in times of war, presidents offer a historical and mythological framework that gives moral justification for the involvement and offers the public a vision and a national identity. We consider statements U.S. presidents have made to illustrate how in times of war the nation's core values can assume a sacred quality and to illustrate the language presidents use to project moral certainty.

> We are Americans, part of something larger than ourselves. For two centuries, we've done the hard work of freedom. And tonight, we lead the world in facing down a threat to decency and humanity. . . . Among the nations of the world, only the United States of America has both the moral standing and the means to back it up. . . . This is the burden of leadership and the strength that has made America the beacon of freedom in a searching world. (George H. W. Bush [1991] upon sending 540,000 troops to the Persian Gulf in 1990)

> All of you . . . have taken up the highest calling of history . . . and wherever you go, you carry a message of hope—a message that is ancient and ever new. In the words of the prophet Isaiah, "To the captives 'come out,' — and to those in darkness, 'be free.'" (George W. Bush [2003] upon sending troops to Iraq in 2003)

> . . . tonight, we are once again reminded that America can do whatever we set our mind to. That is the story of our history, whether it's the pursuit of prosperity for our people, or the struggle for equality for all our citizens; our commitment to stand up for our values abroad, and our sacrifices to make the world a safer place.
>
> Let us remember that we can do these things not just because of wealth or power, but because of who we are: one nation, under God, indivisible, with liberty and justice for all. (Barack Obama [2011] upon announcing the death of Osama bin Laden in 2011)

Critics of war liken the moral certainty characteristic of civil religion to "a kind of fundamentalism" and a "dangerous messianic brand of religion, one where self-doubt is minimal" (Hedges 2002).

Fundamentalism

Anthropologist Lionel Caplan (1987) defines **fundamentalism** as a belief in the timelessness of sacred writings and a belief that the words apply to every setting. In its popular usage, the term *fundamentalism* is applied to a wide array of religious groups around the world, including the religious right in the United States, Orthodox Jews in Israel, and various Islamic groups in the Middle East. Religious groups labeled as fundamentalist are usually portrayed as "fossilized relics . . . living perpetually in a bygone age" (Caplan 1987, 5). Americans frequently use the term *fundamentalism* to explain events involving people in the Middle East, especially political turmoil that threatens the interests of the United States. Such oversimplification misrepresents fundamentalism.

Perhaps the most important characteristic of fundamentalists is their belief that a relationship with God, Allah, or some other supernatural force provides answers to personal and societal problems. In addition, fundamentalists often wish to "bring the wider culture back to its religious roots" (Lechner 1989, 51). In this regard, fundamentalists emphasize the authority, infallibility, and timeless truth of sacred writings as a definitive blueprint for life. This characteristic does not mean that a definitive interpretation of sacred writings actually exists. Indeed, any sacred text has as many interpretations as there are groups that claim it as their blueprint.

Second, fundamentalists usually conceive of history as a "process of decline from an original ideal state," which includes the "betrayal of fundamental principles" (18). They see human history as a cosmic struggle between good and evil: the good embodies the principles outlined in sacred scriptures, and the evil results from countless digressions from those principles. To fundamentalists, truth is not relative, varying across time and place. Instead, truth is unchanging and knowable through the sacred texts.

Third, fundamentalists do not distinguish between what is sacred and what is profane in their day-to-day lives. Religious principles govern all areas of life, including family, business, and leisure. Religiously inspired behavior does not take place only in a church, a mosque, or a temple. For example, religious principles guide the business of producing kosher meats, which are prepared according to strict Jewish dietary laws that are considerably stricter than USDA standards. Kosher laws govern how animals are fed, killed, and processed. Kosher laws prohibit, for example, slaughtering cows with broken bones or that are known to be sick.

ACU photo, Ingrid Barrentine, Childhood photo courtesy of Spc. Simranpreet Singh Lamba

◀ For fundamentalists, religious principles cannot be dismissed out of convenience or because they clash with another set of principles. When U.S. Army Specialist Simranpreet Singh Lamba was a child growing up in India (left) he dreamed of joining the military. When he came to the United States in 2006, he thought he could never join the U.S. military because as a devout Sikh, he could not cut or shave his hair and he wore a turban. His religious beliefs conflicted with military dress codes. But the U.S. military made an exception for Lamba and two other Sikh soldiers. Sikhs value "the principles of justice, equality and truth" and the religion "emphasizes service to others, particularly in the armed forces" (Petrich 2011).

Fourth, fundamentalist religious groups emerge for a reason, usually in reaction to a threat or crisis, whether real or imagined. Consequently, any discussion of a particular fundamentalist group must include some reference to an adversary—that adversary may be capitalism, another religious group, feminists, or secularists.

Finally, fundamentalists believe that the trends toward gender equality are symptomatic of a declining moral order and need to be reversed. In fundamentalist religions, women's rights often become subordinated to ideals that the group considers more important to the well-being of the society, such as the traditional family or the "right to life." Such a priority of ideals is regarded as the correct order of things.

ISLAMIC FUNDAMENTALISM. We cannot apply the term *fundamentalism* to the entire Muslim world, especially when we consider that Muslims make up the majority of the population in at least 45 countries (U.S. Central Intelligence Agency 2012). Religious studies professor John L. Esposito (1986) believes that a more appropriate term is *Islamic revitalism* or *Islamic activism*. While the form of Islamic revitalism may vary from one country to another, it seems to be characterized by a belief that existing political, economic, and social systems have failed; a disenchantment with, and even rejection of, the West; soul-searching; a quest for greater authenticity; and a conviction that Islam offers a viable alternative to secular nationalism, socialism, and capitalism.

Esposito asks, Why has religion become such a visible force in Middle East politics? He believes that Islamic revitalism represents a "response to the failures and crises of authority and legitimacy that have plagued most modern Muslim states" (1986, 53). Those crises can be traced to France and Great Britain's division of the Middle East into nation-states after World War I.

The citizens of these Western-created countries viewed many of the leaders who took control as autocratic, corrupt, and "propped up by Western governments and multinational corporations" (1986, 54). Questions of social justice also arose as oil wealth and modernization policies opened up a vast chasm between the oil-rich countries, such as Kuwait and Saudi Arabia, and the poor, densely populated countries, such as Egypt, Pakistan, and Bangladesh. Western capitalism, which was seen as one of the primary forces behind these trends, seemed blind to social justice, instead promoting unbridled consumption and widespread poverty.

For many people, Islam offers an alternative vision for society. According to Esposito (1986), five beliefs guide Islamic activists who follow many political persuasions, ranging from conservative to militant:

1. Islam is a comprehensive way of life relevant to politics, law, and society.
2. Muslim societies fail when they depart from Islamic ways and follow the secular and materialistic ways of the West.
3. An Islamic social and political revolution is necessary for renewal.
4. Islamic law must replace laws inspired or imposed by the West.
5. Science and technology must be used in ways that reflect Islamic values, to guard against the infiltration of Western values.

Muslim groups differ dramatically in their beliefs about how quickly and by what methods these principles should be implemented. Most Muslims, however, are willing to work within existing political arrangements; they condemn violence as a method of bringing about political and social change. However, some do use violence to try to effect change, and in those cases the concept of *jihad* is often evoked. In thinking about the meaning of *jihad*, it is important to distinguish between religious and political *jihad*. Many Islamic scholars have pointed out that in the religious sense of the word, true *jihad* is the "constant struggle of Muslims to conquer their inner base instincts, to follow the path to God, and to do good in society" (Mitten 2002). But other scholars point out that *jihad* as used by the most radical organizations targets anyone—Muslim or non-Muslim—who stands in the way of their goals.

Secularization

In the most general sense, **secularization** is a process by which religious influences on thought and behavior are gradually removed or reduced. More specifically, it is a process by which some element of society, once part of a religious sphere, separates from its religious or spiritual connection or influences. It is difficult to generalize about the causes and consequences of secularization because they vary across contexts. Americans and Europeans tend to associate secularization with an increase in scientific understanding and in technological solutions to everyday problems of living. In effect, "science and technology give people control over life and death matters once left in the hands of God" (Esposito 1986, 54).

Muslims, in contrast, tend not to attribute secularization to science or to modernization; indeed, many devout Muslims are physical scientists. From a Muslim perspective, secularization is a Western-imposed phenomenon—specifically, a result of exposure to what many people in the Middle East consider the most negative of Western values as reflected in its consumer culture, its violent movies and television programs, and the way its popular culture portrays women and sexualizes young children.

Spc. Leslie Angulo

➤ Here we see orphaned Afghan girls holding up dolls and other toys given to them by U.S. troops. One might argue that Western secular principles governed the design and creation of dolls and their wardrobes. What does it mean to these girls to receive such toys, which contradict fundamental Islamic beliefs about modest dress and other aspects of life?

(Write a Caption)

Write a caption that draws upon the three essential features of religion to describe what is common to people of Muslim faith gathered to pray and fans attending opening ceremonies of the U.S. Tennis Open.

Staff Sgt. Cecilio Ricardo, USAF

Capt. Dan Huvane

Hints: In writing this caption

- review Durkheim's definition of religion, and
- relate the essential features described in that definition to the two scenes.

Critical Thinking

Can you think of a nonreligious ceremony or event in which participation takes on a religious quality?

Key Terms

civil religion
fundamentalism
secularization

Applying Theory: Social Reproduction

MODULE 12.6

Objective

You will learn how the concept of habitus applies to the way the educational system perpetuates social class.

Imagine that a professor projects this photograph onto a screen and asks you to comment about it in writing. What would you write?

Prints & Photographs Division, Library of Congress, LC-DIG-ppmsca-03183

Sociologist Pierre Bourdieu used this line of questioning to document the perceptual schemes that individuals draw upon to think about and react to the world around them. Working-class respondents tend to use plain, concrete language to describe the woman's hand (e.g., "This woman looks like she's got arthritis. Her hands are all knotted. I feel sorry seeing that poor old woman's hands.") Respondents from more advantaged classes tend use abstract, aesthetic language that transcends the situation of the particular woman pictured: "This photograph is a symbol of toil. It puts me in mind of Flaubert's old servant-woman . . . It's terrible that work and poverty are so deforming" (Bourdieu 1984).

Bourdieu was interested in how these perceptual schemes or points of view come into being. He found that the schemes people draw upon are shaped in large part by their social position (Appelrouth and Edles 2007). Bourdieu's work falls under the category of contemporary theoretical synthesis, because he worked to bridge the divide between the individual and society.

Bourdieu maintained that the social positions people come to occupy depend on the amount of economic and cultural capital upon which they can draw. **Economic capital** refers to a person's material resources—wealth, land, money. **Cultural capital** refers to a person's nonmaterial resources, including educational credentials (see Figure 12.6a), the kinds of knowledge acquired, social skills, and aesthetic tastes. Simply consider that children who play soccer gain a kind of cultural capital in that they are exposed to a sport that is considered global in its reach. As a result they have something in common with billions of people around the world. The ability to play soccer and knowledge of its rules opens them to a network of players and fans that extends beyond the community and country in which they live.

Cultural and economic capital are distributed unequally throughout society. When people locate themselves relative to others, they gain a sense of their place in society, and of what is objectively possible.

▾ Figure 12.6a: Distribution of Educational Credentials, Median Income, and Unemployment Rate among U.S. Population 25 Years and Over
The figure represents the distribution of a specific kind of cultural capital—educational credentials. Notice that 15 percent of the population has less than a high school diploma. Among those with less than a high school diploma, note that there is a 14.1 percent unemployment rate, and the median income is $451 per week.

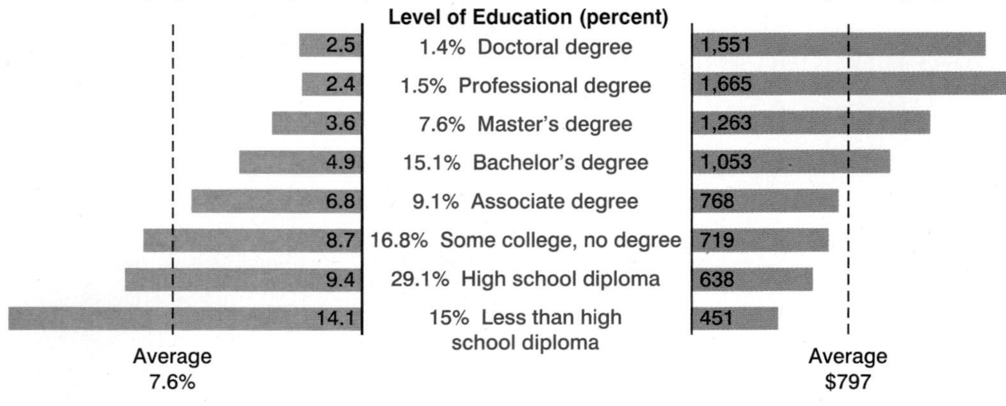

Source: U.S. Bureau of Labor Statistics (2012)

According to Bourdieu's theory, high school dropouts come to know or internalize what is objectively possible for someone with their educational credentials. As a result, they are *likely* to expect an income of $451 per week as reasonable and may even aspire to earn that income. They are also likely to assume that higher incomes are out of reach and to feel that they are just a step away from poverty. Someone with a professional degree, on the other hand, will likely *expect* to live free of poverty and to earn a salary of at least $1,665 per week.

Habitus and Social Reproduction

The **habitus** is a frame of mind that has internalized the objective reality of society. This objective reality becomes the mental filter that structures people's perceptions, experiences, responses, and actions. It is through the habitus that the social world is understood and that people acquire a sense of place and a point of view that informs how they interpret their own and others' actions. The habitus plays a vital role in a process sociologists call **social reproduction**, the perpetuation of unequal relations so that almost everyone, including the disadvantaged, come to view this inequality as normal and legitimate and tend to shrug off or resist calls for change.

➤ Bourdieu believed that the habitus also affects how people physically hold themselves and move about in the world (e.g., posture, facial expressions, gestures). Imagine you are a teacher or school administrator. How might the way each student holds and presents his body affect your perception of academic ability?

Lance Cpl. John Robbart III

Molly Hayden, U.S. Army Garrison Grafenwoehr, Public Affairs Office

According to Bourdieu, no institution does more to ensure the reproduction of inequality than education (Appelrouth and Edles 2007). He argues that the system of education is widely misperceived as meritocratic—that grades are awarded according to demonstrated academic abilities and not family connections or class privilege. We point to tests as fair and objective measures of academic performance. Tests dominate the educational experience in that students' mental energies are organized around taking them and their grades are largely dependent on their test results.

© Corporation for National and Community Service, Office of Public Affairs

⮜ Bourdieu sees the exam as the "clearest expression of academic values" (1984, 142). Tests are treated as a valid measure of knowledge. Thus, test performance is the socially accepted, largely unchallenged way to demonstrate that a specified body of knowledge has been acquired.

Upon close scrutiny, however, educational systems and the tests used to grade and sort students actually perpetuate preexisting inequalities. An honest assessment of tests shows that they are not objective measures, if only because those who create tests are likely to be from academically advantaged backgrounds. Part of doing well on tests involves figuring out what test takers want and see as being important. Bourdieu argues that it is harder for students from disadvantaged backgrounds to figure this out because they draw on different cultural capital to process the material on which they are being tested.

Bourdieu believes that most students who drop out make the decision to do so, not the school system. For these students, studying for tests is what education is. Therefore, it is the prospect of studying for tests, failing tests, and a dislike for the kinds of knowledge one must possess to pass tests that spurs them to "voluntarily" drop out. Bourdieu views examinations as one of the most powerfully effective ways of impressing upon students the dominant culture and its values. Bourdieu argues that we can gain insights about the inequalities perpetuated by the educational system by identifying those groups most likely to drop out by declining to return to school after summer breaks or simply disappearing during the school year. Bourdieu urges us to grasp the significance of the differential educational mortality rate, illustrated in Chart 12.7a.

Chart 12.7a: Probability of Graduating from High School by Race and Ethnicity
The chart shows the percentages of ninth-graders by racial classification who graduate from high school four years after entering ninth grade. It is also a rough indicator of the percentages of high school students who "survive" the system. The survival rate is highest for those classified as Asian/Pacific Islanders and white and lowest for those classified as black.

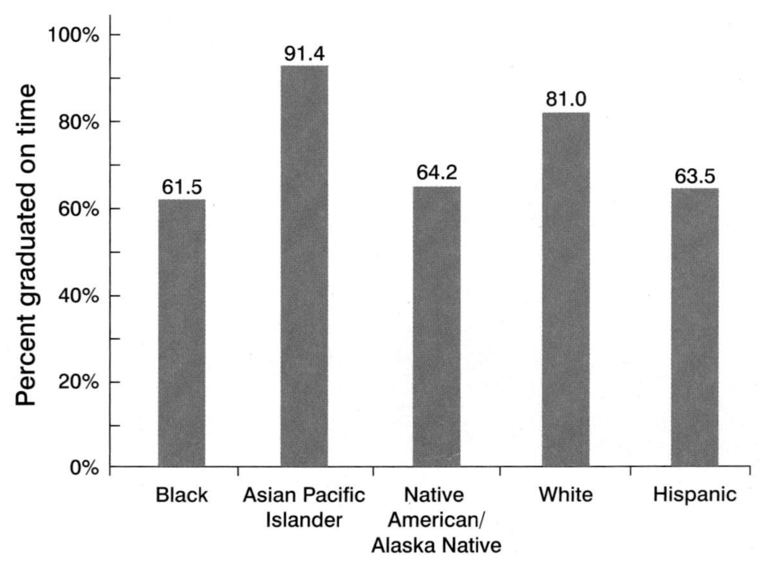

Source: Stillwell (2010)

Bourdieu argues that the act of self-elimination both reflects and sustains the preexisting hierarchy. Thus, to give a full account of the process by which people are sorted, selected, and eliminated within the educational system, we must take into account not only the judgments of educators, which are supported by test scores but also the convictions that disadvantaged individuals hold about themselves.

Critical Thinking

Use Bourdieu's theory to frame your educational experiences.

Key Terms

cultural capital
economic capital
habitus
social reproduction

Summary: Putting It All Together

Although every experience has the potential to educate, sociologists who study education tend to emphasize schooling, a deliberate, planned effort that takes place in a brick-and-mortar or virtual classroom to impart specific skills or information. Symbolic interactionists focus on the role curriculum plays in conveying meaning and a sense of self. From the functionalist perspective, schools perform a number of important functions. The conflict perspective draws our attention to unequal power arrangements between advantaged and disadvantaged groups. The credential society represents one such arrangement that advantages college graduates over those with less education. Here employers give hiring, promotion, and salary advantages to those holding degrees even over those with a proven on-the-job track record.

When sociologists study curriculum they look at racial differences in learning and achievement. Generally speaking, minority students as a group do not do as well as their white and Asian counterparts on standardized testing. The lower scores are a reflection of many factors including family background, neighborhood, and peer group (adolescent society) influences. These factors are connected to concepts of economic and cultural capital—the resources students bring to the classroom and draw upon to negotiate the educational experience, including tests.

Religions (past and present) have three essential features: (1) beliefs about the sacred and the profane, (2) rituals, and (3) a community of worshipers. The religious variety that exists can be attributed to the fact that people create everything encompassed by religion—images of gods, rites, and sacred objects. That is, people play a fundamental role in determining what is sacred and how to act in the presence of the sacred. Consequently, at some level, people worship what they (or their ancestors) have created.

Because some form of religion appears to have existed for as long as humans have, functionalists maintain that religion must serve some vital social functions for the individual and for the group. In this regard, Durkheim observed that whenever any group has a strong conviction, that conviction almost always takes on a religious character. Religious gatherings function as vehicles for affirming convictions and mobilizing the group to uphold them, especially when those convictions are threatened. Conflict theorists inspired by Marx emphasize religion's repressive, constraining, and exploitative qualities. Weber focused on understanding how norms generated by different religious traditions influenced adherents' economic orientations and motivations. Specifically, Weber concluded that a branch of the Protestant tradition—Calvinism—supplied a "spirit" or an ethic that supported the motivations and orientations of capitalism.

Civil religion is an institutionalized set of beliefs about a nation's past, present, and future and a corresponding set of rituals that take on a sacred quality and elicit feelings of patriotism. The dynamics of civil religion are most notable during times of crisis and war and on national holidays. During these times civil religion can take on qualities we associate with fundamentalism.

Go to cengagebrain.com to link to Aplia and CourseMate for the chapter quiz and other activities.

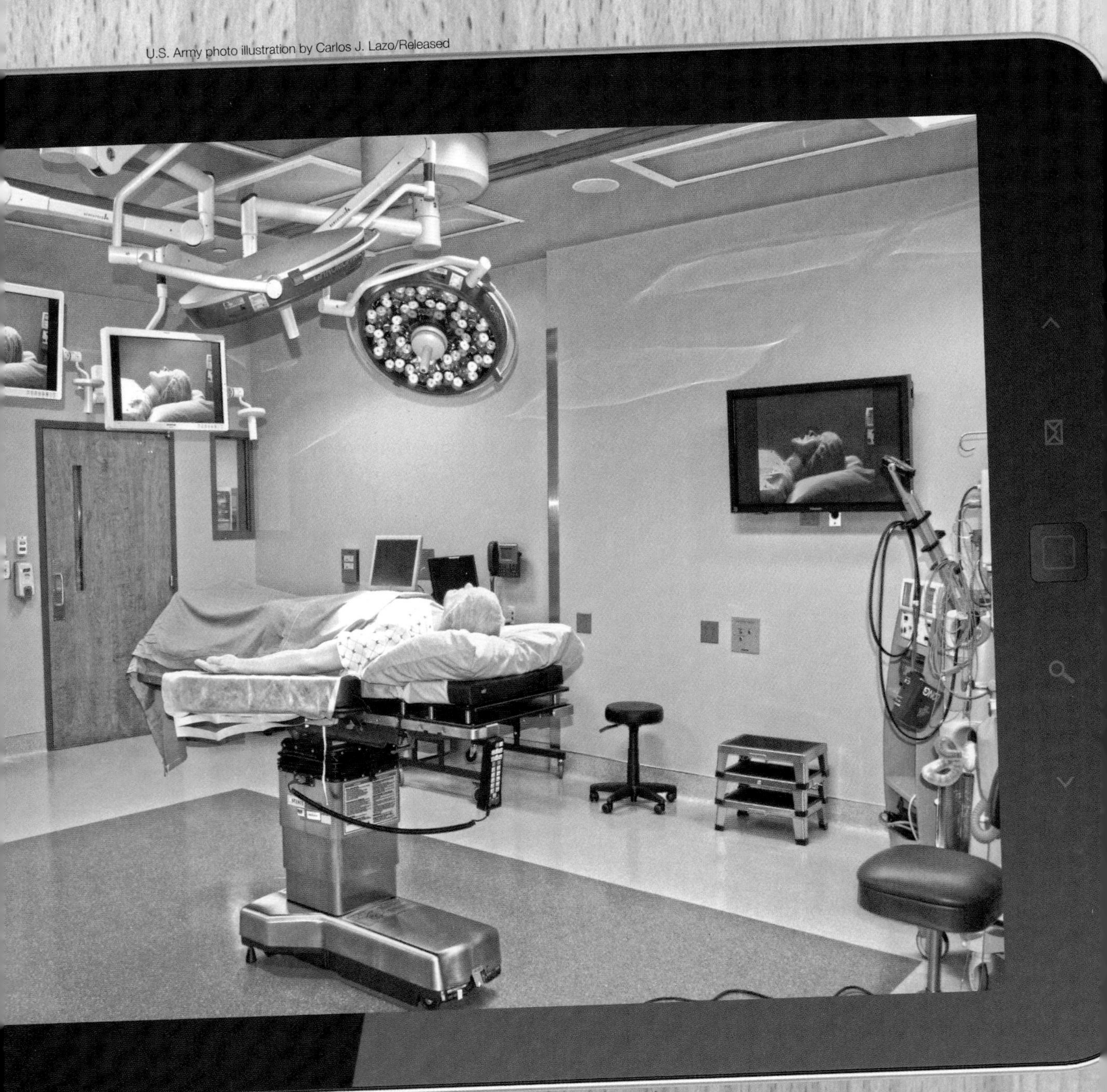

U.S. Army photo illustration by Carlos J. Lazo/Released

CHAPTER 13

SOCIAL CHANGE AND HEALTH CARE

When sociologists study social change, they examine the way human interaction and other activities are transformed over time. In the process, they work to identify the social forces contributing to such change. One of the most significant forces impacting humans—their relationships to machines and each other, their livelihood and lifestyles, and their private and professional lives—is robotics. Robotics is changing virtually every area of life. Simply consider robotics' effect on the way health care is delivered. Remote surgery (telesurgery), for example, allows doctors to operate on patients from another location, making the physical distance between patient and surgeon immaterial (*ScienceDaily* 2010). Robotics is one innovation in a long line of technologies, beginning arguably with the plow, that has fundamentally changed the way people relate to each other and the environment.

Go to Sociology CourseMate on cengagebrain.com to watch a video about how where you live can impact your life expectancy.

MODULE 13.1

Social Change: The Case of Robotics

Objective

You will learn the factors that trigger social change and the consequences of change.

U.S. Air Force

Have you ever been part of something considered cutting edge?

This robot named DieselZilla was built by students at American River College. The robot is helping to plant a tree on Earth Day. Robotic technology may be at a **tipping point**, a situation in which what was once a rare practice or event snowballs into something dramatically more common. The process by which the rare becomes commonplace is at first gradual; changes go largely unnoticed or are dismissed as insignificant. But at some point a critical phase is reached so that the next increment of change "tips" the scale in a dramatic way.

U.S. Navy photo by John F. Williams/Released

Missy Gish

◂ How much closer can this egg be pushed to the edge of the tabletop before it can no longer sustain its position and falls to the floor? The point immediately before that moment is the tipping point. Likewise, there is a tipping point with robots. Here we see spectators fascinated by a robot. Are we at the tipping point, the moment before robots become commonplace?

A robot is an electromechanical machine or agent programmed to do specified tasks with varying degrees of guidance from humans. Robots are more than mechanical or motorized devices because they are designed and programmed to process information about something or someone before responding. Today, robots are used to do a wide array of tasks including mowing lawns (service robots), inspecting suspicious packages for improvised explosive devices (military robots), delivering a drug or antibodies to a specific site in the human body (nanorobots), exploring the sea and outer space (explorer robots), moving artificial limbs (prosthetics robots), and providing customer service and virtual experiences (chatbots and virtual robots).

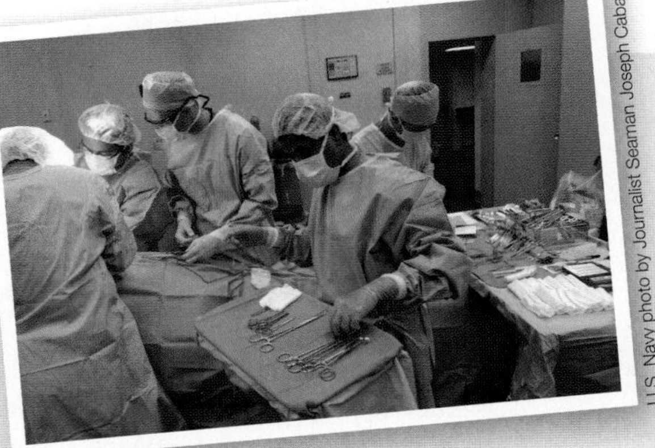
U.S. Navy photo by Journalist Seaman Joseph Caballero

➤ Who in this operating room is in danger of having their job replaced by a robot? The surgical nurse handing instruments to the physician is at risk. A robot known as Gestonurse is programmed to hand surgical instruments to a surgeon and keep track of those instruments, eliminating the chance that one might be sewn into the patient. Another surgical robot known as DaVinci will certainly not eliminate the need for surgeons, but it will reduce the number of surgeons needed to perform complex operations. Likewise, anesthesiologists charged with administering drugs and monitoring vital signs during operations face competition from a robot called McSleepy, who is programmed to do those very tasks without becoming distracted.

Sociologists define **social change** as any significant modification in the way social activities and human relationships are organized. When sociologists study change, they ask at least three key questions: What has changed? What factors trigger change? What are the consequences of the change?

What Has Changed?

Social change is an important sociological topic. In fact, sociology first emerged as an attempt to understand the dramatic social changes accompanying the Industrial Revolution. As a result, sociology offers a variety of concepts that focus our attention on revolutionary changes. Those concepts include industrialization, globalization, rationalization, McDonaldization, urbanization, and the information explosion. In this module we will learn how robotics is embedded in each of these revolutionary changes.

INDUSTRIALIZATION. The Industrial Revolution changed our society into a **hydrocarbon society**—one in which fossil fuels power machines from alarm clocks to photocopiers. In this sense fossil fuels shape virtually every aspect of our personal and social lives. The energy source that lights the night, cools our houses and offices, powers appliances, and charges our mobile phones is likely coal. The energy source that heats our houses, workplaces, and water is likely natural gas; the energy source that enables trains, planes, cars, and buses to move people and goods is most certainly oil. It is fair to say that industrialization wove energy consumption into the fabric of society.

Veterans Affairs and DEKA

◀ This robotic arm is a contemporary illustration of the extent to which we are still a hydrocarbon society. The arm uses sensors to process data that signals the arm and hand to move a certain way, but the sensors and arm cannot work without energy from battery packs that are regularly charged with electricity generated by fossil fuels.

GLOBALIZATION. Sociologists distinguish between global interdependence and globalization. **Global interdependence** describes a situation where human activity transcends national borders. Examples are endless but include travel, trade, texting, phone calls, and outsourcing. Global interdependence also applies to social issues such as water shortages, pollution, unemployment, and drug addiction as they are part of the larger global situation. Because the level of global interdependence is constantly increasing, it is part of a dynamic process known as **globalization**—the ever-increasing flow of goods, services, money, people, information, and culture across political borders.

Sgt. Kevin Stabinsky

▶ Most of us know that it is the hands of workers in foreign countries that sew clothes we wear. Likewise, most of us know our mobile phones are made by workers in Asian factories, most notably factories located in China. The efficiency of these factories is such that there are six billion mobile phones in the hands of people all over the world. Personal robots, projected to be as commonplace as mobile phones, will also be manufactured and assembled by hands in factories located in China, India, Korea, Eastern Europe, and Latin America.

In addition, countries specialize in specific kinds of robots. Korea, for instance, is considered a world leader in creating service robots to take care of the elderly and to teach English. The United States, with the largest military budget in the world, is the leader in creating robots to protect soldiers on the battlefield. European companies are using robots to manufacture solar panels (Burnstein 2009).

RATIONALIZATION. The term *rationalization* refers to changes in the way daily life has come to be organized since industrialization to accommodate large numbers of people. Rationalization is a process whereby thought and action rooted in emotion (such as love, hatred, revenge, or joy), superstition, respect for mysterious forces, and tradition are replaced by thought and action rooted in the most efficient (fastest and/or cheapest) ways to achieve a valued goal. Robotics can be analyzed in the context of rationalization. Consider how the use of rescue robots on the battlefield could change how the military trains its soldiers.

One of the hallmark goals of military training is to establish strong emotional ties among soldiers and instill in them the directive to "leave no soldier behind." Soldiers are willing to put their lives on the line for one another because of the love for those in the unit cultivated during military training.

TATRC

< This search-and-rescue robot is able to lift and carry people and objects weighing up to 500 pounds across all terrains and to grasp fragile objects without damaging them. While the robot represents an efficient tool for saving lives without risking the life of another, how might its presence reduce the need to cultivate strong emotional ties among soldiers?

McDONALDIZATION. This organizational trend refers to a process whereby the principles governing the fast-food industry come to dominate other sectors of the economy (Ritzer 1993). Those principles are efficiency, quantification and calculability, predictability, and control (see Module 4.4). For example, pharmacies, car washes, banks, and liquor stores have adopted drive-thru services to facilitate their goal of efficiently accommodating customers. Of course, drive-thru service is not all there is to McDonaldization. Whatever the service offered—a college degree in 18 months, a medical clinic in a grocery store, matchmaking with success guaranteed in 6 weeks, a prepaid funeral, or the cheapest plane ticket—we can find one or more of the four McDonaldization principles operating.

> Here a student is learning to drive a car using a driving simulator (virtual robot). Learning through simulation has all the features of McDonaldization. It is efficient because the simulator allows the student to experience a range of driving situations in the shortest amount of time. The simulated experience will be the same for everyone (predictability), and the time spent practicing can be easily calculated. Lastly, students can control the pace of their learning.

U.S. Army Photo Files

URBANIZATION. **Urbanization** is a transformative process in which people migrate from rural to urban areas and change the ways they use land, interact, and earn a living. Today, urban populations encompass not only city dwellers but also suburbanites and even residents of small towns that have been pulled in by urban sprawl. Urban sprawl spreads development beyond cities by as much as 40 or 50 miles; puts considerable distance between homes, stores, churches, schools, and workplaces; and makes people dependent on automobiles (Sierra Club 2007). In addition, the automobile and highway system have permitted people to live in places that give them access to more space than they need for a comfortable life. In the past 30 years, the average house size has increased from 1,400 to 2,330 square feet (National Association of Home Builders 2007). Urban sprawl has blurred boundaries between cities, suburbs, and rural towns.

Urbanization and urban sprawl affect the amount of time people spend driving—time they might spend doing other things. Driverless or robotic cars allow people the freedom to work, read, or relax while on the road. In addition, these cars have the potential to make the roads safer and more efficient. Google, which is developing and testing driverless cars, envisions another scenario. "Vehicles would become a shared resource, a service that people would use when needed. You'd just tap on your smart phone, and an autonomous car would show up where you are, ready to drive you anywhere. You'd just sit and relax or do work" (Guizzo 2011).

Chris Caldeira

◂ Many cities already have car share programs in place. They give people 24/7 access to a fleet of fuel-efficient vehicles for an hourly fee of around $7.50. Currently, people have to walk to these cars, which are parked in strategic locations. If Google's vision is realized, people will use a smart phone app to call a driverless car to their location. What might this vision mean for taxicab drivers?

THE INFORMATION EXPLOSION. Sociologist Orrin Klapp (1986) defined the **information explosion** as an unprecedented increase in the amount of stored and transmitted data/information from all media sources (including electronic, print, radio, and television). Arguably, the information explosion began with the invention of the printing press. Today, the information explosion is driven by the Internet, a vast fossil fuel–powered computer network linking billions of computers around the world. The Internet has the potential to give users access to every word, image, and sound that has ever been recorded (Berners-Lee 1996). In addition to the Internet, advances in electronics, miniaturization, digitalization, and software programming give people devices that can collect, access, store, transmit, analyze, and manipulate data and other information in nanoseconds. Specifically, these four technologies have speeded up old ways of doing things and have changed how people and machines learn and process information. It is these four technologies that have contributed to the proliferation of robots. The robot brain consists of computer chips with memory and software programs that allow it to process commands and other prompts. Sensors allow the robot to gather information that it uses to complete tasks for which it is designed.

What Factors Trigger Change?

We usually cannot pinpoint a single factor that triggers a specific social change because change is part of an endless sequence of interrelated events. We can, however, identify key factors that trigger change in general. These triggers include, but are not limited to, technological innovation, revolutionary ideas, conflict, the pursuit of profit, and social movements.

TECHNOLOGICAL INNOVATION. In the broadest sense of the word, **technology** involves using knowledge, tools, applications, and other inventions in ways that allow people to adapt to and negotiate their surroundings. Technology can be material (mobile phones, hammers, and ATMs) or conceptual (ways of organizing human activity such as the assembly line or computer software that powers apps). Technologies change the structure of human activity and social relationships: they save time and can dissolve social and physical boundaries, as when X-rays allow physicians to see into the body. With regard to robotics, a countless number of technological innovations have moved the robot from the realm of science fiction to reality. For example, new innovations in software and sensor technologies allow robots and people to work side by side on a task (Schuster and Winrich 2009).

REVOLUTIONARY IDEAS. In *The Structure of Scientific Revolutions*, philosopher Thomas Kuhn (1975) defines paradigms as the dominant and widely accepted theories and concepts in a particular area or field of study that seem to offer the best way of looking at the world at a particular time. A scientific revolution occurs when enough people in the community reject an existing paradigm in favor of a new paradigm. The new paradigm causes converts to see the world in an entirely new light and to wonder how they could possibly have taken the old paradigm seriously. "When paradigms change, the world itself changes with them" (111).

CONFLICT. Sociologist Lewis Coser (1973) argues that conflict occurs when those who control valued resources strive to protect their interests from those who hope to gain a share. Conflict can be an invigorating force that prevents a social system from becoming stagnant, unresponsive, or inefficient. Conflict—whether it involves military buildups, violent clashes, or public debate—is both a consequence and a cause of change. In general, any kind of change has the potential to trigger conflict between those who stand to benefit from the change and those who stand to lose because of it.

➤ Robotic technologies have roots in conflict. The U.S. military has developed unmanned aerial vehicles (drones), robots that search for improvised explosive devices (see photo), and self-driving trucks and other vehicles. Military robots are controlled by human operators at some remote location.

Private First Class Matthew Clifton, USA

THE PURSUIT OF PROFIT. The capitalist system is a change agent because of the ways it makes profit. Most notably, capitalists must find ways to produce goods and services in the most cost-effective manner, and that includes eliminating labor. One example relates to the manufacturing giant Foxconn, which employs 1.2 million workers in China to make Apple iPads and other electronic devices. Foxconn plans to replace two million human hands with robotic "hands" as early as 2013. Robots will save labor costs of between $175 and $200 per month per employee. In addition, robots are more productive than human counterparts because they work faster, cannot be distracted, can work 24/7, and do not seek higher salaries or better working conditions (Yee and Jim 2011).

SOCIAL MOVEMENTS. Social movements occur when enough people organize to resist a change or to make a change. Usually those involved in social movements work outside the system to advance their cause because the system has failed to respond. To draw attention to their cause and accomplish their objectives, supporters may strike, demonstrate, walk out, or otherwise disrupt the social order.

To date there have been few, if any, major social movements directed against robots outside of protests against military drone strikes, for example, and workers who have likely gone on strike over robotic technologies. But one can readily imagine scenarios in which robotic technologies are used to suppress social movements or to ensure their success. As one example, some engineers are trying to develop what is known as swarm robotics, miniature flying robots that work together like a swarm of ants or bees to accomplish a task such as monitoring environmentally protected areas (e.g., rain forests) or to make sure that loggers or miners follow conservation laws. Swarm robots could also be used to monitor crowd activity and to identify "instigators" behind a social movement (Peck 2012).

What Are the Consequences of Change?

Sociology offers a rich set of concepts, theories, and questions to guide any analysis of social change and its consequences. Three basic questions can be applied to thinking about almost any change: What are the intended and unintended consequences of X? Who benefits from X and at whose expense? Does the emergence of X challenge, sustain, or alter existing meanings? Moreover, the sociological perspective can direct an analysis of change toward many different areas of social life. For example, how might robots change culture, the way people are socialized, the way people interact, the economy, the nature of work, and so on?

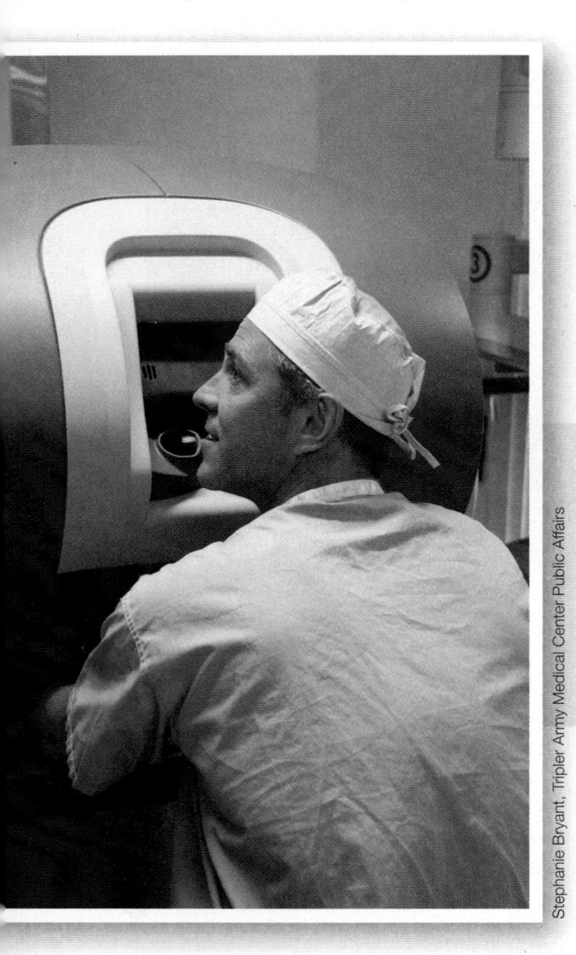

Stephanie Bryant, Tripler Army Medical Center Public Affairs

◀ This surgeon sits at da Vinci Surgical System, a robotic system where the surgeon "looks through two eye holes at a 3-D image of the procedure, meanwhile maneuvering the arms with two foot pedals and two hand controllers" (U.S. Army 2012). Can you think of ways this robotic system might change the surgeon's relationship to the patient?

(Write a Caption)

Write a caption that highlights how a robot that simulates the reactions of a patient is part of the revolutionary changes described in this module.

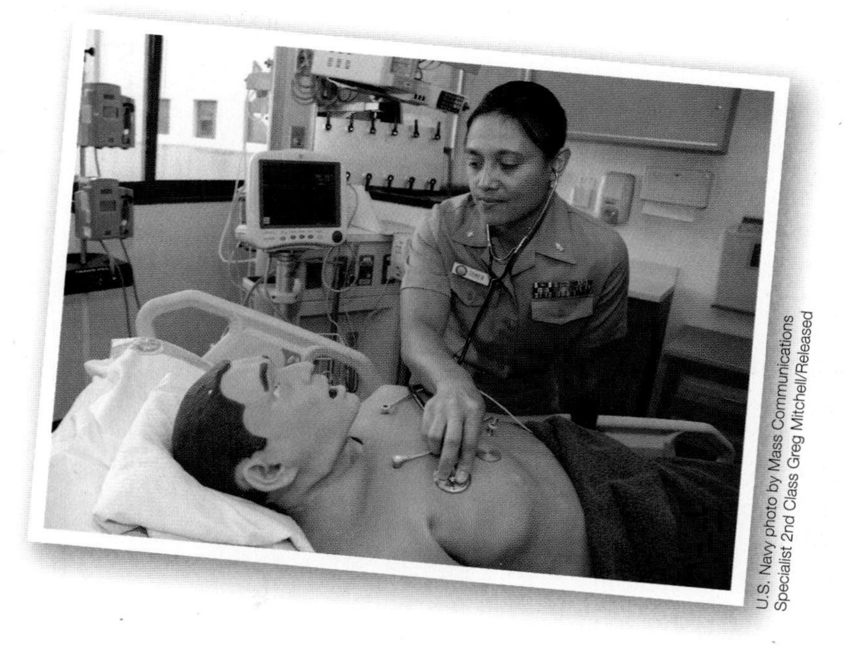

U.S. Navy photo by Mass Communications Specialist 2nd Class Greg Mitchell/Released

Hints: In writing this caption

- review the major revolutionary changes (e.g., industrialization and rationalization),
- consider which best applies to using robots as teaching tools, and
- explain how this robot simulator embodies the identified change.

Critical Thinking

Look over the definition of robots and list up to five ways robots are involved in your life.

Key Terms

global interdependence
globalization
hydrocarbon society
information explosion
social change
technology
tipping point
urbanization

MODULE 13.2

Human Societies and Carbon Footprints

Objective

You will learn about the relationship between the amount of surplus wealth a society is able to create and the carbon footprints its members leave.

Which lifestyle—that of a Buddhist monk or a professional golfer, both living in the United States—creates a larger carbon footprint?

Chris Caldeira

Tim Hipps, FMWRC Public Affairs

A **carbon footprint** is the impact a person makes on the environment by virtue of his or her lifestyle. That impact is measured in terms of fossil fuels consumed or units of carbon dioxide emitted from burning that fuel. We can think of carbon footprints as primary or secondary. The primary carbon footprint is the total amount of carbon dioxide emitted as the result of someone's direct use of fossil fuels to heat and cool a home, run appliances, power an automobile, and so on. The secondary footprint is the total amount of carbon dioxide emitted to manufacture a product for or deliver a service to that person.

Lifestyles and Carbon Footprints

Tim Gutowski (2008a), a professor of mechanical engineering at MIT, and 21 of his students studied 18 different lifestyles in the United States, including a person without a home (homeless), a Buddhist monk, a patient in a coma, and a professional golfer. After extensive interviews they estimated the energy each lifestyle requires. They estimated that none of the 18 lifestyles uses less than 120 gigajoules of energy, even that of a Buddhist monk, whose lifestyle is devoted to simple living, modest dress, and a vegetarian diet. A gigajoule (GJ) is a metric measure of energy consumption. It is a particularly useful measure because it can be applied to different types of energy consumption, such as kilowatts of electricity, liters of heating oil or gasoline, and cubic feet of natural gas. One GJ is equivalent to the energy needed to cook over 2,500 hamburgers or to keep a 60-watt bulb lit for six months (Natural Resources Canada 2010).

Gutowski and his students estimated that the Buddhist monk consumed 120 GJ of energy each year and the professional golfer consumed 8,000 GJ. The Buddhist monk consumes about one-third as much energy as the average American (350 GJ) but double the average amount of energy a typical person on the planet consumes (64 GJ). The Buddhist monk's 120 GJ emits 8.5 metric tons of carbon dioxide each year; the professional golfer's lifestyle (8,000 GJ) produces 566 metric tons of carbon emissions.

Gutowski's study (2008b) has important implications: the United States has a "very energy-intensive system" that in effect sets the lower limits on how much energy any American uses. If even a Buddhist monk uses 120 GJ, then we might conclude that energy use is woven into the fabric of American society (Revkin 2005). We consider the work of sociologist Douglas Massey (2002) and others who were heavily influenced by Gerhard and Jean Lenski (1986) to understand just how energy use becomes woven into the fabric of a society.

Types of Societies

In this module we take a long view and consider six broad types of societies in which humans have lived over their history as a species. Each is defined by the technologies people employ to produce food and extract resources. The six societies are hunting and gathering, pastoral, horticultural, agrarian, industrial, and postindustrial. Each of the six is distinguished by the amount of surplus wealth that society is able to produce. **Surplus wealth** is a situation in which the amount of available food items and other products exceed that which is required to **subsist**, or to meet basic needs for human survival. Of course, the greater the surplus wealth a society creates, the larger its carbon footprint.

HUNTING AND GATHERING SOCIETIES. **Hunting and gathering societies** do not possess the technology that allows them to create more than they need to survive; people subsist on wild animals and vegetation. As the name suggests, hunters and gatherers do not live in a fixed location; they are always on the move, securing food and other subsistence items. A typical hunting and gathering society is composed of 45 to 100 members related by blood or marriage. The institution of family is central to people's lives, and emphasis is placed on group welfare (Massey 2002). The division of labor is simple, and most people engage in activities related to survival. The statuses that matter revolve around gender, age, and kinship. Because almost no surplus wealth exists, there is little inequality.

Sociologist Douglas Massey (2002) predicts that by 2020 the last hunter-gatherers on the planet will cease to exist, ending six million years of dedication to "the most successful and long-persistent lifestyle in the career of our species" (Diamond 1992, 191). In fact, it may already be impossible to find a society that meets all its needs from hunting and gathering.

PASTORAL AND HORTICULTURAL SOCIETIES. About 10,000 to 12,000 years ago, humans began domesticating plants and animals. Instead of searching for and gathering wild grains and vegetation, people planted seeds and harvested crops. Instead of hunting for wild animals, people captured, tamed, and bred them. Domestication offered a predictable food source and allowed people to secure more grain, meat, and milk than they needed to survive. Domestication of animals offered people another source of power—animal muscle—to transport heavy loads, to guard sheep, and so on. Surplus food gave some people time to pursue other activities beyond securing subsistence, such as making vases to store food.

Domestication is the hallmark of two types of societies: pastoral and horticultural. **Pastoral societies** rely on domesticated herd animals to subsist. Pastoralism was adopted by people living in deserts and other places with limited amounts of vegetation. Those able to acquire and manage the largest herds assumed powerful statuses and passed their advantaged positions on to their children.

➤ Today in Afghanistan there are an estimated 1.2 million people considered pastoral. Here the nomadic Kuchi people are shown as they move goats, donkeys, camels, and cattle through the Panjshir Valley to the high country for the summer.

U.S. Air Force Photo/Tech. Sgt. John Cumper

Most pastoral peoples are nomadic, moving their herds when grazing land and/or water sources are depleted. Even though they may be on the move, pastoralists are able to accumulate possessions such as tents, carpets, bowls, and other cultural artifacts, because now they have animals to carry those possessions. In the course of their travels, pastoralists encounter other nomads and settled peoples with whom they trade and/or fight to secure grazing land. The statuses people occupy revolve around gender, age, and kinship, but material possessions and success in conflicts are now also important to determining status.

In contrast to pastoralists, people who live in **horticultural societies** rely on hand tools such as hoes to work the soil and digging sticks to punch holes in the ground into which seeds are dropped. Horticultural peoples grow crops rather than gather food and employ slash-and-burn technology in which they clear land of forest and vegetation to make fields for growing crops and grazing animals. When the land becomes exhausted, people move on, repeating the process. In contrast to pastoralists, horticultural societies are relatively settled. The horticultural system offers a level of predictability and residential stability that gives people the incentive and means to create surplus wealth, including houses, sculptures, and jewelry. The creation of surplus wealth is accompanied by conflict over available resources and by inequality with regard to its distribution.

AGRARIAN SOCIETIES. The invention of the plow 6,000 years ago triggered a revolution in agriculture and marked the emergence of **agrarian societies** built on the cultivation of crops using plows pulled by animals to achieve subsistence. The plow made it possible to cultivate large fields and increase food production to a level that could support thousands to millions of people, many of whom lived in cities and/or were part of empires. Although the plow was a significant invention, planting and harvesting food still depended largely on human and animal muscle.

U.S. Navy photo by Lt. Cmdr. Chuck Bell

USAID/Winrock International

▲ The major agricultural revolution that launched the agrarian societies occurred around 5000 BC with the invention of the scratch plow (still used in some parts of the world); its forward-curving blade cut deep into the soil, bringing nutrients to the surface and turning weeds under. The plow was a great advance over hoes (left) because it allowed farmers and their descendants to replenish the soil year after year (Burke 1978, 9).

Agrarian societies are noted for dramatic inequality; monarchs (kings, queens, or emperors) hold absolute power over their subjects, who for the most part do not question such power. In agrarian societies there are a small number of landowning elites and large numbers of people known as serfs, peasants, or enslaved. There are also small numbers of merchants, traders, and craftspeople. In addition to dramatic inequality, agrarian societies are often engaged in wars to protect and/or expand their territory and to exercise control over resources.

Sgt. Jim McKinney, 11th PAD

◀ One well-known symbol of ancient and medieval empires is the citadel, often located within a town or city and considered the last line of defense for protecting residents. The existence of a citadel offers insights about the kinds of human activity that surplus wealth made possible—including centralized government, roads, systems of writing, militaries, palaces, and much more. The citadel pictured is located in Kirkuk, Iraq, and was built in 884 and 858 BC (Elliott 2009).

The plow is believed to have changed the status of women relative to men in dramatic ways. As surplus wealth and populations increased in size, warfare between peoples fighting over land and resources became commonplace. The invention and proliferation of metal weapons supported military forces created to advance and protect the interests of the political elite. The military excluded women, who spent much of their reproductive life pregnant or nursing, to offset the high death rates and low life expectancy (Boulding 1976).

Women's status was reduced in yet another way. The plow increased the amount of land that could be cultivated, and by extension it increased the amount of food produced. Men operated the plows and managed heavy draft animals in fields away from home. Women's reproductive lives and caretaking responsibilities, in conjunction with the efficient technology men operated, made it difficult for women to outproduce men. As a result, women's share of the production diminished. Elise Boulding (1976) described women's situation as follows:

> The shift of the status of the woman farmer may have happened quite rapidly, once there were two male specializations . . . plowing and the care of cattle. The situation left women with all the subsidiary tasks, including weeding and carrying water to the fields.

INDUSTRIAL SOCIETIES. **Industrial societies** rely on mechanization or machines to subsist. These innovations allowed humans to produce food, extract resources, and manufacture goods at revolutionary speeds and on an unprecedented scale. Mechanization changed everything: most notably, a small percentage of the population grew the food needed to sustain a society that could include hundreds of millions of people. The products of industrialization improved nutrition and living standards, which increased human life expectancy, decreased fertility, and lowered death rates so that the number of older people eventually came to outnumber the young. Under industrialization, population size increased from about 954 million people in 1800 to 2.5 billion by 1950 (Massey 2002).

Chris Caldeira

➤ This line of trucks illustrates the degree of mass production and consumption made possible by industrialization. A large share of the masses, not just an elite few, consumed much more than they need to survive. And, of course, each item has a carbon footprint and, in the case of trucks, leaves a carbon footprint whenever they are driven.

The ability to support unprecedented surplus supported a diverse economy and many institutions, including education, medicine, sports, and government. The mass production of goods allowed people not only to buy products that distinguished them from others but also to buy more products than they needed, creating great social differences. On the one hand, industrialization allowed many more people to experience a high standard of living and social mobility. On the other hand, it created dramatic differences in material wealth between those in the top 10 percent and the bottom 20 percent (Massey 2002).

If we study patterns of conflict in the world, the record shows industrial societies have used their military strength and technological advantages to invade and control what are often referred to as developing countries (R. Robertson 1987). The major military operations that the United States has waged since World War II ended include invasions of Korea, Vietnam, Cuba, Grenada, Panama, Afghanistan, and Iraq.

POSTINDUSTRIAL SOCIETIES. Massey (2002) estimates that humans spent 300,000 generations as hunter-gatherers, 500 generations as agrarians, 9 generations in the industrial era, and only 1 generation in the postindustrial era. Sociologist Daniel Bell, who has been writing about the coming of a **postindustrial society** since the 1950s, described it as a society that relies on intellectual technologies of telecommunications and computers (Bell 1999, xxxvii). According to Bell, this intellectual technology encompasses four interdependent revolutionary innovations: (1) electronics that allow for incredible speed of data transmission and calculations, which can be made in nanoseconds; (2) miniaturization or the drastic size reduction of electronic devices; (3) digitalization, which allows voice, text, image, and data to be integrated and transmitted with equal efficiency; and (4) software, applications that allow people to perform a variety of tasks and generate a variety of simulated experiences.

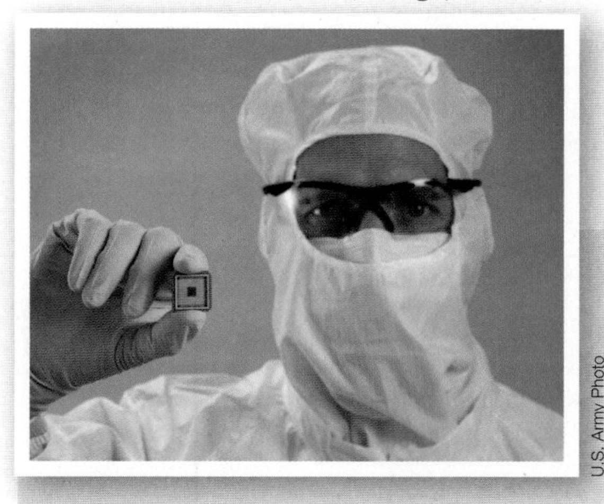

U.S. Army Photo

◂ One example of intellectual technology is the microchip, the brain of many electronic products, including computers, mobile phones, cameras, iPods, and televisions. Microchips run pacemakers and mechanical hearts. They are installed in microwave ovens and cars. Microchips control the deployment of air bags, the movement of artificial limbs, and much more. Microchips have created many new products and services, yet they are ultimately powered by fossil fuels.

According to Bell, postindustrial societies, built upon these intellectual technologies, are distinguished by

- a substantially greater share of the working population employed in service, sales, and administrative support occupations (in the United States, from 29 percent in 1950 to 41 percent in 2010);
- an increased emphasis on education as the avenue of social mobility (in the United States, in 1950, 6 percent of the population had at least a 4-year college degree; in 2010, 28 percent did);
- a recognition that capital is not only financial but also social (that is, access to social networks serves as an important source of information and opportunity);
- the dominance of intellectual technology grounded in mathematics and linguistics that takes the form of what are known as apps;
- the creation of an electronically mediated global communication infrastructure; and
- an economy defined not simply by the production of goods but by applied knowledge and the manipulation of numbers, words, images, and other symbols.

What do these changes mean? In answering this question, keep in mind that Bell does not believe that technology determines the nature of any social change. Rather, it is the ways people choose to use and respond to the technology that shapes social change. Although it is virtually impossible to catalog all the changes associated with the intellectual technologies, they have (1) sped up old ways of doing things; (2) given individuals access to the equivalent of personal libraries, publishing houses, and production studios; (3) changed how people learn; and (4) permitted real-time exchange of information on a global scale.

The unprecedented and immediate access to information and people can be overwhelming. Although computer software and telecommunications technologies have increased the speed of information generation and exchange, keep in mind people must still read, discuss, and contemplate the information to give it meaning. These activities are very slow compared with the speed at which the information is generated.

Relative to other types of societies, the postindustrial society presents its members with a distinct set of challenges that are interpersonal in nature. The communication infrastructure and service economy multiply interactions among people, making interpersonal relationships a primary focus. That focus is complicated because we leave permanent records of our transactions with others. The great difference in the postindustrial society is the tremendous change in the number of people one knows or can potentially know (Bell 1976, 48).

(Write a Caption)

Write a caption linking this turn-of-the-20th-century oil gusher to one of four types of society.

Hints: In writing this caption

- review the four types of society,
- determine which of the four types is built around fossil fuels, and
- describe how oil contributes to surplus wealth and what people do for living.

Prints & Photographs Division, Library of Congress, LC-USZ62-54453

Critical Thinking

Do you hold an occupation that creates products or delivers services that are not needed for subsistence? Explain.

Key Terms

agrarian societies
carbon footprint
horticultural societies
hunting and gathering societies
industrial societies
pastoral societies
postindustrial society
subsist
surplus wealth

Medical Sociology

MODULE

13.3

Objective

You will learn what medical sociology is and the kinds of things medical sociologists study.

Think about your last encounter with the health care system in the United States. Did you come away with a positive or a negative assessment?

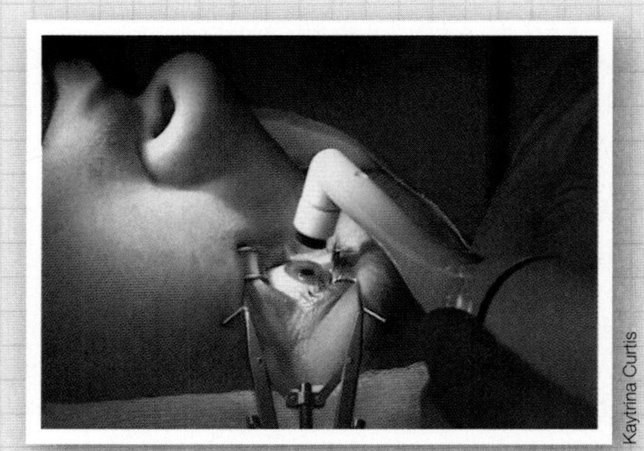

Kaytrina Curtis

According to a nationally representative survey of 15,735 health care consumers in 12 countries (Belgium, Brazil, Canada, China, France, Germany, Luxembourg, Mexico, Portugal, Switzerland, the United Kingdom, and the United States), Americans are among those least satisfied with their system of health care. Almost 40 percent of U.S. respondents gave their system a grade of D or F. Only 22 percent gave the health care system an A (excellent) or B (very good) compared with the 50 percent or more respondents in Luxembourg (69 percent), Belgium (57 percent), Switzerland (52 percent), France (51 percent), and Canada (50 percent) who gave an A or B grade to their systems (Deloitte 2011b). Why such high dissatisfaction? This is the kind of question medical sociologists seek to answer.

Medical sociologists pay special attention to the social context of health care and give special emphasis to the ways in which health, disease, and illness are defined and experienced. They also study how medical care is organized and delivered and relationships among health care providers and other stakeholders.

There are 13,700 sociologists who belong to the American Sociological Association (ASA). Approximately one in five has teaching or research interests that relate to medical sociology (Scelza, Spalter-Roth, and Mayorova 2011; ASA 2012b). As one measure of medical sociology's importance to the discipline, the ASA publishes the *Journal of Health and Social Behavior*. In conjunction with the journal's 50-year anniversary, a special issue titled "Reflection on 50 Years of Medical Sociology" was released. Here we review some key findings from that issue that focus on the social dimensions of medicine. The social dimensions are important because health, disease, and illness are not simply biological experiences. They are also social experiences if only because people give names and assign meaning to biological experiences and establish systems for promoting health and fixing disease and illness.

> Those systems include complementary (integrative) and alternative medicine (CAM). In the United States, approximately 4 in 10 adults and 1 in 9 children use some form of CAM; the most common are deep breathing, meditation, chiropractic, yoga, and diet-based therapies (National Institute of Health 2012).

Chris Caldeira

Key Features of the U.S. System of Health Care

The U.S. health care system is the most expensive in the world. In 2010 the United States spent $2.6 trillion on health care, an amount that is equal to 17.6 percent of the gross domestic product (GDP), or an average of $8,086 per person. Switzerland is next in line, spending 11.2 percent of its GDP, or an average of $4,627 per person. While the United States spends substantially more than other countries, it is not considered a leader in health outcomes, most notably life expectancy, infant mortality, and other measures of well-being (see Figure 13.3a).

Figure 13.3a: Percent of GDP and Per Capita Spending on Health Care

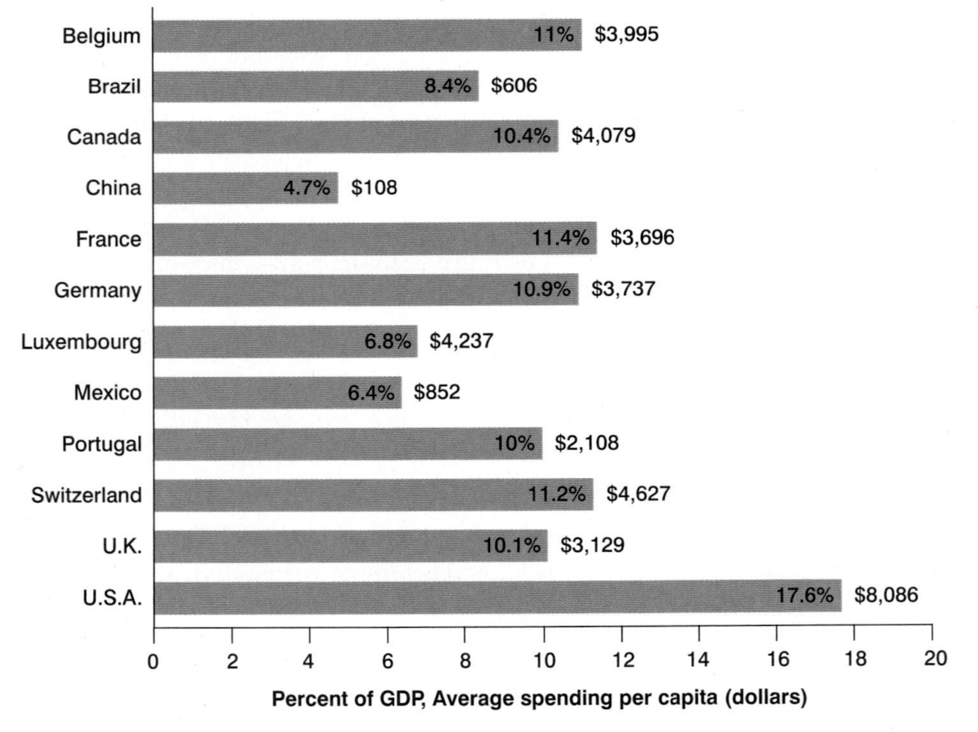

Source of data: Deloitte (2011b, Figure 4)

When compared with the world's wealthiest countries, the United States is the only country that does not provide some kind of universal health insurance coverage (Deloitte 2011b). As a result, 16.3 percent of its population is counted as uninsured (see Figure 13.3b).

▼ Figure 13.3b: Percentage of U.S. Population with Health Insurance by Provider

The most recent data show that 55.3 percent of the U.S. population of 314 million people are covered by employer-sponsored insurance, 14.5 percent of the population are covered by Medicare (for those 65 and over), and almost 16 percent by Medicaid (low-income). About 50 million of the U.S. population are uninsured.

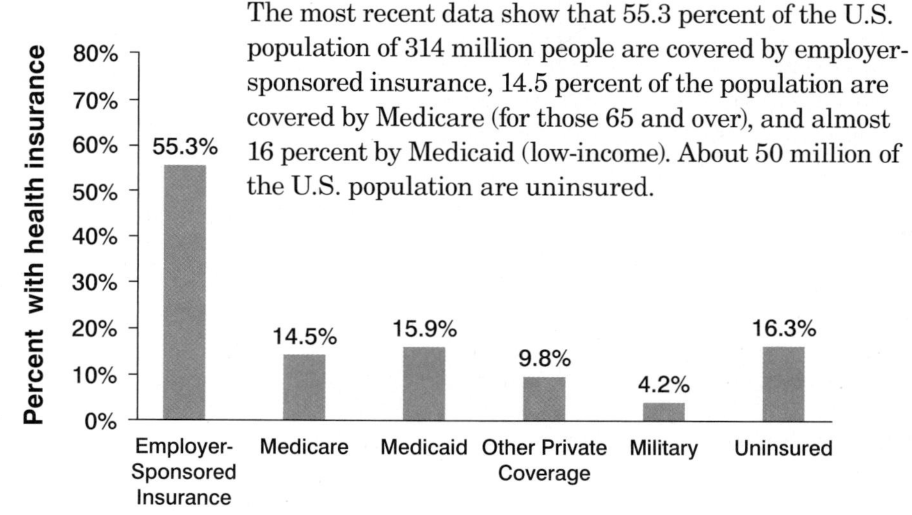

Note: The percentages do not total 100 because some people have more than one type of coverage.
Source of data: U.S. Department of Health and Human Services (2011)

Rising and expensive health care costs have made health care unaffordable for many Americans, even those who are insured. It is important to point out that employer-sponsored health care coverage does not mean that the employer covers all costs of health care services. Typically, employees pay a portion of the monthly premium, deductibles, and co-pays. In addition, employees likely pay a portion of the costs of medical services (20 percent, for example). These out-of-pocket costs are high enough that 25 percent of Americans did not seek medical treatment for an illness or injury, and 19 percent delayed treatment or did not follow a course of treatment because of expense (Deloitte 2011a).

The structure of health care delivery has changed in dramatic ways over the past 50 years. For one, "hospitals no longer function as autonomous units but are large complex corporations that continue to merge into centers of wider, regionally focused health networks" (Wright and Perry 2010).

➤ Chinese Hospital in San Francisco is a 52-bed community hospital that serves the people of Chinatown. Chinese Hospital's mission emphasizes "community ownership and responsiveness" (Chinatown San Francisco 2012). Community hospitals like Chinese Hospital plan with a community's needs in mind, not a region's or investor's needs. Profits are invested back into the hospital.

Chris Caldeira

In addition to changes in hospital reach, a number of technical innovations related to diagnosis and treatment have transformed health care practice and delivery. Examples include genetic and reproductive technologies, imaging (digital mammograms, CT scans, MRIs), and nanotechnologies that deliver medicine and "transform and reorganize human bodies" (Rosich and Hankin 2010).

U.S. Navy photo by Sarah Fortney/Released

< Transplant and other body part replacement options are also among the technologies that have transformed treatment.

Because U.S. health care providers get paid for services rendered and not for health care outcomes, there are strong incentives to use the newest technologies because they are considered state-of-the-art and to also overtest and overprescribe treatments (see Figure 13.3c).

Figure 13.3c: Number of MRI Exams and CT Scans Performed per 1,000 Population in Selected Countries, 2009

The charts show the number of people per 1,000 population who received an MRI exam or CT scan in 2009. Which country has the highest rates? Where does the United States rank? One might wonder whether Greece and the United States have older populations, and whether that fact might help explain the high use. All countries listed have aging populations, however, so that is not an explanation. In addition, the average cost of MRI and CT scans are considerably higher in the United States than in other countries. As one example, the average cost of an MRI in the United States is $1,080 versus $280 in France (Klein 2012).

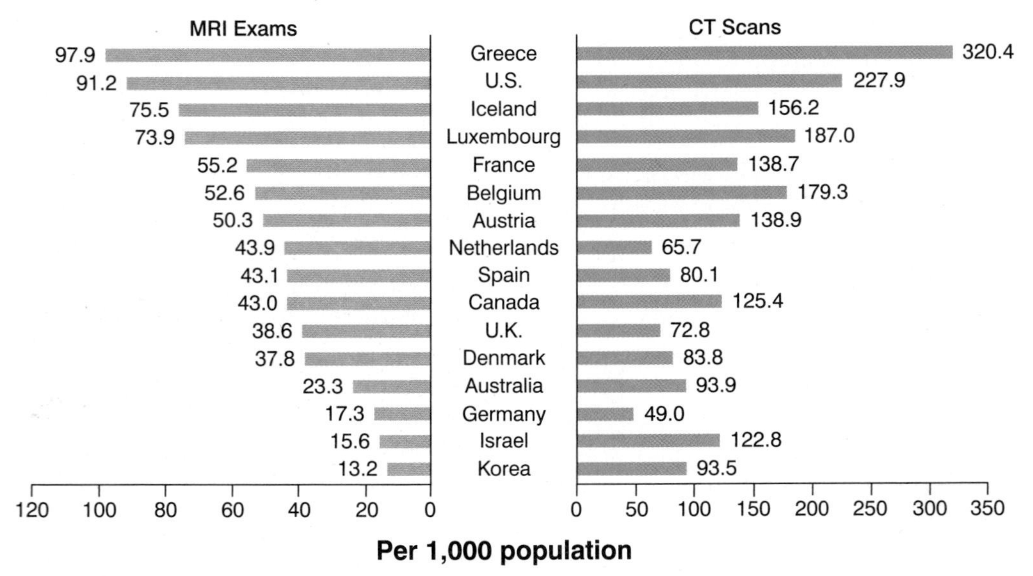

Source of Data: OECD (2011b, slide 25)

One international survey of 100 insurers in 25 countries found that the cost of 22 out of 23 medical services and products ranging from a routine doctor's visit to coronary bypass surgery were highest in the United States. The exception to the rule was cataract surgery, which is highest in Switzerland followed by the United States. The leading explanation for the high cost of health care was, "Providers largely charge what they can get away with, often offering different prices to different insurers, and an even higher price to the uninsured" (Klein 2012).

Until recently physicians have "been the voice of authority" (Rosich and Hankin 2010). In other words, physicians controlled the course of treatment and the flow of information related to diagnosis, prognosis, and treatment. That dominance has been undermined in a number of ways. Most notably other stakeholders—hospitals, pharmaceutical companies, insurance companies, and patients—have taken a visible and more active role in deciding the diagnosis and course of treatment. As a case in point, pharmaceutical companies market drugs directly to consumers. And because patients have access to medical information via the Internet and advertisements, they now assume a larger role in managing preventive care, the course of illness, and treatment.

Insurance companies have also undermined the authority of the doctor by setting the terms of care and treatment. For example, some insurance companies require doctors to handle one health problem per visit so that patients must return a second time (or third) to address another health concern. This, of course, involves multiple co-payments. Finally, many hospitals assign physicians on staff to care for admitted patients, eliminating the involvement of patient's primary care and other doctors.

Health inequalities are deeply rooted and persistent. It is an established fact that those in low-income and disadvantaged populations have worse health outcomes and lower life expectancy than those in advantaged populations. The theory of fundamental causes offers one explanation. According to this theory, health status and socioeconomic status are intertwined so much that those in more advantaged groups have greater access to "an array of resources, such as money, knowledge, prestige, power, and beneficial social connections that protect health no matter what the risk and protective factors are in a given circumstance or given time" (Phelan, Link, and Tehranifar 2010). It should come as no surprise that the most financially well-off have the resources that permit them to achieve a healthier lifestyle (better diet, personal trainers) and to access the best preventive measures and treatment options (including early detection).

Race is also a critical factor in health outcomes. That statement is supported by a considerable amount of research documenting racial disparities with those in minority groups experiencing (relative to whites) lower expectancy, higher rates of disease (most notably cancer, diabetes, and hypertension), and lower disease-specific survival rates (e.g., cancer survival). In addition, while some groups cannot access the health care services they need, other groups overuse services. The major factors contributing to under- and overutilization of medical services include the financial resources of patients (do patients have insurance, does their plan cover a particular treatment of medication?), financial incentives, and the availability of technologies such as MRI and CT scans (see Table 13.3a).

Table 13.3a: Race/Ethnic Profile of the Uninsured in the United States

One reason that minorities have worse health outcomes than those considered white is that they are more likely to be uninsured. Notice, for example, that Hispanics are 16.3 percent of the U.S. population, but they represent 30.7 percent of all those uninsured. In addition, 30.7 percent of people classified as Hispanic are uninsured.

Race/Ethnicity	% of Total U.S. Population	% of All Uninsured	% Not Insured in Each Race/Ethnic Category
White, non-Hispanic	64.5%	46.3%	11.7%
Black	12.8%	16.3%	20.8%
Asian	4.7%	5.2%	18.1%
Hispanic (any race)	16.3%	30.7%	30.7%

Source of data: U.S. Department of Health and Human Services (2011)

Stressful life events and the quality of social relationships affect health outcomes. We know without a doubt that stressful events have an impact on a person's mental and physical health. Such events include the death of a spouse or partner, an abusive relationship, unemployment, chronic discrimination, insufficient income to make ends meet, and caretaking responsibilities for someone with a chronic condition. We also know that health is affected by **social relationships**, the nature, number, and quality of the ties that bind people formally and informally to others. Formal ties are relationships people have with those they interact with in the context of their workplace, school, religious group, and other organizations. Informal relationships include those ties that transcend shared membership in an organization—ties with friends, family, and other relatives. We know that the more social support people have to draw upon, the greater the positive impacts on health and lifestyle (Thoits 2010; Umberson 2010).

Disease, illness, and health are "social constructs as well as medical constructs" (Rosich and Hankin 2010). Disease, health, and illness are not simply biological events; they are social events as well. Perhaps the most dramatic example of this relates to AIDS, which was originally named GRIDS (Gay-Related Immune Deficiency Syndrome). People notice diseases and give them names and in that process convey meanings about the medical condition. The concept of medicalization helps us to see the social processes by which the inevitable problems associated with living and aging become defined as medical problems. Arguably the most vivid examples are from the pharmaceutical industry, which plays an important role in naming physical conditions that can be treated with medicines—not enough eyelashes, menstruation (shortened and fewer cycles), and erectile dysfunction are physical conditions that have been defined as diseases treatable with medication.

When something is medicalized, it becomes something in need of treatment. In this sense a woman's labor to deliver a baby has been medicalized. Rates of caesarean deliveries are considerably higher in some countries such as the United States. These differences in how delivery of a baby is handled offer insights into the social experience of what constitutes risk to baby and mother (see Figure 13.3d). About one-third of women who give birth in Korea, the United States, Australia, and Germany get a C-section. Contrast that to Iceland and the Netherlands, where about 15 percent of women who give birth get a C-section.

▼ **Figure 13.3d: Number of C-Sections per 100 Live Births by Selected Countries**
As you can see, defining a physical condition as a "disease" shapes the experience of illness, the kind of treatment prescribed, and cost of care. Which country listed has the highest number of C-sections per 100 births? Which country has the lowest rate per 100 births?

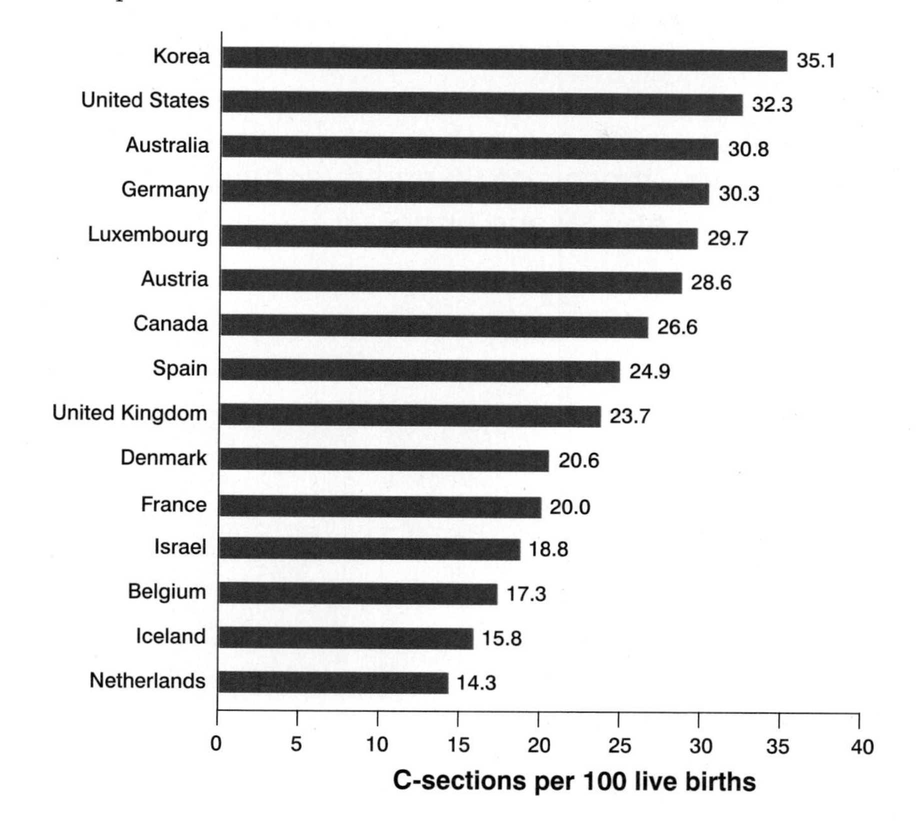

Source of data: OECD (2011b)

(Write a Caption)

Write a caption that describes at least two ways the doctor's authority in the patient–physician relationship has been undermined.

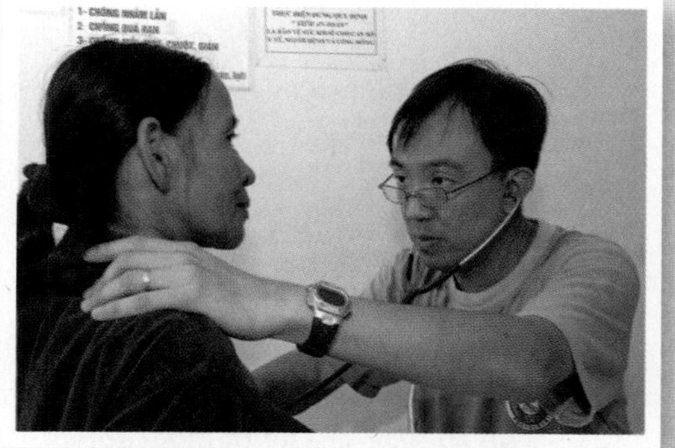

Mass Communication Specialist 3rd Class Bryan M. Ilyankoff

Hints: In writing this caption

- review the ways the health care system has changed, and
- identify trends that reduce physicians' control over the diagnosis and course of treatment.

Critical Thinking

Based on your experience with the health care system in the United States, what grade would you give it? Explain how you arrived at that grade.

Key Term

social relationships

MODULE

13.4 Aging Societies

Objective

You will learn how those age 65 and over have come to outnumber those age 30 and under.

This man is just one of 75.5 million Americans known as baby boomers who are between the ages of 48 and 66. What does it mean to have this many people moving toward retirement?

Chris Caldeira

There is no historical precedent in the United States, or the world, for such a large proportion of a population moving toward older ages. The theory of the demographic transition offers one explanation for how this came to be.

Theory of the Demographic Transition

Sociologists are interested in the factors that affect population size, growth, and age composition. As you can imagine, births and deaths are key factors. The **theory of the demographic transition** postulates that a country's birth and death rates are linked to its level of industrial or economic development. The demographic transition includes three stages: in stage 1, birth and death rates are both high; in stage 2, death rates decrease, causing population size to increase dramatically; and in stage 3, both birth and death rates drop below 20 per 1,000 persons. Since the theory was put forth in the 1930s, a fourth stage has been added (and even a fifth and sixth stage not covered here). The theory centers around a graph depicting historical changes in birth and death rates in Western Europe and the United States, and it helps us to understand the forces behind an aging population (see Figure 13.4a).

Figure 13.4a: Demographic Transition

The figure shows the four stages of the demographic transition. Note that it compares birth and death across time. So at the beginning of stage 1, the birth rate is about 36 births per 1,000 population and the death rate is about 34 deaths per 1,000 population.

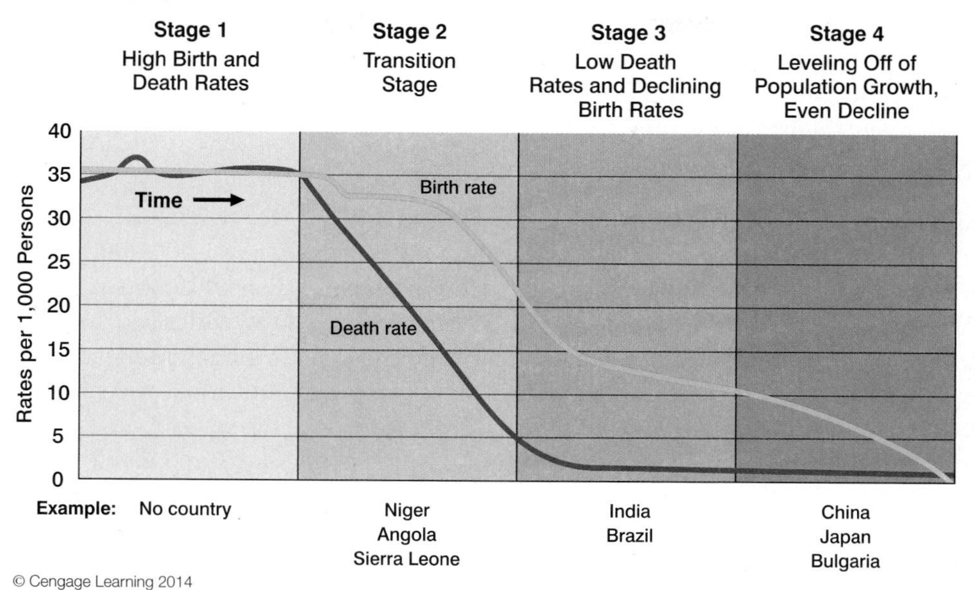

STAGE 1: HIGH BIRTH AND DEATH RATES. For most of human history—the first 2 to 5 million years—populations grew very slowly, if at all. The world population remained at less than 1 billion until around 1850 AD, at which point it began to grow explosively. Demographers speculate that growth until that time was slow because **mortality crises**—violent fluctuations in the death rate caused by war, famine, or epidemics—were a regular feature of life. Stage 1 of the demographic transition is often called the stage of high-potential growth: if something happened to cause the death rate to decline—for example, improvements in agriculture, sanitation, or medical care—the population would increase dramatically. In this stage, life is short and brutal; average life expectancy at birth remained short—perhaps 20 to 35 years—with the most vulnerable groups being women of reproductive age, infants, and children younger than age five. It is believed that women gave birth to large numbers of children and families remained small because one of every three infants died before reaching age one, and two of every three died before reaching adulthood.

STAGE 2: TRANSITION. Around 1650, mortality crises became less frequent, and by 1750 the death rate began to decline slowly. This decline was triggered by a complex array of factors associated with the onset of the Industrial Revolution. The two most important factors were (1) increases in the food supply, which improved the nutritional status of the population and increased its ability to resist diseases; and (2) public health and sanitation measures, including the use of cotton to make clothing and new ways of preparing food. Over a 100-year period, the death rate fell from 50 per 1,000 to less than 20 per 1,000, and life expectancy at birth increased to approximately 50 years of age. As the death rate declined, fertility remained high, thus the **demographic gap**—the difference between the birth and death rates—widened, and the population grew substantially. Accompanying the unprecedented growth in population was urbanization.

Around 1880, birth rates began to drop. The decline was not caused by innovations in contraceptive technology, because the methods available in 1880 had been available throughout history. Instead, the decline in fertility seems to have been associated with several other factors. First, the economic value of children declined; children no longer represented a source of cheap labor but rather became an economic liability to parents. Second, with the decline in infant and childhood mortality, women no longer had to bear a large number of children to ensure that a few survived. Third, a change in the status of women gave them greater control over their reproductive life and made childbearing less central to their life.

STAGE 3: LOW DEATH RATES AND DECLINING BIRTH RATES. Around 1930, both birth and death rates fell to less than 20 per 1,000, and the rate of population growth slowed considerably. Life expectancy at birth surpassed 70 years. The remarkable successes in reducing infant, childhood, and maternal mortality placed accidents, homicides, and suicide among the leading causes of death of young people. The risk of dying from infectious diseases declined, allowing those who would have died of infectious diseases in an earlier era to survive into middle age and beyond, at which point they faced an elevated risk of dying from degenerative and environmental diseases such as heart disease, cancer, and strokes. For the first time in history, people age 50 and older accounted for the largest share of deaths. Before stage 3, infants, children, and young women accounted for the largest share (Olshansky and Ault 1986). As death rates decline, disease prevention becomes important. As a result, people become conscious of the link between health and lifestyle factors such as sleep, nutrition, exercise, and bad habits.

STAGE 4: LEVELING OFF OF POPULATION GROWTH, EVEN DECLINE. Since the demographic transition was first proposed, a fourth stage has been added, in which both birth rates and death rates are low. Birth rates drop to a level below that needed to replace those who die. While death rates are low, there is an increase in lifestyle diseases caused by lack of exercise, poor nutrition, and obesity. Birth rates fall below replacement when the average woman has fewer than two children over the course of her reproductive life. Eastern European countries, Italy, and Japan are examples of countries in which this is the case (see Figure 13.4b).

Katie Englert

◀ Consider that in Japan, the average woman has 1.12 babies; the death rate is 9.0 per 1,000, and the birth rate is 8.0. The number of people age 70–74 exceeds the population age 5–9 (U.S. Census Bureau 2012d).

➤ **Figure 13.4b: Top 10 Countries with Highest Proportion of People Over Age 65, January 2011**

Source: U.S. Census Bureau (2012d)

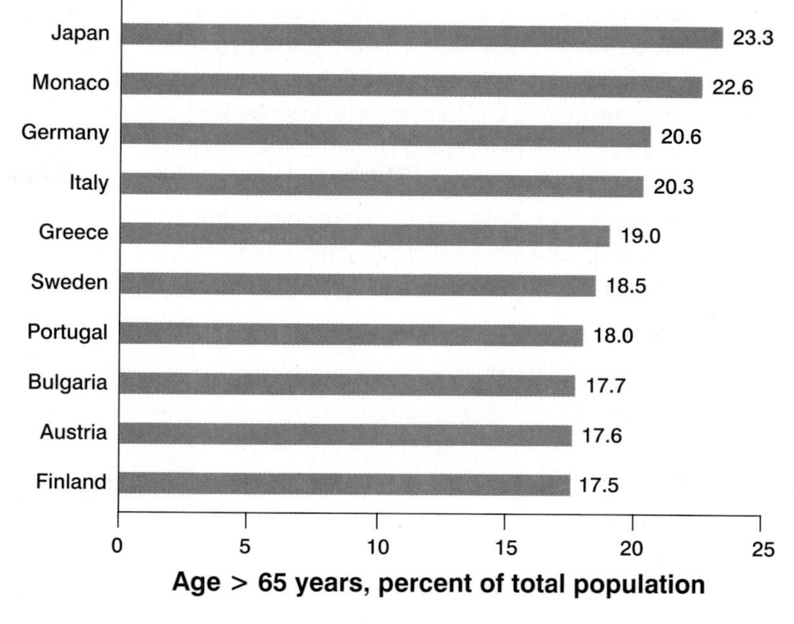

Developing Countries

The theory of the demographic transition forecasted that developing countries would follow the original three-stage model. But their path has deviated in some fundamental ways. Most countries we call developing were once colonies of today's industrialized countries. Colonization created economies in developing countries that supported the colonizing countries' industrialization, but it did not put the developing countries on the path to industrialization—the path that the demographic transition predicts would result in lower birth and death rates.

BIRTH RATES. Birth rates, although still relatively high, are declining in most developing countries. Some important thresholds are associated with declines in fertility. First, less than 50 percent of the labor force is employed in agriculture. (The economic value of children decreases in industrial and urban settings.) Second, at least 50 percent of those between ages 5 and 19 are enrolled in school. (Especially for women, education "widens horizons, sparks hope, changes status concepts, loosens tradition, and reduces infant mortality" [Samuel 1997].) Third, infant mortality is less than 65 per 1,000 live births. (When parents have confidence that their babies and children will survive, they limit the size of their families.) Fourth, 80 percent of the females between ages 15 and 19 are unmarried. (Delayed marriage is important when it is accompanied by delayed sexual activity or protected premarital sex [Berelson 1978].)

DEATH RATES. Death rates in the developing countries have declined at a faster rate than they did in the industrialized countries. That decline has been attributed, in part, to imported Western technologies such as pesticides, fertilizers, immunizations, and antibiotics. Because birth rates were slower to decline, some countries have been caught in a **demographic trap**—the point at which population growth overwhelms the environment's carrying capacity. That is, people's needs for food and shelter exceed the sustainable yield of surrounding forests, grasslands, croplands, or aquifers to the point that people begin to consume the resource base itself (L. Brown 1987).

URBANIZATION. Urbanization in developing economies differs in several major ways from urbanization in developed economies. For one, in the 18th and 19th centuries, millions of Europeans who were pushed off the land were able to migrate to sparsely populated places such as North America, South America, South Africa, New Zealand, and Australia. If these millions had been forced to make their living in the European cities, the conditions there would have been much worse than they actually were (Light 1983). The problem of urbanization in developing countries is compounded by the fact that many people who migrate to the cities come from some of the most economically precarious sections of their countries. In fact, most rural-to-urban migrants are not pulled into the cities by employment opportunities; rather, they are forced to move there because they have no alternatives. When these migrants come to the cities, they face not only unemployment but also a shortage of housing and a lack of services (electricity, running water, and waste disposal).

➤ One distinguishing characteristic of cities in developing countries is the prevalence of slums and squatter settlements, which are much poorer and larger than even the worst slums in industrialized countries.

© David Davis/Shutterstock.com

Sociologist Kingsley Davis uses the term **over-urbanization** to describe a situation in which urban misery—poverty, unemployment, housing shortages, and insufficient infrastructure—is exacerbated by an influx of unskilled, illiterate, and poverty-stricken rural migrants who have come to the cities out of desperation (Dugger 2007).

(Write a Caption)

Chris Caldeira

Sgt. Jesse Stence

Write a caption that highlights two thresholds that are associated with decreases in fertility rates.

Hints: In writing this caption

- review the four thresholds that are associated with lower fertility, and
- consider which of those thresholds is depicted in the photo of a boy scuba diving and the photo of Afghan girls at school.

Critical Thinking

Look up the rates of birth, death, and in- and out-migration for the state in which you live. In recent years, has your state gained or lost population? See if you can find the same information about the city or town in which you live.

Key Terms

demographic gap	mortality crises	theory of the demographic transition
demographic trap	over-urbanization	

Social Movements

MODULE 13.5

Objective

You will learn about types of social movements and some conditions under which people join them.

Is there something about the world you would like to see changed? If so, would you risk imprisonment for that cause?

Tech. Sgt. Anthony Iusi

A **social movement** is formed when a substantial number of people organize to make a change, resist a change, or undo a change to some area of society. For a social movement to form, there must be (1) an actual or imagined condition that enough people find objectionable, (2) a shared belief that something needs to be and can be done about that condition, and (3) an organized effort to attract supporters, articulate the mission, and define a change-making strategy. Usually those involved in social movements work outside the system to advance their cause because the system has failed to address the problem. To draw attention to their cause and accomplish their objectives, supporters may strike, demonstrate, walk out, boycott, go on hunger strikes, riot, or terrorize.

Examples of social movements include the environmental, civil rights, the Tea Party, abortion rights, and pro-life movements. Generally, any social movement encompasses dozens to hundreds of specific groups that have organized to address some conditions they find intolerable. The environmental movement, for example, consists of thousands of different organizations devoted to reducing air and water pollution, preserving wilderness, protecting endangered species, changing lifestyles, and limiting corporate activities that harm the environment.

Types of Social Movements

Social movements can be placed into four broad categories, depending on the scope and type of change being sought: regressive, reformist, revolutionary, and counterrevolutionary. Definitions and examples of each category follow. Keep in mind that the distinctions between the four categories are not always clear-cut. As a result, you might find that some examples fit into more than one category.

REGRESSIVE MOVEMENTS. Regressive, also called reactionary, movements seek to return to an earlier state of being, sometimes considered a golden era. The International Forum on Globalization (2010) represents such an effort. It is a global alliance of 60 organizations in 25 countries that takes issue with the widely held belief that a globalized economy trickles down to even the poorest peoples. This movement seeks to reverse the globalization process by revitalizing local economies, advocating for local food production, and critically examining trade policies that make people dependent on distant sources.

REFORMIST MOVEMENTS. These movements identify a specific feature of society as needing change. The nonprofit organization Polar Bear Sustainability Alliance (2010) focuses on saving the polar bear population and its habitat from extinction. Its goals are to use education and research to protect the world's polar bears, offer educational resources to the public, encourage constructive dialogue, and build an international organization dedicated to saving this population.

REVOLUTIONARY MOVEMENTS. When people take action to make broad, sweeping, and radical structural changes to society's basic social institutions, they are engaged in revolutionary movements (Benford 1992). The Earth Liberation Front (ELF) is an underground eco-defense movement with no formal leadership or membership. Its members anonymously and autonomously engage in economic sabotage, including property destruction and guerrilla warfare, against those seen as exploiting and destroying the natural environment. ELF members have made news for setting fire to SUVs on dealership parking lots and to equipment at logging companies (Goldman 2007; Gillespie 2008).

➤ Radical environmentalists claim to have committed 1,100 acts of arson and vandalism without killing a single person (Goldman 2007).

Major Jason Dickerman

COUNTERREVOLUTIONARY MOVEMENTS. When people join together to maintain a social order that reform and revolutionary movements are seeking to change, they are part of a counterrevolutionary movement. The Global Warming Petition Project (2010) qualifies as such a movement; it seeks to challenge movements demanding that greenhouse gas emissions be reduced. The Petition Project recruits basic and applied scientists to sign a Global Warming Petition urging the U.S. government to reject international agreements to limit greenhouse gas emissions. The movement's website lists the names of 31,487 basic and applied American scientists who have signed the petition.

On the surface, it would appear that social movements form when enough people feel deprived either in an objective or a relative sense. **Objective deprivation** is the condition of those who are the worst off or most disadvantaged—the people with the lowest incomes, the least education, the lowest social status, the fewest job opportunities, and so on. **Relative deprivation** is a social condition that is measured not by objective standards but rather by comparing one group's situation with the situations of groups who are more advantaged. Someone earning an annual income of $100,000 is not deprived in any objective sense. He or she may, however, feel deprived relative to someone making $300,000 or more per year (Theodorson and Theodorson 1969). The research on social movements shows that those who are objectively deprived are less likely than those who are relatively deprived to form or join social movements. The research also shows that people join social movements not to address real or imagined personal deprivations but rather to address larger moral issues, such as mistreatment of animals. The point is that an actual deprivation alone cannot explain why people form or join social movements or how social movements take off.

The Life of a Social Movement

Sociologist Ralf Dahrendorf (1973) offers a three-stage model to capture the life of a social movement. Progression from one stage to the next depends on many things. In the first stage, those without power decide to organize against those with power.

Courtesy of Institute for Research on World-Systems

➤ Dahrendorf maintained that it is "immeasurably difficult to trace the path on which a person . . . encounters other people just like himself, and at a certain point . . . [says] 'Let us join hands'" against those in power to change the system (240).

Such encounters occurred in fall 2009, when thousands of California college students came together to protest a 32 percent hike in tuition imposed as a result of state budget shortfalls. Often, some event—a tuition hike, a shooting, a toxic emission into the air—makes seemingly disconnected and powerless people aware that they share an interest in changing the system. After that event people dare to complain openly and loudly to one another (Ash 1989). At other times, people organize because they have nothing left to lose. They have reached the point when they do not care anymore about what happens if they speak out (Reich 1989).

In the second stage of conflict, those without power organize, provided that they find ways to communicate with one another, they have some freedom to organize, they can garner the necessary resources, and a leader emerges. At the same time, those in power often censor information, restrict resources, and undermine attempts to organize. Resource mobilization theorists maintain that having a core group of sophisticated strategists is key to getting a social movement off the ground. Effective strategists can harness and channel the energies of the disaffected, attract money and supporters, capture media attention, forge alliances with those in power, and develop an organizational structure. Cell phones, text messaging, and the Internet have made organizing easier by allowing interested parties to connect in ways that defy place and time (J. Lee 2003).

In the third stage of conflict, those seeking change enter into direct conflict with those in power. The capacity of the ruling group to stay in power and the amount and kind of pressure exerted from below them affect the speed and depth of change. The intensity of the conflict can range from heated debate to violent civil war. It depends on many factors, including the belief that change is possible and that those in power can control the conflict. If protestors believe that their voices will eventually be heard, the conflict is unlikely to become violent or revolutionary. If those in power decide that they cannot compromise and they mobilize all their resources to thwart protests, two results are possible. First, the protesters may withdraw from the fray because they believe that the sacrifices are too great if they continue. Alternatively, the protesters may decide to meet the enemy head-on, in which case the conflict may become violent.

In spite of protests, California college students were unable to prevent hikes in tuition and fees. What factors do you think prevented them from engaging in perpetual protest until their demands were met?

(Write a Caption)

Write a caption that relates this confrontation between protestors and police to one of three stages in a social movement.

Airman Ryan Ivacic

Hints: In writing this caption

- review the three stages,
- decide which one of the stages this scene best depicts, and
- indicate potential factors that will help to determine whether the protestors will continue to confront authority or retreat.

Critical Thinking

Is there a contemporary or historical social movement with which you are familiar? Under which of the four types of social movements would it fall?

Key Terms

objective deprivation relative deprivation social movement

Applying Theory: Rational Choice Theory and Health Risk

MODULE 13.6

Objective

You will learn that rational choice theory offers one framework for thinking about why people take risks that endanger their health and that of others.

Why do you think these men are taking time to put on protective clothing and gear?

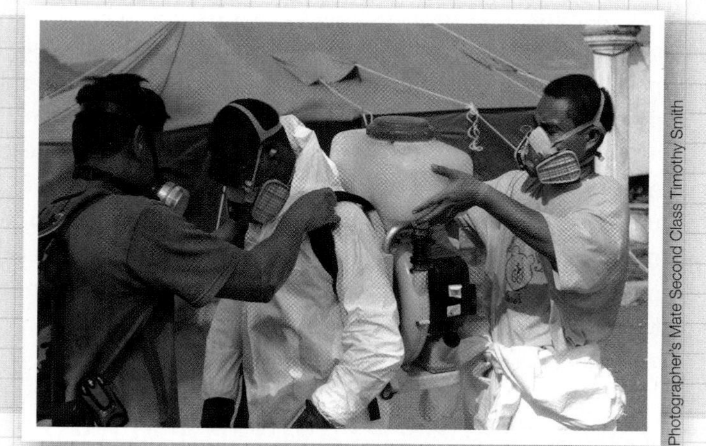

Photographer's Mate Second Class Timothy Smith

If you guessed that the men are protecting themselves from chemicals such as pesticides they are about to spray, you are correct. Can you think of a reason that someone whose job it is to spray pesticides might not wear protective gear?

Among other things, medical sociologists examine the connection between the environment and health risk. In the context of health, the environment includes any external factor that affects the physical and mental well-being of people. Medical sociologists study environmental conditions over which people have little to no direct control such as exposures to chemicals in the workplace, to air pollution, to secondhand smoke, or to pesticides on fruits and vegetables. They also study environments or "lifestyles" people choose (smoking, high-fat diet, failure to protect oneself from hazardous chemicals, and engaging in other risk-taking behaviors).

Courtesy U.S. Army

➤ Medical sociologists study how no-smoking policies at work and other public spaces change people's smoking behavior outside the workplace. They might also study why teenagers take up smoking even though they know the health risks.

Tony Rotundo

There is considerable scientific research that shows an association between pesticide consumption/exposure and a variety of health problems including birth defects, nerve damage, and cancer. While disagreement over what constitutes a safe level of pesticide exposure exists, there is no question that when misused, misapplied, or overused, pesticides are dangerous to human health. So why might a farmer overuse pesticides or apply them in ways that are not recommended by manufacturers?

Sociologist Julia S. Guivant (2003) conducted research with the goal of answering this question. Her findings give support to rational choice theory.

Rational Choice Theories

Rational choice theories are many and diverse. However, all assume that people act rationally. That is, people are viewed as conscious decision makers whose behaviors are largely influenced by weighing the costs and benefits of different possibilities before taking action. For the most part, rational choice sociologists do not seek to predict individual behavior, per se. Rather, they use cost-benefit analysis to explain how a pattern of behavior emerges, is sustained, or gets dismantled. Specifically, a series of rational choices made by many individuals adds up to explain collective actions (Hedström and Stern 2008). Guivant applies rational choice theory to farmers' decisions to overuse, misuse, and misapply pesticides. Her research gives attention to the critical role economic incentives play in directing behaviors in environmentally unfriendly directions with adverse health consequences.

The Case of Brazilian Farmers

In an effort to learn the farmers' point of view, Guivant studied farmers working land along the Cubatão River in southern Brazil. This river, which supplies drinking water to a population of 500,000, is well-known for its pesticide contamination from fungicides, insecticides, and herbicides. (Keep in mind that Brazil ranks among the top five agricultural exporting countries in the world.) Of 1,278 samples of fruits and vegetables produced in this region, 81.2 percent were found to be contaminated with pesticide residues. Serious irregularities were found in 18 percent of these samples, and 5 percent of the samples showed residues from illegal pesticides.

Pesticides and other agricultural technologies were introduced to the region in the 1960s, ending subsistence farming and ushering in a relatively prosperous market-oriented economy that supported a middle class. Even though Brazil has laws on the books mandating professional supervision of pesticide use, the laws were not enforced.

In fact, Guivant found that farmers she studied routinely violated recommended manufacturer's instructions by spraying extra mixtures of pesticides every four days for preventive purposes.

Against this backdrop Guivant studied 48 small- to medium-sized family-run and family-owned farms producing mostly tomatoes and potatoes. In addition, she interviewed agronomists, doctors, sales agents, and bank lenders familiar with pesticide use in the area. Guivant found that farmers held the following beliefs:

1. It is best to eliminate all pests, even harmless ones.
2. One can never apply too much pesticide. The farmers pointed to relatively high profit margins that increased as their pesticide use increased.
3. There are no viable alternatives to pesticides.

Farmers indicated that they were aware of recommendations for using pesticides. Most claimed that they followed guidelines regarding the waiting period between pesticide applications but that other farmers did not. They also reported seeing farmers throw empty pesticide containers in the river, disregard weather conditions when spraying, and fail to use safety equipment. The farmers did not seem to know that pesticides could penetrate the skin, nor that there were ways to reduce pesticide use. While farmers had some knowledge of risk, most did not take steps to protect themselves from exposure. The farmers argued if the risks were real, why then did they not see more deaths among farmers?

➤ Farmers defined cases of eye irritation, dizziness, vomiting, and headaches as not particularly harmful; those symptoms were just things that came along with the job of farming. If such symptoms occur, "the farmer simply waits for them to pass, usually without going to the doctor" (Guivant 2003, 46). Farmers did not take the time to do things like rinse eyes or skin even when irritated.

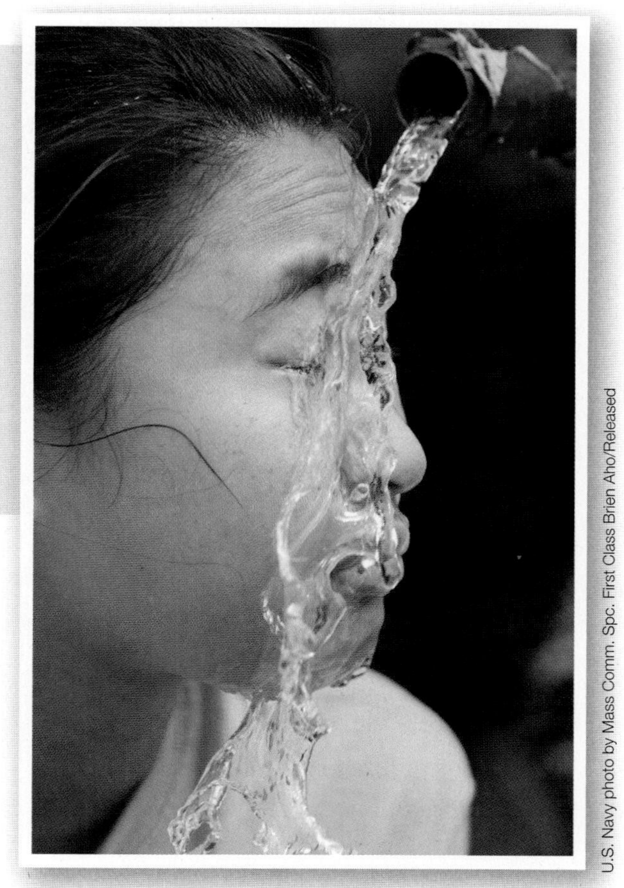

U.S. Navy photo by Mass Comm. Spc. First Class Brien Aho/Released

If the symptoms last for more than one day, the farmer may go to a doctor and stay in the hospital for a couple of days. These occasional symptoms are not enough to stop farmers from using pesticides or from ignoring recommendations regarding usage. However, Guivant did find that interviewees who had been hospitalized tended to use protective gear when mixing and applying pesticides.

When farmers did experience pesticide-induced health problems, they offered one of three explanations absolving pesticides as the cause:

1. A farmer can do nothing to protect himself from fate.
2. A contaminated person is to blame for disregarding the safety recommendations.
3. A contaminated person has some physical weakness or genetic deficiency such as allergies or weak blood. In fact, some farmers reported mixing pesticides barehanded, claiming an immunity to pesticides' effects.

Guivant also found that farmers had a deep mistrust of technicians and sales staff as a source of information about pesticides. The farmers believed that the technicians considered them ignorant and inferior, labels that the farmers outright rejected. The technicians and sales staff in Guivant's study reported difficulties in convincing farmers to use lower and less toxic doses of pesticides. The farmers saw the experts as not having an economic stake in the advice they offered them. That is, it is easy to make recommendations and not have to live with the perceived economic consequences.

Implications for Improving Health Outcomes

The Brazilian farmers' way of applying pesticide not only increases their health risks but also the health risks of those in the surrounding community and of those who consume their products. So any plan to reduce pesticide use would have positive effects on more people than the farmers. Guivant's research has clear implications for disease/illness prevention. First, no pesticide reduction program can be effective without addressing the farmers' fears. An effective program has to allow farmers an opportunity to change their pesticide-dependent method of farming while guaranteeing them the profit margin to which they have become accustomed. Once farmers have proof that they can still turn a profit upon reducing pesticide use, the likelihood that they will change their farming will certainly increase.

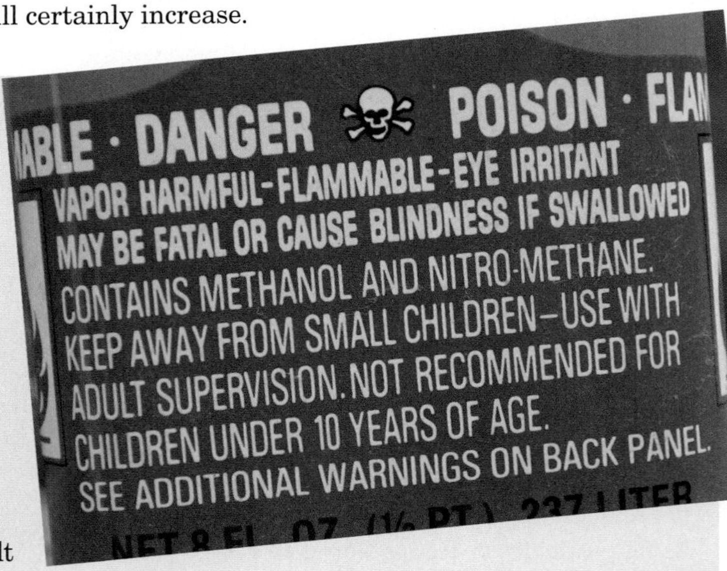

➤ Guivant's research suggests that warning farmers about the health effects of pesticide misuse is not in itself an effective strategy for changing farmers' behavior. Rather, farmers must come to believe that lowering pesticide use and following recommended standards will not result in lost income.

Missy Gish

Critical Thinking

What implications does rational choice theory and Guivant's research have for changing behaviors that increase health risks?

Summary: Putting It All Together

CHAPTER 13

Sociologists define social change as any significant transformation in the way social activities are organized. The discipline of sociology offers concepts to help us identify and describe revolutionary change (e.g., globalization, rationalization). It is worth noting that these key changes could not have occurred without fossil fuels. Social change—whether it be globalization, industrialization, urbanization, or something else—is part of an endless sequence of interrelated events. Even though it is difficult to disentangle the causes of social change, we can identify key factors that trigger change in general. The triggers include technological innovation, revolutionary ideas, conflict, the pursuit of profit, and social movements.

We examined how technological innovations (domestication, the plow, mechanization, digitalization, and so on) created six broad types of human societies, each defined by a signature technology that allowed the society to produce food and exploit resources: hunting and gathering, pastoralism, horticulturalism, agrarianism, industrialism, and postindustrialism. Each society is distinguished by the amount of surplus wealth it is able to produce.

This chapter also considers the characteristics of the U.S. system of health care and changes in the ways medical care is delivered. Characteristics of note include the fact that the U.S. system of health care is the most expensive in the world and that the United States is the only wealthy country that does not provide some form of universal health care coverage. The rising cost of health care has made it unaffordable for many Americans, even those with health insurance. One of the most dramatic changes in the way health care is delivered relates to the authority of the physician. That authority has been diminished as other stakeholders have assumed more visible and aggressive roles in diagnosing and treating illness.

Certainly health care systems will be strained by the aging of populations. The theory of the demographic transition helps us conceptualize the forces behind this unprecedented rise in the proportion of older people relative to the young. The demographic transition includes at least four stages that trace revolutionary changes in the relationship between births and deaths. Births increase the size of population, and deaths decrease it. Life expectancy is a critical factor in shaping the age composition of a society. So even though births have declined, people live longer, increasing the length of time people are on the planet.

We gave special attention to social movements, which are formed when a substantial number of people organize to make a change, resist a change, or undo a change to some area of society. For a social movement to form, there must be (1) an actual or imagined condition that enough people find objectionable, (2) a shared belief that something needs to be and can be done about that condition, and (3) an organized effort to attract supporters, articulate the mission, and define a change-making strategy. Usually those involved in social movements work outside the system to advance their cause, because the system has failed to address the problem.

Go to cengagebrain.com to link to Aplia and CourseMate for the chapter quiz and other activities.

Chris Caldeira

14 CHAPTER

SOCIOLOGY AT THE FOREFRONT

We have learned that sociologists are like inquisitive observers especially fascinated by what they cannot see taking place behind doors and walls of buildings. Driven by curiosity, sociologists have the urge to peer inside and learn about the human activities and relationships hidden from view. Keep in mind that it is not always easy to look inside or enter these structures. Those inside are often reluctant to reveal their lives to outsiders. The six modules in this chapter allow us to see that the sociological perspective provides valuable insights about an extraordinarily wide range of topics—from animal–human relationships to tattoos and body piercing to ageism. In reading each module, we will see "that the first wisdom of sociology is this—things are not what they seem" (Berger 1963, 23).

Go to Sociology CourseMate on cengagebrain.com to watch a video on how wind farms in Scotland promise to provide a renewable energy source.

MODULE

14.1 Animal–Human Relationships

Objective

You will learn that sociology as a discipline excludes animals in its analyses of society but that some sociologists are working to change that.

Do you think horses and other animals have distinct personalities, feel emotions, and possess a consciousness?

Spc. Emily Knitter

Do you think that animals who work for humans as police dogs, as guides to the blind, and as companions feel on some level a responsibility to perform the roles humans have assigned them? If you think so, then you will support those sociologists who are challenging the anthropocentric roots of their discipline.

Animals and Society

Sociologist Amy J. Fitzgerald (2007) maintains that sociologists have neglected human–animal relationships because sociology is largely **anthropocentric**; that is, it puts humans at the center of analysis and excludes animals and their realities from serious consideration. Fitzgerald points out that sociology's intellectual roots are grounded in the belief that humans are unique and vastly different from and superior to animals.

Fitzgerald references the work of two influential theorists—Karl Marx and George Herbert Mead—both of whom made a point of drawing a clear distinction between human and animal realities. Both argued that, unlike animals, humans possess a sense of self, language, and consciousness.

Specifically, Marx saw animals as having no self-awareness. He reasoned that consciousness arises out of social relations with others and argued that animals have "no relations with anything, no relations at all. An animal's relation to others does not exist as a relation" (1845). By "no relations" Marx meant that animals could not take others' expectations into consideration. Instead, animals respond by **instinct**, behavior that is not learned but is a matter of reflex (Fitzgerald 2007).

Lance Cp. Matthew K. Hacker

Department of Defense

➤ Marx distinguished humans and animals according to how the two produce food and build shelter. Birds, for example, do not build fancy dwellings or stock up on worms. Humans, on the other hand, project their food needs beyond the moment and often eat amounts of food beyond what is needed to survive. These different approaches suggest that animals and humans possess a different consciousness about the self.

Like Marx, Mead (1934) saw animals and humans as sharing few, if any, cognitive traits. Mead argued that human interaction revolves around interpretations of reality before a response is made and that animals have "no mind, no thought" (Fitzgerald 2007, 959) because their sounds and moves are immediate (void of interpretation) and thus meaningless. Mead maintained that humans, unlike animals, anticipate the consequences of their actions and take others into account before acting and responding (Fitzgerald 2007).

Research on Animal Consciousness

Because both Marx and Mead used their understanding of human consciousness as the standard against which to compare their understanding of animal consciousness, they concluded that animals were inferior and deficient.

➤ Sociologist Leslie Irvine (2007) asserts that this belief is challenged by a growing body of compelling research showing that animals can feel emotions, are able to role-take, and can communicate (Morgana 2007). There is credible research showing that chimpanzees and gorillas use sign language to communicate with humans; there is also research on a parrot that speaks original thoughts.

Airman 1st Class Devin Doskey

Professor Con Slobodchikoff, who has spent three decades observing prairie dogs, recorded and analyzed the sounds that these animals make to warn other prairie dogs about who is in the area. Slobodchikoff found that variations in the sounds prairie dogs made suggested that they are very specific about who is in their territory, whether it be a human, a hawk, a coyote, or even a "domesticated dog." To further test this idea, Slobodchikoff sent four people on a walk into prairie dog territory. First, each of the four walked though wearing the same colored shirts. Then each walked through another four times, each time wearing one of four shirt colors (blue, yellow, green, and gray). Slobodchikoff discovered that the warning sounds changed depending on the intruder and what he was wearing.

➤ Essentially, prairie dogs were warning, "'Here comes the tall human in the blue,' versus, 'Here comes the short human in the yellow'" (Slobodchikoff 2011).

Rhonda Foley

Researchers have also documented that primates, pigeons, and bottlenose dolphins recognize themselves when gazing into a mirror (Irvine 2007). Recognizing oneself in a mirror suggests that animals can role-take—that is, they can imagine how others are seeing them. From a sociological point of view, when someone can role-take, he or she has developed a sense of self. Thus, if animals do possess a consciousness, then people are obligated to reconsider their relationships with all varieties of animals and reassess the ways in which they use animals (Irvine 2007).

Sociologists who study animal–human relationships are giving attention to the complex bonds humans and animals form. Think about it—people everywhere have relationships with animals. People "eat them, wear them, bless them, help them when they are in need, and train them to help us when we are in need" (Bekoff 2007, xxxi–xxxii).

Canine Companions for Independence

➤ Some sociologists have studied working animals trained to do specific tasks, such as pulling loads, plowing fields, performing search and rescue operations, assisting the disabled, guarding property, serving as research subjects, acting as companions, and providing entertainment. Hundreds of millions of animals around the world work alongside their owners (Bekoff 2007). The dog shown in this photo has been trained to withdraw money from ATMs.

Other topics of interest include learning how animal–human bonds can promote mental and physical health and studying the conditions under which people treat animals as family members. The animal rights movement adds another interesting dimension with emphasis on the strategies activists employ to present their message and the role culture plays in impeding and enabling such efforts.

Critical Thinking

Do you agree with George Herbert Mead's assessment that animals have "no mind, no thought"? Explain.

Key Term

anthropocentric   instinct

Commercialization of Childhood

MODULE 14.2

Objective

You will learn how corporations learn what attracts children to a product.

Would you allow a market researcher to observe and talk to your child while he or she was taking a bath?

Missy Gish

Sociologist Juliet Schor is interested in the **commercialization of childhood**, a process by which children "are being turned into consumers almost from birth, and by adolescence, their social worlds [are] almost totally constructed around cool commodities, brand names, and the latest trendiest commercial music, films and lingo" (2004, 298). This commercialization is especially prevalent in the United States, a country with 4.6 percent of the world's population that consumes almost half the toys produced in the world each year (Statista 2012). As one measure of the extent to which American children influence household consumption, Schor points to corporate research showing that 6- to 12-year-olds shop with an adult two to three times each week and put six items in carts on each occasion.

Corporate Construction of Childhood

Schor draws attention to the **corporate construction of childhood**, a term that refers to the reach, power, and influence of a small number of corporations that sell and/or market most of the things children buy or want their parents to buy for them. Those corporations include Disney, Viacom, MTV Networks, Fox, AOL/Time Warner, Mattel, Hasbro, Nintendo, Sony, Microsoft, McDonald's, Burger King, Coca-Cola Company, and PepsiCo.

To gather data for her book about this process, Schor conducted 25 interviews, shadowed market researchers conducting focus groups with children, sat in on industry meetings, attended professional child marketing conferences, and read as many related publications as she could find. Among other things, Schor describes what she calls a new kind of intrusive research that examines the most intimate details of children's lives and goes far beyond survey questions and focus groups, the traditional venues by which marketers learn about consumer tastes. Schor maintains that marketers began using these intrusive methods in the 1980s because they were not satisfied with the limited information survey questions yield. Only so much can be gleaned from questions like, What shampoo do you use? What do you like about that shampoo? Have you ever heard of brand X? What do you think of it? As Schor describes it, corporations used to survey thousands of potential consumers or actual consumers about their products or their competitor's products. With the advent of caller ID, answering machines, mobile phones, and unlisted numbers, surveys became a less reliable and impractical way to gather data.

This new research "scrutinizes the most intimate details of children's lives" in naturalistic settings. That is, marketers videotape, interview, and observe children as they play, eat, take baths, do homework, and get ready for bed. They do an "in-depth analysis of the rituals of daily life" (22). Studying children in this way allows corporations to learn about the intimate details of children's lives and, based on information gleaned, develop strategies to market their products to them.

➤ Industry researchers patrol aisles watching, even recording, parents while they shop with their children in toy, department, and grocery stores. These researchers use what are known as ethnographic methods to help them build strategies to get children to notice and want their products. Schor learned that industry researchers visit playgrounds, stores, schools, homes, clubs, and any setting that attracts children.

Missy Gish

Ethnographic Methods

Ethnographic methods involve the holistic study of people—in this case children—in the places/settings where they live or carry out the activities of interest. So a company that makes bubble bath, for example, sends researchers to observe and interview children in bathrooms; a company that sells music would observe and interview children in the places where they listen to music, including their bedroom, bathroom, or other private settings; or a company that makes children's clothes would observe and interview children as they choose clothes from a closet or drawer.

Ethnographic research allows the child to set the pace, subject, and direction of the conversation. As one researcher told Schor, the "one shot approach does not work. She needs three visits . . . by the third time they are used to you . . . creating enough trust that the children will be willing to open up" (100). One-on-one sessions that allow the researcher to focus on one child at a time are preferred over focus groups in which one member can dominate the session or participants may censor their responses due to fear of what others may think.

Ethnographic research allows marketers to be there during the "moments of truth" (101) in children's lives—those moments when feelings and emotions surface. Being a part of those moments allows industry researchers to capture the images, metaphors, and symbolic meanings that children keep private but that drive their motivation and behavior.

Schor learned that these researchers pay parents for private time with children in a bedroom, bathroom, or playroom. One researcher recounts sitting on the toilet seat taking notes while the child she was observing showered behind a pulled curtain. "She explained that after a few minutes the kids forget she is in the room. That's when they perform the private behaviors it's her job to discover, such as grabbing the shampoo container and pretending it's a microphone, singing the latest Britney Spears tune and play-acting a super hero" (102).

Some marketing firms plant hidden cameras on children ages 6 to 9 before placing them in a store setting and instructing them to "purchase" a designated number of products.

Delawese Fulton

➤ Cameras record children as they move through the store looking for things to buy. Researchers learn what products children look at, how long they look, to which aisles children are drawn and linger, and what products they put into a shopping cart. After shopping, the researcher reviews the tape and talks to the child about what he or she was doing and thinking.

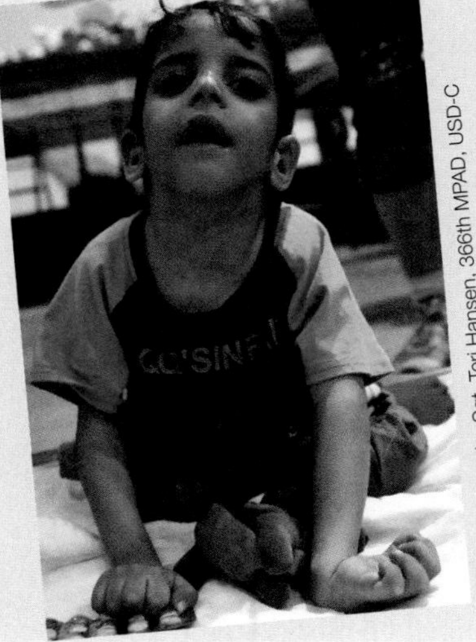

U.S. Army photo by Sgt. Teri Hansen, 366th MPAD, USD-C

◄ Industry researchers use eye-tracking technologies to follow eye movement and other eye activities (pupil dilation, blinks) as children interact with specific products, look through magazines, and watch television shows and commercials. The researchers are interested in how easy it is for the child to see a particular brand on a shelf with other brands, how long the child gazes at a product relative to another, what part of an advertisement loses a child's attention, and at which of two monitors a child looks when two different shows or ads are playing. Eye tracking offers valuable insights as to what product and packaging best captures children's attention.

Critical Thinking

Can you think of a product you simply had to have as a young child? Describe your memories of this product and your need to obtain it.

Key Terms

commercialization of childhood

corporate construction of childhood

ethnographic methods

MODULE 14.3

Tattoos and Body Piercing as Expressions of Identity

Objective

You will learn why tattoos and body piercing have become popular forms of self-expression.

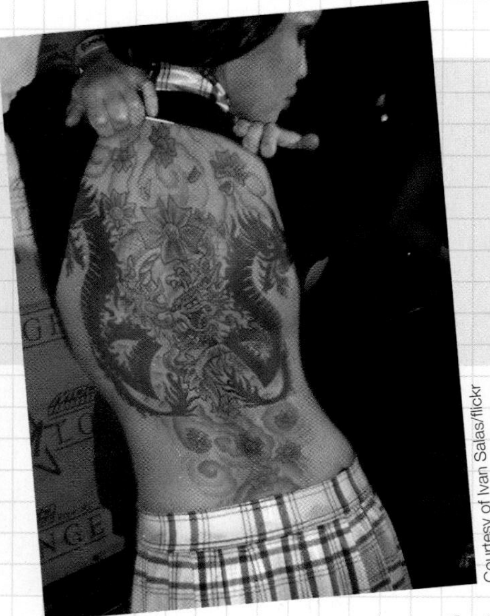

Courtesy of Ivan Salas/flickr

Do you have a tattoo or a body piercing? Why did you decide to modify yourself in this manner?

An estimated 40 percent of those known as Millennials (born between 1971 and 1981) have a tattoo, about 20 percent have 2 to 5 tattoos, and 7 percent have 6 or more. Almost 25 percent have a piercing on their body "in some place other than an earlobe—about 6 times the share of older adults who've done this" (PEW Research Center 2010). The age group representing their parents' generation (ages 40–64) has considerably fewer tattoos (10 percent) and body piercings (6 percent) (PEW Research Center 2007). How do sociologists explain this generational difference in the presentation of self? To answer this question, we will consider the work of sociologist Lauren Langman (2003), who traces body modification in part to the economic instability that has accompanied globalization.

Globalization and Identity

There is no question that globalization has transformed local economies and the global workforce—it has eliminated many occupations, created new ones, moved jobs to new locations, and radically changed the structure of opportunity. In the mid-to-late 1980s, many workers in Europe and the United States lost manufacturing jobs to outsourcing and technological innovations; the reemployed often found themselves underemployed or working part-time. Beginning in the late 1990s, a second wave of outsourcing occurred that moved information and customer service jobs to locations such as India and China. These two waves of economic transformation disrupted not just the affected workers' lives but their families and communities.

Langman (2003) argues that, until recently, the status system in Western capitalist societies was centered around occupation. But a period of prolonged economic instability has supported a proliferation of what Langman calls identity-granting subcultures that offer new ways of gaining status and recognition.

➤ Like masks, which allow people to hide their real identity and assume an alternate one, the Internet supports a variety of identity-granting subcultures that offer members the equivalent of a virtual mask in that they are free to hide their real selves and present themselves as something they are not.

Heather Gunn

Langman maintains that body modification represents another kind of venue by which people acquire an alternate identity. His thesis is this: in a world without job security, predictable careers, or enduring relationships, body modification offers a sense of permanency, connection, and personal empowerment. We develop his thesis in the sections that follow.

Body Modification as a Fashion Statement

Every society has conventions about how to dress. People dress in ways that distinguish themselves from others. Clothes, accessories, hairstyles, jewelry, and other personal products announce to others "who one is and who one is not" (Langman 2003, 238). The fashion we choose presents an identity that locates us inside and outside particular groups. Globalization, and the consumption-oriented societies it has fostered, offers many fashion choices for sale, including some that signal their wearers are outside the mainstream. One fashion choice, body modification—especially that which is visible and extreme—announces contempt for and rejection of the mainstream culture. In its extreme forms, body modification can include splitting tongues, decorating genitalia, implanting studs or earrings in the nose, and saltings (injections of saline solution to grossly enlarge and distort a body part). While these body adornments elicit shock, disgust, and fear among those considered mainstream, they are still sources of status. Such extreme body modifications ensure that one will be noticed—that one has enough power over others to command their gaze. "Whether the gaze invokes disgust by the mainstream, or acceptance by the subculture, to be viewed as different is to gain empowering recognition" (Langman 2003, 242).

For significant numbers of college-age students, tattoos and body piercing have become a rite of passage. Body modification is now so common it is considered ordinary. Still, when done in excess—multiple rings, implants, studs—body modification signals a rejection of mainstream culture and its values.

Symbolic Significance

In Western culture tattoos, scarification, and body piercing have typically been regarded as a regression to so-called primitive practices. According to Langman, those who engage in body modification often identify as modern primitives.

➤ Symbolically, the modern primitive identifies with cultures that consider body modification to be sacred and magical and that treat it as a symbolic means for people to connect with a larger community. The rituals accompanying tattoos and other body modification dramatize that passage into a desired group, such as an ancient warrior class.

Tony Rotundo

Langman argues that "to decorate one's body is to claim authorship. Although this may involve pain, the pain is an expression of agency, recognition and inclusion" (242). Insofar as globalization has caused great pain to many displaced people, the pain of body modification becomes a way of authoring one's own pain rather than having it imposed by impersonal forces. Ironically, it does little to change reality; in fact, one could argue that it is a harmless form of protest that does nothing to change the dominant moral codes or the structure of opportunity.

Critical Thinking

Do you or does someone you know have tattoos or body piercing that might be considered excessive? Does Langman's theory apply to you or this person? Explain why or why not.

Gender Transitions in the Workplace

MODULE 14.4

Objective

You will learn how the two-category conceptions of gender in the context of the workplace are affected when an employee transitions from one gender to another.

Chris Caldeira

How would you react if someone you knew as one gender transitioned to another gender?

Sociologists Kristen Schilt and Catherine Connell (2007) did in-depth interviews with 28 transsexual/transgender people living in Los Angeles, California, and Austin, Texas, to learn the process by which they transitioned in the workplace from their birth gender to their destination gender. The two researchers sought to understand how transmen (female-to-male), transwomen (male-to-female), and their colleagues at work negotiated gender identity.

Schilt and Connell recruited participants from transgender activists and support groups, from personal contacts, and from listservs. Schilt and Connell questioned the recruits about how their sex change affected interactions with coworkers.

Negotiating Cross-Gender Interactions

Transmen and transwomen found that they had to adjust to new gender boundaries, areas where they were once included but now found themselves excluded. Transmen noticed that they were less frequently included in "girl talk" at work. For those transmen who said they always felt like a man, even before their transition, this exclusion came as a relief; now they did not have to notice and talk about things like clothes and changes in hairstyles. But for those transmen who identified as queer, bisexual, or lesbian before the sex change, this exclusion caused them to feel sadness and loss over the new distance between them and their female coworkers.

Transwomen, on the other hand, tended not to participate in "guy talk" (sports and sexual objectification of women) before their surgery and felt that their male colleagues now understood why they were never interested in such matters. One transwoman who worked at a high-tech company remembered that her male boss worried that her taking estrogen might adversely affect her computer programming abilities. Another transwoman who was one of four co-owners of a professional business was forced to sell her share of the company to the three male partners because they were concerned that as a woman she could no longer be a serious business partner, thinking that now she would be overly concerned "with frivolities of appearance" (606).

Both transmen and transwomen recount changing interaction styles to correspond to their new gender. Ellen, a transwoman, found that when asked for "my opinion on a subject from men, I have to remember—do not express it as firmly as I actually believe" (607). She had to come to terms with the fact that muting opinion is a common occurrence among women in the workplace.

Transmen found that their female colleagues expected them to do the heavy lifting now—move office furniture, carry heavy boxes, and change water cooler bottles. They had to come to terms with gender rituals that positioned "women as frail and men as able" (608). Both transmen and transwomen observed that some of their colleagues challenged their destination gender by referencing their birth gender and by criticizing them for retaining too many of their birth gender traits now that they were another gender (transwomen who were considered assertive as men before their sex change were now considered bossy as women).

Negotiating Same-Gender Interactions

Both transmen and transwomen reported that, for the most part, their colleagues included them in same-gender spaces and interactions. This inclusion in same-sex gender spaces often came after an invitation. For example, men would ask a transman "when they were going to start using the men's locker room . . . one set of co-workers even had a mock ceremony in which they presented him with a key to the men's room" (610). Transwomen found themselves "eating lunch with the girls" and chatting with them in the bathroom.

Chris Caldeira

◂ Transmen and transwomen found that during the transition from one gender to the other they had to "gain access to same-gender sanctuaries in the workplace" (610) such as locker rooms and bathrooms.

The transmen and transwomen described experiences where their colleagues treated them as gender apprentices: female coworkers offered to teach transwomen how to put on makeup or how to shop, whereas male coworkers taught transmen how to tie a Windsor knot and give the kind of advice that is typically handed down from father to son about the "right way to be a man." Some transmen appreciated the training, but others found it coercive. Transwomen, on the other hand, welcomed offers to show them how to be a woman. Schilt and Connell speculate that transwomen appreciated the training more than the transmen did because, when they were men, transwomen faced sanctions for imagining and trying out female roles. Their transmen counterparts had more opportunities to imagine and try out being male before their gender transition. This difference suggests that women's cross-gender behavior is more socially accepted than men's cross-gender behavior.

➤ Transwomen, in particular, found that they had more friends now and that they could openly express interest in feminine things—such as babies and children—that they often repressed as men.

Tony Rotundo

Transmen, on the other hand, did not experience the same feelings of liberation because when they were women they had greater leeway to express interest in so-called masculine things. Transmen did, however, find themselves included in certain male rituals like backslapping, which was done "with more force than they probably slapped each other on the back" as a way of "affirming to me that they saw me as a male" (612).

Analysis

Schilt and Connell found that the 28 transmen and transwomen they interviewed actually hoped to create alternate femininities and masculinities—that is, femininities and masculinities that departed from rigid two-category definitions and ideas of what males and females should be and do. However, regardless of their aspirations, their coworkers saw it as their job to "repatriate them" so that they conformed to rigid stereotypes of how men and women "just are." Schilt and Connell wondered whether the existence of a transgender colleague "could lead to an undoing of gender inequality in the workplace, as co-workers and employers could begin to rethink their stereotypes about gender, workplace performance and natural abilities" (603). This undoing did not happen. Why? According to Schilt and Connell it is because there is no interactional transcript to guide people when they interact with those who are not clearly male or female. Furthermore, it is difficult for anyone to operate outside the gender expectations that are constantly reproduced in daily interactions.

Critical Thinking

How do you think you would react if a colleague at work transitioned to another gender? In what ways would your reaction be similar or different from those documented in Schilt and Connell's research?

MODULE

14.5 Ageism and the Anti-Aging Industry

Objective

You will learn the meaning and origins of ageism.

If you were asked to describe one of your grandparents, what would you say?

Lisa Southwick

Chances are that you would have very positive things to say about your grandparents. The positive experiences most of us have stand in stark contrast to the pervasive negative stereotypes regarding the mental abilities and physical characteristics of older people in general.

Professor Glen Hougan (2007) found that students taking his Design for an Aging Population course held deeply ingrained negative stereotypes about the elderly. Hougan required his students to work in teams to create an aging suit that would simulate some of the physiological challenges and experiences associated with aging. The point of this exercise was to sensitize students to the needs of older populations. However, the suits students created actually did little to stimulate innovative product designs; instead, they supported designs that mirrored and reinforced stereotypes about the elderly.

Missy Gish

◄ As Hougan described it, one group of students decided to wear "an 'old man' latex mask that one buys at a joke shop. Their explanation, in all seriousness, was that it enhanced the feeling of being elderly. Another group, using that same principle but a different tactic, incorporated movement restriction devices into clothes they perceived that the elderly wear. Their version of what the elderly wear was a drab, wrinkled, oversized suit jacket and pants one finds in the far reaches of a thrift store" (2007, n.p.).

Hougan also asked his students to write down three words to describe the elderly. Seventy-five percent of the words related to being weak, slow, or feeble. For the course to be effective, the professor had to find a strategy to help his students overcome their prejudices and stereotypes. Hougan did this by capitalizing on students' generally positive relationships with grandparents, asking students to interview them, research their lives, and to design something for them.

Ageism

Ageism is systematic prejudice, stereotyping, and/or discrimination on the basis of age. Ageism applies to any age group but is especially common toward those considered elderly. **Institutionalized ageism** occurs when discrimination based on age is viewed as an accepted, established, and even fair way of doing things, but on close analysis, such practices put a certain age group at a disadvantage relative to those in other age groups. In the context of product design for older people, senior design analyst Gretchen Anderson (2010) maintains that "if we view seniors through the products that are available to them, then they would be viewed as 'cranky, stupid, and tacky.'" The products actually reflect the ageist beliefs held by designers and manufacturers. Anderson offers the example of orthopedic shoes. "The orthopedic shoe, big and bulky, has a style and a color that has not changed in over 40 years." Anderson believes that when most designers "talk about needs of seniors there is a tendency to imagine someone whose eyesight, dexterity, and hearing are so impaired that they are incapable of having an experience."

Of all the -isms—racism, sexism, classism—ageism is the most widespread and the least likely to be challenged. It is the most widespread because everyone, if they live long enough, regardless of race, class, or gender, will eventually be considered "older." Ageism toward older people goes unchallenged on a variety of fronts. As one example, the popularity of birthday cards that make fun of, even ridicule, people sometimes as young as 30 for having lowered sex drive, loss of memory, and declining physical appeal represents one example. Ageism is further reinforced by the ways older people are portrayed in commercials, movies, and sitcoms. Keep in mind that, while the media puts forth these images, it does not create them out of thin air but bases them on audience expectations and perceptions. When Hougan asked his class to name a movie about older people, the students named *Cocoon* (1987) and *Grumpy Old Men* (1993). The title *Grumpy Old Men* speaks for itself, and *Cocoon* features retired men living in a Florida retirement home. They are rejuvenated after taking a swim in a pool that functions as a fountain of youth.

Hougan maintains that if we take a long view, we will see that ageism is rooted in a number of social shifts that have gradually eroded the status of the elderly over time. Those shifts include the following:

- Increased literacy rates have diminished the older persons' role as authorities and keepers of oral traditions. Literate societies have libraries, bookstores, and now online holdings. Increased literacy is also synonymous with formal schooling. In the 1860s the United States became the first country in the world to embrace mass education, making elementary-level education for all the law. Within 60 years all states had passed compulsory attendance laws. Of course, these laws meant that young people spent their days in an age-segregated environment (e.g., schools) isolated from the influence of other age groups, in particular the elderly.

- With industrialization, technical skills became more valued than lived experiences. Moreover, since technical skills are always in need of update, these steady advances put older people who do not grow up with the latest technologies in a position in which they must find ways to master them.
- The increased mobility that accompanied industrialization dispersed families geographically and as a result weakened connections to older relatives. This is significant to ageism because it is interactions with older people that help dispel stereotypes and other misconceptions about them.
- At one time people worked until they were no longer able, at which point they "retired." As early as 1875 the idea that workers retire and enter a new stage in life took root. In that year the American Express railroad company established the first private pension plan in the United States. In subsequent years and decades, other employers began offering pension plans and governments implemented benefit plans.

Social Security Administration

➤ This poster is prompting people to "boldly go" online to use Social Security services, including applying for Social Security. In 1935 the Social Security Act became law, allowing people to retire at age 62 with partial monthly benefits and age 65 with full monthly benefits lasting until death. Retirement can be viewed as a systematic process that removes older people from the workforce (a kind of de facto segregation).

- The numbers of age-segregated environments in which people spend much of their day surrounded by people who are their own age (schools, day care, camps, retirement communities, clubs, neighborhoods) are increasing.
- The biological aging process has come to be perceived as a process that moves people closer to the inevitable—death. While this may seem obvious, death was not always something associated with old age.

Prints & Photographs Division, Library of Congress, LC-DIG-npcc-19247

➤ At one time in the United States death was something people in every age group experienced, but especially those less than 1 year of age. Twenty percent of all deaths in 1900 involved infants less than 1 year old. As late as 1950, 20 percent of deaths (1 in every 5) involved people under age 45. Today, 85 percent of deaths involve someone 55 and older.

- An increasing value is placed not just on youth and beauty but on idealized youth and beauty that can only be achieved through artificial means (surgery, right camera angles, computerized touch-up). When these ideals are juxtaposed against dominant images of old age, it makes aging an especially dreaded process, so much so that an anti-aging industry has emerged.

The Anti-Aging Industry

In the past decade or so there has been an unprecedented increase in the number of products and procedures that are marketed under the label "anti-aging." These products promise to prevent visible signs of aging, prevent physical and mental decline, reverse the biological process of aging, and extend life. There are essentially two branches of what can only be called an anti-aging industry: (1) potions, lotions, and cosmetic treatments that reduce the physical signs of aging (wrinkle-prevention crèmes, hair restoring products), and (2) medical interventions and procedures that intervene to prevent or reverse aging (prescriptions such as Botox and Viagra, cosmetic surgeries) (Vincent, Tulle, and Bond 2008). While there is little systematic research on the social consequences of anti-aging products, there are a number of questions that guide the sociological perspective as it relates to this industry. Sociologist John A. Vincent and his colleagues (2008) have outlined such questions as the following:

- In the event that science can find a way to prolong life beyond the 120 years believed to be the natural limit on human life, what are the potential social consequences of extending life on an already aging population?
- What symbolic meanings of aging and old age does the anti-aging industry reflect and promote? Are there other meanings of aging and old age other than something that is problematic?
- How will anti-aging trends change the way we think about the purpose of medicine and the role it plays in our lives? Will medicine redirect limited resources toward reducing the appearance and effects of aging? What other medical needs will not be met as a result?
- Does the anti-aging enterprise intensify ageism and subject those who are aging to "dissection, manipulation and control" (292)?

Lisa Southwick

➤ Do aging and old age have a value, or are they something to be resisted at any cost? When we consider the important role that grandparents and great-grandparents play in people's lives, we see there is value to stepping up and assuming one's place in the life cycle. One could argue that young people need secure older people in their lives who are not offended by that status. How might the anti-aging movement affect that security if older people seek to be young and not act like grandparents or great-grandparents?

Critical Thinking

Describe someone considered older who has made an impact on your life. In what ways did their older age contribute to that impact?

Key Terms

ageism institutionalized ageism

MODULE

14.6 Environmental Sociology

Objective

You will learn that environmental sociologists focus on how human activities, especially profit-making activities and overconsumption, affect the quality of life on the planet.

Is there a product you routinely buy that is designed to be thrown away after one or two uses? Have you ever considered where those products go after you toss them out?

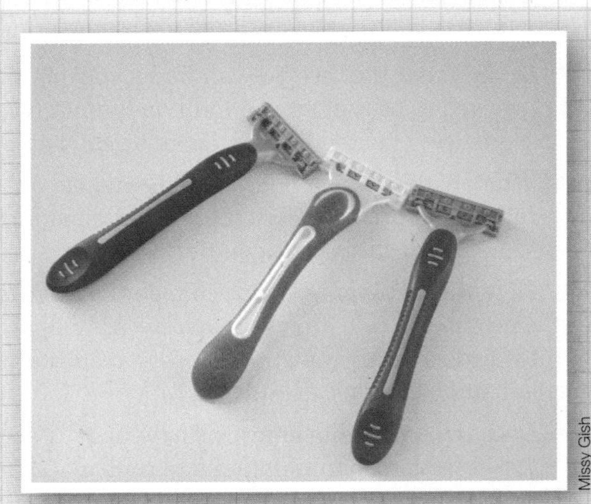
Missy Gish

If you think about how the manufacturing and disposal of consumable products impacts the environment, then you share the interests of sociologists who make the relationship between the environment and human society their focus of study.

Environment and Society

Environmental sociologists seek to identify the social, political, economic, and technological factors that contribute to pollution, overconsumption, and waste, which in turn threaten ecosystems, human life, and other species that share the planet (ASA 2012a). This focus differs from classic sociological inquiry that makes humans the center of study. Environmental sociologists seek to uncover the ways in which human activities influence and affect the natural environment (Dunlap and Catton 1979; Buttel and Humphrey 2002).

Much of the focus of environmental sociology has been on explaining how the consumption patterns of the United States and other industrialized countries are connected to environmental destruction. Sociologists associate environmental problems with the global economy's ever-increasing drive for profit. The term **treadmill of production** captures the ceaseless increases in production and, by extension, consumption that are needed to sustain the global economy's success, which is measured by increased profits (Schnaiberg 1980; Gould, Pellow, and Schnaiberg 2008).

➤ When sociologists think about the cycle of consumption, they don't think about any one consumer, but rather they think about what any one consumer's behavior represents with regard to waste. Consider disposable diapers and the number that any one baby goes through before being potty trained. One estimate puts that number at 8,000 (Environmental Protection Agency 2011). Since the advent of disposable diapers in the 1970s, the average age at which children are potty trained has increased, further increasing the number of diapers any one child uses.

Karl Wiesbaden

This never-ending cycle or treadmill has a devastating impact on the environment because the relentless focus on producing and consuming increases energy consumption, waste, and harmful emissions. Environmental sociologists contend that the damage to the environment is not shared equally; minority and low-income people who consume much less than the wealthy are disproportionately overexposed to environmental pollutants (Stretesky and Lynch 2002; Hooks and Smith 2004; Downey 2005). For example, a disproportionate share of toxic waste dumps and hazardous emissions is found in minority communities within the United States and in poorer nations abroad. This unequal distribution of environmental hazards based on race or socioeconomic status is evidence of **environmental injustice** (Bullard et al. 2007). For example, a study of factors that affect exposure to hazardous waste found that the majority of people living within 1.8 miles of hazardous waste facilities in the United States are minorities (Bullard et al. 2007).

Changing Consumption Habits

In thinking of ways to change consumption habits, we must recognize that energy consumption is woven into the fabric of American society and, to varying degrees, into the fabric of the world's countries. Changing these habits requires a revolution in thinking and behaving. This revolution will not look like the so-called Green Revolution that to date seems to emphasize consuming green products and taking easy steps to change the way we live, as described by the following headlines: "21 Ways to Save the Earth and Make More Money," "365 Ways to Save the Planet Earth," and "Ten Ways to Green Up Your Sex Life" (Friedman 2008). Our current energy system took more than a hundred years of investments to establish (the age of oil in the United States began in 1901); it may take a hundred years of investments to put a clean energy system in place (Friedman 2008). If we consider the total energy consumed from all fossil fuel sources on a global scale (although it is unequally distributed), humans consume the equivalent of 10 million barrels or 420 million gallons per hour—the United States consumes 84 million gallons of that hourly total (Friedman 2008). This is the scale of demand that is only expected to increase in coming years.

Chris Caldeira

⮝ Environmental sociologists are particularly interested in identifying realistic strategies that countries, communities, and individuals are taking to promote sustainable relationships between humans and their natural environment. Examples of successful strategies can be found from the individual level, such as solar-powered laundromats, to the country level, where South Korea is one model of recycling and Sweden a model of energy efficiency (Zumbrun 2008; Williamson 2011).

Critical Thinking

Are there areas of your life in which you are guilty of overconsumption? Explain.

Key Terms

environmental injustice   treadmill of production

Summary: Putting It All Together

CHAPTER 14

These six modules have considered the work of sociologists who have

- challenged the discipline of sociology to broaden its scope of study,
- revealed the processes by which corporations instill a desire to consume,
- connected the presentation of self to globalization and capitalism (profit-generating anti-aging products),
- described how people respond when some change the gender category by which they have been known,
- tackled a form of institutionalized discrimination (aging and gender), and
- challenged us to think about the human relationship to the environment.

After reading the six modules, it may appear that sociology lacks a focus—that it is all over the place. Taken together, however, these modules speak to some important themes. First, it is impossible to compile a comprehensive list of the topics that sociologists study, because almost any topic involving humans is a possible area of study. Sociology is distinguished not by the topics studied but by the perspectives and questions used to frame research and analysis—questions like, What are the intended and unintended consequences of X? Who benefits from a social arrangement and at whose expense? How are meanings assigned, accepted, sustained, and challenged?

Second, while sociologists seek to understand human activity, they are especially drawn to those activities that are hidden from view, that are controversial, or that go largely unchallenged. The discipline of sociology offers theories, concepts, and methods needed to look beyond popular meanings and official interpretations of what is going on around us. Sociologist Peter Berger (1963) points out that sociologists are driven to debunk the social systems they study. The sociological perspective compels sociologists to explore levels of reality that are deep below the surface.

Third, from a sociological point of view, no topic involving human activity is insignificant. Studying those who get tattoos or who transition from one gender to another may seem better suited for talk shows. But sociology is serious in its approach. It challenges those who are curious to look below the surface and entertain the possibility that the proliferation of tattoos may be connected to global economic transformations or that rigid binary definitions and ideas about males and females constrain human potential and experiences. Sociology also challenges us to leave our comfort zones—to open doors that appear closed, to knock on doors that are bolted shut, and to find ways to gain access to those who hide behind them.

Go to cengagebrain.com to link to Aplia and CourseMate for the chapter quiz and other activities.

Glossary

absolute poverty—a situation in which people lack the resources to satisfy the basic needs no person should be without. Absolute poverty is usually expressed as a living condition that falls below a certain threshold or minimum.

absorption assimilation—the process by which a subordinate ethnic, racial, and/or cultural group adapts to the ways of the dominant group, which sets the standards to which they must adjust.

achieved statuses—human-created social categories and characteristics acquired through some combination of personal choice, effort, and ability.

adaptive culture—the part of the culture (nonmaterial) that adjusts to a new product or innovation, specifically to the associated changes that product or innovation promotes.

adolescent status system—a classification system in which participation in some activities results in popularity, respect, acceptance, and praise, and participation in other activities results in isolation, ridicule, exclusion, disdain, and disrespect.

advantaged—a situation in which the symbolic opportunities (positive images) and valued resources are disproportionately held by a particular group relative to another group.

ageism—a systematic prejudice, stereotyping, and/or discrimination on the basis of age.

agents of socialization—significant people, groups, and institutions that act to shape our sense of self and social identity, help us realize our human capacities, and teach us to negotiate the world in which we live.

aging population—a society in which the percentage of the population that is 65 and older is increasing relative to other age groups.

agrarian society—a society that emerged with the invention of the plow 6,000 years ago, triggering a revolution in agriculture. The plow made it possible to cultivate large fields and increase food production to a level that could support thousands to millions of people, many of whom lived in cities and/or were part of empires.

alienation—a state of being in which humans lose control over the social world they have created and are dominated by the forces of their inventions.

altruistic (suicide)—a state in which the ties attaching the individual to the group are so strong that a person's sense of self cannot be separated from the group.

anomic (suicide)—a state in which the ties attaching the individual to the group are disrupted due to dramatic changes in circumstances.

anomie—a state of cultural chaos resulting from structural strain.

anthropocentric—a point of view that puts humans at the center of analysis and excludes animals and their realities from serious consideration.

ascribed statuses—human-created social categories and characteristics that are the result of chance in that people exert no effort to obtain them. Birth order, race, sex, and age qualify as ascribed statuses.

assimilation—a process by which ethnic, racial, and/or cultural distinctions between groups disappear because one group is absorbed, sometimes by force, into another group's culture or because two cultures blend to form a new culture.

attribution theory—a set of concepts and propositions that gives attention to the process by which people explain their behavior and that of others.

audience reference groups—those who are watching, listening, or otherwise giving attention to someone or something.

authoritarian governments—a form of government in which no separation of powers exists; a single person (a dictator), a group (a family, the military, a single party), or a social class holds all power. No official ideology projects a vision of the "perfect" society or guides the government's political or economic policies.

authority—legitimate power, or power that people believe is just and proper. A leader has authority to the extent that people view him or her as being legitimately entitled to it.

back stage—areas out of sight of an audience, where individuals let their guard down and do things that would be inappropriate or unexpected in a front-stage setting.

beliefs—conceptions that people accept as true concerning how the world operates and the place of the individual in relation to others.

biography—all the events and day-to-day interactions from birth to death that make up a person's life.

biological sex—the physiological, including genetic, characteristics associated with being male or female.

bourgeoisie—the owners of the means of production.

brain drain—the emigration from a country of the most educated and most talented people, including those trained to be hospital managers, nurses, accountants, teachers, engineers, political reformers, and other institution builders.

bureaucracy—a completely rational organization—one that uses the most efficient means to achieve a valued goal.

calculability—a feature of McDonaldization that emphasizes numerical indicators by which customers and the service provider can judge the amount of product and the speed of service (e.g., delivery within 30 minutes).

capitalism—an economic system in which the raw materials and the means of producing and distributing goods and services are privately owned.

carbon footprint—the impact a person makes on the environment by virtue of his or her lifestyle. That impact is measured in terms of fossil fuels consumed or units of carbon dioxide emitted from burning that fuel.

carceral culture—a social arrangement under which the society largely abandons physical and public punishment and replaces it with surveillance to control people's activities and thoughts.

caregiver burden—the situation in which caregivers believe that their emotional balance, physical health, social life, and financial status suffer because of their caregiver role.

caregivers—those who provide service to people who, because of physical impairment, a chronic condition, or cognitive impairment cannot do certain activities without help.

case studies—objective accounts intended to educate readers about a person, group, or situation.

caste system—a system of stratification in which people are ranked according to ascribed statuses.

censorship—an action taken to prevent information believed to be sensitive, unsuitable, or threatening from reaching some audience.

charismatic authority—legitimate power that is grounded in exceptional and exemplary personal qualities. Charismatic leaders are obeyed because their followers believe in and are attracted irresistibly to the leader's vision.

church—a group whose members hold the same beliefs regarding what is considered sacred and profane, share rituals, and gather in body or spirit at agreed-on times to share and reaffirm their commitment to those beliefs and practices.

civil religion—an institutionalized set of beliefs about a nation's past, present, and future and a corresponding set of rituals. These beliefs and rituals can take on a sacred quality and elicit intense feelings of patriotism.

claims makers—those who articulate and promote claims and who tend to gain in some way if the targeted audience accepts their claims as true.

class conflict—an antagonism between exploiting and exploited classes.

class system—a system of stratification in which people are ranked on the basis of their achievements related to merit, talent, ability, or past performance.

coercive organizations—organizations that draw in people who have no choice but to participate.

colonialism—a situation in which a foreign power uses superior military force to impose its political, economic, social, and cultural institutions on an indigenous population in order to control their resources, labor, and markets.

color line—a barrier supported by customs and laws separating nonwhites from whites, especially with regard to their roles in the division of labor; it can be traced to the European colonization of Africa.

commercialization of childhood—a process by which children are transformed into consumers beginning at birth, and by adolescence, their lives and activities revolve around commodities, including brand-name products and the latest trendiest commercial music, films, and lingo.

commercialization of sexual ideals—the process of introducing products into the market using advertising campaigns that promise consumers they will achieve a sexual ideal if they buy and use the products.

commodification—a process by which economic value is assigned to things not previously thought of in economic terms such as an idea, a natural resource (water, a view of nature), or a state of being (youth, sexuality).

commodification of sexuality—a process by which companies create products for people to buy with the promise that those products will allow them to express themselves as sexual beings or elicit a sexual response from others.

communitarian utopias— a counterculture in which members withdraw into a separate community in order to live with minimum interference from the larger society, which they view as evil, materialistic, wasteful, or self-centered.

comparison reference groups—groups that provide people with a frame of reference for judging the fairness of a situation in which they find themselves; justifying certain behaviors; and assessing their performance relative to others.

comprehensive dyads—two people who have more than a superficial knowledge of each other's personality and life; they know each other in a variety of ways.

conflict—the major force that drives social change.

conformists—people who have not violated the rules or expectations of a group and are treated accordingly.

conformity (as a response to structural strain)—acceptance of cultural goals and the pursuit of those goals through legitimate means.

consumerism—an ideology that locates the meaning of life in possessions.

content analysis—a method of analysis in which researchers identify themes, sometimes counting the number of times something occurs or specifying categories in which to place observations.

control—a feature of McDonaldization that emphasizes replacing employee labor with nonhuman technologies and/or requiring, even demanding, that employees and customers behave in a certain way.

control variables—variables that researchers hold constant in order to focus on just the relationship between the independent variable and the dependent variable.

core economies—the wealthiest, most highly diversified economies in the world with strong, stable governments.

corporate construction of childhood—a term that refers to the reach, power, and influence of a small number of corporations that sell and/or market most of the things children buy or want their parents to buy for them.

corporate crime—a crime committed by a corporation as a result of the way that it does business competing with other companies for market share and profits.

countercultures—subcultures that challenge, contradict, or outright reject the values of the mainstream culture of which they are a part.

credential society—a situation in which employers use educational credentials as screening devices for sorting through a pool of largely anonymous applicants.

crime—an act that breaks a law.

critical social theory—a sociological perspective that examines human-constructed categories and institutional practices with the aim of shedding light on the central issues, situation, and experiences of groups differently placed within "specific political, social, and historic contexts characterized by injustice" (P. Collins 2006).

cultural anchor—some item of material or nonmaterial culture that elicits broad consensus and support among diverse membership, even in the face of debate and dissent about its meaning.

cultural capital—a person's nonmaterial resources, including educational credentials, the kinds of knowledge acquired, social skills, and acquired aesthetic tastes.

cultural diffusion—the process by which an idea, an invention, or a way of behaving is borrowed from a foreign source and then adapted to the culture by the borrowing people.

cultural diversity—the cultural variety that exists among people who find themselves sharing some physical or virtual space.

cultural lag—a situation in which adaptive culture fails to adjust in necessary ways to a material innovation and its disruptive consequences.

cultural particulars—the *specific* practices that distinguish cultures from one another.

cultural relativism—a point of view advocating that a foreign culture not be judged by the standards of a home culture, and that a behavior or way of thinking must be examined in its cultural context—that is, in terms of that culture's values, norms, beliefs, environmental challenges, and history.

cultural universals—those things all cultures have in common.

culture—the way of life of a people, specifically the shared and human-created strategies for adapting and responding to the social and physical environment.

culture shock—the mental and physical strain that people can experience as they adjust to the ways of a new culture. In particular, newcomers find that many of the behaviors and responses they learned in their home culture, and have come to take for granted, do not apply in the foreign setting.

culture of spectacle—a social arrangement by which punishment for crimes—torture, disfigurement, dismemberment, and execution—is delivered in public settings for all to see.

decolonization—the process of gaining political independence from a colonizing power.

demographic gap—the difference between the birth and death rates.

demographic trap—the point at which population growth overwhelms the environment's carrying capacity.

dependent variable—the behavior to be explained or predicted.

desegregation—the process of ending legally sanctioned racial separation and discrimination including removing legal barriers to interaction and offering legal guarantees of protection and equal opportunity.

deviance—any behavior or physical appearance that is socially challenged and/or condemned because it departs from the norms and expectations of some group.

deviant subcultures—groups that are part of the larger society but whose members share norms and values favoring violation of that larger society's laws.

differential association—a theory that explains how deviant behavior, especially juvenile delinquency, is learned. This theory states that it is exposure to criminal patterns and isolation from anticriminal influences that put people at risk of turning criminal.

disability—a state of being that society has imposed on those with certain impairments because of how inventions have been designed and social activities have been organized to accommodate the shortcomings of only those considered unimpaired.

disciplinary society—a social arrangement that institutionalizes surveillance and fuels an anxiety about being watched—an anxiety that promotes socially desired behavior.

discrimination—the intentional or unintentional unequal treatment of racial or ethnic groups without considering merit, ability, or past performance.

disenchantment—a great spiritual void accompanied by a crisis of meaning in which the natural world becomes less mysterious and revered and becomes the object of human control and manipulation.

dispositional factors—things that people are believed to control, including personal qualities related to motivation, interest, mood, and effort.

division of labor—work that is broken down into specialized tasks, each performed by a different set of workers trained to do that task. The labor and resources needed to manufacture products often come from many locations around the world.

dominant ethnic group—the most advantaged ethnic group in a society; the ethnic group that possesses the greatest access to valued resources, including the power to create and maintain the system that gives it these advantages.

dramaturgical sociology—a sociological perspective that studies social interactions emphasizing the ways in which those involved work to create, maintain, dismantle, and present a shared understanding of reality.

dyad—the smallest group, consisting of two people.

economic capital—a person's material resources—wealth, land, money.

economic systems—the social institutions that coordinate human activity to produce, distribute, and consume goods and services.

education—any experiences that train, discipline, and shape the mental and physical potentials of the maturing person.

efficiency—a feature of McDonaldization that emphasizes using the methods that will achieve a desired end in the shortest amount of time.

egoistic (suicide)—a state in which the social ties attaching the individual to the group are weak.

embodied cultural capital—all that has been consciously and unconsciously instilled in a person through the socialization process.

emotional labor—work that requires employees to display and suppress specific emotions and/or manage customer/client emotions.

emotion work—the process by which people consciously manage their feelings by evoking an expected emotional state or suppressing an inappropriate emotional state.

empire—a group of countries under the direct or indirect control of a foreign power or government that acts to shape the political, economic, and cultural life of the people over which it has power.

endogamy—norms requiring or encouraging people to choose a partner from the same social category as their own—for example, a partner of the same race, sex, ethnicity, religion, or social class.

environmental injustice—unequal exposure to environmental hazards based on race or socioeconomic status.

esteem—the reputation that someone occupying a particular status has acquired based on the opinion of those who know and observe him or her.

ethnic cleansing—an extreme form of forced segregation in which a dominant group uses force and intimidation to remove people of a targeted racial or ethnic group from a geographic area, leaving it ethnically pure, or at least free of the targeted group. Ethnic cleansing also involves the destruction of cultural artifacts associated with the targeted groups, such as monuments, cemeteries, and churches.

ethnic group—people who share, believe they share, or are believed by others to share a national origin; a common ancestry; a place of birth; or distinctive social traits (such as religion, style of dress, or language) that set them apart from other ethnic groups.

ethnic renewal—a situation in which someone discovers an ethnic identity, including the process by which an individual takes it upon himself to find, learn about, and claim an ethnic heritage.

ethnocentrism—a point of view in which people use their home culture as the standard for judging the worth of another culture's ways.

ethnographic methods—a holistic means of studying people in the places/settings where they live or carry out the activities of interest.

ethnomethodology—an investigative and observational approach that focuses on how people make sense of everyday social activities and experiences.

exogamy—norms requiring or encouraging people to choose a partner from a social category other than their own—for example, to choose a partner outside their immediate family who is of the other sex or of another race.

externality costs—hidden costs of using, making, or disposing of a product that are not figured into the price of the product or paid for by the producer. Such costs include those associated with cleaning up the environment and with treating injured and chronically ill workers, consumers, and others.

extreme wealth—the most excessive form of wealth, in which a very small proportion of people in the world have money, material possessions, and other assets (minus liabilities) in such abundance that a small fraction of it, if spent appropriately, could provide adequate food, safe water, sanitation, and basic health care for the 1 billion poorest people on the planet.

false consciousness—with regard to religion, a point of view in which oppressed individuals or groups accept the economic, political, and social arrangements that constrain their chances in life because they are promised compensation for their suffering in the next world.

falsely accused—people who have not broken the rules but are treated as if they have.

family—a social institution that binds people together through blood, marriage, law, and/or social norms.

fatalistic (suicide)—a state in which the ties attaching the individual to the group are so oppressive there is no hope of release.

female infanticide—the targeted abortion of female fetuses because of a cultural preference for males and corresponding low status assigned to females.

femininity—traits believed to be characteristic of females.

feminism—a perspective that examines the larger social, economic, and political context in order to understand the position of women in society relative to men and that advocates equal opportunity.

folkways—norms that apply to the mundane aspects or details of daily life.

formal care—caregiving provided, usually for a fee, by credentialed professionals (whether in the person's home or some other facility).

formal curriculum—the content of the various academic subjects—mathematics, science, English, reading, physical education, and so on.

formal dimension (of organizations)—the official, by-the-book way an organization should operate.

formal education—a deliberate, planned effort to impart specific skills or information; a systematic process (for example, military boot camp, on-the-job training, or smoking cessation classes) in which someone designs the educating experiences.

formal organizations—coordinating mechanisms that bring together people, resources, and technology and then direct human activity toward achieving a specific outcome such as maintaining order in a community (a police department) or fund-raising (Race for the Cure).

formal sanctions—reactions backed by laws, rules, or policies specifying the conditions under which people should be rewarded or punished for specific behaviors.

front stage—areas visible to an audience, where people feel compelled to present themselves in expected ways.

function—the contribution a part makes to maintain the stability of an existing social order.

fundamentalism—a belief in the timelessness of sacred writings and a belief that such writings apply to all kinds of environments.

games—structured, organized activities that involve more than one person; they are characterized by a number of constraints, such as established roles and rules and a purpose toward which all activity is directed.

gender—the socially created and learned distinctions that specify the physical, behavioral, and mental and emotional traits characteristic of males and females.

gendered—a situation in which institutions have an established pattern of segregating the sexes, empowering one sex and not the other, and/or subordinating one sex relative to the other in ways that gender systematically shapes the experiences, constraints, and opportunities of the participating men and women.

gender gap—the disparity in opportunities available to men and women relative to the other.

gender ideals—a standard for masculinity or femininity against which real cases can be compared. A gender ideal is at best a caricature, in that it exaggerates the characteristics that are believed to make someone the so-called perfect male or female.

gender identity—the awareness of being a man or woman, of being neither, or something in between (gender identity also involves the ways one chooses to hide or express that identity).

gender role—the cultural norms that guide people in enacting what is considered to be feminine and masculine behavior.

gender stratification—the extent to which opportunities and resources are unequally distributed between men and women.

generalizability—the extent to which researchers' findings can be applied to the larger population of which their sample was a part.

generalized other—a system of expected behaviors and meanings that transcend the people participating. An understanding of the generalized other is achieved by simultaneously and imaginatively relating the self to many others "playing the game."

generation—a cohort composed of people who are born at a particular time in history and/or separated from other age cohorts by time. A generation is distinguished from others by its cultural disposition (dress, language, preferences for songs, activities, entertainment), posture (walk, dance, the way the body is held), access to resources, and socially expected privileges, responsibilities, and duties.

genocide—the calculated and systematic large-scale destruction of a targeted racial or ethnic group that can take the form of killing an ethnic group en masse, inflicting serious bodily or psychological harm, creating intolerable living conditions, preventing births, "diluting" racial or ethnic lines through rape and forced births, or forcibly removing children to live with another group.

gesture—any action that requires people to interpret its meaning before responding. Language is a particularly important gesture because people interpret the meaning of words before they react. In addition to spoken words, gestures also include nonverbal cues, such as tone of voice, inflection, facial expression, posture, and other body movements or positions that convey meaning.

glass ceiling—a barrier that prevents women from rising past a certain level in an organization, especially for women who work in male-dominated workplaces and occupations. The term applies to women who have the ability and qualifications to advance but who are not well-connected to those who are in a position to advocate for or mentor them.

glass escalator—the largely invisible forces that launch men into positions of power, even within female-dominated occupations, as when management singles out men for special attention and advancement.

global interdependence—a situation in which human interactions and relationships transcend national borders and in which social problems within any one country are shaped by events taking place outside the country, indeed in various parts of the globe.

globalization—the ever-increasing flow of goods, services, money, people, technology, information, and other cultural items across political borders. This flow has become more dense and quick-moving as constraints of space and time break down. As a result of globalization, no longer are people, goods, services, technologies, money, and images fixed to specific geographic locations.

group—two or more people interacting in largely predictable ways who share expectations about their purpose for being. Group members hold statuses and enact roles that relate to the group's purpose.

group think—a phenomenon that occurs when a group under great pressure to take action achieves the illusion of consensus by putting pressure on its members to shut down discussion, dismissing alternative courses of action, suppressing expression of doubt or dissent, dehumanizing those against whom the group is taking action, and/or ignoring the moral consequences of their actions.

habitus—a frame of mind that has internalized the objective reality of society. This objective reality becomes the mental filter that structures people's perceptions, experiences, responses, and actions. It is through the habitus that the social world is understood and that people acquire a sense of place and a point of view that informs how they interpret their own and others' actions.

Hawthorne effect—a phenomenon in which research subjects alter their behavior when they learn they are being observed.

hegemony—a process by which a power maintains its dominance over foreign entities.

heteronormativity—a normative system that presents the gendered heterosexual nuclear family as the ideal and departures from that system as deviant, even threatening.

hidden curriculum—the teaching method, types of assignments, kinds of tests, tone of the teacher's voice, attitudes of classmates, the number of students absent, the frequency of teacher absences, and the criteria teachers use to assign grades. These so-called extraneous factors convey messages to students not only about the value of the subject but also about the values of society, the place of learning in their lives, and their role in society.

hidden ethnicity—for members of an advantaged ethnic group, a sense of self that is based on no awareness of an ethnic identity because their culture is considered normal, normative, or mainstream.

homophobia—an irrational fear held by some heterosexuals that a same-sex person will make a sexual advance toward them. It also refers to a fear of being in close contact with someone of the same sex.

horizontal mobility—a change in one's social situation that does not involve a change in social status.

horticultural societies—societies organized around the use of hand tools such as hoes to work the soil and digging sticks to punch holes in the ground into which seeds are dropped. Horticultural peoples grow crops rather than gather food and employ slash-and-burn technology to make fields for growing crops and grazing animals.

human activity—involves all the things people do with, to, and for one another and what they think and do as a result of others' influence.

hunting and gathering societies—societies that do not possess the technology to create surplus wealth; people subsist on wild animals and vegetation and are always on the move, securing food and other subsistence items. A typical hunting and gathering society comprises 45 to 100 members related by blood or marriage.

hydrocarbon society—a society in which the use of fossil fuels shapes virtually every aspect of people's personal and social lives.

hypothesis—a trial prediction about the relationship between the independent and dependent variables. Specifically, the hypothesis predicts how change in an independent variable brings about change in a dependent variable.

I—the active and creative aspect of the self. It is the part of the self that questions the expectations and rules.

ideal type—a deliberate simplification or caricature in that it exaggerates essential traits of something. Ideal does not mean desirable; an ideal is simply a standard against which real cases can be compared.

ideologies—seemingly commonsense views justifying the existing state of affairs, which, upon close analysis, reflect the viewpoints of the dominant groups and disguise their advantages.

illegitimate opportunity structures—social settings and arrangements that offer people the opportunity to commit particular types of crime.

impairment—a physical or mental condition that interferes with someone's ability to perform a major life activity that the average person can perform without technical or human assistance or without changing the physical environment around them.

imperialistic power—a power that exerts control and influence over foreign entities either through military force or through political policies and economic pressure. Imperialists believe that their cultural, political, or economic superiority justifies control over other entities. In fact, such control is viewed as something that benefits the conquered and the entire planet.

impression management—in social situations, as on a stage, people manage the setting, their dress, their words, and their gestures so that they correspond to an impression they are trying to make.

income—the money a person earns, usually on an annual basis through salary or wages.

independent variable—the variable that explains or predicts the dependent variable.

individual discrimination—behavior that blocks another's opportunities or does harm to life or property.

individuation of social problems—a point of view whereby people tend to view the "problem individuals" as the cause and "fixing" them as the solution rather than looking at the social system in which so-called deviant behavior or appearances are embedded.

industrial societies—societies that rely on mechanization or on externally powered machines to subsist. Mechanization allowed humans to produce food, extract resources, and manufacture goods at revolutionary speeds and on an unprecedented scale.

informal care—caregiving that family members, neighbors, and friends provide in a home setting.

informal dimension (of organizations)—any aspect of an organization's operations that departs from the way the organization should officially operate.

informal sanctions—spontaneous, unofficial expressions of approval not backed by the force of law or official policy.

information explosion—an unprecedented increase in the amount of stored and transmitted data and messages in all media (including electronic, print, radio, and television).

ingroup—the group to which a person belongs, identifies, admires, and/or feels loyalty.

innovation (as a response to structural strain)—the acceptance of cultural goals but the rejection of legitimate means to achieve them.

instinct—behavior that is not learned but part of one's nature and elicited by reflex.

institutionalized ageism—discrimination based on age that is viewed as an accepted, established, and even fair way of doing things. On close analysis, however, such practices put a certain age group at a disadvantage relative to those in other age groups.

institutionalized cultural capital—anything (material or nonmaterial) recognized as important to success in a particular social setting.

institutionalized discrimination—the established, customary way of doing things in society—the unchallenged laws, rules, policies, and day-to-day practices established by a dominant group that keep minority groups in disadvantaged positions.

institutions—relatively stable and predictable social arrangements created and sustained by people that have emerged over time with the purpose of coordinating human activities to meet some need, such as food, shelter, or clothing. Institutions consist of statuses, roles, and groups.

instrumental rational action—result-oriented behavior and practices that emphasize the most efficient methods for achieving some valued goal, regardless of the consequences. In the context of an industrial and capitalist society, *efficient* means the most cost-effective and time-saving way to achieve a goal.

integration—a situation in which two or more racial groups interact in what was once a segregated setting; integration may be court-ordered, legally mandated, or the natural outcome of people crossing the "color line" once legal barriers have been removed.

internalization—a process by which people accept as binding learned ways of thinking, appearing, and behaving.

intersectionality—the interconnections among race, class, gender, sexual orientation, religion, ethnicity, age (generation), nationality, disability, and other social statuses that, taken together, profoundly shape life chances.

intersexed—people with some mixture of male and female primary sex characteristics.

involuntary ethnicity—a umbrella ethnic category created by the government or another dominant group to which people from many different cultures and countries are assigned.

That category becomes the label by which these diverse peoples are known and with which they are forced to identify.

involuntary minorities—those who did not choose to be a part of a country (nor did their ancestors); rather, they were forced to become part of it through enslavement, conquest, or colonization. Those of Native American, African, Mexican, and Hawaiian descent are examples.

iron cage of irrationality—the process by which supposedly rational systems produce irrationalities.

issue—a societal matter that affects many people and that can only be explained by larger social forces that transcend the individuals affected.

language—a symbol system that assigns meaning to particular sounds, gestures, pictures, or specific combinations of letters to convey meaning.

latent dysfunctions—unanticipated disruptions to the existing social order.

latent functions—a part's unanticipated, unintended, and unrecognized effects on an existing social order.

law—a rule governing conduct created by those in positions of power and enforced by entities given the authority to do so, such as police. A law specifies the prohibited behavior, the categories of people to whom the law applies, and the punishment to be applied to violators.

legal-rational authority—legitimate form of power that derives from a system of impersonal and formal rules that specify the qualifications for occupying an administrative or judicial position; the individual holding that position has the power to command others to act in specific ways.

life chances—a critical set of potential opportunities and advantages, including the chance to survive the first year of life, to grow to a certain height, to receive medical and dental care, to avoid a prison sentence, to graduate from high school, to live a long life, and so on.

life expectancy—the average number of years after birth a person can expect to live.

linguistic relativity hypothesis—the idea that "[n]o two languages are ever sufficiently similar to be considered as representing the same social reality. The worlds in which different societies live are distinct worlds, not merely the same world with different labels attached" (Sapir 1949, 162).

lobbyists—people whose job it is to solicit and persuade state and federal legislators to create legislation and vote for bills that favor the interests of the group they represent. Lobbyists can work for corporations, a private individual, or the public interest.

looking-glass self—the way in which a sense of self develops: specifically, people act as mirrors for one another. We see ourselves reflected in others' real or imagined reactions to our appearance and behaviors. We acquire a sense of self by being sensitive to the appraisals that we perceive others to have of us.

manifest dysfunctions—a part's anticipated disruptions to an existing social order.

manifest functions—a part's anticipated, recognized, or intended effects on maintaining order.

masculinity—traits believed to be characteristic of males.

mass media—forms of communication designed to reach large audiences without direct face-to-face contact between those creating/conveying and receiving messages.

master status—a status that takes on such great importance that it overshadows all other statuses a person occupies. That is, it shapes every aspect of life and dominates social interactions.

master status of deviant—an identification that overrides most other statuses a person holds, such that he or she is identified first and foremost as a deviant.

material culture—all the physical objects that people have invented or borrowed from other cultures.

McDonaldization of society—a process whereby the principles governing the fast-food industry come to dominate other sectors of the American economy and society and the world. Those principles are (1) efficiency, (2) quantification and calculation, (3) predictability, and (4) control.

me—the social self or the part of the self that is the product of interaction with others and that knows the rules and expectations; the me is the sense of self that emerges out of role-taking experiences.

means of production—the resources such as land, tools, equipment, factories, transportation, and labor that are essential to the production and distribution of goods and services.

mechanical solidarity—a system of social ties based on uniform thinking and behavior.

mechanisms of social control—strategies people use to encourage, often force, others to comply with social norms.

medicalization—the process of defining a behavior as an illness or medical disorder and then treating it with a medical intervention.

melting pot assimilation—a process by which previously separate groups accept many new behaviors and values from one another, intermarry, procreate, and identify with a blended culture.

militaristic power—a political entity such as a country that believes military strength, and the willingness to use it, is the source of national and even global security. Usually a peace-through-strength doctrine—peace depends on military strength and force—is cited to justify military buildups and interventions on foreign soil.

military-industrial complex—a relationship between those who declare, fund, and manage wars (the Department of Defense, the office of the president, and Congress) and corporations that make the equipment and supplies needed to wage war.

minority groups—subpopulations within a society that are regarded and treated as inherently different from those in the mainstream. They are systematically excluded (whether consciously or unconsciously) from full participation in society and denied equal access to power, prestige, and wealth.

misandry—sexism directed at men that is so extreme that it involves a hatred of those in that category.

misogyny—sexism directed at women that is so extreme that it involves a hatred of those in that category.

modernization—a process of economic, social, and cultural transformation in which a country "evolves" from an underdeveloped to a modern society.

monarchy—a form of government in which the power is in the hands of a leader (known as a monarch) who reigns over a state or territory, usually for life and by hereditary right. Typically, the monarch expects to pass the throne on to someone who is designated as the heir, usually a firstborn son.

moral superiority—the belief that an ingroup's standards represent the only way, leaving no room for negotiation and no tolerance for other ways.

mores—norms that people define as essential to the well-being of a group. People who violate mores are usually punished severely: they may be ostracized, institutionalized, or condemned to die.

mortality crises—violent fluctuations in the death rate caused by war, famine, or epidemics.

mortification—the process by which the self is stripped of all its supports and "shaped and coded" (Goffman 1961).

multinational corporation—an enterprise that owns, controls, or licenses facilities in countries other than the one in which it is headquartered.

mystic—a type of counterculture in which members search for "truth and for themselves" (Yinger 1977) and in the process turn inward.

nature—human genetic makeup or biological inheritance.

negatively privileged property class—people completely lacking in skills, property, or employment, or who depend on seasonal or sporadic employment; they constitute the very bottom of the class system.

negative sanctions—expressions of disapproval for violating norms.

negotiated order—the sum of existing and newly negotiated expectations that are part of any social situation.

neocolonialism—a new form of colonialism where more powerful foreign governments and foreign-owned businesses continue to exploit the resources and labor of the postcolonial peoples.

nonmaterial culture—intangible human creations that include beliefs, values, norms, and symbols.

nonparticipant observation—detached watching and listening by a researcher who only observes and does not become part of group life.

nonprejudiced discriminators—fair-weather liberals or people who accept the creed of equal opportunity but discriminate because they simply fail to consider discriminatory consequences or because discriminating gives them some advantage.

nonprejudiced nondiscriminators—all-weather liberals or people who accept the creed of equal opportunity, and their conduct conforms to that creed.

normative reference groups—groups that provide people with norms that they draw upon or consider when evaluating a behavior or a course of action.

norms—rules and expectations for the way people are supposed to behave, feel, and appear in a particular social situation.

nurture—the interaction experiences that make up every person's life, or more generally, the social environment.

objectified cultural capital—physical and material objects that a person owns outright or has direct access to.

objective deprivation—a state of being characteristic of those who are the worst off or most disadvantaged—the people with the lowest incomes, the least education, the lowest social status, the fewest job opportunities, and so on.

observation—a research strategy that involves watching, listening to, and recording behavior and conversations in context as they happen.

oligarchy—rule by the few, or the concentration of decision-making power in the hands of a few persons who hold the top positions in a hierarchy.

operationalized—for a variable to be operationalized, the researcher must give clear, precise instructions about how to observe or measure it.

organic solidarity—a system of social ties founded on interdependence, specialization, and cooperation.

outgroup—any group to which a person does not belong. Obviously, one person's ingroup is another person's outgroup.

over-urbanization—a situation in which urban misery—poverty, unemployment, housing shortages, and insufficient infrastructure—is exacerbated by an influx of unskilled, illiterate, and poverty-stricken rural migrants who have come to the cities out of desperation.

panethnicity—a broad catchall ethnic category in which people with distinct histories, cultures, languages, and identities are lumped together and viewed as belonging to that category (e.g., Hispanic/Latino).

panopticon—philosopher Jeremy Bentham's design for the most efficient and rational prison—the perfect prison. (*Pan* means a complete view and *optic* means seeing.)

participant observation—a method of research in which the researchers join a group, interact directly with those they are studying, assume a role critical to the group's purpose, and/or live in a community under study.

pastoral societies—societies that rely on domesticated herd animals to subsist. Pastoralism was adopted by people living in deserts and other regions in which vegetation was limited. Domesticating animals allowed people to produce surplus wealth or more milk and meat than they needed to subsist.

patriarchy—an arrangement in which men have systematic power over women in public and private (family) life.

peer group—people who are approximately the same age, participate in the same day-to-day activities, and share a similar overall social status in society.

peer pressure—instances in which people feel directly or indirectly pressured to engage in behavior that meets the approval and expectations of peers and/or to fit in with what peers are doing. That pressure may be to smoke (or not smoke) cigarettes, to drink (or not drink) alcohol, and to engage (or not engage) in sexual activities.

penalties—constraints on a person's opportunities and choices, as well as the price paid for engaging in certain activities, appearances, or choices deemed inappropriate of someone in a particular category.

peripheral economies—economies built around a few commodities or even a single commodity, such as coffee, peanuts, or tobacco, or on a natural resource, such as oil, tin, copper, or zinc.

phenomenology—an analytical approach that focuses on the everyday world and how people actively produce and sustain meaning.

play—voluntary, spontaneous activity with few or no formal rules.

pluralism—a situation in which different racial and ethnic groups coexist in harmony; have equal social standing; maintain their unique cultural ties, communities, and identities; and participate in the economic and political life of the larger society. These groups also possess an allegiance to the country in which they live and its way of life.

political action committees (PACs)—special-interest groups that raise money to be donated to the political candidates who seem most likely to support their economic, social, and/or political needs and interests. There are more than 4,500 registered PACs.

political parties—organizations that try to acquire power to influence social action. Parties are organized to represent people of a certain class or social status or with certain interests.

political system—the institution that regulates the access to and use of power to control access to scarce and valued resources and to make laws, policies, and decisions that affect others' life chances.

positive checks—events that increase deaths, including natural disasters, epidemics of infectious and parasitic diseases, war, and famine.

positively privileged property class—those who monopolize the purchase of the highest-priced consumer goods, have access to the most socially advantageous kinds of education, control the highest executive positions, own the means of production, and live on income from property and other investments.

positive sanctions—expressions of approval for complying with norms.

positivism—the belief that valid knowledge about the world can be derived only from using the scientific method.

postindustrial societies—societies that rely on intellectual technologies of telecommunications and computers, which encompass four interdependent revolutionary innovations: (1) electronics, (2) miniaturization, (3) digitalization, and (4) software.

post-structuralism—a theoretical position within sociology that contends that the behavior-constraining powers of social structures are exaggerated. While post-structuralists recognize the constraining powers of social structures, they emphasize that people reshape and change these structures.

power—the probability that an individual can achieve his or her will, even against opposition.

power elite—those few people who occupy such lofty positions in the social structure of leading institutions that their decisions affect millions, even billions, of people worldwide.

predestination—the belief that God has foreordained all things, including the salvation or damnation of individual souls.

predictability—a principle of McDonaldization that emphasizes the expectation that a service or product will be the same no matter where in the world or when (time of year, time of day) it is purchased.

prejudice—a rigid and, more often than not, unfavorable judgment about a category of people that is applied to anyone who belongs to that category.

prejudiced discriminators—active bigots or people who reject the notion of equal opportunity and profess a moral right, even a duty, to discriminate. They derive significant social and psychological gains from the conviction that anyone from their racial or ethnic group is superior to other such groups.

prejudiced nondiscriminators—timid bigots or people who reject the creed of equal opportunity but refrain from discrimination, primarily because they fear possible sanctions or being labeled as racists. Timid bigots rarely express their true opinions about racial and ethnic groups, and often use code words such as *inner city* or *those people* to camouflage their true feelings.

primary deviants—those whose rule breaking is viewed as understandable, incidental, or insignificant in light of some socially approved status they hold.

primary group—a group characterized by strong emotional ties among members who feel an allegiance to one another.

primary sector—economic activities that generate or extract raw materials from the natural environment. Mining, fishing, growing crops, raising livestock, drilling for oil, and planting and harvesting forest products are examples.

primary sex characteristics—the anatomical traits essential to reproduction.

prison-industrial complex—the corporations and agencies with an economic stake in building and supplying correctional facilities and in providing services.

privileges—special taken-for-granted advantages and immunities or benefits enjoyed by a dominant group relative to minority groups.

profane—everything that is not considered sacred, including things opposed to the sacred (such as the unholy, the irreverent, and the blasphemous) and things that stand apart from the sacred (such as the ordinary, the commonplace, the unconsecrated, and the bodily).

proletariat—those individuals who must sell their labor to the bourgeoisie.

pure deviants—people who have broken the rules and are caught, punished, and labeled as outsiders.

race—human-constructed categories to which people are assigned. People in each category are connected by shared and selected ancestors, history, and physical features and are widely regarded as a distinct racial group.

racial common sense—ideas and assumptions people hold in common about race or a racial group believed to be so obvious or natural they need not be questioned.

racial formation—a theoretical viewpoint in which race is presented, not as a concrete biological category, but as a product of the system of racial classification.

racism—a set of beliefs that uses biological or innate factors to explain and justify inequalities between racial and ethnic groups.

radical activist—a type of counterculture in which members preach, create, or demand a new order with new obligations to others.

rationalization—a process in which thought and action rooted in emotion, superstition, respect for mysterious forces, or tradition is replaced by instrumental rational action or means-to-ends thinking.

rebellion (as a response to structural strain)—the rejection of both the valued goals and the legitimate means of attaining them. Rebellion involves establishing a new set of goals and means of obtaining them.

redlining—systematic and institutionalized practices that deny, limit, or increase the cost of services to neighborhoods because residents are low-income and/or minority. Redlining can involve financial services (loans, checking accounts, credit cards, mortgages), insurance, health care, and grocery stores.

reentry shock—culture shock in reverse that can be experienced upon returning home after living in another culture.

reference group—any group whose standards people take into account when evaluating something about themselves or others, whether it be personal achievements, aspirations in life, or individual circumstances.

relative deprivation—a social condition that is measured not by objective standards but rather by comparing one group's situation with the situations of groups who are more advantaged.

relative poverty—a situation that is measured not by some essential minimum but rather by comparing a particular situation against an average or advantaged situation.

reliability— a standard for assessing an operational definition that emphasizes the ability of a measure to yield consistent results.

religion—a system of shared rituals and beliefs about the sacred and the profane that bind together a community of worshipers.

representative democracy—a form of government in which power is vested in citizens who vote into office those candidates they believe can best represent their interests.

research design—a plan for deciding who or what to study and the method of gathering data.

research methods—various techniques that sociologists and other investigators use to formulate and answer meaningful questions and to collect, analyze, and interpret data.

resocialization—an interactive process during which the affected party reconstructs his or her identity, and by which he or she renegotiates relationships with significant others who must also adjust to the changing person and circumstances.

retreatism (as a response to structural strain)—the rejection of both culturally valued goals and the legitimate means of achieving them.

reverse ethnocentrism—a viewpoint that regards a home culture as inferior to an idealized foreign culture.

ritualism (as a response to structural strain)—the rejection of the cultural goals but a rigid adherence to legitimate means society has in place to achieve them.

rituals—rules that govern how people behave in the presence of the sacred. These rules may take the form of instructions detailing the appropriate day(s) and occasions for worship, acceptable dress, and wording of chants, songs, and prayers.

role—the behavior expected of a status in relation to another status—for example, the role of brother in relation to sister; the role of physician in relation to patient.

role conflict—a predicament in which the roles associated with two or more distinct statuses that a person holds conflict in some way.

role expectations—norms about how a role should be enacted relative to other statuses.

role performance—the actual behavior of the person occupying a role.

role-set—the various role relationships with which someone occupying a status is involved.

role strain—a predicament in which there are contradictory or conflicting role expectations associated with a single status.

role-taking—imaginatively stepping into another person's shoes to view and evaluate the self.

routine—the usual ways of thinking and doing things.

sacred—everything that is regarded as extraordinary and that inspires in believers deep and absorbing sentiments of awe, respect, mystery, and reverence. Sacred things may include objects, living creatures, elements of nature, places, states of consciousness, holy days, ceremonies, and other activities.

sanctions—reactions of approval or disapproval to behavior that departs from or conforms to group norms.

schooling—a program of formal, systematic instruction that takes place primarily in classrooms but also includes extracurricular activities and out-of-classroom assignments.

scientific method—a carefully planned research process with the goal of generating observations and data that can be verified by others.

scientific racism—the use of faulty science to support systems of racial rankings and theories of social and cultural progress that placed whites in the most advanced ranks and stage of human evolution.

secondary deviants—those whose rule breaking is treated as something so significant that it cannot be overlooked or explained away. The secondary deviant is a person who comes to be defined by his or her deviance.

secondary groups—groups that consist of two or more people who interact for a specific purpose. Secondary group relationships are confined to a particular setting and specific tasks. Members relate to each other in terms of specific roles.

secondary sector—economic activities that transform raw materials from the primary sector into manufactured goods such as computers and cars.

secondary sex characteristics—physical traits not essential to reproduction such as breast development, quality of voice, distribution of facial and body hair, and skeletal form, that supposedly result from the action of so-called male (androgen) and female (estrogen) hormones.

secondary sources or archival data—third-party data that have been collected for a purpose not related to the research study.

secret deviants—people who have broken the rules but whose violation goes unnoticed or, if it is noticed, no sanctions are applied.

secularization—a process by which religious influences on thought and behavior are gradually removed or reduced. More specifically, it is a process by which some element of society, once part of a religious sphere, separates from its religious or spiritual connection or influences.

segmentalized dyad—a two-person group in which the parties know little about each other's personality and personal life; and what they do know is confined to a specific situation, such as the classroom, a hair salon, or other specialized setting.

segregation—the physical and/or social separation of people by race or ethnicity. It may be legally enforced (de jure) or socially enforced without the support of laws (de facto).

selective forgetting—a process by which people forget, dismiss, or fail to pass on to their children an ancestral connection to one or more ethnicities.

selective perception—a situation where prejudiced persons notice only the behaviors that support their stereotypes and then use those observations to support the stereotypes they hold.

self-administered survey—a set of questions that respondents read and answer.

self-awareness—a state in which a person is able to observe and evaluate the self from another's viewpoint.

self-fulfilling prophecy—a point of view that begins with a false definition of a situation that is assumed to be accurate. People behave as if that definition were true so that the misguided behavior produces responses that confirm the false definition.

self-referent terms—terms to distinguish the self (including I, me, mine, first name, and last name) and to specify the statuses one holds in society (athlete, doctor, child, and so on).

semiperipheral economies—economies characterized by moderate wealth (but extreme inequality) and moderately diverse system of production and consumption. Semiperipheral economies exploit peripheral economies and are in turn exploited by core economies.

sense of self—self-knowledge derived from stepping outside the self and seeing it from another's point of view and also imagining the effects one's words and actions have on others.

sex—a distinction based on primary sex characteristics.

sexism—the belief that one sex—and by extension, one gender—is innately superior to another, justifying unequal treatment of the sexes.

sexuality—all the ways people experience and express themselves as sexual beings. The study of sexuality considers the range of social activities, behaviors, and thoughts that generate sexual sensations and experiences and that allow for sexual expression.

sexual orientation—"an enduring pattern of emotional, romantic, and/or sexual attractions to men, women, or both sexes. Sexual orientation also refers to a person's sense of identity based on those attractions, related behaviors, and membership in a community of others who share those attractions" (American Psychiatric Association 2009).

sexual scripts—responses and behaviors that people learn, in much the same way that actors learn lines for a play, to guide them in sexual activities and encounters. These scripts are

gendered in that males and females learn different scripts about the sex-appropriate responses and behavioral choices open to them in specific situations.

significant others—people or characters (such as cartoon characters, a parent, or the family pet) who are important in a child's life, in that they greatly influence the child's self-evaluation and way of behaving.

significant symbols—gestures that convey the same meaning to the persons transmitting them and receiving them; gestures or sounds that must be interpreted before a response is made.

situational factors—things believed to be outside a person's control—such as the weather, bad luck, and another's incompetence.

social action—actions people take in response to others.

social change—any significant alteration, modification, or transformation in the way social activities and human relationships are organized.

social class—a person's overall economic and social status in a system of social stratification.

social dynamics—the forces that cause societies to change.

social emotions—feelings that we experience as we relate to other people, such as empathy, grief, love, guilt, jealousy, and embarrassment.

social forces—anything human-created that influences, pressures, or pushes people to behave and think in specified ways.

social inequality—the unequal access to and distribution of income, wealth, and other valued resources.

social interaction—everyday encounters in which people communicate, interpret, and respond to each other's words and actions.

socialism—an economic system in which raw materials and the means of producing and distributing goods and services are collectively owned. That is, public ownership—rather than private ownership—is an essential characteristic of this system.

socialization—the lifelong process by which people learn the ways of the society in which they live. More specifically, it is the process by which humans acquire a sense of self or a social identity, develop their human capacities, learn the culture(s) of the society in which they live, and learn expectations for behavior.

social mobility—movement from one social class to another.

social movement—a phenomenon in which a substantial number of people organize to make a change, resist a change, or undo a change to some area of society.

social network—a web of closely to loosely knit social relationships linking people to one another.

social prestige—a level of respect or admiration for a status apart from any person who happens to occupy it.

social relationships—the nature, number, and quality of the ties that bind people formally and informally to others.

social reproduction—the perpetuation of unequal relations such that almost everyone, including the disadvantaged, comes to view this inequality as normal and legitimate and tends to shrug off or resist calls for change.

social statics—the forces that hold societies together and give them endurance over time.

social status—a human-created and defined position in society such as female, teenager, patient, retiree, sister, homosexual, and heterosexual.

social stratification—the systematic process of categorizing and ranking people on a scale of social worth where one's ranking affects life chances in unequal ways.

social structure—a largely invisible system that coordinates human activities in broadly predictable ways. It shapes relationships and opportunities to connect to others; gives people an identity; puts up barriers to accessing resources and people; and determines the relative ease or difficulty with which those barriers can be broken.

sociological imagination—a perspective that allows us to consider how outside forces, especially our time in history and the place we live, shape our life stories or biographies.

sociological perspective—a conceptual framework for thinking about and explaining how human activities are organized and/or how people relate to one another and respond to their surroundings.

sociology—the scientific study of human activity in society.

solidarity—the system of social ties that acts as a cement connecting people to one another and to the wider society.

special-interest groups—groups consisting of people who share an interest in a particular economic, political, or other social issue and who form an organization or join an existing organization to influence public opinion and government policy.

status group—an amorphous group of persons held together by virtue of a lifestyle that has come to be "expected of all those who wish to belong to the circle" and by the level of social esteem and honor others accord them (Weber 1948, 187).

status set—all the statuses any one person assumes.

status symbols—visible markers of economic and social position and rank.

status value—a situation in which people who possess one characteristic are regarded and treated as more valuable or worthy than people who possess other characteristics.

stereotypes—generalizations about people who belong to a particular category that do not change even in the face of contradictory evidence. Stereotypes give holders an illusion that they know the other group and that they possess the right to control images of the other group.

structural constraints—the established and customary rules, policies, and day-to-day practices that affect a person's life chances.

structuralism—a framework that portrays social structures as transcending those who constructed them. As such, these structures have a coercive/constraining power over thought and behavior.

structural strain—a situation in which there is an imbalance between culturally valued goals and the legitimate means to obtain them. An imbalance exists when (1) the sole focus is on achieving valued goals by any means necessary, (2) people are unsure whether following the legitimate means will lead to success, or (3) there are not enough legitimate opportunities to satisfy demand.

subcultures—groups that share in certain parts of the mainstream culture but have distinctive values, norms, beliefs, symbols, language, and/or material culture that set them apart in some way.

subsist—to meet basic needs for human survival.

surplus wealth—a situation in which the amount of available food items and other products exceed that which is required to subsist.

surveillance—a mechanism of social control that involves monitoring the movements, conversations, and associations of those believed to be or about to be engaged in some wrongdoing and then intervening at appropriate moments.

symbols—anything (a word, an object, a sound, a feeling, an odor, a gesture, an idea) to which people assign a name and a meaning.

sympathetic knowledge—firsthand knowledge gained by living and working among those being studied.

system of oppression—a system that empowers and privileges some categories of people while disempowering other categories. The act of disempowering includes marginalizing, silencing, or subordinating another.

system of racial classification—the systematic process by which people are divided into racial categories that are implicitly or explicitly ranked on a scale of social worth.

technology—knowledge, tools, applications, and other inventions used in ways that allow people to adapt to and exercise control over their surroundings. Technology can be material and/or conceptual.

tertiary sector—economic activities related to delivering services, such as health care or entertainment, and to creating and distributing information, such as books or data.

theocracy—a form of government in which political authority rests in the hands of religious leaders or a theologically trained elite group. The primary purpose of a theocracy is to uphold divine laws in its policies and practices. Thus, there is no legal separation of church and state. Government policies and laws correspond to religious principles and laws.

theory of the demographic transition—a theoretical perspective that postulates that a country's birth and death rates are linked to its level of industrial or economic development and that the so-called developing countries will eventually achieve birth and death rates like those of Western Europe and North America; in fact, to get there they would follow the same pattern as industrialized countries.

this-worldly asceticism—a belief that people are instruments of divine will and that God determines and directs their activities. Calvinists glorified God when they accepted a task assigned to them, carried it out in an exemplary and disciplined fashion, and did not indulge in the fruits of their labor.

Thomas Theorem—a theory stating that before people take action or respond to a situation, they quickly decide on and attach a subjective meaning to the situation, which then affects their response; in other words, "if men define situations as real they are real in their consequences" (Thomas and Thomas 1928, 572).

tipping point—a situation in which what was once a rare practice or event snowballs into something dramatically more common. The process by which the rare becomes commonplace is at first gradual; changes are either unnoticed or noticed by a few and dismissed as insignificant or as a curiosity, but at some point a critical phase is reached so that the next increment of change "tips" the scale in a dramatic way.

total fertility—the average number of children that a woman bears in her lifetime.

total institutions—settings in which people surrender control of their lives, voluntarily or involuntarily, thereby submitting to the authority of an administrative staff and undergoing a program of resocialization cut off from the rest of society.

totalitarianism—a form of government characterized by (1) a single ruling party led by a dictator, (2) an unchallenged official ideology that defines a vision of the "perfect" society and the means to achieve that vision, (3) a system of social control that suppresses dissent, and (4) centralized control over the media and the economy.

tracking—also known as ability grouping; a sifting and sorting mechanism by which students are assigned to separate instructional groups within a single classroom; programs such as college preparatory versus general studies; or advanced placement, honors, or remedial classes.

traditional authority—a form of power grounded in the sanctity of time-honored norms that govern how someone comes to hold a powerful position, such as chief, king, queen, or emperor. Usually the person inherits that position by virtue of being born into a family that has held power for some time.

transgender—the label applied to those who feel that their inner sense of being a man or woman does not match their anatomical sex, so they behave and/or dress to actualize their gender identity.

treadmill of production—a term used to describe the ceaseless increases in production and, by extension, consumption that are needed to sustain the global economy's success, which is measured by increased profits.

triad—a three-person group that is sociologically significant because a third person added to a two-person group (a dyad) significantly alters the pattern of interaction between them.

troubles—individual problems or difficulties that are caused by personal shortcomings related to motivation, attitude, ability, character, or judgment. The resolution of a trouble lies in changing the person in some way.

trust—the taken-for-granted assumption that in a given social encounter others share the same expectations and definitions of the situation and that they will act to meet those expectations.

typificatory schemes—systematic mental frameworks that allow people to place what they observe into preexisting social categories with essential characteristics.

tyranny of the normal—a point of view that measures differences against what is thought to be normal and that assumes those with impairments fall short in other ways.

urbanization—a transformative process in which people migrate from rural to urban areas and change the ways they use land, interact, and earn a living.

urban poor—diverse groups of families and individuals residing in the inner city who are on the fringes of the American occupational system and as a result are in the most disadvantaged position of the economic hierarchy.

utilitarian organizations—organizations that draw in those who, in exchange for money, offer their labor, seek to change status, look to acquire a skill, seek a treatment, request a service, or need a product.

validity—a standard by which operational definitions are assessed that focuses on the extent to which a measure accurately represents what it is intended to measure.

values—general, shared conceptions of what is good, right, desirable, or important.

variable—any behavior or characteristic that consists of more than one category.

vertical mobility—a change in a person's social situation that involves a gain or loss in social status.

virtual integration—a perception that racial integration exists derived from simply seeing other racial groups on television and in advertisements; it gives "the sensation of having meaningful, repeated contact with other racial groups without actually having it" (Lynch 2007).

voluntary organizations—organizations that draw in people who give time, talent, or money to address a human and community need or to achieve some other not-for-profit goal.

wealth—the combined value of a person's income and other material assets such as stocks, real estate, and savings minus debt.

welfare state—a term that applies to an economic system that is a hybrid of capitalism and socialism.

white-collar crime—crimes committed by people whose position of respectability and high social status allows them the opportunity to do so in the course of doing their jobs.

witch hunt—campaign to identify, investigate, and correct behavior that has been defined as undermining a group. The targeted behavior is rarely the real cause but is addressed to make a problem appear as if it is being managed.

References

8Asians.com. 2009. "Oh, You Crazy Test Studying Asians Trying to Get Into College." http://www.8asians.com/2009/06/16/oh-you-crazy-test-studying-asians-trying-to-get-into-college/.

AARP (American Association of Retired Persons). 2001. *AARP Caregiver Identification Study*, February. http://assets.aarp.org/rgcenter/post-import/caregiver.pdf.

Abercrombie, Nicholas, Stephen Hill, and Bryan S. Turner. 1988. *The Penguin Dictionary of Sociology.* New York: Penguin.

Abraham, Carolyn. 2006. "The Smartest Virus in History?" *Globe and Mail (Canada)*, August 12: A9.

Addams, Jane. 1910. *Twenty Years at Hull-House: With Autobiographical Notes*. New York: Macmillan.

———. 1912. *A New Conscience and an Ancient Evil*. Urbana: University of Illinois Press.

Advameg, Inc. 2007. "How Products Are Made." http://www.madehow.com/Volume-2/Bread.html.

Aldrich, Howard E., and Peter V. Marsden. 1988. "Environments and Organizations." In *Handbook of Sociology*, edited by N. J. Smelser, 361–392. Newbury Park, CA: Sage.

Alvarez, Lizette, and Robbie Brown. 2011. "Student's Death Turns Spotlight on Hazing." *New York Times*, November 30. http://www.nytimes.com/2011/12/01/us/florida-am-university-students-death-turns-spotlight-on-hazing.html?pagewanted=all.

American Psychiatric Association (APA). 2009. "What Is Sexual Orientation?" http://www.apa.org/topics/sorientation.html.

American Sociological Association. 2012a. "Environment and Technology." www.asanet.org/sections/environment_awards.cfm.

———. 2012b. "Section Membership Counts." http://www.asanet.org/sections/CountsLastFiveYears.cfm.

Amnesty International. 2008. "Brazil Upholds Indigenous Rights in Key Case." December 15. http://us.oneworld.org/article/359082-brazil-upholds-indigenous-rights-key-case.

Anderson, Gretchen. 2010. Quoted in Glen Hougan, "What Has Design Got to Do With It?" March 28. http://theconference.ca/what-has-design-got-to-do-with-it.

Angier, Natalie. 2002. "The Most Compassionate Conservative." *New York Times Book Review*, October 27: 8–9.

Anspach, Renee R. 1987. "Prognostic Conflict in Life-and-Death Decisions: The Organization as an Ecology of Knowledge." *Journal of Health and Social Behavior* 28(3): 215–231.

Anonymous (NKU student). 2012. Student Interview Paper (anonymous). Soc 100: Introduction to Sociology, taught by Joan Ferrante, Fall Semester.

Appelbaum, Eileen, and John Schmitt. 2009. "Review Article: Low Wage Work in High Income Countries: Labor-Market Institutions and Business Strategy in U.S. and Europe." *Human Relations* 62(12): 1907–1934.

Appelrouth, Scott, and Laura D. Edles. 2007. *Sociological Theory in the Contemporary Era*. Thousand Oaks, CA: Pine Forge Press.

Arenson, Karen W. 2006. "Can't Complete High School? Just Go Right Along to College." *New York Times*, May 30: A1.

Aron, R. 1969. Quoted in *The Sociology of Max Weber*, by Julien Freund. New York: Random House.

Ash, Timothy Garton. 1989. *The Uses of Adversity: Essays on the Fate of Central Europe*. New York: Random House.

Asian American Hotel Owners Association. 2009. "AAHOA History." http://www.aahoa.com/Content/NavigationMenu/AboutUs/History/default.htm.

Associated Press. 2006. "Country Singer Carrie Underwood Graduates." *USA Today*, May 7. http://www.usatoday.com/life/people/2006-05-07-underwood_x.htm.

———. 2009. "Cowboy Churches Rope in New Christians." Faith on msnbc.com, January 8. http://www.msnbc.msn.com/id/28567031/ns/us_news-faith/t/cowboy-churches-rope-new-christians/.

Atwater, Eastwood. 1988. *Adolescence*. Englewood Cliffs, NJ: Prentice Hall.

Aud, S., W. Hussar, G. Kena, K. Bianco, et al. 2011b. *The Condition of Education 2011* (NCES 2011-033). U.S. Department of Education, National Center for Education Statistics. Washington, DC: U.S. Government Printing Office. http://nces.ed.gov/programs/coe/pdf/coe_rmc.pdf.

Bailey, Stanley. 2008. "Unmixing for Race Making in Brazil." *American Journal of Sociology* 114(3):577–614.

Barnet, Richard J. 1990. "Reflections: Defining the Moment." *The New Yorker*, July 16: 45–60.

Barrio, Federico. 1988. "History and Perspective of the Maquiladora Industry in Mexico." In *Mexico In-Bond Industry*, edited by T.P. Lee, 7-13. Tiber, Mexico: ASI.

Barrionuevo, Alexai. 2007. "Globalization in Every Loaf: Ingredients Come From All Over, But Are They Safe?" *New York Times*, June 25: C1.

Barton, Gina. 2005. "Prisoner Sues for the Right to Sex Change." *Milwaukee Journal Sentinel Online,* January 22. http://www.freerepublic.com/focus/f-news/1326882/posts.

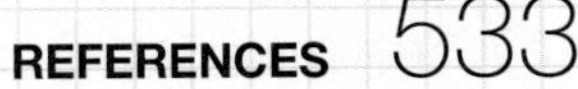

Barton, Len. 1991. "Sociology, Disability Studies and Education: Some Observations." In *Disability Reader: Social Science Perspectives*, edited by T. Shakespeare, Ch. 4. Continuum International Publishing Group.

Baumgartner-Papageorgiou, Alice. 1982. *My Daddy Might Have Loved Me: Student Perceptions of Differences between Being Male and Being Female*. Denver: Institute for Equality in Education.

BBC. 2009. "Women Risk Feet in Fashion's Name." September 7. http://news.bbc.co.uk/2/hi/8241193.stm.

Becker, Howard S. 1963. *Outsiders: Studies in the Sociology of Deviance.* New York: Free Press.

Becker, Howard, and Ruth Hill Useem. 1942. "Sociological Analysis of the Dyad." *American Sociological Review* 7(1): 13–26.

Beilin, H. 1992. "Piaget's Enduring Contribution to Developmental Psychology." *Developmental Psychology* 28:191–204.

Bekoff, Marc. 2007. Preface in *Encyclopedia of Human-Animal Relationships*, edited by M. Bekoff, xxxi–xxxiii. Westport, CT: Greenwood Press.

Bell, Daniel. 1976. "Welcome to the Post-Industrial Society." *Physics Today* (February): 46–48.

———.1999. *The Coming of Post-Industrial Society*. New York: Basic Books.

Bem, Sandra Lipsitz. 1993. *The Lenses of Gender: Transforming the Debate on Sexual Inequality*. Binghamton, NY: Vail-Ballou.

Ben-Ari, Nirit, and Ernest Harsch. 2005. "Sexual Violence, an 'Invisible War Crime.'" *African Renewal* 18, no. 4 (January). http://www.un.org/en/africarenewal/vol18no4/184sierraleone.htm.

Ben-David, Amith, and Yoav Lavee. 1992. "Families in the Sealed Room: Interaction Patterns of Israeli Families during SCUD Missile Attacks." *Family Process* 31(1): 35–44.

Benford, Robert D. 1992. "Social Movements." In *Encyclopedia of Sociology*, edited by E. F. Borgatta and M. L. Borgatta. New York: Macmillan.

Berelson, Bernard. 1978. "Prospects and Programs for Fertility Reduction: What? Where?" *Population and Development Review* 4:579–616.

Berners-Lee, Tim. 1996. "The World Wide Web: Past, Present and Future." http://www.w3.org/People/Berners-Lee/1996/ppf.html.

Berger, Peter L. 1963. *Invitation to Sociology: A Humanistic Perspective*. New York: Anchor.

Berger, Peter, and Thomas Luckmann. 1966. *The Social Construction of Reality.* Garden City, NY: Anchor.

Bernstein, Mary. 2002. "Identities and Politics: Toward a Historical Understanding of the Lesbian and Gay Movement." *Social Science History* 26 (Fall 2002): 3.

Best, Joel. 1989. *Images of Issues: Typifying Contemporary Social Problems.* New York: Aldine de Gruyter.

Beveridge, Andrew. 2007. "Study: Women in Big Cities Bridging Income Gap." Interview with National Public Radio, *Weekend Edition*, August 5. http://www.npr.org/templates/story/story.php?storyId=12513001.

Bianchi, Suzanne M., and Melissa A. Milkie. 2010. "Work and Family Research in the First Decade of the 21st Century." *Journal of Marriage and the Family* 72(June): 705–725.

Biblarz, Timothy J. and Evren Savci. 2010. "Lesbian, Gay, Bisexual, and Transgender Families." *Journal of Marriage and Family* 72(3): 480–497.

Birdsall, Nancy, and Arvind Subramanian. 2004. "Saving Iraq from Its Oil." *Foreign Affairs* 83(4): 77–89.

Blaine, Charley. 2011. "Bookseller Borders Going Out of Business." MSN Money, July 18. http://money.msn.com/exchange-traded-fund/dispatch.aspx?post=8cef1cd2-3da3-4436-9b2d-cd0b1ea01fc7.

Blood, Peter R., ed. 2001. *Afghanistan: A Country Study*. Washington, DC: GPO for the Library of Congress. http://countrystudies.us/afghanistan/index.htm.

Blumer, Herbert. 1969. *Symbolic Interactionism: Perspective and Method*. Berkeley: University of California Press.

Boudon, Raymond, and François Bourricaud. 1989. *A Critical Dictionary of Sociology*, selected and translated by P. Hamilton. Chicago: University of Chicago Press.

Boulding, Elise. 1976. *The Underside of History*. Boulder, CO: Westview.

Bourdieu, Pierre. 1984. *Distinction: A Social Critique of the Judgement of Taste*. Translated by Richard Nice. Cambridge, MA: Harvard University Press.

———. 1986. "The Forms of Capital." In *Handbook of Theory and Research for the Sociology of Education*, edited by J. Richardson, 241–258. New York: Greenwood.

Boxer, Barbara (U.S. Senator). 2007. "Historical Timeline for Women's History." http://boxer.senate.gov/.

Boyd, Danah, and Henry Jenkins. 2006. "MySpace and Deleting Online Predators Act (DOPA)." *MIT Tech Talk*, May 26. http://www.danah.org/papers/MySpaceDOPA.html.

Bradsher, Keith. 2006. "Ending Tariffs Is Only the Start." *New York Times*, February 28: C1.

Brewer, Marilynn B. 1999. "The Psychology of Prejudice: Ingroup Love or Outgroup Hate." *Journal of Social Issues* (Fall). http://findarticles.com/p/articles/mi_m0341/is_3_55/ai_58549254/.

Britton, Dana. 2000. "The Epistemology of the Gendered Organization." *Gender and Society* 14(3): 418–434.

Brown, David. 2010. "A Mother's Education Has a Huge Effect on a Child's Health." *Washington Post*, September 16. http://www.washingtonpost.com/wp-dyn/content/article/2010/09/16/AR2010091606384.html.

Brown, Lester R. 1987. "Analyzing the Demographic Trap." *State of the World 1987: A Worldwatch Institute Report on Progress Toward a Sustainable Society*. New York: Norton.

Brown, DeNeen L. 2005. "The Awakening: Sarah Scantlin's 20-Year Journey from Coma to Silence to Breakthrough."

Washington Post, July 24. http://www.washingtonpost.com/wp-dyn/content/article/2005/07/23/AR2005072301121_pf.html.

Buckley, Alan D. 1998. "Comparing Political Systems." http://homepage.smc.edu/buckley_alan/.

Bullard, Robert, Paul Mohai, Robin Saha, and Beverly Wright. 2007. "Toxic Wastes and Race at Twenty: 1987–2007." http://www.ucc.org/assets/pdfs/toxic20.pdf.

Bullock, Alan. 1977. "Democracy." In *The Harper Dictionary of Modern Thought*, edited by A. Bullock, S. Trombley, and B. Eadie, 211–212. New York: Harper & Row.

Bumiller, Elisabeth. 2009. "Rare Source of Attack on 'Don't Ask, Don't Tell.'" *New York Times*, September 30. http://www.nytimes.com/2009/10/01/us/01pentagon.html.

Burke, James. 1978. *Connections*. Boston: Little, Brown.

Burnstein, Jeff. 2009. "Robotics and the Big Trends." Robotics Online, September 21. http://www.robotics.org/content-detail.cfm/Industrial-Robotics-Feature-Article/Robotics-and-the-Big-Trends/content_id/1709.

Bush, George H. W. 1991. Address before a Joint Session of the Congress on the State of the Union (January 29). http://www.gpo.gov:80/fdsys/pkg/PPP-1991-book1/pdf/PPP-1991-book1-doc-pg74.pdf.

Bush, George W. 2003. "President Bush Announces Major Combat Operations in Iraq Have Ended." Remarks by the president from the USS *Abraham Lincoln* at sea off the coast of San Diego. http://web.archive.org/web/20041020235150/http://www.whitehouse.gov/news/releases/2003/05/iraq/20030501-15.html.

———. 2007. "President's Address to the Nation." January 10. http://georgewbush-whitehouse.archives.gov/news/releases/2007/01/20070110-7.html.

Business-in-Asia.com. 2012. "What Impact Vietnam's New WTO Membership Will Have and What Vietnam Has Committed." Runckel & Associates Consultants. http://www.business-in-asia.com/wto_vietnam_impacts.html.

Buttel, Frederick H., and Craig R. Humphrey. 2002. "Sociological Theory and the Natural Environment." In *Handbook of Environmental Sociology*, edited by Riley E. Dunlap and William Michelson, 33–69. Westport, CT: Greenwood Press.

Cahill, Betsy, and Eve Adams. 1997. "An Exploratory Study of Early Childhood Teachers' Attitudes Toward Gender Roles." *Sex Roles: A Journal of Research* 36(7–8): 517–530.

Caldeira, Chris. 2009. Personal communications with the author, August 10 through September 14.

Campbell, E. Q. 1964. "The Internalization of Norms." *Sociometry* 27:391–412.

Caplan, Lionel. 1987. "Introduction: Popular Conceptions of Fundamentalism." In *Studies in Religious Fundamentalism*, edited by L. Caplan, 1–24. Albany: State University of New York Press.

CBS News. 2011. "Borders to Close, Along with 10,700 Jobs." July 18. http://www.cbsnews.com/2100-201_162-20080539.html.

CBS News Polls. 2005. "Women's Movement Worthwhile." http://www.cbsnews.com/stories/2005/10/22/opinion/polls/main965224.shtml.

Celinska, Katarzyna. 2007. "Individualism and Collectivism in America: The Case of Gun Ownership and Attitudes Toward Gun Control." *Sociological Perspectives* 50(1): 229–247.

Center for Responsive Politics. 2009. "Lobbying—Top Spenders." http://www.opensecrets.org/lobby/top.php?showYear=2009&indexType=s.

———. 2011. "527s: Advocacy Group Spending." http://www.opensecrets.org/527s/index.php.

———. 2012a. "Super PACS." http://www.opensecrets.org/pacs/superpacs.php?cycle=2012.

———. 2012b. "Top PACs." Updated March 5. http://www.opensecrets.org/pacs/toppacs.php?Type=C&cycle=2012.

Centers for Disease Control. 2009. "National Suicide Statistics at a Glance." http://www.cdc.gov/violenceprevention/suicide/statistics/aag.html.

———. 2011a. "Health Disparities and Inequalities Report." MMWR, January 14, 60(Supplement). www.cdc.gov/mmwr/pdf/other/su6001.pdf.

———. 2011b. "Sexual Behavior, Sexual Attraction, and Sexual Identity in the United States: Data from the 2006–2008 National Survey of Family Growth." National Health Statistics Report 36, March 3. http://www.cdc.gov/nchs/data/nhsr/nhsr036.pdf.

Chaliand, Gerard, and Jean-Pierre Rageau. 1995. *The Penguin Atlas of Diasporas*. New York: Penguin Group.

Chambliss, William. 1974. "The State, the Law, and the Definition of Behavior as Criminal or Delinquent." In *Handbook of Criminology*, edited by D. Glaser, 7–44. Chicago: Rand McNally.

Chan, Sewell. 2010. "World Trade Organization Upholds American Tariffs on Tires from China." *New York Times*, December 13. http://www.nytimes.com/2010/12/14/business/global/14trade.html.

Chance, James. 2002. "Tomorrow the World." *New York Review of Books*, November 21: 33–35.

Chanda, Nayan. 2003. "The New Leviathans: An Atlas of Multinationals Throws Unusual Light on Globalization." Review of *Global Inc.: An Atlas of the Multinational Corporation*, by Medard Gabel and Henry Bruner, in *YaleGlobal Online*, November 12. http://yaleglobal.yale.edu.

Chang, Leslie. 2009. *Factory Girls: From Village to City in a Changing China*. New York: Spiegel & Grau.

Che, Yi, Akiko Hayashi, and Joseph Tobin. 2007. "Lessons from China and Japan for Preschool Practice in the United States." *Journal of College Education* 40(1): 7–12.

Chehabi, H. E., and Juan J. Linz, eds. 1998. *Sultanistic Regimes*. Baltimore and London: Johns Hopkins University Press.

Chen, David W. 2004. "What Is a Life Worth?" *New York Times*, June 20: 42K.

Chen, Yi-Fan, and James E. Katz. 2009. "Extending Family to School Life: College Students' Use of the Mobile Phone." *International Journal of Human-Computer Studies* 67(2): 179–191.

ChildStats.gov. 2012a. "Family and Social Environment." *America's Children: Key National Indicators of Well-Being, 2011*. http://www.childstats.gov/americaschildren/tables.asp.

———. 2012b. "Secure Parental Employment." *America's Children: Key National Indicators of Well-Being, 2011*. http://www.childstats.gov/pdf/ac2011/ac_11.pdf.

Chinatown San Francisco. 2012. "Chinese Hospital." http://www.sanfranciscochinatown.com/attractions/chinesehospital.html.

Christian, Julie, Richard Bagozzi, Dominic Abrams, and Harriet Rosenthal. 2012. "Social Influence in Newly Formed Groups: The Roles of Personal and Social Intentions, Group Norms, and Social Identity." *Personality and Individual Differences* 52(2012): 255–260.

Christian Mission Foster Homes for Orphans. 2012. "About CMFO." http://www.orphanage.org/africa/sierraleone/cmfo/.

Churchill, Winston. 1947. "Speech, House of Commons, 11 November 1947." In *Winston S. Churchill: His Complete Speeches, 1897–1963*, edited by Robert Rhodes James, vol. 7, p. 7566. New York: Chelsea House.

Cole, Roberta L. 2002. "Manifest Destiny Adapted for 1990s War Discourse: Mission and Destiny Intertwined." *Sociology of Religion* 63(4): 403–26.

Coleman, James S. 1966. *Equality of Educational Opportunity*. Washington, DC: U.S. Government Printing Office.

Coleman, James S., John W. C. Johnstone, and Kurt Jonassohn. 1961. *The Adolescent Society*. New York: Free Press.

Collins, Elizabeth M. 2011. "From Hospital to Hollywood." *Soldier's Magazine*, January 26. http://www.army.mil/article/50855/From_hospital_to_Hollywood__a_Soldier_s_story/.

Collins, Patricia Hill. 2000a. *Black Feminist Thought*. Boston: Unwin Hyman.

———. 2000b. *Black Feminist Thought: Knowledge, Consciousness and the Politics of Empowerment*, 2nd ed. New York: Routledge.

———. 2006. *From Black Power to Hip Hop: Racism, Nationalism, and Feminism*. Philadelphia: Temple.

Collins, Randall. 1961. "Conflict Theory of Educational Stratification." In *Schools and Society: A Sociological Approach to Education*, edited by Jeanne H. Ballantine and Joan Z. Spade, 34–40. Thousand Oaks, CA; Sage, 2008.

———. 1971. "A Conflict Theory of Sexual Stratification." *Social Problems* 19(1): 3–21.

Congressional Budget Office. 2010. "Average Federal Tax Rates for All Households, by Comprehensive Household Income Quintile, 1979–2007." http://www.cbo.gov/sites/default/files/cbofiles/attachments/Average_rates_3.pdf.

Connidis, Ingrid Arnet, and Julie Ann McMullin. 2002. "Sociological Ambivalence and Family Ties: A Critical Perspective." *Journal of Marriage and the Family* 64(3): 558–567.

Conrad, Peter. 1975. "The Discovery of Hyperkinesis: Notes on the Medicalization of Deviant Behavior." *Social Problems* 23(1): 12–21.

———. 2005. "The Shifting Engines of Medicalization." *Journal of Health and Social Behavior* 46(1): 3–14.

Conrad, Peter, and Joseph W. Schneider. 1992. *Deviance and Medicalization: From Badness to Sickness.* Expanded ed. Philadelphia, PA: Temple University Press.

Cook-Lynn, Elizabeth. 2008. "Deadliest Enemies: Law and the Making of Race Relations on and off Rosebud Reservation, and: Not Without Our Consent: Lakota Resistance to Termination, 1950–59." *Wicazo Sa Review* 23(1) Spring 2008: 155–158. http://muse.jhu.edu/login?uri=/journals/wicazo_sa_review/v023/23.1cook-lynn.html.

Cooley, Charles Horton. 1909. *Social Organization.* New York: Scribner's.

———. 1961. "The Social Self." In *Theories of Society: Foundations of Modern Sociological Theory,* edited by T. Parsons, E. Shils, K. D. Naegele, and J. R. Pitts, 822–828. New York: Free Press.

Coonan, Clifford. 2008. "Western Demand Drains Philippines of 85 Percent of Its Trained Nurses." *Independent News (UK)*, January 25. http://www.independent.co.uk/news/.

Corak, Miles. 2009. "Chasing the Same Dream, Climbing Different Ladders: Economic Mobility in the United States and Canada." Economic Mobility Project: An Initiative of the PEW Charitable Trusts. http://www.pewtrusts.org/our_work_report_detail.aspx?id=56876&category=596.

Cornell, L. L. 1990. "Constructing a Theory of the Family: From Malinowski through the Modern Nuclear Family to Production and Reproduction." *International Journal of Comparative Sociology* 31(1–2): 67–78.

Cornell, Stephen and Douglas Hartmann. 2007. *Ethnicity and Race: Making Identities in a Changing World*, 2nd Edition. Thousand Oaks, CT: Pine Forge.

Corsaro, William A., and Laura Fingerson. 2003. "Development and Socialization in Childhood." In *Handbook of Social Psychology*, edited by J. Delamater, 125–156. New York: Kluwer Academic/Plenum.

Coser, Lewis A. 1973. "Social Conflict and the Theory of Social Change." In *Social Change: Sources, Patterns, and Consequences*, edited by E. Etzioni-Halevy and A. Etzioni, 114–122. New York: Basic Books.

Cote, James. 1997. "A Social History of Youth in Samoa: Religion, Capitalism, and Cultural Disenfranchisement." *International Journal of Comparative Sociology* 38(3–4): 217.

Cowboy Church Directory. 2011. http://cowboychurch.net/dir/ (accessed December 30, 2011).

Crapanzano, Vincent. 1985. *Waiting: The Whites of South Africa*. New York: Random House.

Culbertson, Howard. 2006. "Why Is Ethnocentrism Bad?" Southern Nazarene University. http://home.snu.edu/~HCULBERT/ethno.htm.

Cuzzort, R. P., and E. W. King. 2002. *20th Century Social Thought*. New York: Harcourt.

D'Andrade, Roy. 1984. "Cultural Meaning Systems." In *Culture Theory*, edited by R. A. Shweder and R. A. LeVine, 88–119. Cambridge: Cambridge University Press.

Dahrendorf, Ralf. 1973. "Toward a Theory of Social Conflict." In *Social Change: Sources, Patterns, and Consequences*, 2nd ed., edited by E. Etzioni-Halevy and A. Etzioni, 100–113. New York: Basic Books.

DailyNK. 2008. "Ten Principles for the Establishment of the One-Ideology System." October 10. http://www.dailynk.com/english/.

Daly, Kerry J. 1992. "Toward a Formal Theory of Interactive Socialization." *Qualitative Sociology* 15(44): 395–417.

Darmoni, Stefan. 2005. "Olympic Statistics." http://www.darmoni.net/.

Davis, F. James. 1978. *Minority-Dominant Relations: A Sociological Analysis*. Arlington Heights, IL: AHM.

Davis, Kingsley. 1940. "Extreme Isolation of a Child." *American Journal of Sociology* 45:554–565.

———. 1947. "Final Note on a Case of Extreme Isolation." *American Journal of Sociology* 3(5): 432–437.

———. 1984. "Wives and Work: The Sex Role Revolution and Its Consequences." *Population and Development Review* 10(3): 397–417.

Davis, Kingsley and Wilbert E. Moore. 1945. "Some Principles of Stratification." In *Ideological Theory: A Book of Readings*, edited by L. A. Coser and B. Rosenberg, 413–445. New York: Macmillan.

Davis, Nancy J. 2005. "Taking Sex Seriously: Challenges in Teaching about Sexuality." *Teaching Sociology* 33(1): 16–31.

Davis, Shannon, Theodore N. Greenstein, and Jennifer P. Gerteisen Marks. 2007. "Effects of Union Type on Division of Household Labor: Do Cohabiting Men Really Perform More Housework?" *Journal of Family Issues* 28:1246–1272.

Day, Thomas. 2009. "About Caregiving." http://www.longtermcarelink.net/eldercare/caregiving.htm#caregivers.

Deegan, Mary Jo. 1978. "Women and Sociology: 1890–1930." *Journal of the History of Sociology* 1 (Fall 1978): 11–32.

Defense Contract Management Agency. 2012. "About the Agency." http://www.dcma.mil/about.cfm.

Defense of Marriage Act of 1996, Public Law 104–199, 104th Cong. (September 21, 1996). http://frwebgate.access.gpo.gov/cgi-bin/getdoc.cgi?dbname=104_cong_public_laws&docid=f:publ199.104.pdf.

Deggans, Eric. 2011. "Latest TV Trend? The Black Friend." *Washington Post*, October 28. http://www.washingtonpost.com/lifestyle/style/latest-tv-trend-the-black-best-friend/2011/10/25/gIQAwYw4OM_story.html.

DeGrange, McQuilkin. 1939. "Comte's Sociologies." *American Sociological Review* 4(1): 17–26.

Deloitte. 2011a. "2011 Survey of Health Care Consumers in the United States: Key Findings, Strategic Implications." http://www.deloitte.com/assets/Dcom-UnitedStates/Local%20Assets/Documents/US_CHS_2011ConsumerSurveyinUS_062111.pdf.

———. 2011b. "2011 Survey of Heath Care Consumers Global Report: Key Findings, Strategic Implementations." http://www.deloitte.com/assets/Dcom-UnitedStates/Local%20Assets/Documents/US_CHS_2011ConsumerSurveyGlobal_062111.pdf.

Dennis, Kingsley. 2008. "Viewpoint: Keeping a Close Watch—The Rise of Self-Surveillance and the Threat of Digital Exposure." *Sociological Review* 56(3): 347–357.

DeParle, Jason and Sabrina Tavernise. 2012. "For Women Under 30, Most Births Occur Outside Marriage." *New York Times*, February 17. http://www.nytimes.com/2012/02/18/us/for-women-under-30-most-births-occur-outside-marriage.html?pagewanted=all.

Der Spiegel. 2003. Quoted in "Europe Seems to Hear Echoes of Empires Past," by Richard Bernstein. *New York Times*, April 14. http://www.nytimes.com/.

Deutsch, Francine M. 1999. *Having It All: How Equally Shared Parenting Works*. Cambridge, MA: Harvard University Press.

Dewhurst, Christopher J., and Ronald R. Gordon. 1993. Quoted in "How Many Sexes Are There?" *New York Times,* March 12: A15.

Dhar, Aarti. 2012. "Blow to Misuse of Sex Determination Tests." *The Hindu*. http://www.thehindu.com/health/policy-and-issues/article2801466.ece.

Diamond, Jared. 1992. Quoted in "A Brief History of Human Society: The Origin and Role of Emotion in Social Life: 2001 Presidential Address," by D. S. Massey. *American Sociological Review* 67(1): 1–29.

Doane, Ashley W. 1997. "Dominant Group Ethnic Identity in the United States: The Role of 'Hidden' Ethnicity in Intergroup Relations." *Sociological Quarterly* 38(3): 375–397.

Donegan, E., M. Stuart, J. C. Niland, et al. 1990. "Infection with Human Immunodeficiency Virus Type 1 (HIV-1) Among Recipients of Antibody-Positive Blood Donations." *Annals of Internal Medicine* 113(10): 733–739.

Dorsey, David, and Jana Leon. 2000. "Positive Deviant." *Fast Company*. http://www.fastcompany.com/online/41/sternin.html.

Downey, Liam. 2005. "Assessing Environmental Inequality: How the Conclusions We Draw Vary According to the Definitions We Employ." *Sociological Spectrum* 25(3): 349–369.

Dreger, Alice. 2009. "The Sex of Athletes: One Issue, Many Variables." *New York Times*, October 24. http://www.nytimes.com/2009/10/25/sports/25intersex.html?_r=.

Dreifus, Claudia. 2005. "Declaring with Clarity, When Gender Is Ambiguous." *New York Times,* May 31: D2.

DuBois, W.E.B. 1919 [1970]. "Reconstruction and Africa." In *W.E.B. DuBois: A Reader*, 372–381. New York: Harper & Row.

Dugger, Cecilia. 2005. "Study Finds Small Developing Lands Hit Hardest by 'Brain Drain.'" *New York Times*, October 23: A10.

———. 2007. "UN Predicts Urban Population Explosion." *New York Times*, June 28: A6.

Duhigg, Charles, and Keith Bradsher. 2012. "How U.S. Lost Out on iPhone Work." *New York Times*, January 22: A1+.

Dunlap, Riley E., and William R. Catton Jr. 1979. "Environmental Sociology." *Annual Review of Sociology* 5(1979): 243–273.

Durkheim, Émile. 1897. *Suicide*. New York: Free Press.

———. 1901. *The Rules of Sociological Method and Selected Texts on Sociology and Its Method*. Edited by S. Lukes. Translated by W. D. Halls. Repr., New York: Free Press, 1982.

———. 1915. *The Elementary Forms of the Religious Life*, 5th ed. Translated by J. W. Swain, 1964. Repr., New York: Macmillan.

———. 1933. *The Division of Labor in Society.,* Translated by. G. Simpson. Repr., New York: Free Press, 1964.

———. 1961. "On the Learning of Discipline." In *Theories of Society: Foundations of Modern Sociological Theory*, vol. 2., edited by T. Parsons, E. Shils, K. D. Naegele, 860–865. New York: Free Press.

Dychtwald, Ken, and Joe Flower. 1989. *Age Wage: The Challenges and Opportunities of an Aging America*. Los Angeles: Tarcher.

Dyer, Gwynne. 1985. *War.* New York: Crown.

Dyson, Michael. 2008. Quoted in *Hip Hop: Beyond Beats and Rhymes.* PBS (Independent Lens). http://www.pbs.org/independentlens/hiphop/masculinity.htm.

Ebersole, Luke. 1967. "Sacred." *A Dictionary of the Social Sciences*, edited by J. Gould and W. L. Kolb. New York: UNESCO.

Edlund, Martin. 2005. "The Globalization of Hip Hop." *Slate Magazine*, May 25. http://www.slate.com/articles/arts/music_box/2005/05/the_world_is_phat.html.

Education Encyclopedia. 2009. "Early Childhood Education: International Context." http://www.answers.com/topic/early-childhood-education-international-context.

Ehrenreich, Barbara. 2001. *Nickel and Dimed*. New York: Holt.

Eisenhower, Dwight D. 1961. "Farewell Radio and Television Address to the American People." *American Rhetoric: Top 100 Speeches*. http://www.americanrhetoric.com/speeches/dwightdeisenhowerfarewell.html.

Elliott, Karla P. 2009. "The Citadel, a Kirkuk Province Historical Site." U.S. Army Homepage, February 11. http://www.army.mil/article/16730/The_Citadel__a_Kirkuk_province_historical_site/.

Ellis, Blake. 2011. "Average Student Loan Debt Tops $25,000." CNN Money, November 3. http://money.cnn.com/2011/11/03/pf/student_loan_debt/index.htm.

Engels, Friedrich. 1884. *The Origin of the Family, Private Property, and the State*, edited by E. B. Leacock. New York: International Publisher.

Ennis, Sharon R., Marays Rios-Vargas and Nora G. Albert. "The Hispanic Population: 2010." *2010 Census Briefs*, May 11. http://2010.census.gov/2010census/data/.

Environmental Protection Agency. 2011. "10 Fast Facts on Recycling." Updated October 7, 2011. http://www.epa.gov/reg3wcmd/solidwasterecyclingfacts.htm.

Environmental Working Group. "2011 Farm Subsidy Data Base." http://farm.ewg.org/top_recips.php?fips=00000&progcode=totalfarm®ionname=theUnitedStates (accessed March 29, 2012).

Epstein, Cynthia Fuchs. 2006. "Great Divides: The Cultural, Cognitive, and Social Bases of the Global Subordination of Women." *American Sociological Review* 72(February): 1–22.

Erikson, Kai T. 1966. *Wayward Puritans.* New York: Wiley.

Esposito, John L. 1986. "Islam in the Politics of the Middle East." *Current History* (February): 53–57, 81.

Esser, James K. 1998. "Alive and Well After 25 Years: A Review of Groupthink Research." *Organizational Behavior and Human Decision Processes* 73(2/3): 116–141.

Etzioni, Amitai. 1975. *Comparative Analysis of Complex Organizations*. New York: Free Press of Glencoe.

Evans, Pam. 1991. Quoted in "What Is Prejudice as It Relates to Disability Anti-Discrimination Law?" by David Reubain. *Disability Rights Education and Defense Fund*. http://www.dredf.org/international/paper_ruebain.html.

Export.gov. 2012. "Maquiladoras: Frequently Asked Questions." NAFTA, Mexico. http://infoserv2.ita.doc.gov/ticwebsite/naftaweb.nsf/504ca249c786e20f85256284006da7ab/ab04e16e721ef690852566ff005ece25!OpenDocument.

Eyerman, Ron, and Bryan S. Turner. 1998. "Outline of a Theory of Generations." *European Journal of Social Theory* 1(1): 91–106.

Facebook. 2011. "Data Use Policy | Facebook." http://www.facebook.com/about/privacy/your-info (accessed December 31, 2011).

Fagot, Beverly, Richard Hagan, Mary Driver Leinbach and Sandra Kronsberg. 1985. "Differential Reactions to Assertive and Communicative Acts of Toddler Boys and Girls." *Child Development* 56(6): 1499–1505.

Farrell, James T. 2010. "America." http://www.annabelle.net/topics/author.php?firstname=James_T.&lastname=Farrell.

Federal Bureau of Investigation (FBI). 2009. "Crime Alerts." http://www.fbi.gov/wanted/alert/alert.htm.

Federal Interagency Forum on Aging-Related Statistics. 2000. "Older Americans 2000: Key Indicators of Well-Being." http://www.agingstats.gov/.

Federal Procurement Data System. 2012. "Top 100 Contractors Report." Fiscal Year 2011. https://www.fpds.gov/fpdsng_cms/index.php/reports.

Ferree, Myra Marx. 2010. "Filling the Glass: Gender Perspectives on Family." *Journal of Marriage and Family* 72(June): 420–439.

Fein, Melvyn L. 2011. "On Loss and Losing. A Crucial Nexus." Listed in *Selected Works of Melvyn L. Fein.* http://works.bepress.com/melvyn_fein/.

Felluga, Dino. 2009. "Modules on Foucault: On Panoptic and Carceral Society." *Introductory Guide to Critical Theory*. http://www.cla.purdue.edu/academic/engl/theory/.

Fishbein, Martin. 1963. "The Perception of Non-Members: A Test of Merton's Reference Group Theory." *Sociometry* 26(3): 271–286.

Fitzgerald, Amy J. 2007. "Human Perceptions of Animals: Sociology and Human-Animal Relationships." In *Encyclopedia of Human-Animal Relationships*, edited by M. Bekoff, 955–961. Westport, CT: Greenwood Press.

Forbes. 2011. " The 10 Top-Earning American Idols." http://www.forbes.com/pictures/eeel45ekei/1-carrie-underwood-20-million/ (accessed February 28, 2011).

Fortune Magazine. 2009. "Global 500 (2009)." CNN Money. http://money.cnn.com/magazines/fortune/global500/2009/index.html.

———. 2012. "Global 500 (2011)." CNN Money. http://money.cnn.com/magazines/fortune/global500/2011/snapshots/387.html.

Foucault, Michel. 1977. *Discipline and Punish: The Birth of the Prison*. Paris: Gallimard.

Frank, Nathaniel. 2010. *Gays in Foreign Militaries 2010: A Global Primer*, February. University of California, Santa Barbara: Palm Center. http://www.palmcenter.org/files/FOREIGNMILITARIESPRIMER2010FINAL.pdf.

Freedom House. 2008. "Yemen." http://www.freedomhouse.org/country/yemen.

French, Ron, and Mike Wilkinson. 2007. "Fraud Deepens Michigan Housing Crisis—Metro Detroit's Foreclosure Explosion Linked in Part to Mortgage Scams." *Detroit News*, Wednesday, November 28.

Freund, Julien. 1968. *The Sociology of Max Weber.* New York: Random House.

Friedman, Emily. 2009. "Kids Born to Unwed Moms Hit Record High." *ABC News*, May 13. http://abcnews.go.com/Health/WomensHealth/story?id=7575268&page=1.

Friedman, Thomas. 2008. *Hot, Flat, and Crowded: Why We Need a Green Revolution—and How It Can Renew America*. New York: Farrar, Strauss, Giroux.

Fry, Richard T. 2007. "The Changing Racial and Ethnic Composition of U.S. Public Schools." Pew Hispanic Center Report, August 30. http://www.pewhispanic.org/files/reports/79.pdf.

Fudge, Tom. 2010. "Weight Loss May Become a Medical Specialty." KPBS Public Broadcasting, January 10. http://www.kpbs.org/news/2010/jan/20/weight-loss-may-become-medical-specialty/.

Gabel, Medard, and Henry Bruner. 2003. *Global Inc.: An Atlas of the Multinational Corporation*. New York: New Press.

Gage, Sue-Je Lee. 2007. "The Amerasian Problem: Blood, Duty, and Race." *International Relations* 21(1): 86–102.

Galginaitis, Carol. 2007. "The Lessons of Loss." http://www.clf.uua.org/betweensundays/earlychildhood/LessonsLoss.html.

Gans, Herbert. 1972. "The Positive Functions of Poverty." *American Journal of Sociology* 78:275–289.

Garb, Frances. 1991. "Secondary Sex Characteristics." In *Women's Studies Encyclopedia. Vol. 1: Views from the Sciences*, edited by H. Tierney, 326–327. New York: Bedrick.

Garfinkel, Harold. 1967. *Studies in Ethnomethodology.* Malden, MA: Blackwell.

Gates, William H., Sr., and Chuck Collins. 2003. *Wealth and Commonwealth: Why America Should Tax Accumulated Fortunes*. Boston: Beacon.

Geertz, Clifford. 1984. *American Anthropologist* 86(2): 263–278.

Gerson, Kathleen. 1999. "Review of *Gender Vertigo: American Families in Transition*." *Contemporary Sociology* 28(4): 419–420.

Gerth, Hans, and C. Wright Mills. 1954. *Character and Social Structure: The Psychology of Social Institutions*. London: Routledge & Kegan Paul.

Gharib, Susie. 2009. "Japan's Model Health Care Program." Nightly Business Report. September 17, 2009. http://www.pbs.org/nbr/site/onair/transcripts/japan_has_successful_health_care_strategy_090917/.

Ghaziani, Amin, and Delia Baldassarri. 2011. "Cultural Anchors and the Organization of Differences: A Multi-method Analysis of LGBT Marches on Washington." *American Sociological Review* 76(2): 179–206.

Gibbs, Jack P. 1965. "Norms: The Problem of Definition and Classification." *American Journal of Sociology* 60:586–594.

Gibson, David. 2004. "What Did Jesus Really Look Like?" *New York Times*, February 21: A17.

Gillespie, Elizabeth M. 2008. "Earth Liberation Front Believed Responsible for Arson at Multimillion Dollar Seattle Homes." *Huffington Post*, March 3. http://www.huffingtonpost.com/2008/03/03/earth-liberation-front-be_n_89651.html.

Global Warming Petition Project. 2010. http://www.petitionproject.org/.

Goffman, Erving. 1959. *The Presentation of Self in Everyday Life.* New York: Anchor.

———. 1961. *Asylums: Essays on the Social Situation of Mental Patients and Other Inmates.* New York: Anchor.

Goldman, Russell. 2007. "Environmentalists Classified as Terrorists Get Stiff Sentences." *ABC News*, May 25. http://www.abcnews.go.com/.

Google. 2012. "About Google." http://www.google.com/intl/en/about/index.html.

Gordon, John Steele. 1989. "When Our Ancestors Became Us." *American Heritage* (December): 106–221.

Gordon, Milton M. 1978. *Human Nature, Class and Ethnicity*. New York: Oxford University Press.

Gordon, Steven. 1981. "The Sociology of Sentiments and Emotion." In *Social Psychology: Sociological Perspectives*, edited by M. Rosenberg and R. H. Turner, 562–592. New York: Basic Books.

Gould, Kenneth A., David N. Pellow, and Allan Schnaiberg (2008). *The Treadmill of Production: Injustice and Unsustainability in the Global Economy*. Boulder, CO: Paradigm Publishers.

Grameen. 2005. "Grameen Bank at a Glance." http://www.grameen-info.org.

Granovetter, Mark. 1973. "The Strength of Weak Ties." *American Journal of Sociology* 76(6): 1360–1380.

Greenberg, Jack. 2001. Quoted in "McAtlas Shrugged: An FP Interview with Jack Greenberg." *Foreign Policy* (May/June): 26+.

Gross, Terry. 2009. "Fighting America's 'Financial Oligarchy.'" *Fresh Air*. April 15. http://www.npr.org/templates/transcript/transcript.php?storyId=103122382.

Guđmundsson, Hans Kristján. 2008. "The Gender Challenge in Research Funding—Assessing the European National Scenes: Iceland." http://ec.europa.eu/research/science-society/document_library/pdf_06/iceland-research-funding_en.pdf.

Guivant, Julia S. 2003. "Pesticide Use, Risk Perception and Hybrid Knowledge: A Case Study from Southern Brazil." *International Journal of Sociology of Agriculture and Food* 11(1): 41–51.

Guizzo, Erico. 2011. *Automaton* (blog). "How Google's Self-Driving Car Works." October 18. http://spectrum.ieee.org/automaton/robotics/artificial-intelligence/how-google-self-driving-car-works.

Gutowski, Timothy (and students). 2008a. "Environmental Life Style Analysis (ELSA)." IEEE International Symposium on Electronics and the Environment, May 19–20, San Francisco. http://web.mit.edu/ebm/www/Publications/ELSA%20IEEE%202008.pdf.

———. 2008b. Quoted in "Reducing Your Carbon Footprint." *Congressional Quarterly Researcher* 18(42): 985–1008.

Hacker, Andrew. 1971. "Power to Do What?" In *The New Sociology: Essays in Social Science and Social Theory in Honor of C. Wright Mills*, edited by I. L. Horowitz, 134–146. New York: Oxford University Press.

Hall, Kenji. 2009. "No Outcry about CEO Pay in Japan." *Bloomsberg Businessweek*, February 10. http://www.businessweek.com/globalbiz/content/feb2009/gb20090210_949408.htm.

Hallinan, M. T. 1988. "Equality of Educational Opportunity." In *Annual Review of Sociology*, vol. 14, edited by W. R. Scott and J. Blake, 249–268. Palo Alto, CA: Annual Reviews.

Hamington, Maurice. 2007. "Jane Addams." *Stanford Encyclopedia of Philosophy*, online edition. http://plato.stanford.edu/entries/addams-jane/.

Hannerz, Ulf. 1992. *Cultural Complexity: Studies in the Social Organization of Meaning*. New York: Columbia University Press.

Harris, Gardiner. 2003. "If the Shoe Won't Fit, Fix the Foot? Popular Surgery Raises Concern." *New York Times,* December 7: 24.

Harvard Pediatric. 2010. "Violent Video Games and Young People." *Harvard Mental Health Letter*, October. http://www.health.harvard.edu/newsletters/Harvard_Mental_Health_Letter/2010/October/violent-video-games-and-young-people.

Hasbro Toys. 2010. "G.I. Joe: Chronology." http://www.hasbro.com/gijoe/en_US/.

Hatfield, William. 2006. "White Men Can Jump." In *Sociology: A Global Perspective* (6th edition) by Joan Ferrante, 309–310. Belmont, CA: Wadsworth.

Hedges, Chris. 2002. *War Is the Force That Gives Us Meaning*. New York: Anchor Books/Doubleday.

Hedström, Peter, and Charlotta Stern. 2008. "Rational Choice and Sociology." *The New Palgrave Dictionary of Economics*, 2nd ed. Edited by S. Durlauf and L. Blume. http://www.dictionaryofeconomics.com/dictionary.

Henrik, Kreutz. 2000. "A Pragmatic Theory of Action—Collective Action as Theory of the Result of Individual Harmful Addiction." Abstract. http://www.mtapti.hu/.

Hersh, Seymour. 2004. "Annals of National Security: Torture at Abu Ghraib." *New Yorker*, May 10. http://www.newyorker.com/archive/2004/05/10/040510fa_fact.

Hippocratic Oath. 1943. *Hippocratic Oath: Text, Translation, and Interpretation*. Translated from the Greek by Ludwig Edelstein. Baltimore: Johns Hopkins Press.

Hira, Ron, and Anil Hira. 2008. *Outsourcing America*. New York: American Management Association.

Hirsch, E. D., Jr., Joseph F. Kett, and James Trefil. 1993. *The Dictionary of Cultural Literacy*. New York: Houghton Mifflin.

Hochschild, Adam. 1998. *King Leopold's Ghost*. New York: Houghton Mifflin.

Hochschild, Arlie. 2003a. *The Commercialization of Intimate Life: Notes from Home and Work*. San Francisco and Los Angeles: University of California Press.

———. 2003b. *The Managed Heart: Commercialization of Human Feeling*. Berkeley: University of California Press.

Hodge, Matthew J. 2006. "The Fourth Amendment and Privacy Issues on the 'New' Internet: Facebook.com and Myspace.com." *South Illinois University Law Journal*. https://litigation-essentials.lexisnexis.com/webcd/app?action=DocumentDisplay&crawlid=1&srctype=smi&srcid=3B15&doctype=cite&docid=31+S.+Ill.+U.+L.+J.+95&key=565175060a0dfffc5ae88b5650ce98b1.

Hollingsworth, Leslie Doty. 2005. "Ethical Considerations in Prenatal Sex Selection." *Health and Social Work* 30(2): 126–134.

Hooks, Gregory, and Chad L. Smith. 2004. "The Treadmill of Destruction: National Sacrifice Areas and Native Americans." *American Sociological Review* 69(4): 558–575.

Hougan, Glen. 2007. "Ageism and Design: Connecting to Your Grandparents." http://www.idsa.org/sites/default/files/Hougan-Ageism_and_Design.pdf.

Huffington Post. 2011. "World's Last Typewriter Factory Closes Its Doors (Update)," updated June 26. http://www.huffingtonpost.com/2011/04/26/worlds-last-typewriter-factory-closes_n_853670.html.

Hughes, Everett C. 1984. *The Sociological Eye: Selected Papers*. New Brunswick, NJ: Transaction.

Hull, Anne. 2005. "In Rural Texas, Blessings and Culture Shock." *Washington Post*, September 11: A01.

Humes, Karen R., Nicholas A. Jones and Roberto R. Ramirez. 2011. "Overview of Race and Hispanic Origin: 2010." *2010 Census Briefs*, March. http://www.census.gov/prod/cen2010/briefs/c2010br-02.pdf.

Humphreys, Lee. 2011. "Who's Watching Whom? A Study of

Interactive Technology and Surveillance." *Journal of Communication* 61:575–595.

Hunt, Mary. 2004. "'Survival Debt' Is the Scariest." *Cincinnati Post*, October 26: 7B.

Hyslop, Noreen. 2011. "Cowboy Church Movement Gaining Popularity." *Tulsa (OK) World*, November 26. Reprinted from "'Cowboy Church' Movement Gains Popularity in Southwest," by Noreen Hyslop, *Dexter Daily Statesman*. http://www.tulsaworld.com/news/article.aspx?subjectid=18&articleid=20111126_11_A22_ULNSer484407.

Industrial College of the Armed Forces. 2010. *Final Report: Strategic Materials Industry* (Spring). Washington, DC: National Defense University, Fort McNair.

Innocence Project. 2011. "Dewey Bozella: Exonerated After 26 Years." http://www.innocenceproject.org/Content/Dewey_Bozella_Exonerated_After_26_Years.php.

Instituto Brasileiro de Geografia e Estatística (IBGE). 2010. http://www.ibge.gov.br/english/#sub_populacao.

International Association for the Evaluation of Educational Achievement. 2011. "Trends in International Mathematics and Science Study." http://www.iea.nl/timss_2011.html.

International Center for Alcohol Policies. 2010. "Minimum Age Limits Worldwide." http://icap.org/table/MinimumAgeLimitsWorldwide.

International Centre for Prison Studies. 2012. "Entire World—Prison Population Rates per 100,000 of the National Population." http://www.prisonstudies.org/info/worldbrief/wpb_stats.php?area=all&category=wb_poprate (accessed March 29, 2012).

International Forum on Globalization. 2010. http://www.ifg.org/about.htm.

International Labour Union. 2009. "Recovering from the Crisis: A Global Jobs Pact." http://www.ilo.org/public/libdoc/ilo/2009/109B09_101_engl.pdf.

Irvine, Leslie. 2007. "Sentience and Cognition: Selfhood in Animals." In *Encyclopedia of Human-Animal Relationships*, edited by M. Bekoff, 1311–1314. Westport, CT: Greenwood Press.

Isaacs, Julia B. 2011. "Economic Mobility of Families Across Generations." Executive Summary. Economic Mobility Project, Pew Charitable Trusts. http://www.economicmobility.org/assets/pdfs/EMP_Across_Generations.pdf.

Jackson, Maggie. 2003. "More Sons Are Juggling Jobs and Care for Parents." *New York Times*, June 15. http://www.nyt.com.

Janis, Irving L. 1972. *Victims of Groupthink*. Boston: Houghton Mifflin.

Japan Times. 2003. "Japan: A Developing Country in Terms of Gender Equality." June 14. http://www.japantimes.co.jp.

Jerome, Richard. 1995. "Suspect Confessions." *New York Times Magazine*, August 13: 28–31.

Johansson, S. Ryan. 1987. "Status Anxiety and Demographic Contraction of Privileged Populations." *Population and Development Review* 13(3): 439–470.

Johnston, Norman. 2009. "Prison Reform in Pennsylvania." The Pennsylvania Prison Society. http://www.prisonsociety.org/about/history.shtml.

Jones, Rachel. 2006. "For First Time, More Poor Live in Suburbs Than Cities." National Public Radio, December 6. http://www.npr.org/templates/story/story.php?storyId=6598999.

Judt, Tony. 2004. "Dreams of Empire." *New York Review of Books*, November 4. http://www.nybooks.com/.

Kagan, Jerome. 1989. *Unstable Ideas: Temperament, Cognition, and Self.* Cambridge, MA: Harvard University Press.

Kaiser Family Foundation. 2010. "Generation M²: Media in the Lives of 8- to 18-Year-Olds." January 20. http://www.kff.org/entmedia/mh012010pkg.cfm.

Kaiser, Robert G. 2005. "In Finland's Footsteps: If We're So Rich and Smart, Why Aren't We More Like Them?" *Washington Post*, August 7: B01.

Kang, K. Connie. 1995. *Home Was the Land of Morning Calm: A Saga of a Korean-American Family*. Reading, MA: Addison-Wesley.

Kaplan, E. H., and R. Heimer. 1995. "HIV Incidence among New Haven Needle Exchange Participants: Updated Estimates from Syringe Tracking and Testing Data." *Journal of Acquired Immune Deficiency Syndrome* 10(2): 175–176.

Karen, David. 2005. "No Child Left Behind? Sociology Ignored!" *Sociology of Education* 78(2): 165–169.

Katzenbach, Jon, Fredrick Beckett, Steven Dichter, Marc Feigen, et al. 1997. *Real Change Leaders: How You Can Create Growth and High Performance At Your Company*. New York: Crown.

Kemper, Theodore. 1968. "Reference Groups, Socialization and Achievement." *American Sociological Review* 33(1): 31–45.

Kher, Unmesh. 2002. "Protectionism: Sweet Subsidy." *Time*, February 25. http://www.time.com/time/magazine/article/0,9171,1001897,00.html.

Kilcullen, John. 1996. "Marx on Capitalism." http://philpapers.org/rec/KILMOC.

Kimbrough v. United States. 2007. Certiorari to the United States Court of Appeals for the Fourth Circuit, Supreme Court of the United States No. 06–6330. Argued October 2, 2007; Decided December 10, 2007.

Kipnis, Laura. 2004. "The State of the Unions: Should This Marriage Be Saved?" *New York Times*, January 25: 25.

Kirchner, Mary Beth, and Rebecca Wyker. 2009. "Somewhere Out There." *This American Life* radio show, episode 374. http://www.thisamericanlife.org/radio-archives/episode/374/Somewhere-Out-There.

Kirka, Danica. 2004. "Car Bomb, Mortar Fire Kills 5 U.S. Soldiers." *Lexington-Herald Leader (KY)*, July 9: A3.

Kivisto, Peter, and Dan Pittman. 2007. "Goffman's Dramaturgical Sociology: Personal Sales and Service in a Commodified World." In *Illuminating Social Life: Classical and Contemporary Theory*, 271–290. Thousand Oaks, CA: Pine Forge Press.

Klapp, Orrin E. 1986. *Overload and Boredom: Essays on the Quality of Life in an Information Society*. New York: Greenwood.

Klein, Ezra. 2012. *Wonkblog* (blog). "Why an MRI Costs $1,080 in America and $280 in France." *Washington Post*, March 3. http://www.washingtonpost.com/blogs/ezra-klein/post/why-an-mri-costs-1080-in-america-and-280-in-france/2011/08/25/gIQAVHztoR_blog.html.

Koehler, Nancy. 1986. "Re-Entry Shock." In *Cross-Culture Re-Entry: A Book of Readings,* 89–94. Abilene, TX: Abilene Christian University Press.

Kolb, Bryan. 2007. "How Do Nature and Nurture Consort?" *Vancouver Sun*. October 30. http://www.vancouversun.com/.

Kotbi, Nabil. 2009. Quoted in "Clinic Treats Emotional Toll of Recession" by C. Ferrette. *The Journal News (NY),* February 28. http://www.lohud.com.

Kraft Foods. 2012. "Fact Sheet: Oreo's 100th Birthday." http://www.kraftfoodscompany.com/sitecollectiondocuments/pdf/Oreo_Global_Fact_Sheet_100th_Birthday_as_on_Jan_12_2012_FINAL.pdf.

Krantz, Matt, and Barbara Hansen. 2011. "CEO Pay Soars While Workers' Pay Stalls." *USA Today*, April 4. http://www.usatoday.com/money/companies/management/story/CEO-pay-2010/45634384/1.

Krovatin, Quindlen. 2008. "Red Star Athletes: How China Churns Out Champions," May 26. http://blog.newsweek.com.

Kuhn, Thomas. 1975. *The Structure of Scientific Revolutions.* Chicago: University of Chicago.

Kurdek, Lawrence A. 2005. "What Do We Know about Gay and Lesbian Couples?" *Current Directions in Psychological Science* 14(5): 251–254.

Lam, Peng Er. 2009. "Declining Fertility Rates in Japan: An Ageing Crisis Ahead." *East Asia* 26:177–190.

Lamb, David. 1987. *The Arabs: Journeys Beyond the Mirage*. New York: Random House.

Lambert, Bruce. 1993. "Abandoned Filipinos Sue U.S. Over Child Support." *New York Times*, June 21, 1993. http://www.nytimes.com/1993/06/21/world/abandoned-filipinos-sue-us-over-child-support.html?pagewanted=all.

Langman, Lauren. 2003. "Culture, Identity and Hegemony: The Body in a Global Age." *Current Sociology* 51(3-4): 223–247.

Layton, Lyndsey. 2012. "High School Graduation Rate Rises in U.S." *Washington Post*, March 19. http://www.washingtonpost.com/local/education/high-school-graduation-rate-rises-in-us/2012/03/16/gIQAxZ9rLS_story.html.

Lechner, Frank J. 1989. "Fundamentalism Revisited." *Society* (January/February): 51–59.

Lee, Jennifer. 2003. "Critical Mass: How Protesters Mobilized So Many and So Nimbly." *New York Times*, February 23. http://www.nyt.com/.

Lee, Keehyeung. 2006. "Looking Back at the Politics of Youth Culture, Space, and Everyday Life in South Korea since the Early 1990s." *Cultural Space and Public Sphere in Asia*. Asia's Future Initiative, Seoul, South Korea. http://asiafuture.org/csps2006/50pdf/csps2006_2b.pdf.

Lehrman, Sally. 1997. "WO: Forget Men Are from Mars, Women Are from Venus. Gender." *Stanford Today,* May/June: 47.

Leidner, Robin. 1993. *Fast Food, Fast Talk: Service Work and the Routinization of Everyday Life.* Berkeley, CA: University of California Press.

Leland, John. 2008. "More Men Take the Lead Role in Caring for Elderly Parents." *New York Times*, November 28. http://www.nytimes.com/2008/11/29/us/29sons.html.

Lencioni, Patrick. 2007. *Three Signs of a Miserable Job.* San Francisco: Jossey-Bass.

Lenski, Gerhard, and Jean Lenski. 1986. *Human Societies: An Introduction to Macrosociology*. New York: McGraw-Hill.

Lepkowski, Wil. 1985. "Chemical Safety in Developing Countries: The Lessons of Bhopal." *Chemical and Engineering News* 63:9–14.

Lewin, Tamar. 1990. "For More People in 20s and 30s, Home Is Where the Parents Are." *New York Times*, December 21. http://www.nytimes.com.

Lewis M. Paul. 1998. "Marx's Stock Resurges on 150-Year Tip." *New York Times,* June 27: A17.

Light, Ivan. 1983. *Cities in World Perspective*. New York: Macmillan.

Lilly, Amanda. 2009. "A Guide to China's Ethnic Groups." Washington Post.com, July 8. http://www.washingtonpost.com/wp-dyn/content/article/2009/07/08/AR2009070802718.html.

Lincoln, C. Eric, and Lawrence H. Mamiya. 1990. *The Black Church in the African American Experience*. Durham, NC: Duke University Press.

Lino, Mark. 2011. "Expenditures on Children by Families, 2010." U.S. Department of Agriculture, Center for Nutrition Policy and Promotion. Miscellaneous Publication No. 1528-2010. www.cnpp.usda.gov/publications/crc/crc2010.pdf.

Linton, Ralph. 1936. *The Study of Man: An Introduction*. New York: Appleton-Century-Crofts.

Lloyd, Janice. 2011. "Apples Top Most Pesticide-Contaminated List." *USA Today*, June 13. http://yourlife.usatoday.com/fitness-food/safety/story/2011/06/Apples-top-list-of-produce-contaminated-with-pesticides/48332000/1.

Lockheed Martin. 2012. "Who We Are." http://www.lockheedmartin.com/us/who-we-are.html.

Lockard, C. Brett, and Michael Wolf. 2012. "Employment Outlook: 2010–2020." *Monthly Labor Review* (January). http://www.bls.gov/opub/mlr/2012/01/art5full.pdf.

Lorber, Judith. 2005. "Night to His Day: The Social Construction of Gender." In *The Spirit of Sociology: A Reader*, edited by R. Matson, 292–305. New York: Penguin.

Lourenço, Orlando, and Armando Machado. 1996. "In Defense of Piaget's Theory: A Reply to 10 Common Criticisms." *Psychological Review* 103(1): 143–164.

Love, David A. 2007. "Walgreens Suit Shows Employment Discrimination Still a Problem." *Progressive Media Project*, March 13. http://www.progressive.org/media_mplove031307.

Luce, Edward. 2007. *In Spite of the Gods*. New York: Anchor Books.

Lutfey, Karen, and Jeylan T. Mortimer. 2003. "Development and Socialization through the Adult Life Course." In *Handbook of Social Psychology*, edited by J. Delamater, 193–202. New York: Kluwer Academic/Plenum.

Lyall, Sarah. 2007. "Gay Britons Serve in Military with Little Fuss, as Predicted Discord Does Not Occur." *New York Times,* April 27: A14.

Lynch, Michael. 2007. Review of *By the Color of Our Skin: The Illusion of Integration and the Reality of Race*, by Leonard Steinhorn and Barbara Diggs-Brown. http://findarticles.com/p/articles/mi_m1568/is_7_31/ai_57815517/.

Maddison, Angus. 2001. *The World Economy: A Millennial Perspective*. Development Centre Studies. Paris: OECD.

Madrick, Jeff. 2004. "Economic June: The Earning Power of Women Has Really Increased, Right? Take a Closer Look." *New York Times,* June 16: C2.

Mageo, Jeannette. 1992. "Male Transvestism and Cultural Change in Samoa." *American Ethnologist* 19(3): 443.

———. 1998. *Theorizing Self in Samoa: Emotions, Genders, and Sexualities*. Ann Arbor: University of Michigan Press.

Mahmood, Cynthia K., and Sharon L. Armstrong. 1992. "Do Ethnic Groups Exist?: A Cognitive Perspective on the Concept of Cultures." *Ethnology* 31(1): 1–14.

Marcus, Steven. 1998. "Marx's Masterpiece at 150." *New York Times Book Review,* April 26: 39.

Marger, Martin. 1991. *Race and Ethnic Relations: American and Global Perspectives*. Belmont, CA: Wadsworth.

———. 1992. *Race and Ethnic Relations.* Belmont, CA: Wadsworth.

Markus, Hazel, and Paula Nurius. 1986. "Possible Selves." *American Psychologist* 41(9): 954–969.

Martin, Patricia Yancey. 2004. "Gender as Social Institution." *Social Forces* 82(4): 1249–1273.

Martinez, J.R. 2004. Quoted in "Seeing Past the Scars of Battle," by T. Dwyer. *Washington Post*, October 24. http://www.washingtonpost.com/wp-dyn/articles/A57280-2004Oct23.html.

———. 2011. "Biography for J.R. Martinez." *The Internet Movie Data Base*. http://www.imdb.com/name/nm3210615/bio (accessed December 30, 2011).

Marx, Karl. 1843. Introduction to "A Contribution to the Critique of Hegel's Philosophy of Right." In *Collected Works*, vol. 3, edited by J. O'Malley; translated by A. Jolin and J. O'Malley. New York: Cambridge University Press, 1970. http://www.marxists.org/archive/marx/works/1843/critique-hpr/intro.htm.

———. 1844. *Economic and Philosophic Manuscripts of 1844*. Translated by Richard Hooker. http://www.marxists.org/archive/marx/works/1844/manuscripts/labour.htm.

———. 1845. *The German Ideology* (Part IA Idealism and Materialism). http://www.marxists.org/archive/marx/works/1845/german-ideology/ch01a.htm.

———. 1856. *Class Struggles of France, 1848–1850.* http://www.marxists.org/archive/marx/works/1850/class-struggles-france/index.htm.

———. 1875. "Critique of the Gotha Program." http://www.marxists.org/archive/marx/works/1875/gotha/index.htm.

———. 1887. Quoted in *A Marx Dictionary*, by Terrell Carver. Totowa, NJ: Barnes & Noble.

———. 1888. "The Class Struggle." In *Theories of Society,* edited by T. Parsons, E. Shils, K. D. Naegele, and J. R. Pitts, 529–535. New York: Free Press, 1961.

Marx, Karl, and Friedrich Engels. 1848. *Manifesto of the Communist Party*. http://www.marxists.org/archive/marx/works/1848/communist-manifesto/ch01.htm.

Mason, Margie. 2010. "Mullahs Help Promote Birth Control in Afghanistan." *The Guardian (UK)*, March 2. http://www.guardian.co.uk/world/feedarticle/8970751.

Massey, Douglas S. 2002. "A Brief History of Human Society: The Origin and Role of Emotion in Social Life: 2001 Presidential Address." *American Sociological Review* 67(1): 1–29.

Mattel. 2010. "Barbie: The History." http://corporate.mattel.com/about-us/history/default.aspxp.

Mayor, Susan. 2005. "BMA Calls on G8 Governments to Address 'Brain Drain.'" *British Medical Journal* 330(7506): 1466.

McDonald's. 2012. "Our Story." http://www.mcdonalds.com/us/en/our_story.html.

McDonald's Saudi Arabia. 2007. "Promotions." http://www.mcdonaldsarabia.com.

McIntosh, Peggy. 1992. "White Privilege and Male Privilege: A Personal Account of Coming to See Correspondences through Work in Women's Studies." In *Race, Class and Gender: An Anthology*, edited by M. L. Andersen and P. H. Collins, 70–81. Belmont, CA: Wadsworth.

Mead, George Herbert. 1934. *Mind, Self and Society*. Chicago: University of Chicago Press.

Mead, Margaret. 1928. *Coming of Age in Samoa: A Psychological Study of Primitive Youth for Western Civilisation*. New York: William Morrow.

Medical News Today. 2011. "Fetal Gender Test Determines Sex of Fetus at 7 Weeks Gestation." http://www.medicalnewstoday.com/articles/232502.php.

Megan (posted by). 2009. *Bloginization* (blog). "Cultural Differences: McDonald's in Japan." May 7. http://bucknellorgtheory09.wordpress.com/2009/05/07/cultural-differences-mcdonalds-in-japan/.

Merton, Robert K. 1938. "Social Structure and Anomie." *American Sociological Review* 3(5): 672–682.

———. 1957a. "The Role-Set: Problems in Sociological Theory." *British Journal of Sociology* 8:106–120.

———. 1957b. *Social Theory and Social Structure*. Glencoe, IL: Free Press.

———. 1976. Discrimination and the American Creed. In *Sociological Ambivalence and Other Essays*, 189–216. New York: Free Press.

———. 1997. "On the Evolving Synthesis of Differential Association and Anomie Theory: A Perspective from the Sociology of Science." *Criminology* 35(3): 517–525.

Meyrowitz, J. 2007. "Watching Us Being Watched: State, Corporate, and Citizen Surveillance." Paper presented at the symposium *The End of Television? Its Impact on the World (So Far)*. Annenberg School for Communication, University of Pennsylvania, Philadelphia, February 17–18.

Michels, Robert. 1962. *Political Parties*. Translated by E. Paul and C. Paul. New York: Dover.

Migration Policy Institute. 2007. "New Report Probes Efforts to Protect Migrant Workers." http://www.migrationpolicy.org/news.

Mihesuah, Devan A. 2008. *Big Bend Luck*. Bangor, ME: Booklocker.com.

Milgram, Stanley. 1974. *Obedience to Authority: An Experimental View*. New York: Harper & Row.

———. 1987. "Obedience." In *The Oxford Companion to the Mind*, edited by R. L. Gregory, 566–568. Oxford, UK: Oxford University Press.

Mills, C. Wright. 1956. *The Power Elite*. New York: Oxford University Press.

———. 1959. *The Sociological Imagination*. New York: Oxford.

———. 1963. "The Structure of Power in American Society." In *Power, Politics and People: The Collected Essays of C. Wright Mills*, edited by I. L. Horowitz, 23–38. New York: Oxford University Press.

Mills, Janet Lee. 1985. "Body Language Speaks Louder Than Words." *Horizons* (February): 8–12.

Mitten, David. 2001. Quoted in "Harvard's Muslims Grieving, Wary," by Ken Gewertz. *Harvard Gazette*, September 20. http://news.harvard.edu/gazette/2001/09.20/29-muslims.html.

Moore, Jim. 2004. "The Puzzling Origins of AIDS." *American Scientist* 92:540–547.

Morgana, Aimee. 2007. "Communication and Language: Interspecies Communication—N'Kisi the Parrot: A Personal Essay." In *Encyclopedia of Human-Animal Relationships*, edited by M. Bekoff, 242–248. Westport, CT: Greenwood Press.

Morrow, David J. 1996. "Trials of Human Guinea Pigs." *New York Times*, May 8: C1.

Moss, Michael. 2003. "False Terrorism Tips to F.B.I. Uproot the Lives of Suspects." *New York Times*, June 19: A1.

Murdock, George P. 1945. "The Common Denominator of Cultures." In *The Science of Man in the World Crisis*, edited by Ralph Linton, 123–142. New York: Columbia.

Murphy, Dean E. 2004. "Imagining Life without Illegal Immigrants." *New York Times*, January 11: 16WK.

Murphy, S.L., J. Xu and K.D. Kochanek. 2012. "Deaths: Preliminary Data for 2010." *National Vital Statistics Reports*, January 11, 60(4). http://www.cdc.gov/nchs/data/nvsr/nvsr60/nvsr60_04.pdf.

Murray, Christopher, S. C. Kulkarni, C. Michaud, N. Tomijima, M. T. Bulzacchelli, et al. 2006. "Eight Americas: Investigating Mortality Disparities Across Races, Counties, and Race-Counties in the United States." *PLoS Med* 3(9): e260. doi:10.1371/journal.pmed.0030260.

Naofusa, Hirai. 1999. "Traditional Cultures and Modernization: Several Problems in the Case of Japan." Institute for Japanese Culture and Classics. http://www2.kokugakuin.ac.jp.

The Nation's Report Card. 2011. "Reading Report." http://www.nationsreportcard.gov.

National Alliance for Caregiving and AARP. 2004. *Caregiving in the U.S.* http://www.caregiving.org/data/04finalreport.pdf.

National Association of Home Builders. 2007. http://www.nahb.org/.

National Center for Charitable Statistics. 2009. http://nccs.urban.org/.

National Center for Education Statistics. 2011a. *The Condition of Education 2011*. "Programs and Courses: Undergraduate Fields of Study," Table A-40-1. http://nces.ed.gov/programs/coe/tables/table-fsu-1.asp.

National Center for Education Statistics. 2011b. "Post-Secondary Education." *Digest of Educational Statistics: 2010*, chap. 3. http://nces.ed.gov/pubs2011/2011015_3a.pdf.

———. 2011c. "Table 2. Enrollment in Educational Institutions, by Level and Control of Institution: Selected Years, Fall 1980 through Fall 2010." *Digest of Education Statistics: 2010* (NCES 2011-015). http://nces.ed.gov/programs/digest/d10/tables/dt10_002.asp?referrer=report.

National Conference of State Legislatures. 2012. "Labor and Employment Legislation." Issues and Research. http://www.ncsl.org/issues-research.aspx?tabs=951,69,169.

National Council for Crime Prevention in Sweden. 1985. *Crime and Criminal Policy in Sweden*. Report no. 19. Stockholm: Liber Distribution.

National Institute of Health. 2012. "The Use of Complementary and Alternative Medicine in the United States." National Center for Complementary and Alternative Medicine (NCCAM). Last modified February 28, 2012. http://nccam.nih.gov/news/camstats/2007/camsurvey_fs1.htm.

National Park Service. 2009. "We Shall Overcome: Historic Places of the Civil Rights Movement." http://www.nps.gov/nr/travel/civilrights/.

Natural Resources Canada. 2010. "Gigajoule (GJ)." http://oee.nrcan.gc.ca/commercial/technical-info/tools/gigajoule-definition.cfm.

Nature. 2007. "Holy Cow: Hinduism's Sacred Animal." PBS, August 12. http://www.pbs.org/wnet/nature/holycow/hinduism.html.

Nawotka, Edward. 2004. "The Globalization of Hip-Hop Starts and Ends with 'Where You're At.'" *USA Today*, December 9. http://www.usatoday.com/life/books/reviews/2004-12-09-where-youre-at_x.htm.

Newport, Sally F. 2000. "Early Childhood Care, Work, and Family in Japan: Trends in a Society of Smaller Families." *Childhood Education* 77(2): 68+.

Nicholson, Peter. 2008. Quoted in "More Men Take the Lead Role in Caring for Elderly Parents." *New York Times*, November 28. http://www.nytimes.com/2008/11/29/us/29sons.html.

Nike. 2010. *Corporate Responsibility Report, FY07, 08, 09*. http://www.nikebiz.com/crreport/content/pdf/documents/en-US/full-report.pdf.

Norris, Floyd. 2012. "Manufacturing Is Surprising Bright Spot in U.S. Economy." *New York Times*,

January 5. http://www.nytimes.com/2012/01/06/business/us-manufacturing-is-a-bright-spot-for-the-economy.html.

Norris, Michele. 2002. "Comedy and Race in America: Three Comedians Who Get Serious Laughs from Thorny Issue." *All Things Considered*, December 9–11. http://www.npr.org/programs/atc/features/2002/dec/comedians/.

North Carolina Department of Correction. 2009. "Greene Correctional Institution." http://www.doc.state.nc.us/DOP/prisons/greene.htm.

Northern Kentucky University. 2012. "2012 Freshman Scholarships." http://financialaid.nku.edu/scholarships/freshmen.php.

Nottingham, Elizabeth K. 1971. *Religion: A Sociological View*. New York: Random House.

Novas, Himilce. 1994. "What's in a Name?" In *Everything You Need to Know about Latino History*, 2–4. New York: Dutton Signet.

Noypayak, Walailak. 2001. "Thailand: Experiences in Trade Negotiations in the Tourism Sector." World Trade Organization's Tourism Symposium, Geneva, February 22–23. http://www.wto.org/spanish/tratop_s/serv_s/thailand.doc.

Oakes, Jeannie. 1985. *Keeping Track: How Schools Structure Inequality*. Binghamton, NY: Vail-Ballou.

———. 1986a. "Keeping Track. Part 1: The Policy and Practice of Curriculum Inequality." *Phi Delta Kappan* 67 (September): 12–17.

———. 1986b. "Keeping Track. Part 2: Curriculum Inequality and School Reform." *Phi Delta Kappan* 67 (October): 148–154.

Obama, Barack. 2004. *Dreams from My Father*. New York: Crown.

———. 2009. Quoted in "Michelle Obama May Have White Slave-Owner Ancestor." *The Guardian*, October 9. http://www.guardian.co.uk/world/2009/oct/09/michelle-obama-white-slave-owner.

———. 2011. Transcript: "Osama bin Laden Killed: Barack Obama's Speech in Full." *The Telegraph (UK)*, May 2. http://www.telegraph.co.uk/news/worldnews/barackobama/8487354/Osama-bin-Laden-killed-Barack-Obamas-speech-in-full.html.

Oberschall, Anthony. 2000. "Utopian Visions: Engaged Sociologies for the 21st Century." *Contemporary Sociology* 29(1): 1–13.

OECD. 2010. "A Family Affair: Intergenerational Social Mobility Across OECD Countries." In *Economic Policy Reforms: Going for Growth*, chap. 5. http://www.oecd.org/dataoecd/2/7/45002641.pdf.

———. 2011a. *Education at a Glance 2011*. http://www.oecd.org/.

———. 2011b. *Health at a Glance 2011: OECD Indicators*. Released November 23. http://www.oecd.org/dataoecd/24/8/49084488.pdf.

———. 2011c. *OECD Family Database*. Paris. http://www.oecd.org/document/4/0,3746,en_2649_37419_37836996_1_1_1_37419,00.html.

———. 2011d. "OECD Health Data 2011—Frequently Requested Data." http://www.oecd.org/document/16/0,3746,en_2649_33929_2085200_1_1_1_1,00.html.

Ogawa, Naohiro, and Robert D. Retherford. 1997. "Shifting Costs of Caring for Elderly Back to Families in Japan: Will It Work?" *Population and Development Review* 23(1): 59–94.

Ogawa, Naohiro, Robert D. Retherford, and Rikiya Matsukura. 2009. "Japan's Declining Fertility and Policy Responses." In *Ultra-Low Fertility in Pacific Asia*, edited by G. Jones, P. T. Straughan, and A. Chan, 40–72. New York: Routledge.

Ogbu, John U. 1990. "Minority Status and Literacy in Comparative Prospective." *Daedalus* 119(2): 141–168.

Ogburn, William F. 1968. "Cultural Lag as Theory." In *Culture and Social Change*, 2nd ed., edited by O. D. Duncan, 86–95. Chicago: University of Chicago Press.

Olshansky, S. Jay, and A. Brian Ault. 1986. "The Fourth Stage of the Epidemiologic Transition: The Age of Delayed Degenerative Diseases." *Milbank Quarterly* 64(3): 355–391.

Omi, Michael, and Howard Winant. 1986. *Racial Formation in the United States: From the 1960s to the 1980s.* New York: Routledge & Kegan Paul.

———. 2002. Racial Formation. In *Race Critical Theories*, edited by Philomena Essed and David T. Goldberg, 123–145. London: Blackwell.

Online Newshour. 2001. "Favoring Boys in India." http://www.pbs.org/newshour/bb/asia/july-dec01/india_boys_8-16.html.

———. 2009. "States Face Shortages of Primary Care Doctors," January 6. http://www.pbs.org/newshour/bb/health/jan-june09/doctors_01-06.html.

The Oregonian. 2011. "2011 Oregon High School Graduation Rates." http://schools.oregonlive.com.

Orenstein, Peggy. 2001. "Parasites in Prêt-à-Porter." *New York Times Magazine*, July 1: 31–35.

Painter, Nell Irvin. 2010. *History of White People*. New York: Norton.

Peck, Morgan. 2012. "How Flying Robots Might Prevent Deforestation." *Web 2.0 News*, March 20. http://web2n.com/2012/03/20/how-flying-robots-might-prevent-deforestation/.

Perlez, Jane. 2004. "Asian Maids Often Find Abuse, Not Riches, Abroad." *New York Times*, June 22: A3.

Petrich, Marisa. 2011. "Sikh Soldier Answers Lifelong Calling to Serve." U.S. Army Homepage, June 2. http://www.army.mil/article/58866/.

PEW Research Center. 2007. "As Marriage and Parenthood Drift Apart, Public Is Concerned about Social Impact" (July 1). http://pewresearch.org/pubs/526/marriage-parenthood.

———. 2008. "Inside the Middle Class: Bad Times Hit the Good Life" (April 9). http://pewresearch.org/pubs/793/inside-the-middle-class.

———. 2010. "Millennials: Confident. Connected. Open to Change" (February 24). http://www.pewsocialtrends.org/2010/02/24/millennials-confident-connected-open-to-change/.

———. 2011. "Is College Worth It?" (May 15). http://www.pewsocialtrends.org/2011/05/15/is-college-worth-it/.

Pfuhl, Erdwin, and Stuart Henry. 1993. *The Deviance Process*, 3rd ed. New York: Aldine de Gruyter.

Phelan, Jo C., Bruce G. Link, and Parisa Tehranifar. 2010. "Social Conditions as Fundamental Causes of Health Inequalities: Theory, Evidence, and Policy Implications." *Journal of Health and Social Behavior* 51(1) suppl.: S28–S40.

Piaget, Jean. 1923. *The Language and Thought of the Child.* Translated by M. Worden. New York: Harcourt, Brace & World.

———. 1929. *The Child's Conception of the World.* Translated by J. Tomlinson and A. Tomlinson. Savage, MD: Rowan & Littlefield.

Polar Bear Sustainability Alliance. 2010. The Polar Bear Population Project. http://www.polarbearsinternational.org/sustainability-alliance/.

Population Reference Bureau. 2009. "Women Giving Birth in One Year, Ages 15–19." http://www.prb.org/DataFinder/Topic/Rankings.aspx?ind=55.

Proweller, Amira. 1998. *Constructing Female Identities: Meaning Making in an Upper Middle Class Youth Culture.* Albany: State University of New York.

Public Broadcasting System (PBS). 2008. "About Hip Hop." Independent Lens. http://www.pbs.org/independentlens/hiphop/about_hiphop.htm.

Raag, Tarja, and Christine Rackliff. 1998. "Preschoolers' Awareness of Social Expectations of Gender: Relationships to Toy Choices." *Sex Roles: A Journal of Research* 38(9–10): 685.

Rabin, Roni Caryn. 2008. "TV Ads Contribute to Childhood Obesity, Economists Say." *New York Times*, November 20. http://www.nytimes.com/2008/11/21/health/research/21obesity.html.

Radelet, Michael L., Hugo Adam Bedau, and Constance Putnam. 1994. *In Spite of Innocence: Erroneous Convictions in Capital Cases.* Boston: Northeastern University Press.

Random House Encyclopedia. 1990. "European Imperialism in the 19th Century." New York: Random House.

Redfield, Robert. 1962. "The Universally Human and the Culturally Variable." In *Human Nature and the Study of Society: The Papers of Robert Redfield*, vol. 1, edited by M. P. Redfield, 439–453. Chicago: University of Chicago Press.

Reich, Jens. 1989. Quoted in "People of the Year." *Newsweek*, December 25: 18–25.

Reporters Without Borders. 2010. "Web 2.0 versus Control 2.0," March 18. http://en.rsf.org/web-2-0-versus-control-2-0-18-03-2010,36697.

Republic of Turkey. 2012. "Tourism Statistics." http://www.kultur.gov.tr/EN/belge/2-25287/tourism-statistics.html.

Reuters. 2009. "Wash Less, Line Dry, Donate!" October 21. http://blogs.reuters.com/shop-talk/2009/10/21/wash-less-line-dry-donate/.

Revkin, Andrew C. 2005. "On Climate Change, a Change of Thinking." *New York Times*, December 4. http://www.nytimes.com/2005/12/04/weekinreview/04revkin.html.

Rhode, Deborah L., and Christopher J. Walker. 2008. "Gender Equity in College Athletics: Women Coaches as a Case Study." Standford Journal of Civil Rights and Civil Liberties 4:1–50.

Richtel, Matt. 2011. "As Doctors Use More Devices, Potential for Distraction Grows." *New York Times*, December 14. http://www.nytimes.com/2011/12/15/health/as-doctors-use-more-devices-potential-for-distraction-grows.html?pagewanted=1&_r=1&partner=rss&emc=rss.

Ridgeway, Cecilia. 1991. "The Social Construction of Status Value: Gender and Other Nominal Characteristics." *Social Forces* 70(2): 367–386.

Ritzer, George. 1993. *The McDonaldization of Society.* Newbury Park, CA: Pine Forge Press.

———. 2008. *The McDonaldization of Society*, 5th ed. Thousand Oaks, CA: Pine Forge Press.

Robertson, Ian. 1988. *Sociology*, 3rd ed. New York: Worth.

Robertson, Roland. 1987. "Economics and Religion." *The Encyclopedia of Religion.* New York: Macmillan.

Rodney, Walter. 1973. *How Europe Underdeveloped Africa.* Dar-Es-Salaam: Bogle-L'Ouverture Publications, London and Tanzanian Publishing House. http://www.marxists.org/subject/africa/rodney-walter/how-europe/index.htm.

Roethlisberger, F. J., and William J. Dickson. 1939. *Management and the Worker.* Cambridge, MA: Harvard University Press.

Rokeach, Milton. 1973. *The Nature of Human Values.* New York: Free Press.

Roksa, Josipa, and Tania Levey. 2010. "What Can You Do with That Degree? College Major and Occupational Status of College Graduates over Time." *Social Forces* 89(2): 389–416.

Roosendaal, A. 2010. "Facebook Tracks and Traces Everyone: Like This!" Tilburg Law School Research Paper No. 03/2011. Available at Social Science Research Network. http://ssrn.com/abstract=1717563.

Rose, Stephen J., and Heidi Hartman. 2004. "Still a Man's Labor Market: The Long-Term Earnings Gap." Institute for Women's Policy Research. http://www.iwpr.org/publications/pubs/still-a-mans-labor-market-the-long-term-earnings-gap.

Rosenthal, Robert, and Lenore Jacobson. 1968. *Pygmalion in the Classroom.* New York: Holt, Rinehart & Winston.

Rosich, Katherine J., and Janet R. Hankin. 2010. "Executive Summary: What Do We Know? Key Findings from 50 Years of Medical Sociology." *Journal of Health and Social Behavior* 51(1) suppl.: S1–S9.

Rostow, W. W. 1960. *The Stages of Economic Growth: A Non-Communist Manifesto.* Cambridge, UK: Cambridge University Press.

Royal Family. 2009. "How the Monarchy Works." http://www.royal.gov.uk/Home.aspx.

Sachs, Jeffrey. 2010. Quoted in *The Economist: Democracy in America* (blog). "Interview: Seven Questions for Jeffrey Sachs." January 25. http://www.economist.com/blogs/democracyinamerica/2010/01/seven_questions_jeffrey_sachs.

Sahadi, Jeanne. 2006. "Where Women's Pay Trumps Men's." CNNmoney.com. http://www.money.cnn.com/2006/02/28/commentary/everyday/sahadi/index.htm.

Samuel, John. 1997. "World Population and Development: Retrospect and Prospects." *Development Express.* Ottawa: Canadian International Development Agency.

Salvation Army. 2012. About Us. http://www.salvationarmyusa.org/usn/www_usn_2.nsf/vw-local/About-us.

Sapir, Edward. 1949. "Selected Writings of Edward Sapir." In *Language, Culture and Personality*, edited by D. G. Mandelbaum. Berkeley: University of California Press.

Saslow, Eli. 2007. "Island Hoping: In American Samoa, High School Football Is Seen as the Ultimate Escape." WashingtonPost.com, August 13.

Sato, Minako. 2005. "Cram Schools Cash in on Failure of Public Schools." *Japan Times*, July 28. http://www.japantimes.co.jp.

Satterly, D. J. 1987. "Jean Piaget (1896–1980)." In *The Oxford Companion to the Mind,* edited by R. I. Gregory, 621–622. Oxford, UK: Oxford University Press.

Save the Children. 2007. *State of the World's Mothers*. http://www.savethechildren.org/site/c.8rKLIXMGIpI4E/b.6153061/k.A0BD/Publications.htm.

Scelza, Janene, Roberta Spalter-Roth, and Olga Mayorova. 2011. "A Decade of Change: ASA Membership from 2000–2010." http://www.asanet.org/images/research/docs/pdf/2010_asa_membership_brief.pdf.

Schilt, Kristen, and Catherine Connell. 2007. "Do Workplace Gender Transitions Make Gender Trouble?" *Gender, Work and Organization* 14(6): 596–618.

Schmalz, Jeffrey. 1993. "From Midshipman to Gay-Rights Advocate." *New York Times,* February 4: B1+.

Schmid, Randolph. 1997. "'Multiracial' Category Rejected for Census Forms." *Chicago Sun-Times*, July 9.

Schnaiberg, Allan. 1980. *The Environment: From Surplus to Scarcity*. New York: Oxford University Press.

Schor, Juliet. 2004. *Born to Buy: The Commercialized Child and the New Consumer Culture.* New York: Scribner.

Schuster, George, and Marvin Winrich. 2009. "Robotics Safety." Rockwell Automation. Retrieved April 14. http://www.isa.org/FileStore/Intech/WhitePaper/Robotics_Safety.pdf.

ScienceDaily. 2010. "McSleepy Meets DaVinci: Doctors Conduct First-Ever All-Robotic Surgery and Anesthesia," October 19. http://www.sciencedaily.com/releases/2010/10/101019171811.htm.

Scully, Matthew. 2002. *Dominion: The Power of Man, the Suffering of Animals, and the Call to Mercy.* New York: St. Martin's.

September 11 Victim Compensation Fund. 2001. "Calculating the Losses." *New York Times*, December 21: B6.

Shen, Hua, and Adrian Ziderman. 2009. "Student Loans Repayment and Recovery: International Comparisons." *Higher Education* 57(3): 315–333.

Sher, Lauren. 2011. "Royal Wedding: Prince William and Catherine Middleton Kiss Twice on Balcony as World Watches." *ABC News*, April 29. http://abcnews.go.com/International/Royal_Wedding/royal-wedding-prince-william-kate-middletons-balcony-kiss/story?id=13480895.

Shoup, Anna. 2006. "Afghanistan and the War on Terror: The Taliban." *Online Newshour*. http://www.pbs.org/newshour/indepth_coverage/asia/afghanistan/keyplayers/taliban.html.

Sierra Club. 2007. "Sprawl Overview." http://www.sierraclub.org/sprawl/overview/.

Simmel, Georg. 1950. *The Sociology of Georg Simmel*. Translated, edited, and introduction by Kurt H. Wolff. New York: Free Press.

Simpson, Richard L. 1956. "A Modification of the Functional Theory of Social Stratification." *Social Forces* 35:132–137.

SIPRI. 2012. *SIPRI Yearbook 2011*. http://www.sipri.org/yearbook/2011.

Sklair, Leslie. 2002. *Globalization: Capitalism and Its Alternatives*, 3rd ed. Oxford, UK: Oxford University Press.

Slobodchikoff, Con. 2011. "New Language Discovered: Prairiedogese." Interview with Con Slobodchikoff by Jad Abumrad and Robert Krulwich. National Public Radio, January 11. http://www.npr.org/2011/01/20/132650631/new-language-discovered-prairiedogese.

Smart, Ninian. 1976. *The Religious Experience of Mankind*. New York: Scribner's.

Smith, Adam. 1776. *An Inquiry into the Nature and Causes of the Wealth of Nations*. www2.hn.psu.edu/faculty/jmanis/adam-smith/Wealth-Nations.pdf.

Smith, T., E. Darling, and B. Searles. 2011. "2010 Survey on Cell Phone Use While Performing Cardiopulmonary Bypass." Presented at the 32nd Annual Seminar of The American Academy of Cardiovascular Perfusion, Reno, Nevada, January 27–30. http://prf.sagepub.com/content/26/5/375.abstract.

Sobie, Jan Hipkins. 1986. "The Cultural Shock of Coming Home Again." In *The Cultural Transition: Human Experience and Social Transformation in the Third World and Japan*, edited by M. I. White and S. Pollack, 95–102. Boston: Routledge & Kegan.

Social Issues Reference. 2009. "Preschool—History and Demographics." http://social.jrank.org/pages/519/Preschool-History-Demographics.html.

Soldo, Beth J., and Emily M. Agree. 1988. "America's Elderly." *Population Bulletin* 43(3): 5+.

Son, Eugene. 1998. "G.I. Joe—A Real American FAQ." www.eugeneson.com/.

Sowell, Thomas. 1981. *Ethnic America: A History*. New York: Basic Books.

Spector, Malcolm, and J. I. Kitsuse. 1977. *Constructing Social Problems*. Menlo Park, CA: Cummings.

Spitz, Rene A. 1951. "The Psychogenic Diseases in Infancy: An Attempt at Their Etiologic Classification." In *The Psychoanalytic Study of the Child*, vol. 27, edited by R. S. Eissler and A. Freud, 255–278. New York: Quadrangle.

Spreitzer, Elmer A. 1971. "Organizational Goals and Patterns of Informal Organizations." *Journal of Health and Social Behavior* 12(1): 73–75.

Statista. 2012. "Percentage of Global Toy Market Revenue by Country in 2009 and 2010." http://www.statista.com/statistics/194415/share-of-total-toy-market-revenue-by-country-since-2009/.

Stein, Arlene. 1989. Three Models of Sexuality: Drives, Identities, and Practices." *Sociological Theory* 7(1): 1–13.

Steves, Rick. 2009. *Travel as a Political Act*. New York: Nation Books.

Stillwell, R. 2010. *Public School Graduates and Dropouts from the Common Core of Data: School Year 2007–08* (NCES 2010-341). National Center for Education Statistics, Institute of Education Sciences, U.S. Department of Education. Washington, DC. http://nces.ed.gov/pubsearch/pubsinfo.asp?pubid=2010341.

Strauss, Anselm. 1978. *Negotiations: Varieties, Contexts, Processes, and Social Order*. San Francisco: Jossey-Bass.

Stretesky, Paul B., and Michael J. Lynch. 2002. "Environmental Hazards and School Segregation in Hillsborough County, Florida, 1987–1999." *Sociological Quarterly* 43(4): 553–573.

Stub, Holger. 1982. *The Social Consequences of Long Life*. Springfield, IL: Thomas.

Sumner, William Graham. 1907. *Folkways*. Boston: Ginn.

Sutherland, Edwin H., and Donald R. Cressey. 1978. *Principles of Criminology*, 10th ed. Philadelphia: Lippincott.

Takahashi, Junko. 1999. "Century of Change: Marriage Sheds Its Traditional Shackles." *Japan Times*, December 13. http://www.japantimes.co.jp.

Tampa Bay Times. 2012. "Subsidies and Tariffs: Sweetening the Pot." March 29. http://www.tampabay.com/specials/2008/graphics/sugar-prices/.

Telles, Edward E. 2004. *Race in Another America: The Significance of Skin Color in Brazil*. NJ: Princeton University Press.

Thanh Nien News. 2006. "Cars Mean Traffic Jams in Vietnam's Southern Metro." http://www.thanhniennews.com/2006/Pages/20061171234220.aspx.

Thayer, John E., III. 1983. "Sumo." In *Kodansha Encyclopedia of Japan*, vol. 7, 270–274. Tokyo: Kodansha.

Thebodo, Stacey Woody. 2011. "Surviving Cross-Cultural Re-entry." *Abroad View*. http://abroadviewmagazine.net/.

Theodori, Gene L. 2007. "Campus Cowboys and Cowgirls: A Research Note on College Rodeo Athletes." *Southern Rural Sociology* 22(2): 127–136.

Theodorson, George A., and Achilles G. Theodorson. 1969. *A Modern Dictionary of Sociology*. New York: Harper.

Thoits, Peggy A. 2010. "Stress and Health: Major Findings and Policy Implications." *Journal of Health and Social Behavior* 51: S41–S53.

Thomas, William I. 1923. *The Unadjusted Girl*. Boston: Little, Brown.

Thomas, William I., and Dorothy Swain Thomas. 1928. *The Child in America*. Repr., New York: Johnson, 1970.

Thurow, Roger, and Geoff Winestock. 2005. "Bittersweet: How an Addiction to Sugar Subsidies Hurts Development." aWorldConnected.org (German). http://www.aworldconnected.org/article.php/242.html. Reprinted in Mercy Corps, *Global Envision*, January 26, 2005. http://www.globalenvision.org/library/15/722.

Thurston, Baratunde. 2012. "Interview: Baratunde Thurston Explains 'How to Be Black.'" *Fresh Air*, February 1. http://www.npr.org/2012/02/01/146198412/baratunde-thurston-explains-how-to-be-black.

Tilghman, Andrew. 2012. "Pentagon opens more military jobs to women." *Army Times*, February 9. http://www.armytimes.com/news/2012/02/military-ban-on-women-lifted-for-1-percent-of-military-jobs-020912w/.

Time.com. 2008. "Feminism Poll." http://www.time.com/time/magazine/article/0,9171,988616,00.html.

Toro, Luis Angel. 1995. "'A People Distinct from Others': Race and Identity in Federal Indian Law and the Hispanic Classification in OMB Directive No. 15." *Texas Tech Law Review* 26:1219–1274.

Traina, Cristina. 2005. "Touch on Trial: Power and Right to Physical Affection." *Journal of the Society of Christian Ethics* 25(1): 3–34.

Tuller, David. 2004. "Gentleman, Start Your Engines?" *New York Times*, June 21: E1.

Tumin, Melvin M. 1953. "Some Principles of Stratification: A Critical Analysis." *American Sociological Review* 18:387–394.

Twain, Mark. 1905. *King Leopold's Soliloquy*. Boston: The P. R. Warren Co.

Tyler, Greg. 2011. "Salaries Go Up Even When Income Doesn't." *Sport Digest*, April 11. http://thesportdigest.com/2011/04/salaries-go-up-even-when-income-doesn%E2%80%99t/.

Umberson, Debra. 2010. "Social Relationships and Health: A Flashpoint for Health Policy." *Journal of Health and Social Behavior* 51: S54–S66.

U.S. Army. 2012. "TAMC Head, Neck Docs Use Robot to Increase Patients' Quality of Life." January 31, http://www.army.mil/media/233611.

U.S. Bureau of Labor Statistics. 2010. "Census of Fatal Occupational Injuries (CFOI)—Current and Revised Data." http://www.bls.gov/iif/oshcfoi1.htm#2010.

———. 2011a. "Anesthesiologists." *Occupational Outlook Handbook, 2010–11 Edition*. http://www.bls.gov/oes/current/oes291061.htm.

———. 2011b. "Home Health Aides and Personal and Home Care Aides." *Occupational Outlook Handbook, 2010–11 Edition*. http://www.bls.gov/oco/ocos326.htm#oes_links.

———. 2011c. "Median Weekly Earnings of Full-Time Wage And Salary Workers by Detailed Occupation and Sex, 2010." http://www.bls.gov/cps/cpsaat39.pdf.

———. 2011d. "Occupational Employment and Wages, May 2010: 29-1069 Physicians and Surgeons, All Other." http://www.bls.gov/oes/current/oes291069.htm.

———. 2011e. "Occupational Employment and Wages, May 2010: 53-7081 Refuse and Recyclable Material Collectors." http://www.bls.gov/oes/current/oes537081.htm.

———. 2011f. "Women in the Labor Force: A Databook" (December). http://www.bls.gov/cps/wlf-databook2011.htm.

———. 2012a. "Education Pays." Employment Projections, March 23. http://www.bls.gov/emp/ep_chart_001.htm.

———. 2012b. "Table 2. Median Weekly Earnings of Full-Time

Wage and Salary Workers by Union Affiliation and Selected Characteristics." Economic News Release, January 27. http://www.bls.gov/news.release/union2.t02.htm.

U.S. Census Bureau. 2002. *Demographic Trends in the 20th Century*, by F. Hobbs and N. Stoops. Census 2000 Special Reports, Series CENSR-4. Washington, DC: U.S. Government Printing Office.

———. 2010a. "People with Income Below Specified Ratios of Their Poverty Threshold by Selected Characteristics: 2010." Table 6. http://www.census.gov/hhes/www/poverty/data/incpovhlth/2010/table6.pdf.

———. 2010b. "Profile of U.S. Exporting Companies, 2007–2008." http://www.census.gov/foreign-trade/Press-Release/edb/2008/.

———. 2011a. "Families and Living Arrangements." http://www.census.gov/population/www/socdemo/hh-fam.html.

———. 2011b. "Father's Day: June 19, 2011" (April 20). http://www.census.gov/newsroom/releases/archives/facts_for_features_special_editions/cb11-ff11.html.

———. 2011c. "Household Income for States: 2009 and 2010." American Community Survey Briefs 10/02, September. http://www.census.gov/prod/2011pubs/acsbr10-02.pdf.

———. 2011d. "More Young Adults Are Living in Their Parents' Home, Census Bureau Reports" (November 3). http://www.census.gov/newsroom/releases/archives/families_households/cb11-183.html.

———. 2011e. "Mother's Day: May 8, 2011" (March 17). http://www.census.gov/newsroom/releases/archives/facts_for_features_special_editions/cb11-ff07.html.

———. 2011f. "One-Third of Fathers with Working Wives Regularly Care for Their Children, Census Bureau Reports" (December 5). http://www.census.gov/newsroom/releases/archives/children/cb11-198.html.

———. 2011g. *Overview of Race and Hispanic Origin: 2010*. http://www.census.gov/prod/cen2010/briefs/c2010br-02.pdf.

———. 2011h. "Profile America: Anniversary of Americans with Disabilities Act: July 26" (May 31). http://www.census.gov/newsroom/releases/archives/facts_for_features_special_editions/cb11-ff14.html.

———. 2011i. "Same-Sex Couple Households." American Community Survey Briefs, September 2011. http://www.census.gov/prod/2011pubs/acsbr10-03.pdf.

———. 2011j. *Statistical Abstract of the United States: 2012*. http://www.census.gov/compendia/statab/.

———. 2011k. "Table 60. Interracially Married Couples by Race and Hispanic Origin of Spouses: 1980 to 2010." *Statistical Abstract of the United States*. http://www.census.gov/compendia/statab/2012/tables/12s0059.pdf.

———. 2012a. "A Half-Century of Learning: Historical Census Statistics on Educational Attainment in the United States, 1940 to 2000: Graphs." http://www.census.gov/hhes/socdemo/education/data/census/half-century/files/US.pdf.

———. 2012b. "Current Population Survey Data on Fertility." http://www.census.gov/hhes/fertility/data/cps/index.html.

———. 2012c. *Current Population Survey Interviewing Manual*. www.census.gov/apsd/techdoc/cps/CPS_Manual_Jan2012_Entire.pdf.

———. 2012d. "Definitions." Current Population Survey. http://www.census.gov/cps/about/cpsdef.html.

———. 2012e. *International Data Base*. http://www.census.gov/population/international/data/idb/informationGateway.php.

———. 2012f. "Meat Consumption by Type and Country: 2009 and 2010." *Statistical Abstract of the United States*. http://www.census.gov/compendia/statab/cats/international_statistics.html.

———. 2012g. "Who's Minding the Kids? Child Care Arrangements: Spring 2010." http://www.census.gov/hhes/childcare/data/sipp/2010/tables.html.

U.S. Central Intelligence Agency. 2010. "United States." *World Factbook*. https://www.cia.gov/library/publications/the-world-factbook/index.html.

———. 2012. *World Factbook*. https://www.cia.gov/library/publications/the-world-factbook/.

U.S. Department of Agriculture. 2011a. "Rural Income, Poverty, and Welfare: Poverty Geography." Briefing Rooms, September 17. http://www.ers.usda.gov/Briefing/IncomePovertyWelfare/PovertyGeography.htm.

———. 2011b. "Rural Labor and Education: Farm Labor." http://www.ers.usda.gov/Briefing/LaborAndEducation/FarmLabor.htm.

U.S. Department of Criminal Justice Statistics. 2010. "Prisoners under the Jurisdiction of the Federal Bureau of Prisons, by Adjudication Status, Type of Offense, and Sentence Length, 1990, 2000, and 2009a." http://www.albany.edu/sourcebook/pdf/t600232009.pdf.

U.S. Department of Defense. 1990. "DOD Directive 1332.14." In *Gays in Uniform: The Pentagon's Secret Reports*, edited by K. Dyer, 19. Boston: Alyson.

———. 2011. "Active Duty Military Personnel Strengths by Regional Area and by Country (309a) September 30, 2010". http://siadapp.dmdc.osd.mil/personnel/MILITARY/history/hst1009.pdf.

U.S. Department of Education. 2010. *Digest of Education Statistics*. http://nces.ed.gov/programs/digest/.

———. 2011a. National Center for Education Statistics, Common Core of Data (CCD), "Public Elementary/Secondary School Universe Survey, 2008–09." Figure 28-1 in Aud, S., W. Hussar, G. Kena, K. Bianco, et. al, *The Condition of Education 2011* (NCES 2011-033).

———. 2011b. National Center for Education Statistics. Washington, DC: U.S. Government Printing Office. http://nces.ed.gov/programs/coe/pdf/coe_pcp.pdf.

U.S. Department of Energy. 2011. "Top World Oil Producers, 2009." U.S. Energy Information Administration. http://www.eia.doe.gov/countries.

U.S. Department of Health and Human Services. 2011. "Overview of the Uninsured in the United States: A Summary of the 2011 Current Population Survey." ASPE Issue Brief, September. http://aspe.hhs.gov/health/reports/2011/CPSHealthIns2011/ib.shtml.

———. 2012. "2012 HHS Poverty Guidelines." http://aspe.hhs.gov/poverty/12poverty.shtml.

U.S. Department of Justice. 2007. "Criminal Offenders Statistics." http://bjs.ojp.usdoj.gov/.

———. 2010. "Correctional Populations in the United States, 2009." Bureau of Justice Statistics Bulletin, December. http://bjs.ojp.usdoj.gov/content/pub/pdf/cpus09.pdf.

———. 2011. "Press Release: Rate of Violent Victimization Declined 13 Percent in 2010," September 15. http://www.bjs.gov/content/pub/press/cv10pr.cfm.

U.S. Department of Labor. 2012. "Equal Pay Tool Kit." http://www.dol.gov/wb/equal-pay/.

U.S. Department of Labor, Bureau of Labor Statistics. 2011a. "College Enrollment and Work Activity of 2010 High School Graduates." Economic News Release, April 8. http://www.bls.gov/news.release/hsgec.nr0.htm.

———. 2011b. National Longitudinal Surveys. http://www.nlsinfo.org/ordering/display_db.php3#NLSY79.

———. 2011c. "Nonfatal Occupational Injuries and Illnesses Requiring Days Away from Work, 2010." News Release, November 9. http://www.bls.gov/news.release/archives/osh2_11092011.pdf.

———. 2012a. "The Employment Situation—January 2012." News Release, April 6. http://www.bls.gov/news.release/empsit.nr0.htm.

———. 2012b. "Labor Force Statistics from the Current Population Survey." http://www.bls.gov/cps/tables.htm#weekearn.

———. 2012c. "Union Members Summary." News Release, January 27. http://www.bls.gov/news.release/union2.nr0.htm.

———. 2012d. "Usual Weekly Earnings of Wage and Salary Workers, Fourth Quarter 2011." News Release, January 24. http://www.bls.gov/news.release/pdf/wkyeng.pdf.

U.S. Department of State. 2012. "FY 2012 Executive Budget Summary: Function 150 and Other International Programs." Released February 14, 2011. http://www.state.gov/s/d/rm/rls/ebs/2012/index.htm.

U.S. Department of the Treasury. 2012. "Major Foreign Holders of Treasury Securities." March 15. http://www.ustreas.gov/tic/mfh.txt.

U.S. Federal Reserve. 2011. "Mortgage Debt Outstanding." Updated December 23. http://www.federalreserve.gov/econresdata/releases/mortoutstand/current.htm.

U.S. Federal Reserve. 2012. "Consumer Credit." Updated March 7. http://www.federalreserve.gov/releases/g19/Current/#fn3a.

U.S. General Accounting Office. 2004. "Defense of Marriage Act: Update to Prior Report." http://www.gao.gov/new.items/d04353r.pdf.

U.S. Office of Management and Budget. 1997. "Revisions to the Standards for Classification of Federal Data on Race and Ethnicity." *Federal Register Notice*, October 30. http://www.whitehouse.gov/omb/fedreg/1997standards.html.

United Nations. 1948. "Convention on the Punishment and Prevention of the Crime of Genocide." http://www.ess.uwe.ac.uk/documents/gncnvntn.htm.

———. 2006. *Beyond Scarcity: Power, Poverty, and the Global Water Crisis.* Human Development Report. http://hdr.undp.org/hdr2006/.

———. 2008. *The Least Developed Countries Report 2008: Growth, Poverty and the Terms of Development Partnership*. United Nations Conference on Trade and Development. http://www.unctad.org/en/docs/ldc2008_en.pdf.

———. 2010a. "Beyond the Midpoint." http://content.undp.org/go/newsroom/publications/poverty-reduction/poverty-website/mdgs/beyond-the-midpoint.en.

———. 2010b. "We Can End Poverty 2015." www.un.org/millenniumgoals.

———. 2011a. "Human Development Statistical Annex." Human Development Report 2011. http://hdr.undp.org/en/media/HDR_2011_EN_Tables.pdf.

———. 2011b. "World Population Prospects, the 2010 Revision." Updated June 28. http://esa.un.org/unpd/wpp/Excel-Data/mortality.htm.

———. 2012. UN Data. http://data.un.org/.

United Nations University. 2006. "Richest 2% Own Half the World's Wealth." http://update.unu.edu/archive/issue44_22.htm.

United States African Command. 2008. http://www.africom.mil/.

University of Kentucky (UK) News. 2003. "UK Offers Drive-Thru Flu Shots." http://news.uky.edu.

University of Michigan Health System. 2010. "Television and Children." http://www.med.umich.edu/yourchild/topics/tv.htm (updated August 2010).

Uperesa, Fa' Anofo Lisaclaire. 2010. "Fabled Futures: Development, Gridiron Football, and Transnational Movements in American Samoa." http://gradworks.umi.com/34/20/3420691.html.

UPS. 2010. "Worldwide Facts." http://www.ups.com/content/us/en/about/facts/worldwide.html.

USDA Economic Research Service. 2004. "Rural Income, Poverty, and Welfare." http://www.ers.usda.gov/topics/rural-economy-population/rural-poverty-well-being/geography-of-poverty.aspx.

Van Evera, Stephen. 1990. "The Case against Intervention." *Atlantic Monthly* (July): 72–80.

Vann, Elizabeth. 1995. "Implications of Sex and Gender Differences for Self: Perceived Advantages and Disadvantages of Being the Other Gender." *Sex Roles: A Journal of Research* 33(7–8): 531.

Varadarajan, Tunku. 1999. "A Patel Motel Cartel?" *New York Times*, July 4. http://www.nytimes.com/1999/07/04/magazine/a-patel-motel-cartel.html?pagewanted=all.

Varghese, B., J. E. Maher, T. A. Peterman, B. M. Branson, and R. W. Steketee. 2002. "Reducing the Risk of Sexual HIV Transmission: Quantifying the Per-Act Risk for HIV on the Basis of Choice of Partner, Sex Act, and Condom Use." *Sexually Transmitted Diseases* 29(1): 38–43.

Verkuyten, Maykel. 2005. "Ethnic Group Identification and Group Evaluation Among Minority and Majority Groups." *Journal of Personality and Social Psychology* 88(1): 121–138.

Vincent, John A., Emmanuelle Tulle, and John Bond. 2008. "The Anti-Aging Enterprise: Science, Knowledge, Expertise, Rhetoric, and Values." *Journal of Aging Studies* 22(4): 291–376.

Virginia Racial Integrity Act of 1924. Cited in *Melungeon Studies* (blog), by Dennis Maggard. September 17, 2008. http://melungeon-studies.blogspot.com/2008/09/virginia-racial-integrity-act-of-1924.html.

Wacquant, Loïc J. D. 1989. "The Ghetto, the State, and the New Capitalist Economy." *Dissent* (Fall): 508–520.

Wallerstein, Immanuel. 1984. *The Politics of the World-Economy: The States, the Movements and the Civilizations*. New York: Cambridge University Press.

———. 1990. "Culture as the Ideological Battleground of the Modern World-System." *Theory, Culture, and Society* 7:31–55.

Walmsley, Roy. 2002. "Global Incarceration and Prison Trends." United Nations on Drugs and Crime. http://www.unodc.org/pdf/crime/forum/forum3_Art3.pdf.

Warner, Melanie. 2006. "Salads or No, Cheap Burgers Revive McDonald's." *New York Times*, April 19. http://www.nytimes.com/.

Waters, Mary C. 1990. *Ethnic Options: Choosing Identities in America*. Berkeley: University of California Press.

———. 1994. "Ethnic and Racial Identities of Second Generation Black Immigrants in New York City." *International Migration Review* 28(4): 795–820.

Weber, Max. 1922a. *Economy and Society*, vol. 2. Edited by Guenther Roth and Claus Wittich. Translated by Ephraim Fischof, 1978. Berkeley: University of California Press.

———. 1922b. *The Sociology of Religion*. Translated by E. Fischoff. Boston: Beacon.

———. 1925. *Economy and Society,* part III, chap. 6, pp. 650–678. http://www.faculty.rsu.edu/~felwell/TheoryWeb/readings/WeberBurform.html#Weber.

———. 1947. *The Theory of Social and Economic Organization*. Edited and translated by A. M. Henderson and T. Parsons. New York: Macmillan.

———. 1948. "Class, Status, and Party." In *Essays from Max Weber*, edited by H. Gerth and C. W. Mills. New York: Routledge & Kegan Paul.

———. 1958. *The Protestant Ethic and the Spirit of Capitalism*, 5th ed. Translated by T. Parsons. New York: Scribner's.

———. 1982. "Status Groups and Classes." In *Classical and Contemporary Debates*, edited by A. Giddens and D. Held, 69–73. Los Angeles: University of California.

———. 1994. *Weber: Political Writings* (Cambridge Texts in the History of Political Thought). Edited by Peter Lassman. Translated by Ronald Speirs, xvi. Cambridge, UK: Cambridge University Press.

Weintraub, Jeff, and Joseph Soares. 2005. "Weber on Social Action, Rationality, and Political Ethics." http://Jeffweintraub.blogspot.com/2005_06_01_archive.html.

Weiser, Benjamin. 2003. "Big Macs Can Make You Fat." *New York Times*, January 23: A23.

Wellman, Barry, and Keith Hampton. 1999. "Living Networked On and Off Line." *Contemporary Sociology* 28(6): 648–654.

Wells, Amy Stuart, and Jeannie Oakes. 1996. "Potential Pitfalls of Systematic Reform: Early Lessons from Research on Detracking." *Sociology of Education* 69 (extra issue): 135–143.

Westin, A. F. 2003. "Social and Political Dimensions of Privacy." *Journal of Social Issues* 59:431–453.

Whorf, Benjamin. 1956. *Language, Thought, and Reality: Selected Writings of Benjamin Lee Whorf*, edited by J. B. Carroll. Cambridge: Technology Press of MIT.

Williams, Terry. 1989. *The Cocaine Kids: The Inside Story of a Teenage Drug Ring*. Reading, MA: Addison-Wesley.

Williamson, Lucy. 2011. "South Korea's Enthusiasm for Recycling." BBC News, June 9. http://news.bbc.co.uk/2/hi/programmes/from_our_own_correspondent/9508181.stm.

Wilson, Matthew J. et al. 2011. "NCAA Division I Men's Basketball Coaching Contracts." *Journal of Issues of Intercollegiate Athletics* 4:396–410.

Wilson, William Julius. 1983. "The Urban Underclass: Inner-City Dislocations." *Society* 21:80–86.

———. 1987. *The Truly Disadvantaged*. Chicago: University of Chicago Press.

———. 1991. "Studying Inner-City Social Dislocations: The Challenge of Public Agenda Research" (1990 presidential address). *American Sociological Review* (February): 1–14.

———. 1993. "Another Look at the Truly Disadvantaged." *Political Science Quarterly* 106(4): 639–656.

Wirth, Louis. 1945. "The Problem of Minority Groups." In *The Science of Man*, edited by R. Linton, 347–372. New York: Columbia University Press.

Wiseman, Rosalind. 2003. *Queen Bees and Wannabes*. New York: Three Rivers Press.

Witte, John. 1992. "Deforestation in Zaire: Logging and Landlessness." *Ecologist* 22(2): 58.

Witte, Thomas E. 2005. "The Biggest Man on the Field." *Sportshooter*. October 31. http://www.sportsshooter.com/news/1477.

Wolf, Michael. 2005. "Assembly Required." *Mother Jones*, July/August. http://motherjones.com/politics/2005/07/assembly-required.

Wooddell, George, and Jacques Henry. 2005. "The Advantage of a Focus on Advantage: A Note on Teaching Minority Groups." *Teaching Sociology* 33(3): 301–309.

World Bank. 2008a. "World Bank Updates Poverty Estimates for the Developing World." http://econ.worldbank.org/.

———. 2008b. *World Development Indicators 2008*. Washington, DC: World Bank.

———. 2009. "Understanding Poverty." http://web.worldbank.org/.

———. 2010. "World Bank Updates Poverty Estimates for the Developing World." February 17. http://web.worldbank.org/WBSITE/EXTERNAL/NEWS/0,,contentMDK:21882162~pagePK:34370~piPK:34424~theSitePK:4607,00.html.

———. 2011a. "Gini Index." http://data.worldbank.org/indicator/SI.POV.GINI.

———. 2011b. "International Tourism, Number of Arrivals." http://data.worldbank.org/country/thailand.

World Economic Forum. 2011. *The Global Gender Gap Report 2011: Rankings and Scores.* http://www.weforum.org/reports/global-gender-gap-report-2011.World Watch Institute. 2012. "Global Palm Oil Demand Fueling Deforestation." http://www.worldwatch.org/node/6059.

Wright, Eric R., and Brea L. Perry. 2010. "Medical Sociology and Health Services Research: Past Accomplishments and Future Policy Challenges." *Journal of Health and Social Behavior*, November 51: S107-S119.

Wright, Erik Olin. 2004. "Social Class." *Encyclopedia of Sociological Theory*, edited by George Ritzer. Thousand Oaks, CA: Sage.

———. 2009. "Erik Olin Wright." http://www.ssc.wisc.edu/~wright/.

Yamada, Masahiro. 2000. "The Growing Crop of Spoiled Singles." *Japan Echo* 27(3). http://www.japanecho.com.

Yamamoto, Noriko, and Margaret I. Wallhagen. 1997. "The Continuation of Family Caregiving in Japan." *Journal of Health and Social Behavior* 38 (June): 164–176.

Yee, Lee Chyen, and Clare Jim. 2011. "Foxconn to Rely More on Robots; Could Use 1 Million in 3 Years." Reuters, August 1. http://www.reuters.com/article/2011/08/01/us-foxconn-robots-idUSTRE77016B20110801.

Yeutter, Clayton. 1992. "When Fairness Isn't Fair." *New York Times*, March 24: A13.

Yinger, J. Milton. 1977. "Presidential Address: Countercultures and Social Change." *American Sociological Review* 42(6): 833–853.

Young, Jeffrey R. 2010. "Robot Teachers Are the Latest E-Learning Tool." *Chronicle of Higher Education*, October 31. http://chronicle.com/article/Robot-Teachers-Are-the-Latest/125102/.

Young, T. R. 1975. "Karl Marx and Alienation: The Contributions of Karl Marx to Social Psychology." *Humboldt Journal of Social Relations* 2(2): 26–33.

Zarit, Steven H., Pamela A. Todd, and Judy M. Zarit. 1986. "Subjective Burden of Husbands and Wives as Caregivers: A Longitudinal Study." *Gerontologist* 26:260–266.

Zelditch, Morris. 1964. "Family, Marriage, and Kinship." In *Handbook of Modern Sociology*, edited by Robert E. L. Faris, 680–733. Chicago: Rand McNally.

Zhao, Shanyang, Sherri Grasmuck, and Jason Martin. 2008. "Identity Construction on Facebook: Digital Empowerment in Anchored Relationships." *Computers in Human Behavior* 24:1816–1836.

Zumbrun, Joshua. 2008. "In Pictures: Ten Most Energy-Efficient Countries." *Forbes*, July 7. http://www.forbes.com/2008/07/03/energy-efficiency-japan-biz-energy_cx_jz_0707efficiency_countries.html.

Index

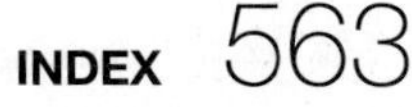